Molecular and Genetic Aspects
of Nitrate Assimilation

Molecular and Genetic Aspects of Nitrate Assimilation

Edited by

JOHN L. WRAY

and

JAMES R. KINGHORN

Plant Molecular Genetics Unit
University of St Andrews

Oxford   New York   Tokyo
OXFORD SCIENCE PUBLICATIONS
1989

Oxford University Press, Walton Street, Oxford OX2 6DP
Oxford New York Toronto
Delhi Bombay Calcutta Madras Karachi
Petaling Jaya Singapore Hong Kong Tokyo
Nairobi Dar es Salaam Cape Town
Melbourne Auckland
and associated companies in
Berlin Ibadan

Oxford is a trade mark of Oxford University Press

Published in the United States
by Oxford University Press, New York

British Library Cataloguing in Publication Data
Molecular and genetic aspects of nitrate
assimilation
1. Plants. Nitrogen. Fixation
I. Wray, John L.   II. Kinghorn, James R.
581.1'33
ISBN 0-19-857696-X

Library of Congress Cataloging in Publication Data
Molecular and genetic aspects of nitrate assimilation/edited by John
L. Wray and James R. Kinghorn.
Includes bibliographies and index.
1. Nitrates—Metabolism.   2. Plants—Assimilation.
3. Fungi—Metabolism.   4. Nitrates—Metabolism—Genetic aspects.
5. Nitrates—Metabolism—Regulation.   I. Wray, John Langford, 1942–
II. Kinghorn. James R.
QK898. N57M65   1989   581.19'214—dc19   88-26844
ISBN 0-19-857696-X

Set by Cotswold Typesetting Ltd, Cheltenham and Gloucester
Printed in Great Britain by
St Edmundsbury Press, Bury St Edmunds, Suffolk

Preface

This book has its origins in the Second International Symposium on Nitrate Assimilation—Molecular and Genetic Aspects, organized by J. L. Wray in St Andrews, Scotland in 1987. This Symposium drew together 24 invited speakers and some 60 other workers in the higher plant and microbial field to discuss the recent advances in the biochemistry, genetics, and molecular biology of nitrate assimilation, the first time that this topic had been considered as an integrated whole since the First International Symposium held in Gatersleben, German Democratic Republic, in 1982. Subsequently, we asked invited speakers to provide a review of their area of expertise and we commissioned an additional review (Chapter 15). This book is the result.

The contributed reviews have been organized into six broadly based parts entitled *Nitrate Uptake, Nitrate Reduction, Nitrite Reduction, Regulation of Nitrate Assimilation, Applied Aspects,* and *Comparative Aspects.* We say broadly based since it will become apparent to the reader that there is some overlap between these different parts. This was unavoidable and instances where it occurs are indicated below.

Nitrate Uptake contains two chapters (Chapters 1 and 2) which consider the molecular and genetic aspects of nitrate uptake in higher plants. These chapters reveal how little we know but set the scene for the way forward. Later chapters include discussions of what little is known in molecular and genetic terms of nitrate uptake in cyanobacteria (Chapter 4), yeasts (Chapter 5), and algae (Chapter 8). The best characterized system at present is clearly that in *Aspergillus nidulans* where a nitrate uptake gene, *crnA*, has been identified and its molecular cloning recently described (Chapter 6).

Nitrate Reduction contains twelve chapters (Chapters 3–14) reflecting the level of interest in, and amount of information derived about, this second step in nitrate assimilation. The first chapter (Chapter 3) may seem an odd one to include in a book on nitrate assimilation, since the preponderance of evidence seems to suggest that the *Escherichia coli* strains commonly used in bacterial genetics laboratories use nitrate mainly in a dissimilatory mode as a terminal electron acceptor under anaerobic conditions, and not as a nitrogen source for growth. The reason for its inclusion, of course, is that, as indicated in Chapter 14, the structure of the *E. coli* molybdenum cofactor is likely to be similar, if not identical, to that of other micro-organisms and also higher plants, suggesting similarity in biosynthetic pathway and genetics. Inclusion of Chapter 3 thus allows the reader to compare the current position of molybdenum cofactor genetics in algae, fungi, and higher plants with that in *E. coli* and to determine whether some of the lessons learnt, and techniques used, with *E. coli* may not also be applied to these other organisms. Chapters 4–8 review nitrate reduction in micro-organisms including a discussion of the structure of nitrate reductase and, where available, its genetics and molecular biology. The position with respect to higher plants is considered in Chapters 9–13, which discuss the structure (Chapter 9), immunology (Chapter 10), and genetics (Chapters 11, 12 and 13) of nitrate reductase and some of the recent developments in its molecular biology (Chapters 12 and 13).

Nitrite Reduction contains four chapters which consider nitrite reduction in *E. coli* (Chapter 15) and higher plants (Chapters 16–18). Chapter 15 discusses the fact that nitrite reduction in *E. coli* is largely a dissimilatory process and indicates that the genetics of nitrite assimilation in enteric bacteria is an essentially unexplored field. Chapter 16 reviews the biochemistry of nitrite reduction in higher plants and describes the first steps towards the establishment of a genetic basis for higher plant nitrite reduction. Chapter 17 is an elegant treatise on the structure and reaction mechanism of spinach nitrite reductase, whilst Chapter 18 discusses the recent molecular cloning and nitrate regulation of the spinach nitrite reductase structural gene. Nitrite reduction is also considered in Chapters 4, 5, 6, and 8 with respect to cyanobacteria, yeasts, filamentous fungi, and algae respectively.

Regulation of Nitrate Assimilation is considered in three chapters (Chapters 19–21). Chapters 19 and 20 review the regulation of nitrate assimilation in the filamentous fungi *A. nidulans* and *Neurospora crassa* and describe the involvement, in molecular terms, of positive-acting regulatory genes whose products mediate nitrate and ammonium regulation. Chapter 21 describes the molecular cloning and nitrate regulation of the higher plant nitrate reductase apoprotein gene. Currently there is no evidence that higher plants possess regulatory genes of the type identified in *A. nidulans* and *N. crassa*. Some discussion of regulation in cyanobacteria, yeasts, algae, and higher plants is also presented in Chapters 4, 5, 8, and 9 respectively.

Applied Aspects are considered in two chapters (Chapters 22 and 23) which discuss some of the ways in which the basic molecular and genetic information available has been, and likely will be, exploited in a biotechnological way in fungi (Chapter 22) and higher plants (Chapter 23).

Finally, *Comparative Aspects*, in a single chapter (Chapter 24), draws together and compares amino acid sequence relationships between bacterial, fungal, and plant nitrate reductase and nitrite reductase proteins.

St Andrews   J. L. W.
                                                                               J. R. K.

Contents

List of contributors   xi

Nitrate Uptake

1 Molecular aspects of nitrate uptake in higher plants   3
   C.-M. Larsson and B. Ingemarsson

2 The genetics of nitrate uptake in higher plants   15
   Roger M. Wallsgrove, Hiroshi Hasegawa, Alan C. Kendall, and Janice C. Turner

Nitrate Reduction

3 Involvement of the chlorate resistance loci and the molybdenum cofactor in the biosynthesis of the *Escherichia coli* respiratory nitrate reductase   27
   David H. Boxer

4 Biochemistry, regulatory aspects, and genetics of nitrate assimilation in cyanobacteria   40
   Alexandra J. Andriesse, Peter J. Weisbeek, and Gerard A. van Arkel

5 Nitrate assimilation in yeasts   51
   Charles R. Hipkin

6 Genetic, biochemical, and structural organization of the *Aspergillus nidulans crnA–niiA–niaD* gene cluster   69
   James R. Kinghorn

7 Structure–function relationships of algal nitrate reductases   88
   Larry P. Solomonson and Michael J. Barber

8 Genetics and regulatory aspects of nitrate assimilation in algae   101
Emilio Fernández and Jacobo Cárdenas

9 Structure and regulation of nitrate reductase in higher plants   125
Wilbur H. Campbell

10 Immunology of nitrate reductase with special reference to higher plants   155
Brian A. Notton

11 Biochemical and somatic cell genetics of nitrate reduction in *Nicotiana*   166
Andreas J. Muller and Ralf R. Mendel

12 Molecular genetics of nitrate reduction in *Nicotiana*   186
Michel Caboche, Isabelle Cherel, Fabienne Galangau, Marie-Angele Grandbastien, Christian Meyer, Therese Moureaux, Frederique Pelsy, Pierre Rouze, Herve Vaucheret, Francoise Vedele, and Michel Vincentz

13 Molecular genetics of nitrate reductase in barley   197
A. Kleinhofs, R. L. Warner, J. M. Lawrence, J. M. Melzer, J. M. Jeter, and D. A. Kudrna

14 Chemistry and biology of the molybdenum cofactor   212
K. V. Rajagopalan

Nitrite Reduction

15 Physiology, biochemistry, and genetics of nitrite reduction by *Escherichia coli*   229
J. A. Cole

16 Molecular and genetic aspects of nitrite reduction in higher plants   244
John L. Wray

17 Structure and function of spinach ferredoxin-
nitrite reductase   263
Lewis M. Siegel and James O. Wilkerson

18 Molecular cloning of spinach nitrite reductase   284
E. Back, W. Burkhart, M. Moyer, L. Privalle, and S. Rothstein

Regulation of Nitrate Assimilation

19 Regulation of nitrate assimilation in *Aspergillus
nidulans*   299
Claudio Scazzocchio and Herbert N. Arst, Jr.

20 Genetics, regulation, and molecular studies of
nitrate assimilation in *Neurospora crassa*   314
George Marzluf and Ying-Hui Fu

21 Molecular analysis of nitrate regulation of
nitrate reductase in squash and *Arabidopsis*   328
Nigel M. Crawford and Ronald W. Davis

Applied Aspects

22 Fungal biotechnology and the nitrate
assimilation pathway   341
Shiela E. Unkles

23 Plant biotechnology and nitrate assimilation   364
*D. Cammaerts, R. Dirks, I. Negrutiu, I. Famelaer, S.
Hinnisdaels, D. Debeys, J. Veuskens, and M. Jacobs*

Comparative Aspects

24 Amino acid sequence relationships between
bacterial, fungal, and plant nitrate reductase
and nitrite reductase proteins   385
James R. Kinghorn and E. I. Campbell

Index   405

Contributors

A. J. Andriesse
Department of Molecular Cell Biology, University of Utrecht, Padualaan 8, 3584 CH Utrecht, The Netherlands

H. N. Arst, Jr.
Department of Bacteriology, Royal Postgraduate Medical School, Hammersmith Hospital, Ducane Road, London, W12 0HS, UK

E. Back
Ciba-Geigy Corporation, Research Triangle Park, North Carolina 27709-2257, USA

M. J. Barber
Department of Biochemistry, University of South Florida, College of Medicine, Tampa, Florida 33612, USA

D. H. Boxer
Department of Biochemistry, Medical Sciences Institute, University of Dundee, Dundee, DD1 4HN, UK

W. Burkhart
Ciba-Geigy Corporation, Research Triangle Park, North Carolina, 27709-2257, USA

M. Caboche
Laboratoire de Biologie Cellulaire, INRA 78 000, Versailles, France

D. Cammaerts
Laboratory of Plant Genetics, Free University of Brussels, Paardenstraat 65, B-1640 St Genesius-Rode, Belgium

E. I. Campbell
Plant Molecular Genetics Unit, University of St Andrews, St Andrews, Fife, KY16 9TH, UK

W. H. Campbell
Department of Biological Sciences, Michigan Technological University, Houghton, Michigan 49931, USA

J. Cárdenas
Departmento de Bioquímica, Biología Molecular y Fisiología, Facultad de Ciencias Universidad de Córdoba, E-14071 Córdoba, Spain

I. Cherel

Laboratoire de Biologie Cellulaire, INRA 78 000, Versailles, France

J. A. Cole

Department of Biochemistry, University of Birmingham, Birmingham B15 2TT, UK

N. M. Crawford

Department of Biology, C-016, University of California at San Diego, La Jolla, California 92093, USA

R. W. Davis

Department of Biochemistry, Stanford Medical School, Stanford, California 94305, USA

D. Debeys

Laboratory of Plant Genetics, Free University of Brussels, Paardenstraat 65, B-1640 St Genesius-Rode, Belgium

R. Dirks

Laboratory of Plant Genetics, Free University of Brussels, Paardenstraat 65, B-1640 St Genesius-Rode, Belgium

I. Famelaer

Laboratory of Plant Genetics, Free University of Brussels, Paardenstraat 65, B-1640 St Genesius-Rode, Belgium

E. Fernández

Departmento de Bioquimíca, Biología Molecular y Fisiología, Facultad de Ciencias Universidad de Cordóba, E-14071 Cordóba, Spain

Y.-H. Fu

Department of Biochemistry, The Ohio State University, 484 West 12th Avenue, Columbus, Ohio 43210, USA

F. Galangau

Laboratoire de Biologie Cellulaire, INRA 78 000, Versailles, France

M.-A. Grandbastien

Laboratoire de Biologie Cellulaire, INRA 78 000, Versailles, France

H. Hasegawa

Department of Biochemistry, AFRC Institute of Arable Crops Research, Rothamsted Experimental Station, Harpenden, Herts., AL5 2JQ, UK

S. Hinnisdaels

Laboratory of Plant Genetics, Free University of Brussels, Paardenstraat 65, B-1640, St Genesius-Rode, Belgium

C. R. Hipkin
Plant and Microbial Metabolism Research Group, School of Biological Sciences, University College of Swansea, Singleton Park, Swansea, SA2 8PP, W. Glamorgan, UK

B. Ingemarsson
Department of Botany, University of Stockholm, S-106 91 Stockholm, Sweden

M. Jacobs
Laboratory of Plant Genetics, Free University of Brussels, Paardenstraat 65, B-1640, St Genesius-Rode, Belgium

J. M. Jeter
Department of Agronomy and Soils and Program in Genetics and Cell Biology, Washington State University, Pullman, WA 99164–6420, USA

A. C. Kendall
Department of Biochemistry, AFRC Institute of Arable Crops Research Rothamsted Experimental Station, Harpenden, Herts., AL5 2JQ, UK

J. R. Kinghorn
Plant Molecular Genetics Unit, University of St Andrews, St. Andrews, Fife, KY16 9TH, UK

A. Kleinhofs
Department of Agronomy and Soils and Program in Genetics and Cell Biology, Washington State University, Pullman, WA 99164-6420, USA

D. A. Kudrna
Department of Agronomy and Soils and Program in Genetics and Cell Biology, Washington State University, Pullman, WA 99164-6420, USA

C. M. Larsson
Department of Botany, University of Stockholm, S-106 91 Stockholm, Sweden

J. M. Lawrence
Department of Agronomy and Soils and Program in Genetics and Cell Biology, Washington State University, Pullman, WA 99164-6420, USA

G. A. Marzluf
Department of Biochemistry, The Ohio State University, 484 West 12th Avenue, Columbus, Ohio 43210, USA

J. M. Melzer
Department of Agronomy and Soils and Program in Genetics and Cell Biology, Washington State University, Pullman, WA 99164-6420, USA

R. R. Mendel
Zentralinstitut fur Genetik und Kulturpflanzenforschung, Akademie der Wissenschaften der DDR, DDR-4325 Gatersleben, German Democratic Republic.

C. Meyer

Laboratoire de Biologie Cellulaire, INRA 78 000, Versailles, France

T. Moureaux

Laboratoire de Biologie Cellulaire, INRA 78 000, Versailles, France

M. Moyer

Ciba-Geigy Corporation, Research Triangle Park, North Carolina 27709-2257, USA

A. J. Müller

Zentralinstitut für Genetik and Kulturpflanzenforschung, Akademie der Wissenchaften der DDR, DDR-4325 Gatersleben, GDR

I. Negrutiu

Laboratory of Plant Genetics, Free University of Brussels, Paardenstraat 65, B-1640 St Genesius-Rode, Belgium

B. A. Notton

Department of Agricultural Sciences, University of Bristol, Institute of Arable Crops Research, Long Ashton Research Station, Bristol, BS18 9AF, UK

F. Pelsy

Laboratoire de Biologie Cellulaire, INRA 78 000, Versailles, France

L. Privalle

Ciba-Geigy Corporation, Research Triangle Park, North Carolina 27709-2257, USA

K. V. Rajagopalan

Department of Biochemistry, Duke University Medical Centre, Durham, NC 27710, USA

S. Rothstein

Ciba-Geigy Corporation, Research Triangle Park, North Carolina 27709-2257, USA

P. Rouze

Laboratoire de Biologie Cellulaire, INRA 78 000, Versailles, France

C. Scazzocchio

Institut de Microbiologie, Batiment 409, Universit Paris XI, 91405 Orsay Cedex, France

L. M. Siegel

Department of Biochemistry, Duke University Medical School, Durham, NC 27710, USA

L. P. Solomonson

Department of Biochemistry, University of South Florida, College of Medicine, Tampa, Florida 33612, USA

J. C. Turner
Department of Biochemistry, AFRC Institute of Arable Crops Research, Rothamsted Experimental Station, Harpenden, Herts., AL5 2JQ, UK

S. E. Unkles
Plant Molecular Genetics Unit, University of St Andrews, St Andrews, Fife, KY16 9TH, UK

G. A. van Arkel
Department of Molecular Cell Biology, University of Utrecht, Padualaan 8, 3584 CH Utrecht, The Netherlands

H. Vaucheret
Laboratoire de Biologie Cellulaire, INRA 78 000, Versailles, France

F. Vedele
Laboratoire de Biologie Cellulaire, INRA 78 000, Versailles, France

J. Veuskens
Laboratory of Plant Genetics, Free University of Brussels, Paardenstraat 65, B-1640 St Genesius-Rode, Belgium

M. Vincentz
Laboratoire de Biologie Cellulaire, INRA 78 000, Versailles, France

R. M. Wallsgrove
Department of Biochemistry, AFRC Institute of Arable Crops Research, Rothamsted Experimental Station, Harpenden, Herts., AL5 2JQ, UK

R. L. Warner
Department of Agronomy and Soils and Program in Genetics and Cell Biology, Washington State University, Pullman, WA 99164-6420, USA

P. J. Weisbeek
Department of Molecular Cell Biology, University of Utrecht, Padualaan 8, 3584 CH Utrecht, The Netherlands

J. O. Wilkerson
Department of Biochemistry, Duke University Medical School, Durham, NC 27710, USA

J. L. Wray
Plant Molecular Genetics Unit, University of St Andrews, St Andrews, Fife, KY16 9TH, UK

Nitrate Uptake

1. Molecular aspects of nitrate uptake in higher plants

C.-M. Larsson and B. Ingemarsson

Introduction

The study of nitrate assimilation in higher plants is gradually assuming a more molecular approach, in common with other fields of experimental biology. While progress is being made concerning the molecular biology of nitrate reduction, which is clearly manifested in this volume, few studies devoted to the molecular biology of nitrate uptake have appeared in the literature so far. The main deficiency in this regard is that the nitrate transport protein, which we must assume is present in plant membranes, has not yet been identified. In fact, information on the molecular nature of transport proteins for inorganic anions in higher plants is scarce, and seems to be limited to the phosphate translocator of the inner chloroplast envelope, a 29–30 kDa protein which has been described in both C_3 (Flügge and Heldt 1976; Flügge 1982) and C_4 (Thompson *et al.* 1987) species.

The present communication attempts to consider some molecular aspects of the membrane nitrate transport system. By 'molecular' we mean the characteristics of information transfer from the genes, as well as the assembly and properties of the mature transport protein. In fact, due to the lack of any substantial body of molecular knowledge, emphasis is placed on what physiological and biochemical examination tells us (and does *not* tell us) about the molecular biology of nitrate transport. Some experimental approaches to the eventual identification of nitrate transport proteins will also be presented.

Location and functional characteristics of nitrate uptake: a multitude of distinct nitrate transporters?

Kinetics

The nitrate ion has to enter the cells against an electrical potential gradient over the plasmalemma (inside negative) ranging from approximately 70 to 250 mV, depending on tissue and species. Since under many conditions the internal nitrate concentration has to be increased relative to the anticipated concentration at electrochemical equilibrium in order to sustain appreciable rates of nitrate reduction, the existence of a transport system capable of utilizing metabolically generated energy during the transport process may be inferred. When the net uptake rate is examined over a range of external nitrate concentrations, uptake data normally obey simple Michaelis–Menten kinetics (reviewed by Clarkson 1986; Jackson *et al.* 1986). This kind of kinetic interpretation of net uptake data might be confused by variable rates of concomitant nitrate efflux. However, short-term measurements of nitrate influx using nitrate labelled with the short-lived ($t_{1/2} = 10$ min) isotope ^{13}N confirm the presence of a saturable transport component (Lee and Drew 1986). There is thus little doubt that nitrate uptake is mediated by a transport protein, most probably located in the plasmalemma.

Concentration kinetics of net ion uptake often indicate the existence of more than one, and typically two saturable components, one with a high affinity and low V_{max}, the other with low affinity and high V_{max}. To these, a non-saturable component might be added. Two saturable components were observed in *Arabidopsis thaliana* (Doddema and Telkamp 1979); essentially biphasic kinetics were also observed in *Phaseolus vulgaris* by Breteler and Nissen (1982), although the authors themselves resolved four different phases. The occurrence of several phases can be interpreted as indicating the presence of a number of different transport systems or, alternatively, a single system that undergoes sharp changes in its mode of operation at certain external concentrations (see e.g. Nissen 1974; Epstein 1976; Borstlap 1981; Kochian and Lucas 1982, for arguments favouring and rejecting different opinions).

There is considerable controversy concerning the physiological interpretation of concentration kinetics; a thorough molecular and genetic examination of nitrate transporters might aid in elucidating this much debated problem. The *A. thaliana* mutant B1 is apparently affected in nitrate uptake at high external concentrations, whereas the uptake kinetics at low nitrate concentrations are similar to those observed in the wild type (Doddema and Telkamp 1979). This finding suggests a mutation in a specific uptake system which is operative mainly at higher external nitrate concentrations. Recent re-examination of the B1 mutant has revealed, however, that uptake of the nitrate analogue, chlorate, is impaired at both high and low concentrations, and that potassium and chloride (but not sulphate) uptake is also reduced relative to the wild type (Scholten and Feenstra 1986). Whether these results indicate that the mutational lesion in the B1 mutant is of a more general type, or whether a mutation in the nitrate uptake system also affects other transport processes, cannot as yet be concluded.

Mechanism: nitrate/anion antiport or nitrate/cation symport?

Microelectrode measurements of the membrane potential in fronds of *Lemna* have revealed a transient depolarization of the membrane following nitrate addition (Novacky *et al.* 1978). Other anions cause the same effect, e.g. phosphate (Ullrich-

Eberius *et al.* 1981) and sulphate (Lass and Ullrich-Eberius 1984). Collectively, these responses have been interpreted as indicating symport of nitrate (or any of the other anions) and protons, where protons are transported in excess of electrical neutrality, thus causing depolarization of the membrane. The subsequent increased rate of proton extrusion is thought to aid in restoration of the membrane potential.

The opposite phenomenon was recorded in maize roots, where nitrate caused a hyperpolarization of the membrane (Thibaud and Grignon 1981). Measurements of the external pH shift indicated formation of one hydroxyl ion for each pair of nitrate ions taken up. This stoichiometry was unaffected by application of the proton pump inhibitor diethylstilbestrol (DES). Based on these observations, the authors proposed a nitrate/hydroxyl ion antiport being involved in nitrate uptake.

The results of Thibaud and Grignon (1981) have been confirmed by Ullrich (1987) for both maize and oats. The hyperpolarization was, however, usually preceded by a minute depolarization (also noted by Thibaud and Grignon 1981), whereas soybean roots generally responded with a rather marked depolarization, resembling the previous results with *Lemna*. Ullrich (1987) argued that these observations were best explained by the nitrate/proton symport hypothesis if one assumed that the kinetics of recovery of the membrane potential could differ between species. At present we cannot safely conclude whether we should distinguish between different populations of nitrate transporters on the basis of their co-transport properties. It is also possible that the response of the membrane potential to nitrate depends on the physiological state of the tissue (discussed by Jackson *et al.* 1986). Only a molecular examination can tell us whether this is because there are several types of nitrate transporters, or because a single type of transporter possesses different modes of operation.

Tissue distribution of plasmalemma nitrate transporters

Based partly on the fate of the absorbed nitrate, Morgan *et al.* (1985) suggested that nitrate was preferentially absorbed and reduced in the root epidermis at low external concentrations, whereas at higher concentrations, nitrate was preferentially taken up by the cortical cells. Support for this model was obtained in recent experiments where the nitrate reductase of maize roots was localized using immunological methods. At low external nitrate (200 μM), 100 per cent of enzyme-linked immunosorbent assay (ELISA)-detectable nitrate reductase (NR) protein was found in the epidermis. Following pretreatment with 20 mM nitrate, however, about one-quarter of the total NR protein was detected in the cortex and stele, possibly as a result of increased uptake in the cortical region (Rufty *et al.* 1986). The relative abundance of NR protein in the different regions corresponded well with the measured *in vitro* enzyme activities. Whether this behaviour with respect to external nitrate concentration indicates differential location of the high- and low-affinity uptake systems cannot be unequivocally concluded, since the actual nitrate concentration in the cortical apoplast is hard to assess.

A different mode of plasmalemma nitrate transport is seen in the stele, where nitrate is transported in the reverse direction, i.e. out of the symplast to the apoplast. Available data (reviewed by Jackson *et al.* 1986) suggest the presence of a protein involved in loading of nitrate into the xylem. Whether this protein has any resemblance at all with 'uptake-transporters' for nitrate is unknown.

The nature of the nitrate uptake system that must be assumed to be present in leaf cells is largely unknown, although the distribution of nitrate assimilation in different leaf tissues (notably in C_4 plants) as well as physiological factors affecting leaf nitrate reduction have been investigated in great detail (reviewed by Jackson *et al.* 1986). The main difficulty is that the apoplastic environment of the leaf cell is not readily accessible for observation or perturbation *in situ*. There seems to be little systematic research on the nitrate uptake properties of leaf protoplasts, although techniques are at hand for the isolation of relatively intact protoplasts which retain the ability to metabolize nitrate (Rathnam 1978; Reed and Canvin 1982; Martinoia *et al.* 1986).

Intracellular transmembrane transport of nitrate

Considerable amounts of nitrate can be accumulated in the vacuoles (Martinoia *et al.* 1981, 1986; Granstedt and Huffaker 1982) from which it can be retrieved or exchanged for nitrate in the cytoplasm (see Clarkson and Lüttge 1984; Clarkson 1986, for a discussion on cytoplasm–vacuole interrelations). This necessitates a nitrate transporter being present in the tonoplast, which possibly is different from the plasmalemma nitrate transporters. Furthermore, it is possible that NR is enclosed in more or less well-defined subcellular compartments. Histochemical and immunochemical analysis of soybean cotyledons suggest that the enzyme is clustered and contained in cytoplasmic vesicles (Vaughn and Duke 1981; Vaughn *et al.* 1984). Immunolocalization has recently indicated a partial (Roldán *et al.* 1987) or exclusive (Kamachi *et al.* 1987) location of NR in the chloroplasts of spinach leaf tissue. The NR of several microalgae also seems to be exclusively located in the chloroplast (Lopez-Ruiz *et al.* 1986). If NR is separated from the bulk cytoplasm by a membrane, then yet another transmembrane transport system might be needed to support nitrate reduction.

Physiological studies of nitrate uptake and translocation in maize show that nitrate entering the root from the external medium is readily reduced, whereas nitrate previously accumulated in the root may be lost by efflux to the ambient solution or translocated to the shoot, but not reduced to any great extent (MacKown *et al.* 1983). The presence of a symplastic transport system, radially traversing the root from epidermis to stele and closely communicating with the vacuoles, has been suggested. In such a system nitrate could be moved from storage compartments to the xylem without contacting existing nitrate reductase or causing induction of the enzyme (Rufty *et al.* 1986). The type of barrier that would separate nitrate in the transport compartment from the rest of the symplast is obscure.

Activity of the nitrate uptake system

The expression of the nitrate uptake system has been studied almost exclusively as the rate of nitrate uptake by intact plants, excised roots, or other plant tissues. The rate of uptake depends, however, on a number of environmental and internal factors, of which the molecular properties of the nitrate uptake system probably constitutes only a part. Particular attention has been paid to the relationships between plant N status and the uptake properties, notably in experiments where the nitrate supply has been subject to changes in the short or long terms: in the most extreme cases, plants have been completely deprived of nitrate and the uptake system subsequently 'induced' by nitrate addition.

Short- and long-term effects of nitrate shortage

Clarkson (1986) outlined a series of possible events following the transfer of N-sufficient plants to N-limited conditions. De-repression of the nitrate uptake system may cause a transient expansion of the uptake system, thus increasing V_{max}, whereas a combination of factors, including decay of carrier and inhibition of growth, reduced the uptake capacity during prolonged starvation. An example of an initial increase in nitrate uptake capacity is found in barley, where 10-day-old seedlings transferred to N-free medium for another 3 days considerably increased their net uptake rate (Lee and Rudge 1986). This effect could be attributed to an increased maximum influx rate (I_{max}) rather than to decreased efflux, as determined with [^{13}N]nitrate as a tracer (Lee and Drew 1986).

During long-term acclimatization to sub-optimal nutrient supply, decline of growth is likely to affect the uptake properties. Experimentally, the relationship between nitrate availability, uptake properties, and growth can be studied in cultures maintained and adapted to different relative rates of addition of nitrate-N (Ingestad 1982). V_{max} for nitrate uptake in *Lemna* and *Pisum* cultured at a range of relative rates of N addition decreased with decreased relative N addition rate (P. Oscarson, B. Ingemarsson, and C.-M. Larsson, unpublished). Measurements of [^{13}N]nitrate fluxes in these species (Oscarson *et al.* 1987; Ingemarsson *et al.* 1987*a,b*), indicate that these differences in net uptake capacity are probably related to different influx capacities, i.e. I_{max} values. In these experiments it appears that growth, i.e. N demand, plays an essential role in the regulation of nitrate uptake. A possible control of nitrate uptake by demand is also observed when roots that have been chilled relative to the shoot subsequently gradually increase their specific nitrate uptake rate, which largely offsets the inhibition caused by low temperature (Clarkson *et al.* 1986). A similar 'compensatory' increase in nitrate uptake is seen in the untreated part of the root when a remote part is deprived of external nitrate (Drew and Saker 1975), suggesting that the nitrate uptake system responds to the overall nitrogen demand of the entire plant.

To what extent do these differences in the apparent activity of the nitrate uptake system reflect events at the molecular level? The answer is not to be found in literature which is chiefly concerned with physiological control systems. Control of net uptake through modulation of efflux has been put forward by Deane-Drummond (1984, 1986), but is not supported in experiments employing [^{13}N]nitrate as an influx/efflux tracer (Lee and Clarkson 1986; Lee and Drew 1986; Oscarson *et al.* 1987; Ingemarsson *et al.* 1987*a,b*). Control of nitrate uptake by the internal nitrate level or by nitrogenous intermediates has been suggested (reviewed by Clarkson 1986; Jackson *et al.* 1986). Regulation of uptake capacity *via* the number of active carriers was considered by Lee and Drew (1986); however, the authors also mentioned other possibilities to explain their data, e.g. factors affecting transmembrane ion and potential gradients.

Induction

One of the salient features of nitrate uptake in higher plants is that the ability to take up nitrate is absent or low in plants that have been deprived of nitrate for some time; upon re-addition of nitrate the uptake system is apparently induced, and full activity is usually seen 2–10 h (sometimes longer) after addition of nitrate (reviewed by Clarkson 1986; Jackson *et al.* 1986). Several reports have shown that inhibitors of translation and transcription block the apparent induction (Jackson *et al.* 1973; Tompkins *et al.* 1978; Clarkson 1986), which strongly supports *de novo* synthesis of at least certain polypeptides of importance for nitrate uptake. Application of these substances is likely to have far-ranging effects on metabolism, including nitrate metabolism. Functional NR, however, does not appear to be a prerequisite for operation of the uptake system (Doddema *et al.* 1978). The generally held view is that nitrate itself is the inducer (Jackson *et al.* 1986) while the nitrate analogue, chlorate, does not induce the uptake system (McClure *et al.* 1986).

In several cases, the pattern of induction has been reported to depend on the initial external nitrate concentration (Morgan *et al.* 1985; Mäck and Tischner 1986; Goyal and Huffaker 1986), which may partly be related to the differential responses of several uptake systems. It should be noted that the formation of nitrate has been reported in plants cultured in the absence of nitrate (Funkhouser and Garey 1981) or in the presence of less oxidized N sources (Lee and Drew 1986; Aslam *et al.* 1987; Lips *et al.* 1987) even when care has been taken to avoid bacterial contamination. This observation is important in attempts to assess whether a 'constitutive' uptake system, distinct from the 'inducible' system is present in plant roots, since it is possible that oxidation of reduced inorganic N might lead to the formation of inducing amounts of nitrate within the tissue. There are also some recent data that indicate virtually zero net nitrate uptake in the early stage of induction (Mäck and Tischner 1986; Goyal and Huffaker

1986), which would not seem easily compatible with the presence of a constitutive system. However, a lower K_m for nitrate uptake in *Arabidopsis* starved for nitrate for 7 days (Doddema and Otten 1979) and in *Hordeum* (measured as $^{13}NO_3^-$ influx) grown on ammonium as the sole N source (Lee and Drew 1986), is compatible with the existence of a constitutive system, different from the inducible system. Clearly, the apparent inducibility of the uptake system offers a most promising starting point for a search for nitrate transporters. McClure *et al.* (1987) observed nitrate-induced ^{35}S-labelling of five polypeptides in maize, one of which, a 31 kDa protein, was found in a membrane fraction enriched in tonoplast and/or endoplasmic reticulum, while the other polypeptides were soluble. No nitrate-induced protein labelling was detected in a plasmalemma-enriched fraction. The appearance of the 31 kDa protein was associated with enhanced nitrate uptake. The authors proposed that the protein might be involved in regulation of nitrate uptake, and also indicated the possibility that the 31 kDa protein actually participates in nitrate transport, although no direct evidence is available at present.

Work in our laboratory, in collaboration with the Department of Plant Physiology, University of Lund, has involved SDS–PAGE of protein from plasmalemma vesicles enriched on aqueous two-phase polymer systems (cf. Larsson 1985), in some cases after labelling with [^{35}S]methionine. The experiments have been performed with nitrogen-depleted or ammonium-grown *Lemna* and *Pisum*, subsequently treated with ammonium (control) or nitrate. Preliminary studies have not revealed any nitrate-specific changes in the plasmalemma proteins of *Lemna*, nor in the ^{35}S-labelling of plasmalemma proteins in *Pisum* (B. Ingemarsson, P. Kjellbom, P. Oscarson, Ch. Larsson, and C.-M. Larsson, unpublished). However, recently Dhugga *et al.* (1988*b*) reported nitrate-induced [^{35}S]-methionine labelling of plasmalemma proteins in maize roots, not previously subjected to any N in the medium, which could be distinguished from labelled proteins induced by chloride by application of two-dimensional electrophoresis. The proteins were of several different sizes: 165, 95, 70, and 40 kDa. The participation of any of these proteins in nitrate transport remains, however, obscure.

The further search for nitrate transporters

Physiological examination of nitrate uptake has possibly left us with a large number of distinct transporters, differing from each other in physiological characteristics such as kinetics, mechanism, location, and inducibility (as discussed above). Identification of these transporters, employing a molecular approach, is obviously necessary in order to gain an insight into the regulation of nitrate uptake.

If we accept that transport proteins for nitrate are present in the plasmalemma and tonoplast (at least), the first objective would be to define systems in which membranes can be fractionated according to their origin. Due to the overlapping density of different biomembranes, density gradient centrifugation might be less suitable than other separation techniques, e.g. free-flow electrophoresis (Auderset *et al.* 1986) or aqueous two-phase polymer systems (Larsson 1985; see Bérczi and Møller 1986 for comparison between density gradient centrifugation and two-phase partitioning). The relative enrichment of various membranes can be analysed by using available markers, e.g. silicotungstic or phosphotungstic staining, K^+-stimulated Mg^{2+}-dependent ATPase activity, glucan synthetase II, or light-induced absorbance changes (LIAC) for plasmalemma; Cl^--stimulated NO_3^--sensitive ATPase for tonoplast (reviewed by Larsson 1985).

We also need to be able to identify the transport protein. A number of alternative procedures are at hand, the most straightforward of which seems to be to compare the protein composition of membranes isolated from plants with an apparently repressed uptake system with that of induced plants, employing electrophoretic techniques, preferably after incorporation of a radioisotope into the proteins (see McClure *et al.* 1987; Dhugga *et al.* 1988*b*). A related approach is to use mutants deficient in nitrate transport capacity. Since nitrate and chlorate seemingly share uptake sites, such mutants could be screened and selected for according to their chlorate resistance (McClure *et al.* 1986). Other selection techniques are described in Chapter 2.

Another possibility involves labelling membrane proteins in the absence and presence of an agent that protects the transport protein, notably the substrate or a substrate analogue. This technique was recently applied in the search for a sucrose carrier in plant tissue (Pichelin-Poitevin *et al.* 1987), and also used in the identification of the phosphate translocator of chloroplast envelopes (Flügge and Heldt 1976). A promising compound, phenylglyoxal, which reacts with arginine residues, was shown to inhibit net nitrate uptake in maize roots, and to be somewhat specific in its action (Dhugga *et al.* 1988*a*). Arginine and lysine have previously been shown to be involved in anion binding, e.g. in the erythrocyte chloride carrier (Jay and Cantley 1986). It might also be possible to tag the analogue if a compound can be found that irreversibly binds to the transporter.

The introduction of immunological methods for identification of membrane proteins would also be of great value in the search for nitrate transporters. thus, antibodies raised against plasmalemma vesicles or separated plasmalemma proteins prepared from induced and non-induced roots could be screened for, e.g. their inhibitory action on nitrate transport in protoplasts. It should also be possible to apply immunoabsorption techniques to isolate antibodies against nitrate-induced plasmalemma proteins.

Concluding remarks

Although information on the molecular nature of nitrate uptake is scarce at the moment, one anticipates that current efforts in many laboratories will produce a considerable expansion of our knowledge in the near future. Identification and isolation of nitrate transporters would permit the production of antibodies and subsequently allow differentiation, quantification, and localization of nitrate transporters, as well as provide a means of identification of relevant genes and mRNAs. An interesting aspect of evolutionary significance is the extent of homology, not only between nitrate transporters showing differences in their physiological properties, but also with transporters for other anions, e.g. phosphate and sulphate. For comparison, extensive homology has been found for transport proteins of such diverse biological origin and function as the yeast H^+-ATPase, the K^+-ATPase of *Escherichia coli*, the (Na^+,K^+)-ATPase of kidney, and the Ca^{2+}-ATPase of muscle, indicating a common ancestral transport protein (Serrano *et al.* 1986). The identification of methods to answer these and related questions would be of obvious benefit for both fundamental and applied research on nitrate nutrition.

Acknowledgements

Financial support from the Swedish Natural Science Research Council is acknowledged.

References

Aslam, M., Rosichan, J. L., and Huffaker, R. C. (1987) Comparative induction of nitrate reductase by nitrate and nitrite in barley leaves. *Plant Physiology* **83**, 579–84.

Auderset, G., Sandelius, A. S., Penel, C., Brightman, A., Greppin, H., and Morré, J. (1986). Isolation of plasma membrane and tonoplast fractions from spinach leaves by preparative free-flow electrophoresis and effect of photoinduction. *Physiologia Plantarum* **68**, 1–12.

Bérczi, A. and Møller, I. M. (1986). Comparison of the properties of plasmalemma vesicles purified from wheat roots by phase partitioning and by discontinuous sucrose gradient centrifugation. *Physiologia Plantarum* **68**, 59–66.

Borstlap, A. C. (1981). Invalidity of the multiphasic concept of ion absorption in plants. *Plant, Cell and Environment* **4**, 189–95.

Breteler, H. and Nissen, P. (1982). Effect of exogenous and endogenous nitrate concentration on nitrate utilization by dwarf bean. *Plant Physiology* **75**, 1099–103.

Clarkson, D. T. (1986). Regulation of the absorption and release of nitrate by plant cells: a review of current ideas and methodology. In *Fundamental, ecological and agricultural aspects of nitrogen metabolism in higher plants* (eds. H. Lambers, J. J. Neeteson and I. Stulen), pp. 3–27. Martinus Nijhoff Publishers, Dordrecht.

Clarkson, D. T. and Lüttge, U. (1984). Mineral nutrition: vacuoles and tonoplasts. *Progress in Botany* **46**, 56–67.

Clarkson, D. T., Hopper, M. J., and Jones, L. H. P. (1986). The effect of root temperature on the uptake of nitrogen and the relative size of the root system in *Lolium perenne*. I. Solutions containing both NH_4^+ and NO_3^-. *Plant, Cell and Environment* **9**, 535–45.

Deane-Drummond, C. E. (1984). Mechanism of nitrate uptake into *Chara corallina* cells.

Lack of evidence for obligatory coupling to proton pump and a new NO_3^-/NO_3^- exchange model. *Plant, Cell and Environment* **7**, 317–23.

Deane-Drummond, C. E. (1986). Nitrate uptake into *Pisum sativum* L. cv Feltham First seedlings: commonality with nitrate uptake into *Chara corallina* and *Hordeum vulgare* through a substrate cycling model. *Plant, Cell and Environment* **9**, 41–8.

Dhugga, K. S., Waines, J. G., and Leonard, R. T. (1988*a*). Nitrate absorption by corn roots. Inhibition by phenylglyoxal. *Plant Physiology* **86** 759–63.

Dhugga, K. S., Waines, J. G., and Leonard, R. T. (1988*b*). Correlated induction of nitrate uptake and membrane polypeptides in corn roots. *Plant Physiology* **87** 120–5.

Doddema, H. and Otten, H. (1979). Uptake of nitrate by mutants of *Arabidopsis thaliana*, disturbed in uptake or reduction of nitrate. III. Regulation. *Physiologia Plantarum* **45**, 339–46.

Doddema, H. and Telkamp, G. P. (1979). Uptake of nitrate by mutants of *Arabidopsis thaliana*, disturbed in uptake or reduction of nitrate. II. Kinetics. *Physiologia Plantarum* **45**, 332–8.

Doddema, H., Hofstra, J. J., and Feenstra, W. J. (1978). Uptake of nitrate by mutants of *Arabidopsis thaliana*, disturbed in uptake or reduction of nitrate. I. Effect of nitrogen source during growth on uptake of nitrate and chlorate. *Physiologia Plantarum* **43**, 343–50.

Drew, M. C. and Saker, L. R. (1975). Nutrient supply and the growth of the seminal root system in barley. II. Localized, compensatory increases in lateral root growth and rates of nitrate uptake when nitrate supply is restricted to only part of the root system. *Journal of Experimental Botany* **26**, 79–90.

Epstein, E. (1976). Kinetics of ion transport and the carrier concept. In *Encyclopedia of plant physiology*, New series, Vol. 2B (eds. U. Lüttge and M. G. Pitman), pp. 70–94. Springer-Verlag, Berlin.

Flügge, U. I. (1982). Biogenesis of the chloroplast phosphate translocator. *FEBS Letters* **140**, 273–6.

Flügge, U. I. and Heldt, H. W. (1976). Identification of a protein involved in phosphate transport of chloroplasts. *FEBS Letters* **68**, 259–62.

Funkhouser, E. A. and Garey, A. S. (1981). Appearance of nitrate in soybean seedlings and *Chlorella* caused by nitrogen starvation. *Plant and Cell Physiology* **22**, 1279–86.

Goyal, S. S. and Huffaker, R. C. (1986). The uptake of NO_3^-, NO_2^-, and NH_4^+ by intact wheat (*Triticum aestivum*) seedlings. I. Induction and kinetics of transport systems. *Plant Physiology* **82**, 1051–61.

Granstedt, R. C. and Huffaker, R. C. (1982). Identification of the leaf vacuole as a major nitrate storage pool. *Plant Physiology* **70**, 410–13.

Ingemarsson, B., Oscarson, P., af Ugglas, M., and Larsson, C.-M. (1987*a*). Nitrogen utilization in *Lemna*. II. Studies of nitrate uptake using $^{13}NO_3^-$. *Plant Physiology* **85**, 860–4.

Ingemarsson, B., Oscarson, P., af Ugglas, M., and Larsson, C.-M. (1987*b*). Nitrogen utilization in *Lemna*. III. Short-term effects of ammonium on nitrate uptake and nitrate reduction. *Plant Physiology* **85**, 865–7.

Ingestad, T. (1982). Relative addition rate and external concentration; Driving variables used in plant nutrition. *Plant, Cell and Environment* **5**, 443–53.

Jackson, W. A., Flesher, D., and Hageman, R. H. (1973). Nitrate uptake by dark-grown corn seedings. Some characteristics of apparent induction. *Plant Physiology* **51**, 120–7.

Jackson, W. A., Pan, W. L., Moll, R. H., and Kamprath, E. J. (1986). Uptake, translocation, and reduction of nitrate. In *Biochemical basis of plant breeding*, Vol. 2 (ed. C. A. Neyra), pp. 73–108. CRC Press, Boca Raton.

Jay, D. and Cantley, L. (1986). Structural aspects of the red blood cell anion exchange protein. *Annual Review of Biochemistry* **55**, 511–38.

Kamachi, K., Amemiya, Y., Ogura, N., and Nakagawa, N. (1987). Immuno-gold localization of nitrate reductase in spinach (*Spinacia oleracea*) leaves. *Plant and Cell Physiology* **28**, 333–8.

Kochian, L. V. and Lucas, W. J. (1982). A re-evaluation of the carrier–kinetic approach to ion transport in roots of higher plants. *What's New in Plant Physiology* **13**, 45–8.

Larsson, C. (1985). Plasma membranes. In *Modern methods of plant analysis*, Vol. 1 (ed. H. F. Linskens and J. F. Jackson), pp. 85–104. Springer-Verlag, Berlin.

Lass, B. and Ullrich-Eberius, C. I. (1984). Evidence for proton/sulfate contransport and its kinetics in *Lemna gibba* G1. *Planta* **161**, 53–60.

Lee, R. B. and Clarkson, D. T. (1986). Nitrogen-13 studies of nitrate fluxes in barley roots. I. Compartmental analysis from measurements of ^{13}N efflux. *Journal of Experimental Botany* **37**, 1753–67.

Lee, R. B. and Drew, M. C. (1986). Nitrogen-13 studies of nitrate fluxes in barley roots. II. Effect of plant N-status on the kinetic parameters of nitrate influx. *Journal of Experimental Botany* **37**, 1768–79.

Lee, R. B. and Rudge, K. A. (1986). Effects of nitrogen deficiency on the absorption of nitrate and ammonium by barley plants. *Annals of Botany* **57**, 471–86.

Lopez-Ruiz, A., Verbelen, J. P., Roldán, J. M., and Diez, J. (1985). Nitrate reductase of green algae is located in the pyrenoid. *Plant Physiology* **79**, 1006–10.

Lips, S. H., Soares, M. I. M., Kaiser, J. J., and Lewis, O. A. M. (1987). K$^+$ modulation of nitrogen uptake and assimilation in plants. In *Inorganic nitrogen metabolism* (ed. W. R. Ullrich, P. J. Aparicio, P. J. Syrett and F. Castillo), pp. 233–9. Springer-Verlag, Berlin.

Mäck, G. and Tischner, R. (1986). Nitrate uptake and reduction in sugar-beet seedlings. In *Fundamental, ecological and agricultural aspects of nitrogen metabolism in higher plants* (eds. H. Lambers, J. J. Neeteson and I. Stulen), pp. 33–6. Martinus Nijhoff Publishers, Dordrecht.

MacKown, C. T., Jackson, W. A., and Volk, R. J. (1983). Partitioning of previously accumulated nitrate to translocation, reduction, and efflux in corn roots. *Planta* **157**, 8–14.

Martinoia, E., Heck, U., and Wiemken, A. (1981). Vacuoles as storage compartments for nitrate in barley leaves. *Nature* **289**, 292–4.

Martinoia, E., Schramm, M. J., Kaiser, G., Kaiser, W. M., and Heber, U. (1986). Transport of anions in isolated barley vacuoles. *Plant Physiology* **80**, 895–901.

McClure, P. R., Omholt, T. E., and Pace, G. M. (1986). Anion uptake in maize roots: interactions between chlorate and nitrate. *Physiologia Plantarum* **68**, 107–12.

McClure, P. R., Omholt, T. E., Pace, G. M., and Bouthyette, P. Y. (1987). Nitrate-induced changes in protein synthesis and translation of RNA in maize roots. *Plant Physiology* **84**, 52–7.

Morgan, M. A., Jackson, W. A., and Volk, J. A. (1985). Concentration-dependence of the nitrate assimilation pathway in maize roots. *Plant Science* **38**, 185–91.

Nissen, P. (1974). Uptake mechanisms: inorganic and organic. *Annual Review of Plant Physiology* **25**, 53–79.

Novacky, A., Fischer, E., Ullrich-Eberius, C. I., Lüttge, U., and Ullrich, W. R. (1978). Membrane potential changes during transport of glycine as a neutral amino acid and nitrate in *Lemna gibba* G1. *FEBS Letters* **88**, 264–7.

Oscarson, P., Ingemarsson, B., af Ugglas, M., and Larsson, C.-M. (1987). Short-term studies of nitrate uptake in *Pisum* using ^{13}NO$_3^-$ *Planta* **170**, 550–5.

Pichelin-Poitevin, D., Delrot, S., M'Batchi, B., and Everat-Bourbouloux, A. (1987).

Differential labeling of membrane proteins by N-ethylmaleimide in the presence of sucrose. *Plant Physiology and Biochemistry* **25**, 597–607.

Rathnam, C. K. M. (1978). Malate and dihydroxyacetone phosphate-dependent nitrate reduction in spinach leaf protoplasts. *Plant Physiology* **62**, 220–3.

Reed, A. and Canvin, D. T. (1982). Light and dark controls of nitrate reduction in wheat (*Triticum aestivum* L.) protoplasts. *Plant Physiology* **69**, 508–13.

Roldán, J. M., Romero, F., López-Ruiz, A., Diez, J., and Verbelen, J. P. (1987). Immunological approaches to inorganic nitrogen metabolism. In *Inorganic nitrogen metabolism* (eds. W. R. Ullrich, P. J. Aparicio, P. J. Syrett and F. Castillo), pp. 94–8. Springer-Verlag, Berlin.

Rufty, T. W., Thomas, J. F., Remmler, J., Campbell, W. H., and Volk, R. J. (1986). Intercellular localization of nitrate reductase in roots. *Plant Physiology* **82**, 675–80.

Scholten, H. J. and Feenstra, W. J. (1986). Uptake of chlorate and other ions in seedlings of the nitrate-uptake mutant B1 of *Arabidopsis thaliana. Physiologia Plantarum* **66**, 265–9.

Serrano, R., Kielland-Brandt, M. C., and Fink, G. R. (1986). Yeast plasma membrane ATPase is essential for growth and has homology with $(Na^+ + K^+)$, K^+- and Ca^{2+}-ATPases. *Nature* **319**, 689–93.

Thibaud, J. B. and Grignon, C. (1981). Mechanism of nitrate uptake in corn roots. *Plant Science Letters* **22**, 279–89.

Thompson, A. G., Brailsford, M. A., and Beechey, R. B. (1987). Identification of the phosphate translocator from maize mesophyll chloroplasts. *Biochemical and Biophysical Research Communications* **143**, 164–9.

Tompkins, G. A., Jackson, W. A., and Volk, R. J. (1978). Accelerated nitrate uptake in wheat seedlings. Effects of ammonium and nitrite pretreatments and of 6-methylpurine and puromycin. *Physiologia Plantarum* **43**, 166–71.

Ullrich, W. R. (1987). Nitrate and ammonium uptake in green algae and higher plants: Mechanism and relationship with nitrogen metabolism. In *Inorganic nitrogen metabolism* (ed. W. R. Ullrich, P. J. Aparicio, P. J. Syrett and F. Castillo), pp. 32–8. Springer-Verlag, Berlin.

Ullrich-Erberius, C. I., Novacky, A., Fischer, E., and Lüttge, U. (1981). Relationship between energy-dependent phosphate uptake and the electrical membrane potential in *Lemna gibba* G1. *Plant Physiology* **67** 797–801.

Vaughn, K. C. and Duke, S. O. (1981). Histochemical localization of nitrate reductase. *Histochemistry* **72**, 191–8.

Vaughn, K. C., Duke, S. O., and Funkhouser, E. A. (1984). Immunochemical characterization and localization of nitrate reductase in norflurazon-treated soybean cotyledons. *Physiologia Plantarum* **62**, 481–4.

2. The genetics of nitrate uptake in higher plants

Roger M. Wallsgrove, Hiroshi Hasegawa, Alan C. Kendall, and Janice C. Turner

Introduction

Some of the basic characteristics of nitrate uptake in plants have been dealt with in the previous chapter. We intend here to review what is known about variation and inheritance of nitrate uptake properties in plants, and to outline our approach to the selection of nitrate uptake-deficient mutants. It will soon become obvious that virtually nothing is known about 'the genetics of nitrate uptake in higher plants'—it is hoped that isolation and characterization of uptake mutants will help identify the gene(s) controlling this process.

The molecular details of nitrate uptake by higher plant roots are not yet well understood. Nitrate uptake is known to be 'inducible', with a distinct lag between initial exposure of roots to nitrate and the start of net uptake (Jackson *et al.* 1972). This suggests the synthesis or activation of a specific carrier or carriers. The kinetics of net uptake by such carriers have been interpreted in several ways. Typically, an apparent 'dual phase' response to increasing nitrate concentration is found, and this has been ascribed to dual uptake systems (e.g. Huffaker and Rains 1978), or to a single multi-phase system (Nissen 1974). Whatever the explanation for the observed kinetics, depletion studies have recorded linear rates of nitrate uptake until the concentration in the medium is reduced to 20–30 μM in several plant species (Edwards and Barber 1976; Frota and Tucker 1978; Asher and Edwards 1978).

Species and cultivar variation

Despite the similar responses shown in such depletion studies, species differences in kinetic properties have been reported (Table 2.1). Some caution is needed in evaluating such data, which comes from different groups using different methods, though there is an apparently lower affinity for nitrate in barley roots than in the other species. A direct comparison of nitrate uptake and utilization was made between two wheat genotypes by Rao *et al.* (1977). The varieties Anza and UC44-111 differed markedly in nitrate content and *in vitro* nitrate reductase activity. Despite this, *in vivo* NO_3^- reduction rates were very similar, and the minor differences detectable in short-term uptake rates were not found in the long term.

Table 2.1. Nitrate uptake from nutrient solution by different crop plants (from Cregan and van Berkum 1984)

Species	Temperature (°C)	pH	Concentration for half maximum uptake (μM)
Barley	30	6.0	110
Perennial ryegrass	25	6.2	33
Maize	25	6.0	21

Another detailed study using maize revealed significant differences in long-term nitrate uptake between different inbred lines (Chevalier and Schrader 1977). Hybrids between these lines were studied to monitor the inheritance of such characteristics, but the authors were forced to conclude that 'no apparent relationship existed between inbred parents and their F_1 progeny in ability to remove NO_3^- from solution' (Chevalier and Schrader 1977). Most notably, crossing the two inbred lines with the lowest uptake produced a hybrid with the highest uptake! There is, therefore, little evidence for the heritability of differing uptake traits, and nothing that would help us to identify any gene or genes affecting nitrate uptake.

Studies with mutants

If 'classical' genetics provides no clues, the use of defined mutants may be of help. Tracing biochemical pathways *via* mutants is a well-established technique in microbiology, and is being increasingly exploited in higher plants (see e.g. King 1984; Bright *et al.* 1984).

With a single exception, such an approach has not been used to study nitrate uptake in plants, whereas in filamentous fungi selection for chlorate resistance produced *Aspergillus nidulans* mutants deficient in nitrate (and chlorate) uptake. The affected gene was designated *crnA* (Chapter 6) and its expression found to be under the control of the *areA* gene that mediates nitrogen metabolite represssion (Chapters 6 and 9). Somewhat earlier, the same approach identified the one known higher plant NO_3^- uptake mutant. One line of chlorate-resistant *Arabidopsis thaliana*, designated B1, proved to have reduced nitrate uptake (Oostindier-Braaksma and Feenstra 1973). Kinetic studies on nitrate uptake in

Arabidopsis gave evidence for a dual-phase response, a high affinity Phase I (K_m^1 40 μM, saturated at 0.1 mM NO_3^-) and a lower affinity Phase II (K_m^1 25 mM, 'V_{max}' 200-fold higher than Phase I) (Doddema and Telkamp 1979). Mutant line B1 appeared to lack Phase II (kinetics of uptake at low [NO_3^-] being identical to the wild-type), and although this mutation in a single nuclear gene was expressed both in whole plants and in tissue culture, the phenotype was not apparent in protoplasts (Scholten and Feenstra 1986). Some characteristics of this line suggest that the mutation may not affect a specific nitrate carrier. Chlorate uptake was reduced, as expected, but lower uptake of both Cl^- and K^+ is harder to explain.

At Rothamsted we have been successful in selecting barley (*Hordeum vulgare*) mutants with defects or alterations in a wide variety of biochemical pathways (see Bright *et al.* 1984), including root amino acid uptake (Bright *et al.* 1979, 1983*a*). A programme to select mutants with defects in nitrate uptake was started as part of a larger project to identify the gene(s) involved in the uptake process. In particular, we were interested in identifying mutants with reduced uptake from low NO_3^- concentrations ('Phase I', by analogy with *Arabidopsis*). We had previously used a chlorate-resistance screen which produced nitrate reductase-deficient (*cnx*-type) lines (Bright *et al.* 1983b) but no uptake-deficient mutants.

Selection of uptake-deficient barley mutants

The starting material, as for previous selection programmes, was the M_2 generation of azide-treated barley. This chemical mutagenesis has proved to give good mutation rates coupled with good fertility; selecting in the M_2 generation yields homozygous mutants and thus gives us the ability to find recessive mutations in this diploid species.

As chlorate resistance did not appear to be an effective screen, methods of directly selecting uptake mutants were sought. In the first scheme devised, barley seedlings were grown in bulk liquid culture with 50 μM KNO_3 as sole nitrogen source (Fig. 2.1). Wild-type plants had no apparent nutrient stress symptoms under these conditions, whereas at nitrate concentrations below 20 μM they grew slowly, had pale leaves, and prominent anthocyanin production. After 15–17 days, we looked for individual seedlings showing such 'nitrogen stress', and transferred them to individual culture pots. Their ability to take up nitrate was measured by following nitrate depletion from the medium, using either HPLC determination of NO_3^-, or an ion-specific electrode (see below). All plants with low or zero uptake were then potted into soil.

A second, more labour-intensive, procedure was also used. Seedlings were grown for 6–7 days in 0.5 mM $CaSO_4$, and then transferred individually to 10 ml 250 μM KNO_3 in 0.5 mM $CaSO_4$. After 24 h, the nitrate concentration was measured with a nitrate electrode (Russell ion-selective electrode; ECM 202 meter, EDT Research Ltd). The electrode response to nitrate concentration is

shown in Fig. 2.2a. Wild-type plants reduced the nitrate concentration below detectable limits in 24 h under these conditions. Any seedlings that failed to do so were returned to $CaSO_4$ prior to a secondary screen 7 days later, in which plants were 'induced' for 24 h in 250 μM KNO_3 before transfer to 10 ml of the same solution as before. Nitrate depletion was then monitored over 3 h (see Fig. 2.2b for the uptake curves for two parent variaeties). All plants still showing low uptake at this stage were recovered and either potted into soil or grown on in liquid culture. Inconsistent results with this procedure were traced to variation in seed reserve depletion—plants took up little NO_3^- before the endosperm reserves were exhausted. Delaying the primary screen or physically removing the endosperm before screening overcame this problem, and expressing results on a root length or seedling fresh weight basis allowed more accurate detection of variant individual plants.

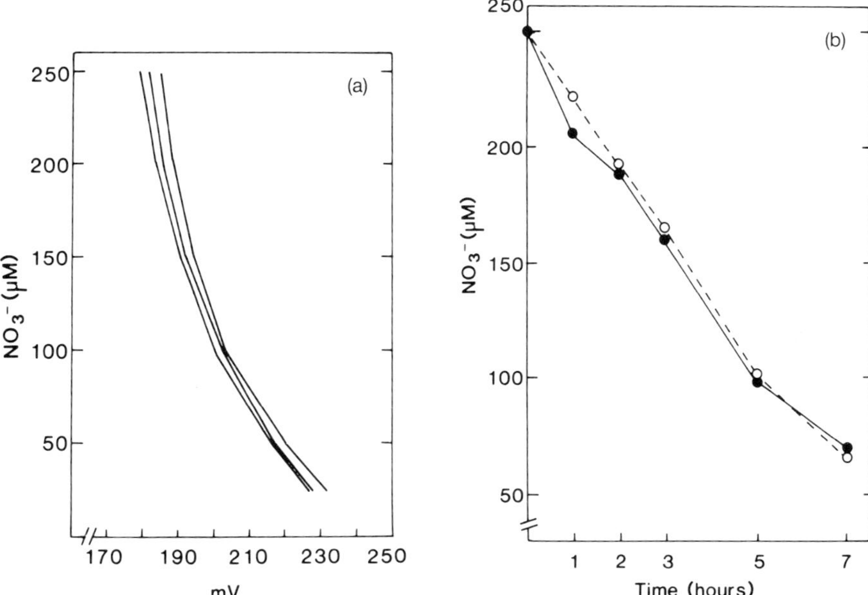

Fig. 2.2. (a) Response of the nitrate electrode to changes in nitrate concentration. The three curves represent data from three separate calibrations. (b) Time course of nitrate uptake by a seedling of two barley cultivars. Solid line—cv. Apex; dotted line—cv Patty.

Overall results from both selection procedures are given in Table 2.2. It is clear that the electrode screening method has been much more productive. In part this is because of the difficulty in recognizing 'uptake' mutants in the bulk tanks against a background of other mutants with visible effects—chlorophyll deficiency, morphological abnormalities, and other deficiencies. With electrode screening such obviously abnormal plants are discarded immediately. Some nitrate uptake mutants might also have escaped detection because their slow growth rates were not immediately obvious against such a varied background.

Unfortunately, many of the selected plants did not survive to maturity, or did not produce seeds. We at first attempted to grow all selected plants in soil or liquid culture in the presence of high concentrations (> 10 mM) of NO_3^-, assuming that any defect in uptake from lower concentrations could be compensated for by 'Phase II'-type uptake, or diffusion. As discussed later, this strategy may have

Table 2.2. Frequency of nitrate uptake deficient or reduced nitrate uptake seedlings in the M_2 generation

Variety	No. screened	Primary screen		Secondary screen	
		Selected	Frequency ($\times 10^{-3}$)	Selected	Frequency ($\times 10^{-4}$)
Bulk liquid culture method					
Apex	9247	49	5.3	7	7.57
Klaxon	15790	26	1.65	4	2.53
Patty	23520	82	3.49	4	1.70
Total	48557	157	3.23	15	3.09
Electrode screening					
Apex	1577	21	13.23	8	50.73
Klaxon	8922	52	5.83	10	11.21
Patty	30213	227	7.51	27	8.94
Totals	40712	300	7.37	45	11.05

been seriously wrong. Of those lines that did produce seed, not all gave rise to progeny with the mutant phenotype, and other lines gave rise to segregating M_3 populations, some individuals having low uptake, some wild-type, and some intermediate. The variation seen in Fig. 2.3 is typical, in this case M_3 seedlings from a line designated RNu (Rothamsted Nitrate uptake) 8639. Insufficient seeds are available from any given line for a reasonable estimate of the ratio of wild-type to mutant phenotype, but in all the lines so far tested less than half of the M_3 progeny carry the low uptake phenotype. For some lines M_4 seed (derived from low-uptake M_3 parents) is becoming available, and analysis of the characteristics of these plants may shed more light on the nature of the inheritance of the mutations. We obviously hope to find one or more pure-breeding uptake-deficient lines for full genetic analysis and characterization. If the M_4 generation is still segregating, it may indicate that homozygous mutations in the particular gene(s) are lethal, and that we initially selected a heterozygous plant. These possibilities are all based on the unproven assumption that we are dealing with stable mutations in single nuclear genes.

A further interesting problem has come to light with some of our selected lines. Uptake-deficient M_3 individuals of line RNu8633 (Fig. 2.4) were grown in liquid culture with either 1 mM KNO_3, 1 mM $(NH_4)_2SO_4$ or 1 mM urea as nitrogen source. As is obvious from Fig. 2.5, growth in nitrate was severely inhibited compared to growth on either of the other media. The individual plant RNu8633/7 failed to survive on nitrate, whereas RNu8633/8 fully recovered after rescue by transfer to urea-containing medium. Nitrate toxicity had been previously found in nitrate reductase-deficient barley (Bright *et al.* 1983b). Brief

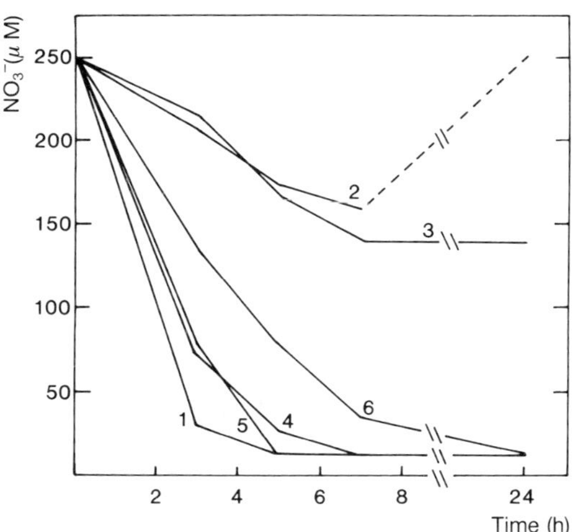

Fig. 2.3. The variation in uptake characteristics in individual M_3 seedlings (1–6) from mutant line RNu8693.

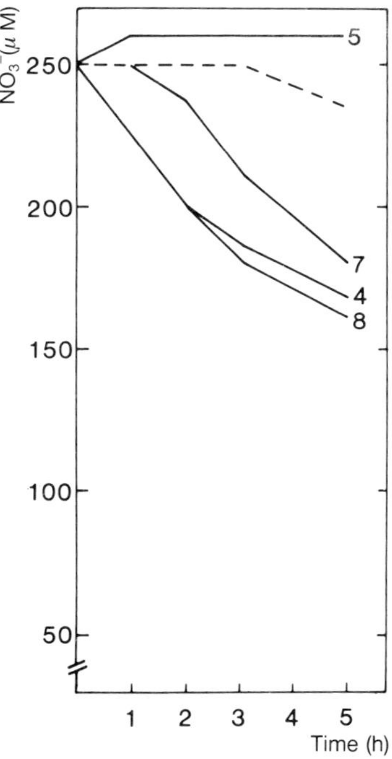

Fig. 2.4. Uptake characteristics of individual 'low-uptake' M_3 seedlings (4, 5, 7, and 8) of mutant line RNu8633 and the M_2 parent line (dashed line).

Fig. 2.5. The growth of RNu8633 M_3 seedlings in the presence of different nitrogen sources (1 mM). Left, plant grown on urea as a sole nitrogen source; centre, plant grown on nitrate as a sole nitrogen source; right, plant grown on ammonium as a sole nitrogen source. Upper picture: plants removed from growth medium. Lower picture: plants growing in pots.

exposure (24 h) to nitrate induced readily detectable nitrate reductase activity in RNu8633 M_3 seedlings, assayed either *in vivo* or *in vitro*, so this line is not deficient in nitrate reductase. So far we have no clear idea of the cause of such nitrate toxicity, which was not found in some of the other lines. As a precaution, newly-selected M_2 lines are being routinely grown in liquid culture with urea. Some of our earlier lines may have been lost because of similar problems with growth in nitrate.

Conclusions

Very little is known about the genetics of nitrate uptake in higher plants. Conventional breeding studies have not found a clear relationship between parent and progeny in nitrate uptake properties. One gene has been identified in *Arabidopsis* that affects uptake at high NO_3^- concentrations but not at low concentrations, and mutation in this gene also reduces Cl^- and K^+ uptake. Direct screening of mutated barley seedlings for reduced uptake has been successful in selecting several 'uptake-deficient' lines, but none of these is yet genetically stable and, therefore, has not so far been characterized. When (if) stable lines are available, the physiology and genetics of such mutants should prove invaluable in the search to understand nitrate uptake in plants and to identify the gene(s) involved in this process.

Acknowledgements

We gratefully acknowledge the assistance of Julian Franklin in designing and maintaining liquid culture systems.

References

Asher, C. J. and Edwards, D. G. (1978). Critical external concentrations for nutrient deficiency and excess. *Proceedings of the 8th International Colloquium on Plant Analysis and Fertiliser Problems, Auckland 1978.* Vol. 1, pp. 13–28. N.Z. DSIR Information series No. 134. Wellington, N.Z.

Bright, S. W. J., Featherstone, L. C., and Miflin, B. J. (1979). Lysine metabolism in a barley mutant resistant to S(2-aminoethyl)cysteine. *Planta* **146**, 629–33.

Bright, S. W. J., Kueh, J. S. H., and Rognes, S. E. (1983a). Lysine transport in two barley mutants with altered uptake of basic amino acids in the root. *Plant Physiology* **72**, 821–4.

Bright, S. W. J., Norbury, P. B., Franklin, J., Kirk, D., and Wray, J. L. (1983b). A conditional-lethal *cnx* type nitrate reductase-deficient barley mutant. *Molecular and General Genetics* **189**, 240–4.

Bright, S. W. J. *et al.* (1984). Manipulation of key pathways in photorespiration and amino acid metabolism by mutation and selection. In: *The genetic manipulation of plants and its application to agriculture* (eds P. J. Lea and G. R. Stewart), pp. 141–69. Oxford University Press.

Chevalier, P. and Schrader, L. E. (1977). Genotypic differences in nitrate absorption and partitioning of N among plant parts in maize. *Crop Science* **17**, 897–901.

Cregan, P. B. and van Berkum, P. (1984). Genetics of nitrogen metabolism and physiological/biochemical selection for increased grain crop productivity. *Theoretical and Applied Genetics* **67**, 97–111.

Doddema, H. and Telkamp, G. P. (1979). Uptake of nitrate by mutants of *Arabidopsis thaliana*, disturbed in uptake or reduction of nitrate. *Physiologa Plantarum* **45**, 332–8.

Edwards, J. H. and Barber, S. A. (1976). Nitrogen uptake characteristics of corn roots at low N concentrations—influenced by plant age. *Agronomy Journal* **68**, 17–19.

Frota, J. N. E. and Tucker, T. C. (1978). Absorption rates of ammonia and nitrate by red kidney beans under salt and water stress. *Soil Science Society of America Journal* **42**, 753–6.

Huffaker, R. C. and Rains, D. W. (1978). Factors influencing nitrate acquisition by plants: assimilation and fate of reduced nitrogen. In *Nitrogen in the environment*, Vol. 2 (eds. D. M. Nielson and J. G. MacDonald), pp. 1–43. Academic Press, New York.

Jackson, W. A., Volk, R. J., and Tucker, T. C. (1972). Apparent induction of nitrate uptake in nitrate-depleted plants. *Agronomy Journal* **64**, 518–21.

King, P. J. (1984). From single cells to mutant plants. *Oxford Surveys of Plant Cell and Molecular Biology* **1**, 7–32.

Nissen, P. (1974). Uptake mechanisms: inorganic and organic. *Annual Review of Plant Physiology* **25**, 53–79.

Oostindier-Braaksma, F. J. and Feenstra, W. J. (1973). Isolation and characterisation of chlorate resistant mutants of *Arabidopsis thaliana*. *Mutation Research* **19**, 175–85.

Rao, K. P., Rains, D. W., Qualset, C. D., and Huffaker, R. C. (1977). Nitrogen nutrition and grain protein in two spring wheat genotypes differing in nitrate reductase activity. *Crop Science* **17**, 283–6.

Scholten, H. J. and Feenstra, W. J. (1986). Expression of the mutant character of chlorate-resistant mutants of *Arabidopsis thaliana* in cell culture. *Journal of Plant Physiology* **123**, 45–54.

Nitrate Reduction

3. Involvement of the chlorate resistance loci and the molybdenum cofactor in the biosynthesis of the *Escherichia coli* respiratory nitrate reductase

David H. Boxer

Introduction

Escherichia coli appears to be unable to utilize nitrate or nitrite as a nitrogen source for growth (Kobayashi and Ishimoto 1973) and there have been no substantiated reported of an assimilatory nitrate reductase (NR) in the bacterium. High levels of NR (nitrate:benzylviologen oxidoreductase) activity are, however, found when the bacterium is grown anaerobically in the presence of nitrate. This activity is associated with the membrane fraction and is catalysed by the respiratory (dissimilatory) NR, a terminal respiratory complex which catalyses the reduction of nitrate to nitrite, deriving the necessary electrons from the quinone pool with concomitant transmembrane proton translocation leading to proton motive force generation across the cytoplasmic membrane (Garland *et al.* 1975). Nitrate can act as the terminal electron acceptor for a number of respiratory pathways employing a variety of electron donors including NADH, D-lactate, glycerol-3-phosphate, fumarate, and formate (Ingledew and Poole 1984).

Structure of nitrate reductase

Nitrate reductase consists of three subunits α, β, and γ with approximate molecular weights of 155, 50 and 20 kDa, respectively (MacGregor *et al.* 1974; Enoch and Lester 1975). The enzyme contains one atom of molybdenum and about 18 atoms each of iron and acid-labile sulphide per 235 kDa (Morpeth and Boxer 1985). The molybdenum, present as the molybdenum cofactor (MoCo), and the majority of the non-haem iron appear to be associated with the α subunit (Chaudry and MacGregor 1983*a*). The site for nitrate reduction is thought to be at the molybdenum centre and therefore would be located on this subunit: this is in accord with molybdenum-electron paramagnetic spectroscopic analysis (George *et al.* 1985) and chemical modification studies (Graham and Boxer 1980*a*). The α-subunit is located at the cytoplasmic face of the membrane (Boxer and Clegg 1975; MacGregor and Christopher 1978), the side of the membrane at which the nitrate is reduced (Kristjannson and Hollocher 1979). Little is known concerning the function of the β-subunit. No prosthetic groups have been ascribed to this polypeptide, which is located at the cytoplasmic face of the membrane (Graham and Boxer 1980*b,c*). The β-subunit appears to be involved in the maintenance of the quaternary structure of the enzyme in its isolated form (DeMoss 1977). The γ-subunit is the apoprotein of cytochrome b_{556} which is found in the more intact preparations of the enzyme (Chaudry and MacGregor 1983*b*). It is highly hydrophobic in nature and readily forms aggregates, which probably explains why some workers have failed to detect the subunit in certain cytochrome-bearing preparations of the enzyme. It is the only subunit for which a complete amino acid sequence (deduced from the DNA sequence) is available (Sodergren and DeMoss 1988). *In situ* surface labelling studies locate this subunit, at least in part, at the periplasmic face of the membrane (Boxer and Clegg 1975) which, taken with the locations of the α- and β-subunits, indicate that the enzyme spans the cytoplasmic membrane. Hydropathy plots for the γ-subunit suggest that it may also span the membrane (Sodergren and DeMoss 1988).

Mechanism for energy conservation in nitrate respiration

Exploiting the permeability properties of viologen dyes and a mutant strain unable to synthesize haem, Garland and co-workers (Jones and Garland 1977; Jones *et al.* 1980) showed that quinol was oxidized at the periplasmic surface of the membrane by the cytochrome b_{556} of the NR complex. This oxidation liberates two protons at the periplasm and the two electrons so generated are passed from cytochrome b_{556} across the membrane *via* the enzyme to the nitrate reduction site, at which two cytoplasmic protons are consumed for each nitrite produced (Fig. 3.1). This mechanism explains the observed stoichiometry of two protons electrogenically translocated per two electrons passing through the nitrate reductase complex. Detailed kinetic studies on the purified enzyme confirmed the presence of two distinct active sites for the ubiquinol-1-dependent nitrate reduction reaction and highlighted the cytochrome as the site of quinol oxidation (Morpeth and Boxer 1985).

Genetics of nitrate reductase

The genes encoding the NR polypeptides form an operon, *nar* (originally *chlC*), which is located at 27 min on the chromosome (Bonnefoy-Orth *et al.* 1981;

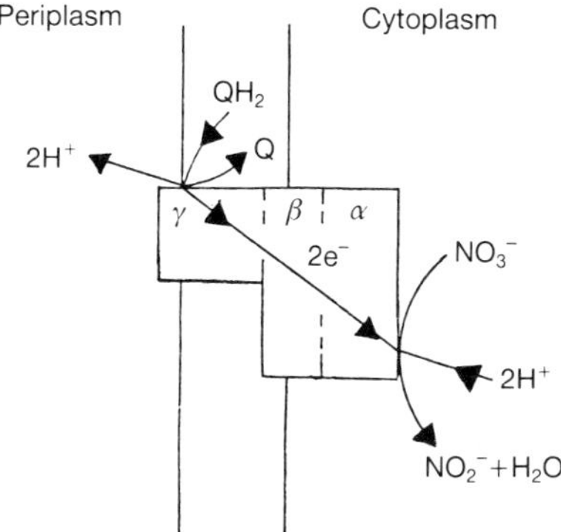

Fig. 3.1. Proton translocation catalysed by respiratory nitrate reductase. The vertical lines represent the membrane. The broken lines represent uncertainties in the positions of the subunits.

Edwards *et al.* 1983; Rondeau *et al.* 1984). The genes encoding the α-, β-, and γ-subunits are designated *narG*, *narH*, and *narI* respectively. A fourth gene (*narJ*), of unknown function, has also been placed in this operon (Sodergren and DeMoss 1988). The genes are arranged in the order *GHJI*, with *narG* in the promoter-proximal position. The promoter region has been sequenced (Li and DeMoss 1987). The use of gene fusions to the *nar* operon has demonstrated that this operon is regulated at the level of transcription (Fimmel and Haddock 1979; Chippaux *et al.* 1981; Stewart 1982). Expression is inducible by anaerobiosis and nitrate: the anaerobic induction of the *nar* operon requires the function of the gene product of the pleiotropic anaerobic regulator *fnr* (Lambden and Guest 1976; Stewart 1982).

Several other genes, *narL*, *narK* and *narX*, are involved with NR or its control and are located close to the *narGHJI* operon. Nitrate induction requires the *narL* gene product, which apparently functions as a positive regulator (Stewart 1982; Stewart and MacGregor 1982). The functions of *narK* and *narX* are unclear (Stewart and Parales 1988).

A second NR structural gene in *E. coli* has recently been reported (Bonnefoy *et al.* 1987) and its product characterized (Iobbi *et al.* 1987). This second enzyme appears respiratory in function, but its role and significance are presently unknown.

The chlorate-resistant mutants

In addition to mutants which lack NR activity due to lesions at the *nar* locus, a second general class of NR-deficient mutants, the chlorate-resistant mutants may be defined. These are distinguished from *nar* mutants by their pleiotropic loss of the activities of all known molybdoenzymes, including NR and map to many different loci on the *E. coli* chromosome. Because the mutations are pleiotropic in nature they cannot define lesions at structural genes of the molybdoenzymes. The chlorate-resistant mutants are so called because they are able to grow anaerobically in the presence of chlorate. Wild-type *E. coli* is sensitive to high levels of potassium chlorate (15 mM) during anaerobic growth, due to the presence of 'chlorate reductase' activity, which converts chlorate to toxic chlorite. The lack of all cellular molybdoenzyme functions results in the loss of chlorate reductase activity, and allows growth in the presence of chlorate (Stewart and MacGregor 1982). Purified NR is an efficient chlorate reductase (Morpeth and Boxer 1985).

Selection of mutants with chlorate has been used extensively to isolate strains deficient in NR (Venables and Guest 1968; Pinchinoty *et al.* 1969; Glaser and DeMoss 1968), although they normally exhibit pleiotropic loss of molybdo-enzymes rather than the specific loss of NR (Glaser and DeMoss 1971; Dubourdieu *et al.* 1976). Mutants defective in the structural genes for NR have been isolated at low frequency by such procedures, especially when lower concentrations of chlorate were used (see below). Seven loci in *E. coli* have been reported to give rise to resistance to chlorate: *chlA*, *chlE*, and *chlD* have been mapped to the 17–18 min region of the chromosome (Venables and Guest 1968); *chlB* to 86 min (Casse 1970); *chlC* and *chlF* to 27 min (Puig *et al.* 1969; Guest 1969; Glaser and DeMoss 1971); *chlG* to 0 min (Glaser and DeMoss, 1972). The *chlF* mutants have been lost without further characterization, but they may correspond to *narL*, *narK*, or *narX*. The *chlC* (*narGHJI*) mutants exhibit resistance to only low levels of chlorate (Stewart and MacGregor 1982) and were subsequently demonstrated to encode the NR polypeptides (see above). This class of mutants possesses normal activities for all molybdoenzymes except nitrate reductase.

The involvement of the *E. coli* chlorate resistance loci with the MoCo was first postulated by Dubourdieu *et al.* (1976). Since all *E. coli* molybdoenzymes are thought to contain the MoCo (Chapter 14), the pleiotropic effects of mutations in these genes are readily explained by lack of this functional cofactor: this explanation appears satisfactory, except for the *chlB* locus (see below). All of the *chl* loci have been cloned (Reiss *et al.* 1987) and their molecular genetic study is underway. Some of the earlier work is compromised by the use of undefined point mutations and the use of non-isogenic control strains. A well characterized set of Mu *c*ts insertion mutants has been isolated by Stewart and MacGregor (1982) and most workers now use these strains.

The *chlA* locus is reported to possess three complementation groups (Reiss *et al.* 1987) although recent work in our laboratory has revealed a fourth (S. L. Rivers and D. H. Boxer, unpublished observations). MoCo activity is absent from a series of *chlA* insertion mutants, consistent with this locus encoding functions for cofactor biosynthesis (Miller and Amy 1983). Johnson and Rajagopalan (1987*a*), pursuing the observation that extracts from a *chlA* mutant gave a positive result in the *Neurospora crassa nit-1* complementation assay for MoCo (Amy 1981; Miller and Amy 1983) demonstrated that the *chlA* extract contained an enzyme which converted an inactive precursor in the *nit-1* preparation into functional MoCo. A *chlM* gene also at the *chlA* locus is required for the converting activity (Johnson and Rajagopalan 1987*b*). This directly shows that the *chlA* locus encodes molybdopterin biosynthetic activity.

We have constructed *lacZ*-gene fusions at the *chlA* locus in which β-galactosidase expression, which is placed under *chlA* control, is low under aerobic growth conditions but is increased by about 20-fold during anaerobiosis. This anaerobic induction is independent of the anaerobic regulator *fnr*. Mobilization of the gene fusions into bacteriophage λ (*lac*$^+$)transducing phage and construction of λ(*chlA'*::*'lacZ*) lysogens in a *chlA*$^+$ background leads to a drop in the overall level of expression. The high expression of the *chlA'*::*'lacZ*/*chlA*$^+$ merodiploid strains is restored if other mutations leading to loss of MoCo activity are introduced, indicating that *chlA* expression is repressed by MoCo or a closely related substance (K. P. Baker and D. H. Boxer, unpublished observations). This is in accord with the molybdopterin biosynthetic function of the locus. The anaerobic regulation is independent of cofactor repression.

The *chlB* locus has been reported to contain three complementation groups (Venables 1972). The available clones, however, contain only 1.1 kb of DNA (Reiss *et al.* 1987) and complement all *chlB* mutants examined to date. Analysis of β-galactosidase expression from *chlB'*::*'lacZ* gene fusions, indicates that *chlB* expression is low, constitutive, and uninfluenced by secondary mutations in other *chl* genes (P. W. Whitty and D. H. Boxer, unpublished observations). Part of the DNA sequence of the *chlB* region has been reported (Reiss and Klingmüller 1987). Mutants at this locus are unique in that, despite having lost the activities of the molybdoenzymes, they still possess MoCo activity (Miller and Amy 1983; Saracino *et al.* 1986). The active product of the *chlB* locus appears to be protein FA (see below; Rivière *et al.* 1975). None of the other chlorate resistance loci is required for protein FA activity (Low *et al.* 1988).

The *chlD* mutants can be suppressed by growth in media containing high molybdate concentrations (Glaser and DeMoss 1971). The *chlD* region has been cloned and partially sequenced (Johann and Hinton 1987). Homology of a *chlD* open reading frame with membrane proteins of transport systems led Johann and Hinton (1987) to propose that this locus encodes a molybdate transport function. This would be consistent with the low levels of nitrate reductase activity normally found in *chlD* mutants. Detailed genetic analysis of the *chlD* locus has not been reported. Expression of *chlD* is repressed with high (>10 nM) molybdate in the growth medium (Miller *et al.* 1987) and is unaffected by secondary mutations at any of the other chlorate resistance loci.

Complementation analysis of *chlE* mutants with clones of the *chlE* region revealed two complementation groups (Reiss *et al.* 1987) in agreement with earlier classical genetic analysis (Venables 1972). Johnson and Rajagopalan (1987*b*) have also distinguished two classes of mutant, *chlE* and *chlN*, at the locus on the basis of *in vitro* MoCo complementation assays. Both *chlE* and *chlN* encode functions required for molybdopterin biosynthesis.

The *chlG* mutants share some features with the *chlD* class of mutants, in that the blockage in molybdoenzyme activity is not usually complete and the phenotype can be somewhat restored by growth in the presence of additional molybdate (Stewart and MacGregor 1982). Strain background also appears to influence the *chlG* phenotype (Buxton and Drury 1984). A problem with this locus is that few alleles are available since most mutageneses produce few *chlG* mutants. The region has been cloned as a 1.0 kb DNA fragment (Reiss *et al.* 1987). Apparently wild-type amounts of molybdopterin are synthesized by *chlG* mutants when normally grown. Active MoCo is only found in the *nit-1* complementation assay of *chlG* extracts if large amounts of molybdate are added to the complementation mixture. This is interpreted as the presence of demolybdo-cofactor in the cell extract (Miller and Amy 1983).

In vitro *complementation of chlorate-resistant mutants*

A route towards the understanding of the biochemistry of the chlorate resistance gene products was provided by Azoulay *et al.* (1967) who showed that the mixing of soluble fractions from a *chlA* and a *chlB* mutant, under anaerobic conditions, gave rise to active NR. Subsequently, it has been found that other *E. coli* molybdoenzymes can also be activated in this manner (Terrière *et al.* 1981; Giordano *et al.* 1984). Appreciable quantities of NR protein, presumably intermediates in the biosynthesis of NR, are found in the *chl* mutants (Graham *et al.* 1980). The complementation process has been used to analyse the biochemical defects in the *chl* mutants, and works well with *chlB* extracts mixed with extracts of any of the other *chl* mutants, indicating that the NR precursor in *chlB* mutants is readily activated (MacGregor and Schnaitman 1973, 1974; Saracino *et al.* 1986). Active protein FA is present in all the extracts added to the *chlB* fraction (Low *et al.* 1988) which would account for these observations.

The complementation approach has allowed the purification of some putative active products of some of the *chl* loci. Fractionation of a *chlA* extract led to the isolation of protein FA which, when added to a *chlB* extract, restored NR activity (Rivière *et al.* 1975). Genetic evidence is consistent with protein FA being a product of the *chlB* locus (Low *et al.* 1988). The cloned *chlB* locus present in a high copy number plasmid leads to the over-production of protein FA activity (P. W. Whitty and D. H. Boxer, unpublished observations). The protein FA activity is uninfluenced by growth conditions, including the presence of tungstate, which suggests that it does not contain molybdenum (Low *et al.* 1988).

A second protein, protein PA, has been identified in a similar manner to protein FA but by fractionating in this case the *chlB* soluble fraction and obtaining a purified preparation which, when added to the *chlA* extract, restores NR activity (Grillet and Giordano 1983). This 70 kDa protein is presumed to be a product of the *chlA* locus. Protein PA has been detected immunologically in various *chl* mutants and its synthesis appears constitutive (Giordano *et al.* 1985). Unlike protein FA, however, its activity is lost following growth in the presence of tungstate. The activity of purified protein PA is low and it was suggested that a component required for its activity was lost during the purification (Grillet and Giordano, 1983). Later work showed that the activity of protein PA could be greatly increased if exposed to preparations containing the active MoCo (Giordano *et al.* 1987). It is thought that protein PA functions as a carrier of MoCo upon which its activity depends. The assignment of protein PA as a *chlA* gene product, therefore, requires further investigation since its complementing activity with *chlA* extracts may arise from the presence of the cofactor which is absent from the *chlA* strain. MoCo has previously been shown to be bound to a protein in *E. coli* extracts, but the reported size of the binding protein is rather less than that quoted for protein PA (Johnson *et al.* 1984). Protein PA and protein FA participate in the *in vitro* activation of several *E. coli* molybdoenzymes in *chl* extracts (Giordano *et al.* 1984).

MoCo involvement in the activation by *in vitro* complementation of molybdoenzymes from numerous sources is well documented (Rajagopalan 1988). However, the activation in this manner of an *E. coli* apomolybdoenzyme has only recently been achieved despite the relatively advanced state of knowledge of *E. coli* molybdoenzymes and their genetics. NR and trimethylamine-N-oxide reductase activation in extracts from tungstate-grown, *chlB* mutants requires, along with protein FA, a source of active MoCo (Saracino *et al.* 1986; Silvestro *et al.* 1986).

Chlorate resistance genes and the biosynthesis of nitrate reductase

The ability to activate inactive NR precursors *in vitro* and the comparative abundance of these precursors in the chlorate-resistant strains have allowed progress to be made towards a molecular understanding of the steps involved in NR biosynthesis with respect to cofactor acquisition and provided a link between the genetics and the biochemistry of the system. Dialysis or gel filtration of the *chlB* soluble fraction to remove low molecular weight material leads to the loss of NR activation on addition of protein FA. The activation can, however, be restored by the addition of heat-treated (100°C for 8 min) undialysed *chlB* crude extract (Boxer *et al.* 1987) from the wild type or any of the chlorate-resistant strains, even those lacking active MoCo. Clearly some heat-stable, low molecular weight substance distinct from the active MoCo is required for the protein FA-dependent activation of the NR in the *chlB* strain. This substance (X) awaits molecular characterization. The addition of ATP increases the efficiency of the activation but X is distinct from ATP. As stated above, *chlB* mutants possess MoCo activity in soluble fractions and this prompted MacGregor and Schnaitman (1974) to propose that the *chlB* product (protein FA) is required for the insertion of the cofactor into aponitrate reductase. We have purified the inactive NR precursor from a *chlB* mutant and elemental analysis showed that it contained stoichiometric amounts of molybdenum (D. C. Low, F. F. Morpeth and D. H. Boxer, unpublished observations). Protein FA, therefore, is not involved in the association of MoCo with aponitrate reductase. Similar analysis showed that the inactive NR precursor in a *chlA* mutant did not contain molybdenum. The NR precursors characterized from both the *chlA* and *chlB* mutants contained normal amounts of iron and acid-labile sulphide. The polypeptides of the enzyme precursors were indistinguishable from those in the fully functional, mature enzyme.

The above findings indicate a multistep pathway for the acquisition of prosthetic groups in the biosynthesis of NR (Fig. 3.2). Iron–sulphur centres are acquired by the aponitrate reductase at an early stage by an unknown mechanism, to form the demolybdoenzyme precursor present in the *chlA* mutant MoCo, which requires *chlA*, *chlE*, *chlD*, and *chlG* gene products for its synthesis, would then be bound to the complex. Protein PA is probably involved at this stage, either in stabilizing the cofactor or assisting in its insertion into the demolybdoenzyme. These steps thus lead to the formation of the form of the enzyme present in the *chlB mutant*. The conversion of this apparently cofactor-sufficient yet inactive form of the enzyme into the active enzyme requires protein FA, the *chlB* active product, and substance X. This rather surprising step, therefore, does not involve free, active MoCo. The lack of MoCo involvement is in agreement with the insensitivity of this stage in the enzyme's biosynthesis to growth in the presence of tungstate which normally inhibits molybdate- and MoCo-requiring reactions. In contrast, the *in vitro* restoration of activity to the NR precursor present in extracts from the *chlB* mutant grown with tungstate does require MoCo, presumably because tungstate inhibits the formation of the molybdenum-containing enzyme precursor (Saracino *et al.* 1986).

As discussed earlier, many of the gene–gene product relationships for the chlorate resistance loci remain to be elucidated. It is clear, however that the *chlA* and *chlB* loci are required for the synthesis of the carbon skeleton of MoCo, presumably from a pterin precursor (Rajagopalan 1988). The active product of the *chlG* locus is likely to act at the level of molybdate–molybdopterin association (Miller and Amy 1983). These, along with *chlD* (molybdate transport) are shown in Fig. 3.2.

None of the steps in the maturation pathway is confined to NR biosynthesis except, possibly, those involved with iron–sulphur centre synthesis. This pathway is almost certainly valid for the biosynthesis of all *E. coli* molybdo-enzymes. *E. coli*, therefore, contains a great deal of genetic information, in addition to that for the structural genes, for the biosynthesis of molybdoenzymes. Such genetic complexity would complicate the expression of cloned molybdoen-zymes in heterologous hosts. A similar situation is found in dinitrogenase biosynthesis where, although the MoCo is different, the genetic and biochemical complexity is comparable (Dixon 1988).

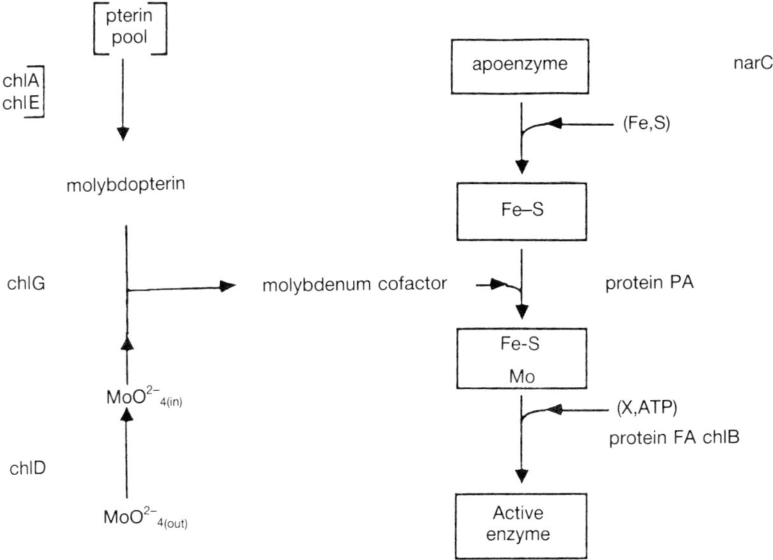

Fig. 3.2. Biosynthesis of respiratory nitrate reductase: cofactor acquisition. The nitrate reductase structural operon (*nar GHJI*) is indicated as *narC*. Apoenzyme refers to the protein (all subunits) part of the enzyme before the attachment of any of the prosthetic groups. It is not known whether the subunits acquire iron–sulphur centres before, during or after their assembly to form the multi-subunit apoenzyme characteristic of both the mature, active enzyme and the demolybdo-form which is found in the *chlA* mutant. Protein PA, since it complements *in vitro* the defect in *chlA* mutants, has been proposed as a product of the *chlA* locus, however this has not been directly established and its activity may be due to molybdenum cofactor bound to the protein. The *chlA* locus encodes molybdopterin-synthesizing functions and is so indicated in the figure. Nevertheless, protein PA may be the product of a further gene at the *chlA* locus.

Acknowledgements

I thank Gérard Giordano for many long discussions on this topic. Work from my laboratory referred to here is supported by grants from the Science and Engineering Research Council.

References

Amy, N. K. (1981). Identification of the molybdenum cofactor in chlorate-resistant mutants of *Escherichia coli*. *Journal of Bacteriology* **148**, 274–82.

Azoulay, E., Puig, J., and Pichinoty, F. (1967). Alteration of respiratory particles by mutation in *Escherichia coli* K-12. *Biochemical and Biophysical Research Communications* **27**, 270–4.

Bonnefoy, V., Burini, J.-F., Giordano, G., Pascal, M.-C., and Chippaux, M. (1987). Presence in the 'silent' terminus region of the *Escherichia coli* K-12 chromosome of cryptic gene(s) encoding a new nitrate reductase. *Molecular Microbiology* **1**, 143–50.

Bonnefoy-Orth, V., Lepelletier, M., Pascal, M.-C., and Chippaux, M. (1981). Nitrate reductase and cytochrome *b* (nitrate reductase) structural genes as parts of the nitrate reductase operon. *Molecular and General Genetics* **181**, 535–40.

Boxer, D. H. and Clegg, R. A. (1975). A transmembrane location for the proton-translocating, reduced ubiquinone-nitrate reductase segment of the respiratory chain of *Escherichia coli*. *FEBS Letters* **60**, 54–7.

Boxer, D. H., Low, D. C., Pommier, J., and Giordano, G. (1987). Involvement of a low-molecular-weight substance in *in vitro* activation of the molybdoenzyme respiratory nitrate reductase from a *chlB* mutant of *Escherichia coli*. *Journal of Bacteriology* **169**, 4678–85.

Buxton, R. S. and Drury, L. S. (1984). Identification of the *dye* gene product mutational loss of which alters envelope protein composition and also affects sex factor F expression in *Escherichia coli* K-12. *Molecular and General Genetics* **194**, 241–7.

Casse, F. (1970). Mapping of the gene *chlB* controlling membrane-bound nitrate reductase and formic hydrogen-lyase activities in *Escherichia coli* K-12. *Biochemical and Biophysical Research Communications* **39**, 429–36.

Chaudry, G. R. and MacGregor, C. H. (1983*a*). *Escherichia coli* nitrate reductase subunit A: its role as the catalytic site and evidence for its modification. *Journal of Bacteriology* **154**, 387–94.

Chaudry, G. R. and MacGregor, C. H. (1983*b*). Cytochrome *b* from *Escherichia coli* nitrate reductase. Its properties and association with the enzyme complex. *Journal of Biological Chemistry* **258**, 5819–27.

Chippaux, M., Bonnefoy-Orth, V., Ratouchnaik, J., and Pascal, M.-C. (1981). Operon fusions in the nitrate reductase operon and study of the control gene *nir*R in *Escherichia coli*. *Molecular and General Genetics* **182**, 477–9.

DeMoss, J. A. (1977). Limited proteolysis of nitrate reductase purified from membranes of *Escherichia coli*. *Journal of Biological Chemistry* **252**, 1696–701.

Dixon, R. (1988). Genetic regulation of nitrogen fixation in *The nitrogen and sulphur cycles* (eds J. A. Cole and S. J. Ferguson), pp. 417–38. Cambridge University Press, Cambridge, UK.

Dubourdieu, M., Andrade, E., and Puig, J. (1976). Molybdenum and chlorate-resistant mutants in *Escherichia coli* K-12. *Biochemical and Biophysical Research Communications* **70**, 766–73.

Edwards, E. S., Rondeau, S. S., and DeMoss, J. A. (1983). *chlC*(*nar*) operon of *Escherichia coli* includes structural genes for α- and β-subunits of nitrate reductase. *Journal of Bacteriology* **153**, 1513–20.

Enoch, H. G. and Lester, R. L. (1975). The purification and properties of formate dehydrogenase and nitrate reductase from *Escherichia coli*. *Journal of Biological Chemistry* **250**, 6693–705.

Fimmel, A. L. and Haddock, B. A. (1979). Use of *chlC*-lac fusions to determine regulation of gene *chlC* in *Escherichia coli*. *Journal Bacteriology* **138**, 726–30.

Garland, P. B., Downie, J. A., and Haddock, B. A. (1975). Proton-translocation and respiratory nitrate reductase of *Escherichia coli*. *Biochemical Journal* **152**, 547–59.

George, G. N., Bray, R. C., Morpeth, F. F., and Boxer, D. H. (1985). Complexes with halides and other anions of the molybdenum centre of nitrate reductase from *Escherichia coli*. *Biochemical Journal* **227**, 925–31.

Giordano, G., Saracino, L., and Grillet, L. (1985). Identification in various chlorate-resistant mutants of a protein involved in the activation of nitrate reductase in the soluble fraction of a *chlA* mutant of *Escherichia coli* K-12. *Biochimica et Biophysica Acta* **839**, 181–90.

Giordano, G., Violet, M., Medani, C.-L., and Pommier, J. (1984). A common pathway for the activation of several molybdoenzymes in *Escherichia coli* K-12. *Biochimica et Biophysica Acta* **798**, 216–25.

Giordano, G., Santini, C.-L., Saracino, L., and Iobbi, C. (1987). Involvement of a protein with molybdenum cofactor in the *in vitro* activation of nitrate reductase from a *chlA* mutant of *Escherichia coli* K-12. *Biochimica et Biophysica Acta* **914**, 220–32.

Glaser, J. H. and DeMoss, J. A. (1968). Comparison of nitrate reductase mutants of *Escherichia coli* selected by alternative procedures. *Molecular and General Genetics* **116**, 1–10.

Glaser, J. H. and DeMoss, J. A. (1971). Phenotypic restoration by molybdate of nitrate reductase activity in *chlD* mutants of *Escherichia coli*. *Journal of Bacteriology* **108**, 854–60.

Graham, A. and Boxer, D. H. (1980*a*). Implication of the α-subunit of *Escherichia coli* nitrate reductase catalytic activity. *Biochemical Society Transactions* **8**, 329–30.

Graham, A. and Boxer, D. H. (1980*b*). Arrangement of respiratory nitrate reductase in the cytoplasmic membrane of *Escherichia coli*: location of the β-subunit. *FEBS Letters* **113**, 15–20.

Graham, A. and Boxer, D. H. (1980*c*). The membrane location of the β-subunit of nitrate reductase from *Escherichia coli*. *Biochemical Society Transactions* **8**, 331.

Graham, A., Jenkins, H. E., Smith, N. H., Mandrand-Berthelot, M.-A., Haddock, B. A., and Boxer, D. H. (1980). The synthesis of formate dehydrogenase and nitrate reductase proteins in various *fdh* and *chl* mutants of *Escherichia coli*. *FEMS Microbiology Letters* **7**, 145–51.

Grillet, L. and Giordano, G. (1983). Identification and purification of a protein involved in the activation of nitrate reductase in the soluble fraction of a *chlA* mutant of *Escherichia coli* K-12. *Biochimica et Biophysica Acta* **749**, 115–24.

Guest, J. R. (1969) Biochemical and genetic studies with nitrate reductase C-gene mutants of *Escherichia coli*. *Molecular and General Genetics* **105**, 285–97.

Ingledew, J. W. and Poole, R. K. (1984). The respiratory chains of *Escherichia coli*. *Microbiological Reviews* **48**, 222–71.

Iobbi, C., Santini, C.-L., Bonnefoy, V., and Giordano, G. (1987). Biochemical and immunological evidence for a second nitrate reductase in *Escherichia coli* K-12. *European Journal of Biochemistry* **168**, 451–9.

Johann, S. and Hinton, S. M. (1987). Cloning and nucleotide sequence of the *chlD* locus. *Journal of Bacteriology* **169**, 1911–16.

Johnson, M. E. and Rajagopalan, K. V. (1987*a*). *In vitro* system for molybdopterin synthesis. *Journal of Bacteriology* **169**, 110–16.

Johnson, M. E. and Rajagopalan, K. V. (1987*b*). Involvement of *chlA, E, M* and *N* loci in *Escherichia coli* molybdopterin synthesis. *Journal of Bacteriology* **169**, 117–25.

Johnson, J. L., Hainline, B. E., Rajagopalan, K. V., and Arison, B. H. (1984). The pterin component of the molybdenum cofactor: structural characterisation of two fluorescent derivatives. *Journal of Biological Chemistry* **259**, 5414–22.

Jones, R. W. and Garland, P. B. (1977). Sites and specificity of the reaction of bipyridilium compounds with anaerobic respiratory systems of *Escherichia coli*. Effects of permeability barriers imposed by the cytoplasmic membrane. *Biochemical Journal* **164**, 199–211.

Jones, R. W., Lamont, A., and Garland, P. B. (1980). The mechanism of proton translocation driven by the respiratory nitrate reductase complex in *Escherichia coli*. *Biochemical Journal* **190**, 79–94.

Kobayashi, M. and Ishimoto, M. (1973). Aerobic inhibition of nitrate assimilation in *Escherichia coli*. *Zeitschrift für Allgemeine Mikrobiologie* **13**, 405–13.

Kristjansson, J. and Hollocher, T. C. (1979). Substrate binding site for nitrate reductase of *Escherichia coli* is on the inner aspect of the membrane. *Journal of Bacteriology* **137**, 1227–33.

Lambden, P. R. and Guest, J. R. (1976). Mutants of *Escherichia coli* K-12 unable to use fumarate as an anaerobic electron acceptor. *Journal of General Microbiology* **97**, 145–60.

Li, S. and DeMoss, J. A. (1987). Promoter region of the *nar* operon of *Escherichia coli*: nucleotide sequence and transcription initiation signals. *Journal of Bacteriology* **169**, 4614–20.

Low, D. C., Pommier, J., Giordano, G., and Boxer, D. H. (1988). Biosynthesis of molybdoenzymes in *E. coli*: *chlB* is the only chlorate-resistance locus required for protein FA activity. *FEMS Microbiology Letters* **49**, 331–6.

MacGregor, C. H. and Schnaitman, C. A. (1973). Reconstitution of nitrate reductase activity and formation of membrane particles from cytoplasmic extracts of chlorate-resistant mutants of *Escherichia coli*. *Journal of Bacteriology* **114**, 1164–76.

MacGregor, C. H. and Schnaitman, C. A. (1974). Nitrate reductase in *Escherichia coli*: properties of the enzyme and *in vitro* reconstitution from enzyme-deficient mutants. *Journal of Supramolecular Structure* **2**, 715–27.

MacGregor, C. H. and Christopher, A. R. (1978). Assymetric distribution of nitrate reductase subunits in the cytoplasmic membrane of *Escherichia coli*: evidence derived from surface labelling studies with transglutaminase. *Archives of Biochemistry and Biophysics* **185**, 204–13.

MacGregor, C. H., Schnaitman, C. A., Mormansell, D. E., and Hodgkins, M. G. (1974) Purification and properties of nitrate reductase from *Escherichia coli*. *Journal of Biological Chemistry* **249**, 5321–7.

Miller, J. B. and Amy, N. K. (1983). Molybdenum cofactor in chlorate-resistant and nitrate reductase-deficient insertion mutants of *Escherichia coli*. *Journal of Bacteriology* **155**, 793–801.

Miller, J. B., Scott, D. B., and Amy, N. K. (1987). Molybdenum-sensitive transcriptional regulation of the *chlD* locus of *Escherichia coli*. *Journal of Bacteriology* **169**, 1853–60.

Morpeth, F. F. and Boxer, D. H. (1985). Kinetic analysis of respiratory nitrate reductase from *Escherichia coli* K-12. *Biochemistry* **24**, 40–6.

Pinchinoty, F., Puig, J., Chippaux, M., Bigliardi-Rouvier, J., and Gendre, J. (1969). Réchèrches sur des mutants bacteriens ayant perdu les activities catalytiques liées a la nitrate reductase A. II. Comportment envers le chlorate et le chorite. *Annales de l'Institut Pasteur* **116**, 400–32.

Puig, J., Azoulay, E., Pinchinoty, F., and Gendre, J. (1969). Genetic mapping of the *chlC* gene of the nitrate reductase A system in *Escherichia coli* K-12. *Biochemical and Biophysical Research Communications* **35**, 659–62.

Reiss, J. and Klingmüller, W. (1987). Direct selection of recombinant plasmids with chlorate. *FEMS Microbiological Letters* **43**, 201–6.

Reiss, J., Kleinhofs, A. and Klingmüller, W. (1987). Cloning of seven differently complementing DNA fragments with *chl* functions from *Escherichia coli* K-12. *Molecular and General Genetics* **206**, 352–5.

Rivière, C., Giordano, G., Pommier, J., and Azoulay, E. (1975). Membrane reconstitution in *chl-r* mutants of *Escherichia coli* K-12. VIII. Purification and properties of the FA factor, the product of the *chlB* gene. *Biochimica et Biophysica Acta* **389**, 219–35.

Rondeau, S. S., Hsu, P.-Y., and DeMoss, J. A. (1984). Construction *in vitro* of a cloned *nar* operon from *Escherichia coli*. *Journal of Bacteriology* **159**, 159–66.

Saracino, L., Violet, M., Boxer, D. H., and Giordano, G. (1986). *In vitro* activation of respiratory nitrate reductase of *Escherichia coli* K-12, grown in the presence of tungstate: involvement of molybdenum cofactor. *European Journal of Biochemistry* **158**, 483–90.

Silvestro, A., Pommier, J., and Giordano, G. (1986). Molybdenum cofactor: a compound in the *in vitro* activation of both nitrate reductase and trimethyl-N-oxide reductase activities in *Escherichia coli* K-12. *Biochimica et Biophysica Acta* **872**, 243–52.

Sodergren, E. J. and DeMoss, J. A. (1988). *narI* Region of *Escherichia coli* nitrate reductase (*nar*) operon contains two genes. *Journal of Bacteriology* **170**, 1721–9.

Stewart, V. (1982). Requirement of Fnr and NarL functions for nitrate reductase expression in *Escherichia coli* K-12. *Journal of Bacteriology* **151**, 1320–5.

Stewart, V. and MacGregor, C. H. (1982). Nitrate reductase in *Escherichia coli* K-12: involvement of *chlC*, *chlE* and *chlG* loci. *Journal of Bacteriology* **151**, 788–99.

Stewart, V. and Parales, J., Jr. (1988). identification and expression of genes *narL* and *narX* of the *nar* (nitrate reductase) locus of *Escherichia coli* K-12. *Journal of Bacteriology* **170**, 1589–97.

Terrière, C., Giordano, G., Medani, C.-L., Boxer, D. H., Haddock, B. A., and Azoulay, E. (1981). Precursor forms of formate dehydrogenase in *chlA* and *chlB* mutants of *Escherichia coli*. *FEMS Microbiology Letters* **11**, 287–93.

Venables, W. A. (1972). Genetic studies with nitrate reductase-less mutants of *Escherichia coli*. I. Fine structure analysis of the *narA*, *narB* and *narE* loci. *Molecular and General Genetics* **103**, 127–40.

Venables, W. A. and Guest, J. R. (1968). Transduction of nitrate reductase loci of *Escherichia coli* by phages P1 and λ. *Molecular and General Genetics* **103**, 127–40.

4. Biochemistry, regulatory aspects, and genetics of nitrate assimilation in cyanobacteria

Alexandra J. Andriesse, Peter J. Weisbeek, and Gerard A. van Arkel

Introduction

Cyanobacteria (or blue-green algae) are a very divergent group of unicellular and filamentous micro-organisms which constitute one of the largest subgroups of Gram-negative prokaryotes. They have very few growth requirements, needing only carbon dioxide and nitrogen in more than trace amounts.

The cyanopbacteria are unique in the prokaryotic world because of their capacity to carry out a plant-like oxygen-evolving photosynthesis utilizing the thylakoids, which are localized on intracytoplasmic membranes. The organization of these membranes is similar to that of the chloroplast photosynthetic membranes, except that they are not stacked like chloroplast thylakoids, but are situated in several layers in the periphery of the cell. Attached to these membranes are the phycobilisomes, complexes of proteins, which are the accessory pigments of photosystem II.

One class of the phycobiliproteins, the phycocyanins, have a blue colour which contributes to the characteristic colour of most cyanobacteria. Under conditions of nitrogen starvation these proteins are broken down by a special protease (Foulds and Carr 1977; Lau *et al.* 1977) and the amino acids so obtained are used to synthesize other proteins. Upon prolonged starvation cultures change in colour from blue-green to yellow-green.

The nitrogen content of healthy cyanobacteria is usually within the range 4–9 per cent on a dry weight basis. They can readily utilize inorganic nitrogen compounds such as nitrate, nitrite, and ammonium salts, as well as several forms of combined organic nitrogen.

Regulation of nitrogen assimilation is mediated by products of the metabolism of ammonium, in which an amino group is coupled to a carbon skeleton. The initial product of ammonium metabolism is mainly glutamine, the formation of which is catalysed by glutamine synthetase in an ATP-dependent reaction (Ohmori and Hattori 1978). The second major product is glutamate, which is formed from glutamine and α-ketoglutarate by glutamate synthase (Meeks *et al.* 1978; Lea and Miflin 1979). In *Anacystis nidulans*, a unicellular non-nitrogen-fixing cyanobacterium, glutamate is formed directly from ammonium and α-ketoglutarate, probably via glutamate dehydrogenase (Meeks *et al.* 1978). However, Flores *et al.* (1983*a*) reported that no glutamate dehydrogenase activity could be detected in cell-free extracts.

Nitrate reduction and carbon assimilation are thus tightly coupled together by ammonium assimilation. To study nitrate assimilation separately from ammonium assimilation, cells are treated with the glutamine synthetase inhibitor L-methionine-DL-sulphoximine (MSX; cf. Ramos *et al.* 1982), which significantly inhibits ammonium assimilation while nitrate uptake and reduction remain fully active.

In the absence of combined nitrogen certain cyanobacteria are able to grow with the dinitrogen molecule as the sole nitrogen source; the nitrogenase system reduces molecular nitrogen to ammonium. Nitrogen fixation is coupled to the nitrate reduction and carbon assimilation via ammonium (Ramos and Guerrero 1983).

Nitrate uptake

Uptake of nitrate is the first step in its assimilation. Because cyanobacteria are relatively impermeable to nitrate when it is present at low concentrations, the uptake must be mediated by an active transport system. Nitrate uptake is not a consequence of nitrate reductase function since, first, whole cells have a much higher affinity for nitrate than isolated nitrate reductase (NR) (Flores *et al.* 1983*a*) and, second, utilization of nitrate by whole cells is energy dependent, whereas *in vitro* reduction takes place without any added energy supply (Ohmori *et al.* 1977; Flores *et al.* 1983*a,b*; Boussiba *et al.* 1984*a*). This active transport system is driven by photosynthetically generated energy. Nitrate transport is therefore a light-dependent phenomenon; nitrate uptake is accelerated by light (Ohmori *et al.* 1977; Tischner and Schmidt 1984), while in the dark no significant uptake of nitrate is observed (Rai *et al.* 1981). Uncouplers of photosynthetic and oxidative phosphorylation (Ohmori *et al.* 1977) and specific inhibitors of ATPases (Rai *et al.* 1981) markedly inhibit nitrate uptake. The presence of ATP significantly enhances uptake in all cases. Addition of ammonium to cells which are taking up nitrate results in a complete cessation of nitrate uptake, but only when glutamine synthetase is active (Ohmori *et al.* 1977; Rai *et al.* 1981; Rai and Singh 1982; Bagchi *et al.* 1985*a*). The inhibitory action of ammonium on nitrate uptake is probably based on competition for ATP between the nitrate uptake processes and the assimilation of ammonium via glutamine synthetase. Ammonium assimilation may therefore reduce the availability of energy for nitrate uptake when both nitrogen sources are present (Ohmori *et al.* 1977; Ohmori and Hattori 1978; Rai *et al.* 1981). In addition, the prevention of ammonium incorporation into carbon skeletons by MSX releases nitrate uptake from its carbon dioxide requirement (Flores *et al.* 1983*c*).

Ammonium also represses the synthesis of the nitrate uptake system; ammonium-grown cells are unable to take up nitrate immediately, while cells grown in nitrate medium exhibit a negligible lag period (Rai *et al.* 1981; Bagchi and Singh 1984; Bagchi *et al.* 1985*a*). Cells grown on molecular nitrogen need additional nitrate for activation of the nitrate uptake system (Bagchi and Singh 1984).

Nitrite uptake

The uptake of nitrite is mediated by passive influx, which is concentration-dependent, and an active high-affinity system, which involves nitrite-binding proteins (Singh and Kashyap 1986; Flores *et al.* 1987). Like nitrate transport, nitrite transport is light-dependent, although photosystem I alone is not sufficient to drive the transport of nitrite, so there is also an involvement of a proton gradient, resulting from photophosphorylation (Singh and Kashyap 1986). Nitrite uptake is effectively inhibited by nitrate, which acts as a competitive inhibitor (Madueño *et al.* 1987): a common permease for nitrate and nitrite transport into *Anacystis* has been suggested.

Ammonium does not play any regulatory role in the nitrite uptake process at physiological pH values (Ohmori *et al.* 1977; Singh and Kashyap 1986), but at higher pH values there is significant inhibition of nitrite uptake, probably caused by products of ammonium assimilation affecting the active component of the nitrite uptake system (Flores *et al.* 1987). The transport system may be constitutively synthesized, but genetic control of its synthesis is inferred from the observation that nitrite transport is repressed in cells grown on reduced nitrogen sources (Singh and Kashyap 1986).

Uptake of ammonium

The uptake of ammonium is energy-dependent (Boussiba *et al.* 1984*a*; Rai *et al.* 1984; Kashyap and Singh 1985) and two distinct systems have been reported (Kashyap and Singh 1985; Singh *et al.* 1986), both of which seem to be specific for ammonium, but differing in their affinity for this compound. The low-affinity system is more dependent on energy generated through ATP hydrolysis, whereas the high-affinity system is probably more dependent on the proton gradient of the membrane (Kashyap and Singh 1985). The two physiologically different transport systems also correspond to two phases of ammonium transport activity (Singh *et al.* 1986). During the first minute after ammonium addition there is a rapid initial uptake; this is followed by a slower secondary uptake, which is associated with product formation (Boussiba *et al.* 1984*a,b*; Rai *et al.* 1984).

The addition of relatively high concentrations of nitrate together with ammonium results in an inhibition of ammonium uptake. Both uptake processes seem to compete, with ammonium uptake being favoured over nitrate uptake (Ohmori *et al.* 1977).

Purification and properties of nitrate reductase

Attempts have been made to purify the cyanobacterial NR for about 20 years. One of the earlier reports was that of Hattori and Myers (1967) who prepared subcellular fractions containing active photosystem I from *Anabaena cylindrica* by sonication or by acetone precipitation. NR was solubilized from the particles obtained by acetone precipitation by treatment with Triton X-100 and sonication. When the subcellular fractions were used, it appeared that, in contrast to plants, this nitrate reductase accepts electrons from ferredoxin that is reduced either by the action of photosystem I or by NADPH and NADP-reductase. NADPH cannot donate electrons directly to the *Anabaena* nitrate reducing system (Hattori and Myers 1967; Hattori 1970).

Also in *Anacystis nidulans*, NR is associated with pigment-containing particles (Manzano *et al.* 1976). As in the *Anabaena* system, ferredoxin, reduced either by NADPH/ferredoxin-NADP reductase or by a light-dependent reaction, can provide electrons for the reduction of nitrate. Reduced pyridine nucleotides by themselves are not effective as electron donors.

Manzano *et al.* (1978) used an affinity chromatography step with ferredoxin–Sepharose for further purification of NR. Binding of NR to the ferredoxin–Sepharose required the ferredoxin to be in the reduced state. Elution took place with buffer free of the reducing agent dithionite. The procedure resulted in a 72-fold purification of the enzyme, with a 67 per cent recovery.

Purification to homogeneity of NR from *Plectonema boryanum* was obtained by Mikami and Ida (1984). Their procedure led to a 19 000-fold increase in specific activity with a yield of 15 per cent. The most effective steps were heat treatment at 70°C for 10 min, chromatography on DEAE–Toyopearl and chromatography on hydroxyapatite. The purified preparation was unstable when diluted in buffers of low ionic strength, but on storage at -20°C in 50 mM tris–HCl (pH 7.5), 0.5 M NaCl, 1 mM EDTA no loss of enzyme was observed after 1 month. The molecular weights measured by SDS–PAGE and sedimentation equilibrium were 83 kDa and 85 kDa, respectively, indicating that the enzyme consists of a single polypeptide chain. The enzyme contains one atom of molybdenum and one $[Fe_4S_4]$ or two $[Fe_2S_2]$ iron–sulphur clusters per molecule.

The pH optimum depends on the system in which the reaction is carried out (Hattori and Myers 1967; Manzano *et al.* 1976; Ida and Mikami 1983; Mikami and Ida 1984). In the ferredoxin-dependent assay, which is normally carried out under aerobic conditions, maximal activity is found in the pH range 8–9, which is the same as for the methylviologen-linked activity in an anaerobic assay. Using an aerobic system with dithionite the pH optimum value moves to 10.0–10.5, which is probably due to a modification of the enzyme (Mikami and Ida 1984). Rapid inactivation occurs in the presence of high concentrations of dithionite. This effect is even more apparent when methylviologen is replaced by ferredoxin in the assay; under these conditions hardly any activity can be detected (Hattori and Myers 1967; Manzano *et al.* 1976). The native activity of the enzyme can be restored by incubation in buffer with cyanate or azide in the presence of dithionite or reduced ferrodoxin (Mikami and Ida 1986).

Purification and properties of nitrite reductase

In *Anacystis nidulans*, nitrite reductase (NiR) is bound to pigment-containing particles. The enzyme can easily be solubilized by sonic disintegration or prolonged vibration treatment (Guerrero *et al.* 1974; Manzano *et al.* 1976). Up to now no report has been published describing purification to homogeneity of a cyanobacterial NiR. Méndez and Vega (1981) purified the enzyme from *Anabaena* sp. 7119 to a 763-fold increase in specific activity and a yield of 15 per cent, but this preparation was not pure.

Enzyme preparations are remarkably stable; no significant loss of activity is observed after storage for several months at $-20°C$, storage for 1 week at $2°C$, or drastic treatments such as boiling or strong acid conditions (Hattori and Uesugi 1968; Guerrero *et al.* 1974; Méndez and Vega 1981). The activity does decrease after repetitive freezing–thawing of the enzyme solution or after prolonged dialysis against low-ionic strength buffers.

Hattori and Uesugi (1968) determined a molecular weight of 52 kDa for the *Anabaena cylindrica* enzyme, while the *Anabaena* sp. 7119 enzyme (Méndez and Vega, 1981) has a molecular weight of 68 kDa and has been shown to be spherical.

Ferredoxin, reduced either by NADPH in the presence of NADP-reductase or chemically by dithionite or photochemically by photosystem I upon illumination, is a good electron carrier for NiR. Effective substitutes are methylviologen, benzylviologen, and diquat, but no reaction takes place in the presence of the flavin nucleotides FAD or FMN, which are known to function as prosthetic groups of several oxidation–reduction enzymes. The pyridine nucleotides NADH and NADPH are only effective together with NADPH-diaphorase and ferredoxin (Hattori and Myers 1966; Hattori and Uesugi 1968; Guerrero *et al.* 1974; Méndez *et al.* 1981).

Both NiR and NR contain an iron–sulphur centre as a prosthetic group; absorption spectra further indicate that NiR is a haemo-protein and NR is not (Méndez and Vega 1981; Mikami and Ida 1984).

Regulation of nitrate assimilation

NR and NiR are adaptive enzymes (Ohmori and Hattori 1970; Guerrero *et al.* 1974; Stevens and Van Baalen 1974; Herrero *et al.* 1981, 1985; Méndez *et al.*

1981; Ramos and Guerrero 1983; Avissar 1985; Herrero and Guerrero 1986). In all cyanobacteria studied ammonium represses the synthesis of both reductases. This influence can only be exerted when ammonium is metabolized through glutamine synthetase; if this enzyme is inhibited, no represssion takes place (Herrero *et al.* 1981; Herrero and Guerrero 1986).

There are two views about the regulation of nitrate reduction by de-repression. First, according to Herrero *et al.* (1981, 1985), there is a difference in regulation by de-repression between unicellular non-nitrogen-fixing and filamentous nitrogen-fixing cyanobacteria. In the latter strains NR is synthesized '*de novo*' when cells are removed from ammonium-containing medium and placed in nitrate-containing medium. In the unicellular cyanobacteria NR is synthesized not only after transfer to nitrate medium, but also after transfer to medium with no bound nitrogen. Therefore, in the nitrogen-fixing filamentous strains nitrate behaves as a nutritional inducer, whereas in unicellular strains its function is not obligatory (Herrero *et al.* 1981, 1985).

Second, according to Bagchi *et al.* (1985*b*) and Avissar (1985) there is de-repression in filamentous nitrogen-fixing cyanobacteria when the cells are grown on molecular nitrogen. Although this seems to conflict with the first view, they also find that under de-repressed conditions (i.e. in ammonium-free medium) the NR activity is higher in nitrate-grown cells than in cells grown with molecular nitrogen. There is a large, but matching pool of free apoprotein in both cells grown with nitrate and those grown on molecular nitrogen. The amount of cofactor controls the NR activity, as is evident from reconstruction experiments with the molybdenum cofactor (MoCo) isolated from *Escherichia coli* (Bagchi *et al.* 1985*b*). Nitrate is not required for '*de novo*' synthesis of the free apoprotein, but is needed for '*de novo*' synthesis of MoCo, after which full activity is obtained (Avissar 1985).

NiR activity in the filamentous, nitrogen-fixing strain *Anabaena* sp. 7119 (Méndez *et al.* 1981), as well as in the unicellular non-nitrogen-fixing strain *Anacystis nidulans* (Herrero and Guerrero 1986), increases after transfer from ammonium-containing medium to medium containing nitrate or nitrite or to medium with no bound nitrogen. Enzyme activity is optimal in the presence of nitrate as the nitrogen source. NiR activity is also present in tungsten-treated cells (cells with inactivated nitrate reductase) (Herrero and Guerrero 1986). The stimulatory effect is, therefore, due to nitrate itself and not (only) to the nitrite resulting from intracellular reduction. Herrero and Guerrero (1986) showed that NR and NiR activities develop simultaneously. This finding, combined with the fact that both enzymes are mainly regulated by ammonium repression, favours the conclusion that the synthesis of both enzymes in the nitrate reduction pathway are controlled by a common mechanism.

Use of nitrate reduction mutants

Several nitrate reduction mutants of *Nostoc muscorum* (Singh and Sonie 1977; Singh *et al.* 1977; Pandey and Singh 1984) have been isolated and used to study the regulation of nitrogen assimilation. Experiments with spontaneous chlorate-resistant mutants deficient in nitrate reduction or nitrogen fixation or both, and mutants dependent on tungsten instead of molybdenum for growth on nitrate or molecular nitrogen, indicate a common genetic determinant in the regulation and synthesis of active NR and nitrogenase. The precise character of the mutations has not been determined.

For genetic studies of cyanobacteria *Anacystis nidulans* R2 (*Synechococcus* PCC 7942) is often used. It has a chromosome of about 2300 Mda and plasmids of 33 and 5.3 Mda, respectively (Van den Hondel *et al.* 1980). This strain is naturally competent; no special treatment is needed for transformation. There are two basic systems for genetic manipulation of *A. nidulans* R2. The first is based on recombination with the genome, which takes place with high frequency. Vectors which make use of this property have been constructed (J. van der Plas, manuscript in preparation). The second transformation system is based on vectors derived from the smaller plasmid (Van den Hondel *et al.* 1980; Kuhlemeier *et al.* 1981). To prevent recombination with the latter, a derivative of *A. nidulans* R2 is used, the small-plasmid-cured strain *A. nidulans* R2 Spc (Kuhlemeier *et al.* 1983).

Nitrate reduction mutants of *A. nidulans* R2 are obtained after chemical or transposon mutagenesis (Kuhlemeier and Van Arkel 1988). These mutants grow slowly and display a characteristic yellow colour on nitrate medium. After transfer to nitrite medium they turn green and grow normally. The NR activity is < 1 per cent of that in the wild type. With the aid of these mutants, wild-type NR genes have been cloned by phenotypic complementation (Kuhlemeier *et al.* 1984*a,b*). Every gene can transform several of the mutants to the wild-type phenotype; this has allowed the definition of three complementation groups. At present the mutants and the corresponding wild-type genes are being analysed in physiological, biochemical, and genetic experiments.

Concluding remarks

Nitrate assimilation in cyanobacteria is mainly regulated by ammonium or by ammonium assimilation products. In short-term experiments ammonium represses the uptake of nitrate; it does not play any regulatory rôle in the nitrite uptake. In long-term experiments ammonium represses the synthesis of both NR and NiR. Both enzymes are ferredoxin-dependent; neither NADH nor NADPH can donate its electrons directly to the enzymes.

NR from *Plectonema boryanum* has been purified to homogeneity. It consists of a single polypeptide of about 85 kDa, with a non-haem iron–sulphur centre and a molybdenum cofactor. NiR has not yet been purified to homogeneity, but experiments indicate that this enzyme also consists of a single polypeptide with an iron–sulphur centre, probably of the haem-type.

For both enzymes of the nitrate reduction pathway much work can still be done in the fields of physiology, biochemistry, and genetics, especially in comparative experiments among various cyanobacterial strains. An interesting aspect of this research is the supposed difference in regulation by de-repression of nitrate reduction genes between unicellular non-nitrogen-fixing cyanobacteria and filamentous nitrogen-fixing strains.

In our laboratory three wild-type genes involved in the reduction of nitrate to nitrite have been cloned and are being analysed. At present we are also working on the cloning of the structural gene for NR from *Anacystis nidulans*, using an expression library and antibodies raised against the purified NR from *Plectonema boryanum*. Our aim is to study the interaction between NR and ferredoxin at the levels of gene expression and protein activity to elucidate the regulation of the distribution of photosynthetically generated reducing power over a large array of enzymes.

References

Avissar, Y. J. (1985). Induction of nitrate assimilation in the cyanobacterium *Anabaena variabilis*. *Physiologia Plantarum* **63**, 105–8.

Bagchi, S. N. and Singh, H. N. (1984). Genetic control of nitrate reduction in the cyanobacterium *Nostoc muscorum*. *Molecular and General Genetics* **193**, 82–4.

Bagchi, S. N., Rai, U. N., Rai, A. N., and Singh, H. N. (1985*a*). Nitrate metabolism in the cyanobacterium *Anabaena cycadeae*: regulation of nitrate uptake and reductase by ammonia. *Physiologia Plantarum* **63**, 322–6.

Bagchi, S. N., Rai, A. N., and Singh, H. N. (1985*b*). Regulation of nitrate reductase in cyanobacteria. Repression–derepression control of nitrate reductase apoprotein in the cyanobacterium *Nostoc muscorum*. *Biochimica et Biophysica Acta* **838**, 370–3.

Boussiba, S., Resch, C. M., and Gibson, J. (1984*a*). Ammonia uptake and retention in some cyanobacteria. *Archives of Microbiology* **138**, 287–92.

Boussiba, S., Dilling, W., and Gibson, J. (1984*b*). Methylammonium transport in *Anacystis nidulans* R-2. *Journal of Bacteriology* **160**, 204–10.

Flores, E., Ramos, J. L., Herrero, A., and Guerrero, M. G. (1983*a*). Nitrate assimilation by cyanobacteria. In *Photosynthetic prokaryotes: cell differentiation and function* (ed. G. C. Papageorgiou and L. Packer), pp. 363–87. Elsevier Biomedical, New York, Amsterdam, Oxford.

Flores, E., Guerrero, M. G., and Losada, M. (1983*b*). Photosynthetic nature of nitrate uptake and reduction in the cyanobacterium *Anacystis nidulans*. *Biochimica et Biophysica Acta* **722**, 408–16.

Flores, E., Romero, J. M., Guerrero, M. G., and Losada, M. (1983*c*). Regulatory interaction of photosynthetic nitrate utilization and carbon dioxide fixation in the cyanobacterium *Anacystis nidulans*. *Biochimica et Biophysica Acta* **725**, 529–32.

Flores, E., Herrero, A., and Guerrero, M. G. (1987). Nitrite uptake and its regulation in the cyanobacterium *Anacystis nidulans*. *Biochimica et Biophysica Acta* **896**, 103–8.

Foulds, I. J. and Carr, N. G. (1977). A proteolytic enzyme degrading phycocyanin in the cyanobacterium *Anabaena cylindrica*. *FEMS Microbiology Letters* **2**, 117–9.

Guerrero, M. G., Manzano, C., and Losada, M. (1974). Nitrite photoreduction by a cell-free preparation of *Anacystis nidulans*. *Plant Science Letters* **3**, 273–8.

Hattori, A. (1970). Solubilization of nitrate reductase from the blue-green alga *Anabaena cylindrica*. *Plant and Cell Physiology* **11**, 975–8.

Hattori, A. and Myers, J. (1966). Reduction of nitrate and nitrite by subcellular preparations of *Anabaena cylindrica*. I. Reduction of nitrite to ammonia. *Plant Physiology* **41**, 1031–6.

Hattori, A. and Myers, J. (1967). Reduction of nitrate and nitrite by subcellular preparations of *Anabaena cylindrica*. II. Reduction of nitrate to nitrite. *Plant and Cell Physiology* **8**, 327–37.

Hattori, A. and Uesugi, I. (1968). Purification and properties of nitrite reductase from the blue-green alga *Anabaena cylindrica*. *Plant and Cell Physiology* **9**, 689–99.

Herrero, A. and Guerrero, M. G. (1986). Regulation of nitrite reductase in the cyanobacterium *Anacystis nidulans*. *Journal of General Microbiology* **132**, 2463–8.

Herrero, A., Flores, E., and Guerrero, M. G. (1981). Regulation of nitrate reductase levels in the cyanobacteria *Anacystis nidulans*, *Anabaena* sp. strain 7119 and *Nostoc* sp. strain 6719. *Journal of Bacteriology* **145**, 175–80.

Herrero, A., Flores, E., and Guerrero, M. G. (1985). Regulation of nitrate reductase cellular levels in the cyanobacteria *Anabaena variabilis* and *Synechocystis* sp. *FEMS Microbiology Letters* **26**, 21–5.

Ida, S. and Mikami, B. (1983). Purification and characterization of assimilatory nitrate reductase from the cyanobacterium *Plectonema boryanum*. *Plant and Cell Physiology* **24**, 649–58.

Kashyap, A. K. and Singh, D. P. (1985). Ammonium transport in unicellular cyanobacterium *Anacystis nidulans*. *Journal of Plant Physiology* **121**, 319–30.

Kuhlemeier, C. J. and Van Arkel, G. A. (1988). Host–vector systems for gene cloning in cyanobacteria. *Methods in Enzymology* **153**, 199–215.

Kuhlemeier, C. J., Borrias, W. E., Van den Hondel, C. A. M. J. J., and Van Arkel, G. A. (1981). Construction and characterization of two hybrid plasmids capable of transformation to *Anacystis nidulans* and *Escherichia coli*. *Molecular and General Genetics* **184**, 249–54.

Kuhlemeier, C. J. *et al.* (1983). A host-vector system for gene cloning in the cyanobacterium *Anacystis nidulans* R2. *Plasmid* **10**, 156–63.

Kuhlemeier, C. J., Logtenberg, T., Stoorvogel, W., Van Heugten, H. A. A., Borrias, W. E., and Van Arkel, G. A. (1984*a*). Cloning of nitrate reductase genes from the cyanobacterium *Anacystis nidulans*. *Journal of Bacteriology* **159**, 36–41.

Kuhlemeier, C. J., Teeuwsen, V. J. P., Janssen, M. J. T., and Van Arkel, G. A. (1984*b*). Cloning of a third nitrate reductase gene from the cyanobacterium *Anacystis nidulans* R2 using a shuttle cosmid library. *Gene* **31**, 109–16.

Lau, R. H., MacKenzie, M. M., and Doolittle, W. F. (1977). Phycocyanin synthesis and degradation in the blue-green bacterium *Anacystis nidulans*. *Journal of Bacteriology* **132**, 771–8.

Lea, P. J. and Miflin, B. J. (1979). Photosynthetic ammonia assimilation. In *Encyclopedia of plant physiology*, New series, Vol. 6 (eds M. Gibbs and E. Latzko), pp. 445–56. Springer-Verlag, Berlin, Heidelberg, New York.

Madueño, F., Flores, E., and Guerrero, M. G. (1987). Competition between nitrate and nitrite uptake in the cyanobacterium *Anacystis nidulans*. *Biochimica et Biophysica Acta* **896**, 109–12.

Manzano, C., Candau, P., Gomez-Moreno, C., Relimpio, A. M., and Losada, M. (1976). Ferredoxin-dependent photosynthetic reduction of nitrate and nitrite by particles of *Anacystis nidulans*. *Molecular and Cellular Biochemistry* **10**, 161–9.

Manzano, C., Candau, P., and Guerrero, M. G. (1978). Affinity chromatography of *Anacystis nidulans* ferredoxin-nitrate reductase and NADP reductase on reduced ferredoxin–sepharose. *Analytical Biochemistry* **90**, 408–12.

Meeks, J. C., Wolk, C. P., Lockau, W., Schilling, N., Shaffer, P. W., and Chien, W.-S. (1978). Pathways of assimilation of $[^{13}N]N_2$ and $^{13}NH_4^+$ by cyanobacteria with and without heterocysts. *Journal of Bacteriology* **134**, 125–30.

Méndez, J. M. and Vega, J. M. (1981). Purification and molecular properties of nitrite reductase from *Anabaena* sp. 7119. *Physiologia Plantarum* **52**, 7–14.

Méndez, J. M., Herrero, A., and Vega, J. M. (1981). Characterization and catalytic properties of nitrite reductase from *Anabaena* sp. 7119. *Zeitschrift für Pflanzenphysiologie* **103**, 305–15.

Mikami, B. and Ida, S. (1984). Purification and properties of ferredoxin-nitrate reductase from the cyanobacterium *Plectonema boryanum*. *Biochimica et Biophysica Acta* **791**, 294–304.

Mikami, B. and Ida, S. (1986). Reversible inactivation of ferredoxin-nitrate reductase from the cyanobacterium *Plectonema boryanum*. The role of superoxide anion and cyanide. *Plant and Cell Physiology* **27**, 1013–21.

Ohmori, K. and Hattori, A. (1970). Induction of nitrate and nitrite reductases in *Anabaena cylindrica*. *Plant and Cell Physiology* **11**, 873–8.

Ohmori, M. and Hattori, A. (1978). Transient change in the ATP pool of *Anabaena cylindrica* associated with ammonia assimilation. *Archives of Microbiology* **117**, 17–20.

Ohmori, M., Ohmori, K., and Strotmann, H. (1977). Inhibition of nitrate uptake by ammonia in a blue-green alga, *Anabaena cylindrica*. *Archives of Microbiology* **114**, 225–9.

Pandey, K. D. and Singh, P. K. (1984). Isolation and characterisation of nitrate reductase mutants and regulation of nitrate reductase and nitrogenase in the cyanobacterium *Nostoc muscorum*. *Molecular and General Genetics* **195**, 180–5.

Rai, A. K. and Singh, S. (1982). Regulation of nitrate uptake in *Nostoc muscorum* by glutamine synthetase. *FEMS Microbiology Letters* **14**, 303–6.

Rai, A. K., Kashyap, A. K., and Gupta, S. L. (1981). ATP dependent uptake of nitrate in *Nostoc muscorum* and inhibition by ammonium ions. *Biochimica et Biophysica Acta* **674**, 78–86.

Rai, A. N., Rowell, P., and Stewart, W. D. P. (1984). Evidence for an ammonium transport system in free-living and symbiotic cyanobacteria. *Archives of Microbiology* **137**, 241–6.

Ramos, J. L. and Guerrero, M. G. (1983). Involvement of ammonium metabolism in the nitrate inhibition of nitrogen fixation in *Anabaena* sp. strain ATCC 33047. *Archives of Microbiology* **136**, 81–3.

Ramos, J. L., Guerrero, M. G., and Losada, M. (1982). Photoproduction of ammonia from nitrate by *Anacystis nidulans* cells. *Biochimica et Biophysica Acta* **679**, 323–30.

Singh, D. P. and Kashyap, A. K. (1986). Nitrite transport in the unicellular cyanobacterium *Anacystis nidulans*. *Biochemie und Physiologie der Pflanzen* **181**, 623–31.

Singh, D. T., Modi, D. R., and Singh, H. N. (1986). Evidence for glutamine synthetase and methylammonium (ammonium) transport system as two distinct primary targets of methionine sulfoximine inhibitory action in the cyanobacterium *Anabaena doliolum*. *FEMS Microbiology Letters* **37**, 95–8.

Singh, H. N. and Sonie, K. C. (1977). Isolation and characterization of chlorate-resistant mutants of the blue-green alga *Nostoc muscorum*. *Mutation Research* **43**, 205–12.

Singh, H. N., Sonie, K. C., and Singh, H. R. (1977). Nitrate regulation of heterocyst differentiation and nitrogen fixation in a chlorate-resistant mutant of the blue-green alga *Nostoc muscorum*. *Mutation Research* **42**, 447–52.

Stevens, S. E. and Van Baalen, C. (1974). Control of nitrate reductase in a blue-green alga. The effects of inhibitors, blue light, and ammonia. *Archives of Biochemistry and Biophysics* **161**, 146–52.

Tischner, R. and Schmidt, A. (1984). Light mediated regulation of nitrate assimilation in *Synechococcus leopoliensis. Archives of Microbiology* **137**, 151–4.

Van den Hondel, C. A. M. J. J., Verbeek, S., Van der Ende, A., Weisbeek, P. J., Borrias, W. E., and Van Arkel, G. A. (1980). Introduction of transposon Tn901 into a plasmid of *Anacystis nidulans:* preparation for cloning in cyanobacteria. *Proceedings of the National Academy of Sciences, USA* **77**, 1570–4.

5. Nitrate assimilation in yeasts

Charles R. Hipkin

Introduction

Few microbial physiologists or molecular biologists would deny the considerable advantages of using yeasts as experimental organisms. Quite apart from any reasoning based on their obvious biotechnological importance, one has only to consider the ease with which they can be cultured and manipulated in the laboratory and the fact that they represent unicellular models of heterotrophic, eukaryotic organisms. It is surprising, therefore, that the study of nitrate assimilation in yeasts has not been popular and that there is a relative paucity of information on this subject in the literature. This review will attempt to bring together much of the relevant research on nitrate assimilation in yeasts which has occurred over the last 30 years. Information obtained from relatively recent research in the author's own laboratory over the last 7 years will serve as a structural framework.

The occurrence of nitrate assimilation in yeasts

One probable reason for the apparent neglect of yeasts in studies of nitrate metabolism has been the tendency to equate yeast with *Saccharomyces cerevisiae*, which, alas, is not able to assimilate nitrate or nitrite. Nevertheless, many other species of yeasts, of diverse origins and ecology, can assimilate nitrate and some of these belong to well known genera of biotechnological (e.g. *Hansenula, Candida*) and biomedical (e.g. *Cryptococcus*) importance. At least 18 genera (Table 5.1) are known to contain species which assimilate nitrate and in some, such as *Hansenula*, it is a feature of the whole genus (Kregar-van Rij 1984). Reports in the older literature stating that, in general, yeasts are unable to assimilate nitrate (Lilly and Barnett 1951) are no longer accepted.

A relatively small number of yeasts belong to an interesting class which are unable to assimilate nitrate but can apparently assimilate nitrite. These include species of *Debaryomyces* (*D. pseudopolymorpha, D. hansenii*) and *Sporopachydermia* (*S. cereana, S. lactativora*). These organisms have been isolated from various sources including human patients (Kregar-van Rij, 1984) where nitrite produced

Table 5.1. Distribution of the ability to assimilate nitrate in yeasts.

(a) Genera containing species that assimilate nitrate.

Ambrosiozyma	*Cryptococcus*	*Pachysolen*	*Sterigmatomyces*
Brettanomyces	*Dekkera*	*Rhodosporidium*	*Trichosporon*
Bullera	*Filobasidium*	*Rhodotorula*	*Wickerhamiella*
Candida	*Hansenula*	*Sporidiobolus*	
Citeromyces	*Leucosporidium*	*Sporobolomyces*	

(b) Genera containing species that assimilate nitrite (but not nitrate)

Debaryomyces   *Sporopachydermia*

(c) Some genera notable for their inability to assimilate nitrate

Kluyveromyces   *Pichia*   *Saccharomyces*   *Schizosaccharomyces*

by bacterial activity (Tannenbaum *et al.* 1976) from ingested nitrate may be a source of nitrogen for growth.

The distribution of the ability to assimilate nitrate (or nitrite) in yeasts shows little obvious taxonomic or ecological pattern. Detailed, systematic studies with yeasts may, nevertheless, provide insights into the evolution of the nitrate-assimilating ability and its physiological ecology in heterotrophic organisms. It is perhaps significant that many yeasts able to assimilate nitrate also show great versatility in their use of a large number of organic nitrogen compounds (Large 1986).

The nitrate assimilation pathway in yeasts

Earlier investigations with yeasts such as *Hansenula anomala* (Silver 1957; Pichinoty and Metenier 1967), *Candida utilis* (Sims *et al.* 1968; Burn *et al.* 1974; Choudary and Ramanda Rao 1976*a*), and *Torulopsis nitratophila* (Rivas *et al.* 1973, 1974) confirmed the similarity between these and other nitrate-assimilating eukaryotes with respect to the general details of the pathway, enzymology, and regulation of nitrate assimilation, at least at the physiological level. Figure 5.1 presents a view of the whole process from the uptake and reduction of nitrate-nitrogen via nitrate reductase (NR) and nitrite reductase (NiR) to the incorporation of this nitrogen into organic form as glutamate. Figure 5.1 also shows the probable involvement of the pentose phosphate pathway in the generation of reduced pyridine nucleotide for nitrate and nitrite reduction and this aspect is discussed later. Evidence suggests that in nitrogen-starved and nitrate-adapted cultures of yeasts the limiting reactions of nitrate assimilation occur at the level of nitrate uptake and/or NR activity (Ali and Hipkin 1985, 1986).

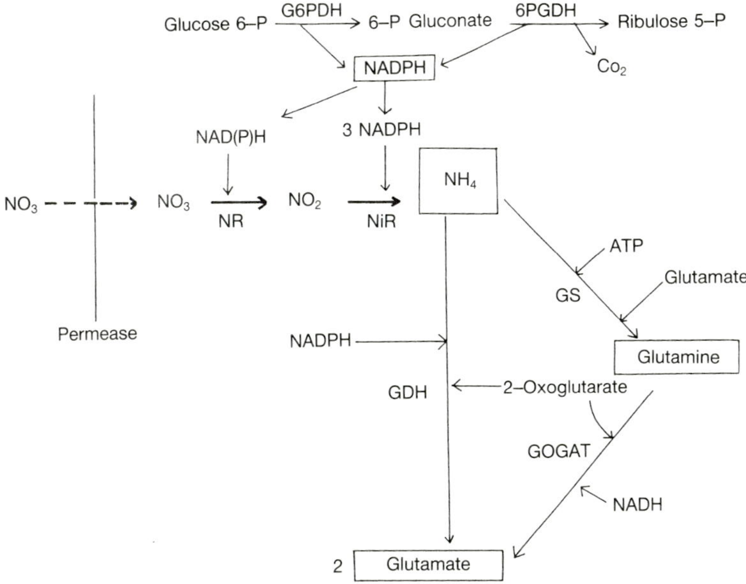

Fig. 5.1. Nitrate assimilation in yeasts. Ammonium, glutamate and glutamine can be envisaged as the end products of nitrate assimilation. There is good evidence for a NADH-dependent GOGAT activity in *Candida nitratophila* which is inhibited by azaserine but not by amino-oxyacetate (unpublished results). NR = nitrate reductase; NiR = nitrite reductase; GS = glutamine synthetase; GDH = glutamic dehydrogenase (biosynthetic); GOGAT = glutamate synthase; G6PDH = glucose-6-phosphate dehydrogenase; 6PGDH = 6-phosphogluconate dehydrogenase.

Nitrate and nitrite uptake

Nitrate uptake has not been studied in any detail in yeasts. In *Candida nitratophila* the kinetics of nitrate uptake suggests a carrier-mediated process (A. Al-Kubisi and C. R. Hipkin, unpublished observation). Moreover in yeasts such as *Sporobolomyces roseus* and *Candida* spp, development of the ability to take up nitrate is prevented by inhibitors of protein synthesis such as cycloheximide and 6-methyl purine. Once developed, nitrate uptake requires a suitable source of assimilatory carbon, such as glucose, and is inhibited by anaerobiosis and antimycin A. Nevertheless these organisms do not accumulate nitrate and there is no equivocal evidence that they possess an active uptake system for nitrate. Notably, tungstate-treated cultures which contain little NR activity do not take up nitrate, which suggests a close coupling between uptake and reduction of nitrate allowing transport along a chemical diffusion gradient (Ali and Hipkin 1985, 1986). However, since nitrate has to be transported against an electrochemical gradient one might expect the operation of an electrogenic porter. Eddy and Hopkins (1985) obtained evidence for a nitrate–proton symport in *Candida utilis*, where entry of nitrate is accompanied by two proton equivalents with the exit of one equivalent of K^+ to maintain charge balance.

Nitrite uptake has been studied in *Candida nitratophila* (A. Al-Kubisi and C. R. Hipkin, unpublished observations) but in few other species. Recent studies have shown that although nitrate-grown, nitrogen-replete cultures of *Candida nitratophila* do not assimilate nitrate or ammonium in the absence of a suitable carbon source, they do take up and reduce nitrite. Ammonium formed stochiometrically from nitrite under these conditions is not assimilated further, but excreted into the medium. Energy is apparently available for uptake and reduction of nitrite but not for nitrate or ammonium assimilation in carbon-deficient cultures. At present the mechanism underlying this discrimination is not understood. Studies with *Candida* also indicate that nitrite and nitrate are taken up via the same uptake system. Thus, nitrite uptake is inhibited competitively by nitrate and nitrate uptake is inhibited by nitrite. For most species of yeasts, nitrite is probably not an important source of nitrogen in the natural environment and there would be little selection pressure to evolve a separate uptake system for it.

Nitrate reductase and nitrite reductase

Earlier studies with *Candida utilis* suggested that the enzymatic reduction of nitrate to ammonium was catalysed by a multi-enzyme complex (nitrosome) with NR and NiR activities (Sims *et al.* 1968). This has not been substantiated and it is now generally accepted that these enzyme activities reside in separate proteins although it is not clear where NR and NiR activities are located in the cell. Pichinoty and Metenier (1967) claimed evidence for a mitochondrial location for NR in *Hansenula anomala* but this has been refuted by Minagawa and Yoshimoto (1983) who suggest a cytoplasmic location for the enzyme.

The NRs of a number of yeasts have been studied but only the enzyme from *Hansenula anomala* (Silver 1957; Zauner and Dellweg 1983; Minagawa and Yoshimoto 1983, 1984), *Rhodotorula glutinis* (Guerrero and Gutierrez 1977; Ito and Suzuki 1978) and *Candida nitratophila* (Hipkin *et al.* 1986; Kay *et al.* 1987) have received detailed attention. The use of affinity chromatography using blue gel (Solomonson 1975) allowed the NRs of these yeasts to be purified to homogeneity (Guerrero and Gutierrez 1977; Zauner and Dellweg 1983; Hipkin *et al.* 1986), and studies with purified enzyme showed yeast NR to resemble NR from other eukaryotic sources (Table 5.2) in being a multimeric molybdoflavoprotein containing a cytochrome of the *b*-type (cytochrome b_{557}). The NR of *Hansenula anomala* differs in its relatively small subunit size, 52 kDa, compared to about 100 kDa for NR subunits from other sources. Another notable feature includes the high turnover number ($> 1000 \text{ s}^{-1}$) of *Candida nitratophila* NR. However, in other properties, such as quarternary structure and thermodynamics, the *Candida* enzyme resembles that of *Chlorella* (Howard and Solomonson 1982; Hipkin *et al.* 1986; Kay *et al.* 1986; Chapter 7). Like *Chlorella* NR, *Candida* NR exists as a tetramer at high concentrations but dissociates into a dimeric form after dilution

Table 5.2. The physical and biochemical properties of NR from yeasts (*Candida nitratophila*, *Hansenula anomala*, and *Rhodotorula glutinis*) compared with the properties of NR from a filamentous fungus (*Neurospora crassa*), an alga (*Chlorella vulgaris*), and spinach leaf (*Spinacea oleracea*)

Organism	*Candida nitratophila*	*Hansenula anomala*	*Rhodotorula glutinis*	*Neurospora crassa*	*Chlorella vulgaris*	*Spinacea oleracea*
Molecular weight (kDa)						
subunit	97	52	118	115	100	110–120
dimer	200	$-^a$	230	228 (290)	200	200–250
tetramer	365	215	–	–	360	–
Stokes radius (nm)	8.5	6.5	7.0	7.0	8.1	6.0
$S_{20,W}$	10.25	8.0	7.9	8.0	10.0	8.1
Pyridine nucleotide specificity	NADH; NADPH	NADPH; NADH	NADPH; NADH	NADPH	NADH	NADH
Specific activity (μkat mg^{-1})	7.0 (NADH–NR)	0.8 (NADPH–NR)	2.5 (NADPH–NR)	2.1	1.7	0.8
Turnover number (s^{-1})	1200 (dimer)	180	570	475	330 (dimer)	160
Optimum pH	7.0	–	7.5	7.5	7.9	7.5
K_m NO$_3$ (μM)						
NADH–NR	120	–	45	–	110	180
NADPH–NR	110	500	125	200	–	–
K_m NADH (μM)	17	80	160	–	–	4.6
K_m NADPH (μM)	30	25	20	62 (22)	–	–
Spectral absorption maxima (nm)	280; 413 (ox) 423; 525; 557 (red)	–	278; 412 (ox) 423; 527; 557 (red)	413 (ox) 423; 528; 557 (red)	279; 413 (ox) 423; 527; 557 (red)	280; 413 (ox) 423; 527; 557 (red)
References	Hipkin *et al.* 1986	Zauner and Dellweg 1983	Guerrero and Guttierez 1977	Garrett and Amy 1978; Pan and Nason 1978; Horner 1983. (Values in parentheses)	Solomonson 1979; Solomonson *et al.* 1975 Howard and Solomonson 1982	Notton and Hewitt 1979; Hewitt and Notton 1980; Fido and Notton 1984; Fido 1987

^{a}–, data not available (or not applicable).

and exhibits a large negative (-174 mV) mid-point potential for its haem. Unlike the NR of *Chlorella* and filamentous fungi, however, *Candida* NR exhibits significant nitrate-reducing activity with either NADH or NADPH as an electron donor, although the V_{max} with NADH is higher than that with NADPH. Bi-specific NR activity has been described for other yeasts (e.g. Table 5.2 and Rivas *et al.* 1973; Choudary and Ramanda-Rao 1976*a*; Jones *et al.* 1987) and the results of a survey of the pyridine nucleotide specificities of a number of yeast NRs (C. R. Hipkin, unpublished observations) is shown in Table 5.3. With the exception of *Sporobolomyces* NR, which is specific for NADPH (Ali and Hipkin 1985), all the NR activities are bi-specific, although many exhibit a higher V_{max} with either NADH or NADPH. Maximum NAD(P)H-NR activity in yeast extracts usually requires the addition of FAD to assays or preparation buffers, suggesting that the flavin component of the enzyme is loosely bound and easily lost during extraction (Silver 1957; Rivas *et al.* 1973; Choudary and Ramanda-Rao 1976*a*; Zauner and Dellweg 1983; Hipkin *et al.* 1986; Jones *et al.* 1987).

Table 5.3. Pyridine nucleotide specificities of nitrate reductase and nitrite reductase from yeasts

Organism	Nitrate reductase		Nitrite reductase	
	NADH	NADPH	NADH	NADPH
Brettanomyces lambicus	+ +	+	ND	ND
Candida nitratophila	+ +	+	−	+
Candida utilis	+ +	+	−	+
Citeromyces matritensis	+	+	−	+
Dekkera intermedia	+ +	+	ND	ND
Hansenula anomala	+	+ +	−	+
Hansenula canadensis	+	+ +	−	+
Pachysolen tannophilus	+	+ +	ND	ND
Rhodosporidium toruloides	+	+	−	+
Rhodotorula glutinis	+	+ +	−	+
Sporobolomyces roseus	−	+	ND	ND

Information presented in the table summarizes data obtained by C. R. Hipkin (unpublished) and A. Al Kubisi and C. R. Hipkin (unpublished). + indicates NR activity with the specified pyridine nucleotide; + + indicates which of the specified pyridine nucleotides supported the higher activity (unless both supported similar levels of activity); − indicates that no activity was detected with the specified pyridine nucelotide; ND indicates not determined.

Following classical studies on *Neurospora crassa* NR (Evans and Nason 1952; Nason and Evans 1953), Silver (1957) showed that the NR of *Hansenula anomala* contained essential sulphydryl groups, as evidenced by the enzyme's sensitivity to para-chloromercuribenzoate, and similar results have been obtained with the NR from *Candida nitratophila* (Hipkin *et al.* 1986). Silver also showed that yeast NR was inhibited by metal binding agents such as azide and cyanide, which presumably blocked the molybdenum active site. Later Rivas *et al.* (1973) described a redox inactivation of *Torulopsis* NR by reduced methylviologen which was reversible after enzyme oxidation by ferricyanide or oxygen. Pichinoty and Metenier (1966) had previously observed a similar reversible redox inactivation of *Hansenula anomala* NR in their anaerobic assay system. Yeast NR is also inactivated rapidly in the presence of NAD(P)H and cyanide (Rivas *et al.* 1974; Hipkin *et al.* 1986). This type of inactivation is also reversible after oxidation with ferricyanide and has been described for several other eukaryotic NR systems (e.g. Solomonson *et al.* 1973; Garrett and Greenbaum 1973).

There have been few studies of yeast NiR (Rivas *et al.* 1973; Choudary and Ramanda Rao 1984; Jones *et al.* 1987; A. Al-Kubisi and C. R. Hipkin, unpublished observations). The yeast enzyme has not been purified sufficiently for detailed study but investigations show that the enzyme's activity is linked specifically to NADPH (Table 5.3). The NiR of *Candida nitratophila* (A. Al Kubisi and C. R. Hipkin, unpublished observations) is inhibited by sulphite but is not subject to a ferricyanide-reversible, reductive cyanide inactivation. Similar results have been obtained by Rivas *et al.* (1974) with *Torulopsis* NR.

Incorporation of reduced inorganic nitrogen into amino nitrogen

Ammonium, glutamate, and glutamine are end-products of nitrate assimilation and as such play a central role in its regulation. Knowledge of the major pathways of ammonium assimilation in yeasts is, therefore, very important if we are to understand how inorganic nitrogen assimilation is controlled in these organisms. Figure 5.1 shows two alternative pathways of ammonium assimilation which could operate in yeasts—the NADPH-glutamate dehydrogenase (GDH) pathway and the glutamine synthetase (GS)/glutamate synthase (GOGAT) pathway. Sims and Folkes and their colleagues showed by a very thorough ^{15}N-labelling study that about 75 per cent of the inorganic nitrogen assimilated by *Candida utilis* entered organic nitrogen metabolism as glutamate and most of the remainder entered via glutamine (Sims and Folkes 1964; Sims *et al.* 1968; Folkes and Sims 1974). They concluded that the major route for ammonium assimilation was via the biosynthetic GDH reaction. It should be noted, however, that glutamate synthase activity has been detected in a number of yeast species (Johnson and Brown 1974; Roon *et al.* 1974; Van Andel and Brown 1977; Casper *et al.* 1985) including *Candida nitratophila* (C. R. Hipkin, Z. Hamoudi, E. Marjot, and A. C. Cannons, unpublished observations) and that the GS/GOGAT pathway could be important for ammonium assimilation under conditions of nitrogen limitation.

The regulation of nitrate assimilation

Viewing nitrate assimilation as a whole (Fig. 5.1), one can readily identify points where there could be rapid, fine control of enzyme or permease activity as well as points of long-term, coarse control of enzyme or permease synthesis. In practice, the rapid inhibition of nitrate assimilation which occurs when ammonium is added to yeast cultures (Burn *et al.* 1974; Cocucci and Cocucci 1981; Ali and Hipkin 1985, 1986) is a striking example of a rapid, fine control. With species such as *Sporobolomyces roseus* and *Rhodotorula glutinis*, nitrate assimilation ceases completely until all of the added ammonium has been taken up from the medium. With other species, such as *Candida nitratophila* and *Candida utilis*, inhibition of nitrate uptake is incomplete so that these cultures assimilate ammonium and nitrate together (Fig. 5.2) but at rates slower than either would have been assimilated alone (Ali and Hipkin 1986). Nitrite assimilation in *Candida nitratophila* is also inhibited by ammonium as long as a carbon source such as glucose is available. This suggests that ammonium must be assimilated before inhibition takes place. It is significant that glutamine is more effective than ammonium as an inhibitor of nitrate and nitrite uptake (A. Al Kubisi, Z. Hamoudi, and C. R. Hipkin, unpublished observations), although, like ammonium, it does not inhibit nitrate or nitrite uptake in *Candida nitratophila* completely.

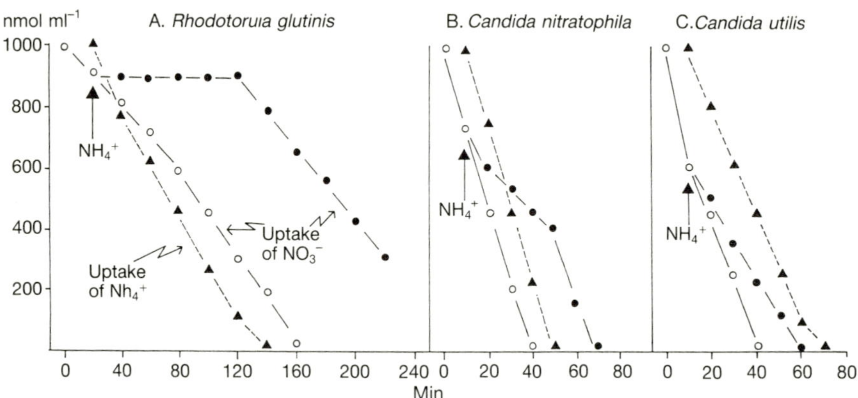

Fig. 5.2. The effect of ammonium on nitrate uptake by yeasts. Nitrate uptake was followed in cultures in the presence (●–●) or absence (○–○) of ammonium. Ammonium uptake was also followed (▲–▲). Arrows indicate addition of ammonium. Culture conditions: glucose (55 mM) as carbon source, cell density 10^8 cell ml.

At present it is not clear at which level the rapid inhibition of nitrate assimilation by ammonium (or glutamine) is effected. Likely targets for control include the nitrate uptake system and/or NR activity. However, since it is not possible yet to study nitrate uptake separate from nitrate reduction in yeasts it is difficult to study the regulation of nitrate uptake directly. A rapid, reversible inactivation of NR activity, like that described for algae (Guerrero *et al.* 1981) would eventually abolish a nitrate concentration gradient resulting in the inhibition of the facilitated diffusion of nitrate. There is evidence for *in vivo* inactivation of NR in yeasts. For example, crude enzyme preparations from

Hansenula anomala contain partially inactive NR which can be activated by ferricyanide under aerobic conditions (Minagawa and Yoshimoto 1983). However, the *in vivo* inactive NR in this organism does not appear to be cyanide inactivated (Minagawa and Yoshimoto 1984). *In vivo* inactivation of NR in *Rhodotorula glutinis* after addition of ammonium or hydroxylamine to nitrate-adapted cultures has also been demonstrated (Ito and Suzuki 1978), but the rates of *in vivo* NR inactivation described by these authors are not fast enough to explain the ammonium inhibition of nitrate uptake. Generally, losses of NR activity in yeast cultures that occur after the addition of ammonium take place over periods of hours (Ali and Hipkin 1985, 1986; Hipkin and Cannons 1987). At present the physiological significance of *in vivo* inactive NR is not clear. Another possibility is that nitrate assimilation becomes limited by the availability of reduced pyridine nucleotides when excess ammonium is supplied. The co-ordinated regulation of nitrate assimilation and NADPH production is discussed later.

The regulation of the synthesis of NR and NiR in plants and fungi provides one of the best examples of coarse control in eukaryotes. Several studies have shown that, in general, NR and NiR activities in yeasts are highest when organisms are grown with nitrate as a nitrogen source but low or absent after growth with reduced forms of nitrogen such as ammonium, glutamate, or glutamine (Silver 1957; Burn *et al.* 1974; Choudary and Ramanda Rao 1976*a,b*; Ali and Hipkin 1985, 1986; Cannons *et al.* 1986). Growth with ammonium nitrate results in the expression of intermediate levels of NR activity (Silver 1957; Choudary and Ramanda Rao 1976*a*; Hipkin *et al.* 1986; Hipkin and Cannons 1987) while growth with L-phenylalanine and D-amino acids, particularly D-norleucine, also allows the expression of significant levels of NR activity in *Candida utilis* (Choudary and Ramanda Rao 1976*a*).

Models have been developed from genetic and biochemical studies with mutants of *Aspergillus nidulans* (Chapters 6, 19) and *Neurospora crassa* (Chapter 20) which explain how the control of NR and NiR activity might be effected through the regulation of enzyme synthesis at the transcriptional level (Cove 1979; Wiame *et al.* 1985). The apparent induction of NR and NiR synthesis in *Aspergillus* cultures grown with nitrate as the sole source of nitrogen is explained by a well-known and elegant model in which the NR molecule itself plays an autogenous role (Cove 1979; Chapters 6, 19). In the absence of nitrate, the model predicts that the NR molecule binds to and inactivates the product of a regulatory gene which promotes the expression of the structural genes for NR and NiR (Cove and Pateman 1969) but probably not the nitrate permease (Brownlee and Arst 1983). This model explains why, in wild-type strains of *Aspergillus* and *Neurospora* (to some extent), NR and NiR activities are not expressed in the absence of nitrate even when reduced nitrogen is absent (Cove 1966; Dantzig *et al.* 1978). However, in some filamentous fungi (Morton 1956), algae (Syrett and Hipkin 1973; Hipkin *et al.* 1983) and yeasts (Ali and Hipkin 1985, 1986; Hipkin and Cannons 1987) NR activity is expressed and *sustained* in ammonium-grown cultures after a period of nitrogen starvation, i.e. in the absence of nitrate. NiR activity also develops in nitrogen-free cultures of yeasts such as *Candida nitratophila, Candida utilis,* and *Hansenula anomala* (A. Al Kubisi and C. R. Hipkin, unpublished observations). Thus, in yeasts, it appears that NR and NiR synthesis could be regulated in part by de-repression. Alternative explanations for the appearance of NR and NiR activity in nitrogen-free cultures of yeasts are not strong. For instance, trace levels of nitrate in putatively nitrogen-free medium could induce low levels of NR activity but would lead only to an ephemeral appearance of NR activity since nitrate is assimilated rapidly by nitrogen-starved cultures (Ali and Hipkin 1985, 1986). Wiame *et al.* (1985) point out that the results of nitrogen starvation experiments may be misleading, since the starvation effect may result from induction, as is perhaps true for arginase in *Saccharomyces.* This is not a good argument against the use of nitrogen starvation experiments in the study of NR and NiR synthesis in yeasts, however, since although nitrogen-replete cultures of yeasts may store arginine, which may become available as an inducer of arginase during nitrogen starvation, they do not accumulate nitrate or nitrite intracellularly (A. H. Ali and C. R. Hipkin, unpublished observations). Thus unless one argues that yeasts and algae are capable of nitrification, which is unlikely (Syrett and Hipkin 1973; Hipkin *et al.* 1980), the nitrogen starvation effect remains legitimate evidence for de-repression. In *Candida nitratophila* the appearance of NR activity in nitrogen-starved cultures correlates with the appearance of NR protein (Fig. 5.3) which can be demonstrated by rocket immunoelectrophoresis (Cannons *et al.* 1986) and *in vivo* labelling (Cannons and Hipkin 1987). It also correlates with the appearance of NR mRNA as demonstrated by *in vitro* translation (Cannons and Hipkin 1987).

These results indicate that a coarse control of nitrate assimilation in yeasts is effected at the level of transcription and that there may be significant differences between yeasts (i.e. *Candida*) and filamentous fungi (i.e. *Aspergillus* and *Neurospora*) in the way in which NR and NiR structural genes are regulated (Hipkin and Cannons 1987). Alternatively, the apparent differences between these organisms may reflect only the different affinities of different NRs for the product of the pathway specific regulatory gene (*vis à vis* Cove's model). In yeasts, such as *Candida nitratophila,* a relatively low affinity between NR and the regulatory protein would allow a low but significant level of NR to accumulate in the absence of nitrate.

Addition of nitrate to ammonium-grown, nitrogen-replete (represssed), or nitrogen-starved (de-repressed) cultures of yeasts results in a marked increase in NR and NiR activities which is prevented by inhibitors of protein synthesis (Chaudary and Ramanda Rao 1976*b*; Ali and Hipkin 1985, 1986; Jones *et al.* 1987). Sorger and Davies (1973) argued that in *Neurospora,* nitrate is required for translation of NR mRNA but not for the transcription of the NR structural gene. In contrast Chaudary and Ramanda Rao (1976*b*) suggested that in the yeast *Candida utilis,* nitrate was required for transcription but not translation of NR

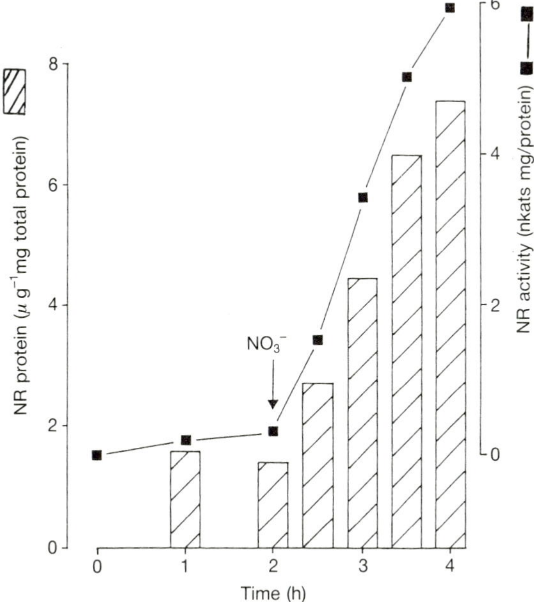

Fig. 5.3. Appearance of NR in Candida nitratophila. A culture previously grown with ammonium was transferred to nitrogen-free medium for 2 h before receiving 50 mM KNO_3 (arrow). NR activity (■–■) was measured in cell-free extracts and NR protein (□) was estimated by rocket immunoelectrophoresis.

mRNA. The increase in NR activity that occurs after the addition of nitrate to nitrogen-starved cultures of *Candida nitratophila* correlates with an increase in NR protein (Fig. 5.3), NR catalytic activity and, initially, NR mRNA (Hipkin and Cannons 1987; Cannons and Hipkin 1987). However, it is not clear whether the increase in NR mRNA levels induced by nitrate, which is seen in *in vitro* translation studies (Cannons and Hipkin 1987), results from a stimulation of NR mRNA formation by transcription or a stimulation of NR accumulation resulting from a post-transcriptional effect of nitrate on mRNA turnover. The fact that nitrate appears to stimulate an increase in the catalytic activity of NR suggests that it plays some regulatory role at a post-translational stage. Calculations based on the specific activity of pure yeast NR (Table 5.1), the amount of NR protein *in vivo* and its catalytic activity in crude extracts (Cannons *et al.* 1986) suggest that < 25 per cent of the total NR population in nitrogen-starved *Candida nitratophila* cells is in an active state. The inactive protein in this population may exist as unassembled NR subunits and the post-translational nitrate effect may involve a stabilization of the active, quaternary protein *in vivo* (Hipkin *et al.* 1985).

NR and NiR activities are lost from nitrate-adapted cultures of yeasts when they are resuspended in nitrogen-free medium, glucose (carbon)-free medium or medium containing a reduced nitrogen source such as ammonium, glutamate and glutamine (Ali and Hipkin 1985; Jones *et al.* 1987). Losses of NR activity that occur when nitrate-adapted cultures of *Candida nitratophila* are transferred to ammonium medium correlate with a decrease in NR protein (Fig. 5.4). *Candida* cultures resuspended in ammonium nitrate, however, do not show a decrease in NR activity initially, although the presence of ammonium may antagonize stimulatory effects of nitrate. Initially, such cultures assimilate ammonium and nitrate together (Fig. 5.2). Thus, initial decreases in NR protein that occur in

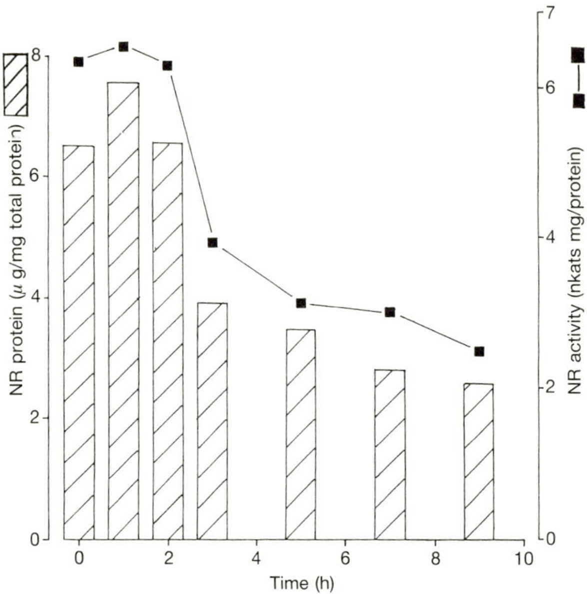

Fig. 5.4. Loss of NR in *Candida nitratophila*. A nitrate grown culture was transferred to fresh medium containing 10 mM ammonium as a nitrogen source. NR activity (■–■) was measured in cell-free extracts and NR protein (□) was estimated by rocket immunoelectrophoresis.

cultures containing reduced nitrogen may result more from inducer exclusion than repression. Similar results are obtained when the effects of ammonium and ammonium nitrate on NiR activity are followed (A. Al Kubisi and C. R. Hipkin, unpublished observations). Repression is more evident with *Candida nitratophila* when glutamine is added to nitrate-adapted cultures, even when nitrate is also present. This can be shown when NR synthesis is followed by *in vivo* labelling NR protein with [^{35}S]methionine (D. A. Kau, A. C. Cannons, and C. R. Hipkin, unpublished observations). In these experiments ammonium-grown cells are transferred to nitrate medium for 2 h so that they contain NR mRNA and high levels of NR protein and NR activity. Further incubation of these cultures with nitrate and [^{35}S]methionine results in an increase in NR activity and NR protein and incorporation of [^{35}S]methionine into *de novo* synthesized NR. However, there is no incorporation of label into NR if nitrate is removed or if ammonium replaces nitrate. In contrast, a substantial amount of NR protein is labelled in the presence of ammonium nitrate. These experiments demonstrate the importance of distinguishing between 'ammonium repression' and inducer exclusion. Under the experimental conditions described here, nitrate is apparently essential for the continued expression of preformed NR mRNA in yeasts. When glutamine is added to nitrate-adapting cultures with nitrate there is little incorporation of label into NR protein. This dramatic effect is good evidence that glutamine plays a prominent role in the regulation of NR in yeasts. Since these cultures are able to take up some nitrate in the presence of glutamine it is unlikely that the glutamine effect is the result of inducer exclusion. It is more likely that in the presence of excess glutamine the production of NR mRNA is represssed and rapid turnover results in its disappearance and thus the cessation of NR synthesis.

In filamentous fungi such as *Aspergillus* and *Neurospora*, the nitrate permease, NR and NiR are subject to nitrogen metabolite represssion (Arst and Cove 1973; Marzluf 1981; Brownlee and Arst 1983; Wiame *et al.* 1985; Chapters 19, 20). In this model the product of a wide domain regulatory gene is required for the expression of genes coding for several enzymes and permeases otherwise represssed in cultures grown with ammonium or glutamine (Chapters 19, 20). Nitrogen metabolite (catabolite) repression in yeasts is not as clearly understood but has been studied in *Saccharomyces* (Cooper 1982; Wiame *et al.* 1985). Significantly, there is evidence which suggests that it is a negatively acting control circuit with the product of the regulatory gene *(gdh CR)* acting as a repressor after it has been activated by glutamine.

The generation of reduced pyridine nucleotide for nitrate assimilation

The necessity for four molecules of NAD(P)H for every molecule of nitrate reduced to ammonium imposes a heavy energetic demand on the heterotrophic cell (Fig. 5.1). Earlier studies by Osmond and ap Rees (1969) indicated that the pentose phosphate pathway might be involved in this process. These authors showed that in *Candida utilis* the activities of glucose 6-phosphate dehydrogenase and transketolase were 2.5 times greater in extracts of cells grown on nitrate medium compared to cells grown on a complex, reduced nitrogen medium. Furthermore, their experiments showed that all treatments that resulted in higher activities of these enzymes also increased $^{14}CO_2$ production from $[1-^{14}C]$glucose. Later Bruinenberg *et al.* (1983*a*) carried out a theoretical analysis of NADPH production and consumption in yeasts. They calculated that when yeasts grow on ammonium as the nitrogen source at least 2 per cent of the glucose metabolized has to be oxidized completely via the pentose phosphate pathway, but this figure increases to 20 per cent when nitrate is the nitrogen source. Bruinenberg *et al.* (1983*b*) also showed that the activities of glucose 6-phosphate dehydrogenase, 6-phosphogluconate dehydrogenase, transketolase, and transaldolase were higher in nitrate cultures than in ammonium cultures of *Candida utilis*. Recent experiments with *Candida nitratophila* show that glucose 6-

phosphate dehydrogenase and 6-phosphogluconate dehydrogenase are regulated in coordination with NR and NiR (A. Al Kubisi and C. R. Hipkin, unpublished observations).

Further studies of the regulation of the supply of reduced pyridine nucleotide by the initial reactions of the pentose phosphate pathway for nitrate assimilation may reveal elegant and unexpected control mechanisms. It is notable that Hankinson and Cove (1974) showed a link between the regulation of nitrate assimilation enzymes and the enzymes of the pentose phosphate pathway in their studies with *Aspergillus*. A link between nitrate assimilation and the pentose phosphate pathway has also been demonstrated for unicellular algae (Hipkin and Cannons 1985).

Genetics of nitrate assimilation in yeasts

Very little is known about the genetics of nitrate assimilation in yeasts except for some preliminary studies (Jones *et al.* 1989) which have been carried out with *Hansenula wingei* (= *Hansenula canadensis*). Since this yeast is haploid, heterothallic, and has a well-defined life cycle (Crandall and Iovannisci 1980) it is a favourable organism for genetic study. Jones *et al.* (1989) isolated a number of different classes of spontaneous, chlorate-resistant mutants of this organism, one of which was unable to utilize nitrate as a nitrogen source but could use nitrite. Genetic analysis of this class revealed a number of complementation groups which were all NR negative and NiR positive. All mutants in this class were also able to grow on hypoxanthine as sole nitrogen source but some, nevertheless, appeared to be defective in molybdenum cofactor. Thus with some, NR activity and growth with nitrate could be restored in the presence of high concentrations of molybdate while others gave negative results in NADPH-NR reconstitution assays using *Neurospora crassa nit*-1. Another class of mutants was unable to use nitrate or nitrite. Analysis of this class indicated three inter-allelic complementation groups possibly representing a complex regulatory locus controlling NR and NiR.

Attempts were made to isolate mutants defective in nitrate assimilation in *Rhodotorula glutinis*. This organism is highly resistant to chlorate but NR mutants were isolated on the basis of their inability to grow on nitrate as a sole nitrogen source (F. Campbell, J. L. Wray, and J. R. Kinghorn, unpublished results).

Concluding remarks

Despite the lack of attention paid to yeast nitrate assimilation over the last 30 years, recent studies have allowed significant advances in our understanding of this process in species of *Candida* and *Hansenula*. Consequently, it is now possible to identify important areas of research for the future. In view of the high activity of some yeast NRs and their versatility with respect to pyridine nucleotide usage, further studies of the thermodynamic properties and structure of yeast NRs may provide useful information relevant to projects involving NR protein engineering. Further research on the regulation of NR synthesis in yeasts will also be worthwhile. Preliminary molecular biological studies have cleared a way for such investigations and despite the lack of knowledge of NR genetics for yeasts, which has been a stumbling block in the past, current techniques in molecular biology will allow research at gene level. Investigations into a possible negative control of NR repression and the nature of the de-repression of NR synthesis in yeasts are obvious areas of interest. Further studies on post-transcriptional control and enzyme turnover are also required. Finally, it should be noted that little is known about nitrate uptake and the provision of reducing power for nitrate assimilation in heterotrophic nitrate assimilators. Yeasts are ideal organisms for such studies.

References

Ali, A. H. and Hipkin, C. R. (1985). Nitrate assimilation in the basidiomycete yeast *Sporobolomyces roseus*. *Journal of General Microbiology* **131**, 1867–74.

Ali, A. H. and Hipkin, C. R. (1986). Nitrate assimilation in *Candida nitratophila* and other yeasts. *Archives of Microbiology* **144**, 263–7.

Arst, H. N., Jr. and Cove, D. J. (1973). Nitrogen metabolite repression in *Aspergillus nidulans*. *Molecular and General Genetics* **126**, 111–41.

Brownlee, A. G. and Arst, H. N., Jr. (1983). Nitrate uptake in *Aspergillus nidulans* and involvement of the third gene of the nitrate assimilation gene cluster. *Journal of Bacteriology* **155**, 1138–46.

Bruinenberg, P. M., Van Dijken, J. P., and Scheffers, A. W. (1983*a*). A theoretical analysis of NADPH production and consumption in yeasts. *Journal of General Microbiology* **129**, 953–64.

Bruinenberg, P. M., Van Dijken, J. P., and Scheffers, A. W. (1983*b*). An enzymic analysis of NADPH production and consumption in *Candida utilis*. *Journal of General Microbiology* **129**, 965–71.

Burn, V. J., Turner, P. R., and Brown, C. M. (1974). Aspects of inorganic nitrogen assimilation in yeasts. *Antonie van Leeuwenhoek* **40**, 93–102.

Cannons, A. C. and Hipkin, C. R. (1987). Evidence for the transcriptional control of nitrate reductase in *Candida nitratophila* from *in vitro* translation studies. *European Journal of Biochemistry* **164**, 383–7.

Cannons, A. C., Ali, A. H., and Hipkin, C. R. (1986). Regulation of nitrate reductase synthesis in the yeast *Candida nitratophila*. *Journal of General Microbiology* **132**, 2005–11.

Casper, P., Bode, R., and Birnbaum, D. (1985). Untersuchungen zur Regulation der Ammoniumassimilation von *Candida maltosa*. *Journal of Basic Microbiology* **25**, 95–101.

Choudary, V. P. and Ramanda Rao, G. (1976*a*). Regulatory properties of yeast nitrate reductase *in situ*. *Canadian Journal of Microbiology* **22**, 35–42.

Choudary, V. P. and Ramanda Rao, G. (1976*b*). Molecular basis of nitrate reductase induction in *Candida utilis*. *Biochemical and Biophysical Research Communications* **72**, 598–602.

Choudary, V. P. and Ramanda Rao, G. (1984). Molecular analysis of inorganic nitrogen assimilation in yeasts. *Archives of Microbiology* **138**, 183–6.

Cocucci, M. and Cocucci, S. M. (1981). Metabolic adaptation of the yeast *Rhodotorula gracilis* growing on ammonium or nitrate as nitrogen source. *Annali di Microbiologia* **31**, 89–101.

Cooper, T. G. (1982). Nitrogen metabolism in *Saccharomyces cerevisiae*. In *The molecular biology of the yeast Saccharomyces* (eds J. N. Strathern, E. W. Jones, and J. R. Broach), pp. 39–99, Cold Spring Harbor Laboratory Press, NY.

Cove, D. J. (1966). The induction and repression of nitrate reductase in the fungus *Aspergillus nidulans*. *Biochimica et Biophysica Acta* **113**, 51–6.

Cove, D. J. (1979). Genetic studies of nitrate assimilation in *Aspergillus nidulans*. *Biological Reviews* **54**, 291–327.

Cove, D. J. and Pateman, J. A. (1969). Autoregulation of the synthesis of nitrate reductase in *Aspergillus nidulans*. *Journal of Bacteriology* **97**, 1374–8.

Crandall, M. J. and Iovannisci, D. M. (1980). Genetics of *Hansenula wingei*. In *Current developments in yeast research 1981. Proceedings of the 5th International Symposium of Yeast* (eds G. Steward and I. Russell), pp. 149–55. Pergamon Press, London.

Dantzig, A. H., Zurowski, W. K., Ball, T. M., and Nason, A. (1978). Induction and repression of nitrate reductase in *Neurospora crassa*. *Journal of Bacteriology* **133**, 671–9.

Eddy, A. A. and Hopkins, P. G. (1985). The putative electrogenic nitrate-proton symport of the yeast *Candida utilis*. *Biochemical Journal* **231**, 291–7.

Evans, H. J. and Nason, A. (1952). The effect of reduced triphosphopyridine nucleotide on nitrate reduction by purified nitrate reductase. *Archives of Biochemistry and Biophysics* **39**, 234–5.

Fido, R. J. (1987). Purification of nitrate reductase from spinach (*Spinacea oleracea* L.) by immunoaffinity chromatography using a monoclonal antibody. *Plant Science* **50**, 111–15.

Fido, R. J. and Notton, B. A. (1984). Spinach nitrate reductase: Further purification and removal of 'nicked' sub-units by affinity chromatography. *Plant Science Letters* **37**, 87–91.

Folkes, B. F. and Sims, A. P. (1974). The significance of amino acid inhibition of NADP-linked glutamate dehydrogenase in the physiological control of glutamate synthesis in *Candida utilis*. *Journal of General Microbiology* **82**, 77–95.

Garrett, R. N. and Amy, N. K. (1978). Nitrate assimilation in fungi. *Advances in Microbial Physiology* **18**, 1–65.

Garrett, R. N. and Greenbaum, P. (1973). The inhibition of the *Neurospora* nitrate reductase complex by metal binding agents. *Biochimica et Biophysica Acta* **302**, 24–32.

Guerrero, M. G. and Gutierrez, M. (1977). Purification and properties of the NAD(P)H:nitrate reductase of the yeast *Rhodotorula glutinis*. *Biochimica et Biophysica Acta* **482**, 272–85.

Guerrero, M. G., Vega, J. M., and Losada, M. (1981). The assimilatory nitrate reducing system and its regulation. *Annual Review of Plant Physiology* **32**, 169–204.

Hankinson, O. and Cove, D. J. (1974). Regulation of the pentose phosphate pathway in the fungus *Aspergillus nidulans*. *Journal of Biological Chemistry* **249**, 2344–53.

Hewitt, E. J. and Notton, B. A. (1980). Nitrate reductase systems in eukaryotic and prokaryotic organisms. In *Molybdenum and molybdenum-containing enzymes* (ed. M. Coughlan), pp. 273–325. Pergamon Press, Oxford.

Hipkin, C. R. and Cannons, A. C. (1985). Inorganic nitrogen assimilation and the regulation of glucose 6-phosphate dehydrogenase in unicellular algae. *Plant Science* **41**, 155–60.

Hipkin, C. R. and Cannons, A. C. (1987). The regulation of nitrate reductase in yeasts. In

Inorganic nitrogen metabolism (eds W. R. Ullrich, P. J. Aparicio, P. J. Syrett, and F. Castillo), pp. 89–93. Springer-Verlag, Berlin.

Hipkin, C. R., Al-Bassam, B. A., and Syrett, P. J. (1980). The role of nitrate and ammonium in the regulation of the development of nitrate reductase in *Chlamydomonas reinhardtii*. *Planta* **150**, 13-18.

Hipkin, C. R., Thomas, R. J., and Syrett, P. J. (1983). Effects of nitrogen deficiency on nitrate reductase, nitrate assimilation and photosynthesis in unicellular marine algae. *Marine Biology* **77**, 101–5.

Hipkin, C. R., Hermann-Smith, J. A., and Syrett, P. J. (1985). The role of nitrate, photosynthesis and protein turnover in the formation of nitrate reductase in the *nit A* mutant of *Chlamydomonas*. *Biochimica et Biophysica Acta* **838**, 191–6.

Hipkin, C. R., Ali, A. H., and Cannons, A. C. (1986). Structure and properties of assimilatory nitrate reductase from the yeast *Candida nitratophila*. *Journal of General Microbiology* **132**, 1997–2003.

Horner, R. D. (1983). Purification and comparison of *nit-1* and wild-type NADPH-nitrate reductases of *Neurospora crassa*. *Biochimica et Biophysica Acta* **744**, 7–15.

Howard, W. D. and Solomonson, L. P. (1982). Quaternary structure of assimilatory NADH : nitrate reductase from *Chlorella*. *Journal of Biological Chemistry* **257**, 10243–50.

Ito, N. and Suzuki, Y. (1978). Inactivation of nitrate reductase of the yeast, *Rhodotorula glutinis*. *Physiologia Plantarum* **44**, 15–20.

Johnson, B. and Brown, C. M. (1974). The enzymes of ammonia assimilation in *Schizosaccharomyces* spp. and in *Saccharomyces ludwigii*. *Journal of General Microbiology* **85**, 169–72.

Jones, C. P., Wray, J. L., and Kinghorn, J. R. (1987). The role of nitrogen sources in the regulation of nitrate reductase and nitrite reductase levels in the yeast *Hansenula wingei*. *Journal of General Microbiology* **133**, 2767–72.

Jones, C. P., Wray, J. L., and Kinghorn, J. R. (1989). Isolation and preliminary characterisation of mutants of the yeast *Hansenula wingei* defective in nitrate assimilation. In preparation.

Kay, C. J., Solomonson, L. P., and Barber, M. J. (1986). Thermodynamic properties of the heme prosthetic group in assimilatory nitrate reductase. *Journal of Biological Chemistry* **261**, 5799–802.

Kregar-van Rij, N. J. W. (ed.) (1984). *The yeasts. A taxonomic study* (3rd edn). Elsevier Science Publishers, Amsterdam.

Large, P. J. (1986). Degradation of organic nitrogen compounds by yeast. *Yeast* **2**, 1–34.

Lilly, V. G. and Barnett, H. L. (1951). *Physiology of the fungi*. McGraw-Hill, New York.

Marzluf, G. A. (1981). Regulation of nitrogen metabolism and gene expression in fungi. *Microbiological Reviews* **45**, 437–61.

Minagawa, N. and Yoshimoto, A. (1983). Assimilatory nitrate reductase of *Hansenula anomala*: Its electron donors and cellular distribution. *Agricultural and Biological Chemistry* **47**, 125–7.

Minagawa, N. and Yoshimoto, A. (1984). The *in vivo* inactive nitrate reductase from *Hansenula anomala*. *Agricultural and Biological Chemistry* **48**, 1907–9.

Morton, A. G. (1956). A study of nitrate reduction in mould fungi. *Journal of Experimental Botany* **7**, 97–112.

Nason, A. and Evans, H. J. (1953). Triphosphopyridine nucleotide–nitrate reductase in *Neurospora*. *Journal of Biological Chemistry* **202**, 655–73.

Notton, B. A. and Hewitt, E. J. (1979). Structure and properties of higher plant nitrate reductase, especially *Spinacea oleracea*. In *Nitrogen assimilation of plants* (eds E. J. Hewitt and C. V. Cutting), pp. 227–44. Academic Press, London.

Osmond, C. B. and ap Rees, T. (1969). Control of the pentose phosphate pathway in yeast. *Biochimica et Biophysica Acta* **184**, 35–42.

Pan, S. S. and Nason, A. (1978). Purification and characterization of homogeneous assimilatory reduced nicotinamide adenine dinucleotide phosphate–nitrate reductase from *Neurospora crassa. Biochimica et Biophysica Acta* **523**, 297–313.

Pichinoty, F. and Metenier, G. (1966). Contribution a l'étude de la nitrate-reductase assimilatrice d'une levure. *Annales de l'Institut Pasteur (Paris)* **111**, 282–313.

Pichinoty, F. and Metenier, G. (1967). Regulation de la biosynthèse et localization de la nitrate-reductase d'*Hansenula anomala. Annales de l'Institut Pasteur (Paris)* **112**, 701–11.

Rivas, J., Guerrero, M. G., Paneque, A., and Losada, M. (1973). Characterization of the nitrate-reducing system of the yeast *Torulopsis nitratophila. Plant Science Letters* **1**, 105–13.

Rivas, J., Tortolero, M., and Pancque, A. (1974). Metal components of the nitrate-reducing system from the yeast *Torulopsis nitratophila. Plant Science Letters* **2**, 283–8.

Roon, R., Even, H. L., and Larimore, F. (1974). Glutamate synthase: properties of the reduced nicotinamide adenine dinucleotide-dependent enzyme from *Saccharomyces cerevisiae. Journal of Bacteriology* **118**, 89–95.

Silver, W. S. (1957). Pyridine nucleotide-nitrate reductase from *Hansenula anomala*, a nitrate reducing yeast. *Journal of Bacteriology* **73**, 241–6.

Sims, A. P. and Folkes, B. F. (1964). A kinetic study of the assimilation of [^{15}N]ammonia and the synthesis of amino acids in an exponentially growing culture of *Candida utilis. Proceedings of the Royal Society (London)* **159**, 479–502.

Sims, A. P., Folkes, B. F., and Bussey, A. H. (1968). Mechanisms involved in the regulation of nitrogen assimilation in microorganisms and plants. In *Recent aspects of nitrogen metabolism in plants* (eds E. J. Hewitt and C. V. Cutting), pp. 91–114. Academic Press, London.

Solomonson, L. P. (1975). Purification of NADH-nitrate reductase by affinity chromatography. *Plant Physiology* **56**, 853–5.

Solomonson, L. P. (1979). Structure of *Chlorella* nitrate reductase. In *Nitrogen assimilation of plants* (eds E. J. Hewitt and C. V. Cutting), pp. 199–205. Academic Press, London.

Solomonson, L. P., Jetschmann, C. and Vennesland, B. (1973). Reversible inactivation of the nitrate reductase of *Chlorella vulgaris* Beijerinck. *Biochimica et Biophysica Acta* **381**, 32–43.

Solomonson, L. P., Lorimer, G. H., Hall, R. L., Borchers, R., and Bailey, J. L. (1975). Reduced nicotinamide adenine dinucleotide-nitrate reductase of *Chlorella vulgaris. Journal of Biological Chemistry* **250**, 4120–7.

Sorger, G. J. and Davies, J. (1973). Regulation of nitrate reductase of *Neurospora* at the level of transcription and translation. *Biochemical Journal* **134**, 673–85.

Syrett, P. J. and Hipkin, C. R. (1973). The appearance of nitrate reductase activity in nitrogen-starved cells of *Ankistrodesmus braunii. Planta* **111**, 57–64.

Tannenbaum, R. S., Weisman, M. and Fett, D. (1976). The effects of nitrate intake on nitrite formation in human saliva. *Food and Cosmetics Toxicology* **14**, 549–52.

Van Andel, J. G. and Brown, C. M. (1977). Ammonia assimilation in the fission yeast *Schizosaccharomyces pombe*, 972. *Archives of Microbiology* **111**, 265–70.

Wiame, J.-M., Grenson, M., and Arst, H. N., Jr. (1985). Nitrogen catabolite repression in yeasts and filamentous fungi. *Advances in Microbial Physiology* **26**, 1–88.

Zauner, E. and Dellweg, H. (1983). Purification and properties of the assimilatory nitrate reductase from the yeast *Hansenula anomala. European Journal of Applied Microbiology and Biotechnology* **17**, 90–5.

6. Genetic, biochemical, and structural organization of the *Aspergillus nidulans* *crnA–niiA–niaD* gene cluster

James R. Kinghorn

Introduction

Over the past 25 years or so, much has been discovered concerning the nitrate assimilation pathway in the ascomycetous filamentous fungus *Aspergillus nidulans* so that now the pathway is perhaps the best genetically characterized of any eukaryotic organism. That genetic studies were first initiated in *A. nidulans* (Cove and Pateman 1963) is perhaps a trivial reason for this. A more accurate explanation is that *A. nidulans* is amenable to intensive and rigorous genetic analyses since both heterokaryon and diploid phases exist which aid the identification of functional genetic units through complementation studies. In addition, these life cycle phases permit the determination of allele dominance/recessivity relationships which in turn provide valuable information on the identification and the nature of regulatory genes and their control circuits. Such functional tests are supported by the availability of both parasexual and sexual cycles which assist the rapid assignment of mutant alleles to individual linkage groups (chromosomes) as well as the determination of distances, in recombination terms, between mutant alleles.

Although genetic aspects of nitrate assimilation are best seen in *A. nidulans*, contributions have been made using certain other filamentous fungi, especially *Neurospora crassa* (Chapter 20) and also, but to a lesser degree, other ascomycetes such as *A. amstelodami*, *A. oryzae*, *A. niger*, *Cephalosporium acremonium*, *Fulvia fulva*, *Fusarium oxysporum*, *Gibberella fujikuroi*, *Penicillium chrysogenum*, *Septoria nodorum*, as well as the basidiomycete *Ustilago maydis* (Chapter 22).

Genetic organization

The genetic components of the pathway in *A. nidulans* have been reviewed before (Cove 1970, 1979; Pateman and Kinghorn 1976; Scazzocchio 1980; Arst 1981,

1984; Dunn-Coleman *et al.* 1984; Arst and Scazzocchio 1985) but since the *A. nidulans* work is regarded as the 'guiding light' among nitrateophiles, no nitrate assimilation text would be complete without a fleeting reference to the general genetic framework. For this reason I shall very briefly re-tell the old and well-told story.

Mutants defective in nitrate assimilation can be selected on the basis of resistance to chlorate and are easily assigned by growth tests (Table 6.1) and by the use of the whole panoply of *A. nidulans* genetics, to one of a number of gene loci. Growth responses outlined in Table 6.1 show that most chlorate-resistant mutants have a dual growth phenotype, i.e. poor growth on nitrate as well as resistance to chlorate. This particular pleiotropic phenotype is rare in microbial genetics, a notable exception being certain pyrimidine auxotrophs which are resistant to 5-fluoro-orotic acid (Chapter 22). The double phenotype makes the selection of mutants via resistance and their subsequent characterization on the basis of their growth defects very powerful and attractive for study. Mutants in all nitrate assimilation genes can be isolated via chlorate resistance, with the exception of internal mutants in the *niiA* gene which have to be identified by the more laborious study of their inability to utilize nitrite (or nitrate) as the sole source of nitrogen.

The enzymes nitrate reductase (NR) (EC 1.6.6.3) and nitrite reductase (NiR) (EC 1.6.6.4) are encoded by *niaD* and *niiA* structural genes, respectively (Cove 1979 and references therein; Fig. 6.1; Table 6.1). NR shares a common molybdenum cofactor (MoCo) with xanthine/hypoxanthine dehydrogenase (purine hydroxylase I). Mutations within a number of *cnx* genes result in impairment of the biosynthesis of this cofactor and consequently *cnx* mutants concomitantly lack NR and hypoxanthine dehydrogenase activities (see Chapter 19 for further discussion of *A. nidulans cnx* genes). More recently a third gene, *crnA*, which is thought to encode an uptake system involved in nitrate transport, had been identified (Tomsett and Cove 1979; Brownlee and Arst 1983; Chapter 19).

The genetic organization is unusual for eukaryotic organisms in that the three genes are tightly linked in a cluster in the order *crnA–niiA–niaD* (Arst *et al.* 1979; Tomsett and Cove 1979; Fig. 6.2). Although eukaryotic genes of related function are normally unlinked, and are often scattered throughout the genome, there are, in fact, a few other examples of gene clusters in *A. nidulans*, such as those involved in utilization of proline (Arst 1984 and references therein), quinic acid (Grant *et al.* 1988) and alcohol (Pateman *et al.* 1985; Gwynne *et al.* 1987). Furthermore, the *niiA* and *niaD* genes appear to be tightly linked in other fungi such as *A. niger*, *P. chrysogenum*, and *U. maydis* (Chapter 22) but the general significance of the linkage of these genes must take into account that the *niiA* and *niaD* genes are non-contiguous, and, in fact, on different chromosomes, in *N. crassa* (Tomsett and Garrett 1980; Chapter 20). It was not entirely clear whether the *A. nidulans* nitrate gene cluster (and the other gene clusters for that matter) formed an operon-like structure, in which co-ordinately regulated, functionally related

Table 6.1. *A. nidulans* mutants defective in nitrate assimilation

Gene mutation[a]	Chlorate resistance[b]	Utilization of sole nitrogen source[c]						Summary of the role of these genes (see text for references and also Chapter 19)
		Nitrate	Nitrite	Ammonium	Hypoxanthine	Proline	Glutamate	
crnA⁻	R	+	+	+	+	+	+	Probably encodes a product for nitrate uptake in young cells
niiA⁻	S	−	−	+	+	+	+	The nitrite reductase structural gene
niaD⁻	R	−	+	+	+	+	+	The nitrate reductase structural gene
cnxA-J⁻	R	−	−	+	−	+	+	At least six loci are thought to be involved in the synthesis of the molybdenum cofactor required for nitrate reductase and hypoxanthine dehydrogenase activities
*areA*ʳ	R	−	−	+	−	−	−	Control gene, whose product is required for nitrogen metabolite (or ammonium) repression of the nitrate assimilation systems
nirA⁻	R	−	−	+	+	+	+	Control gene required for nitrate induction of nitrate assimilation systems

[a] − or r superscripts denote loss-of-function mutations. In the case of the *areA* and *nirA* genes other allele forms exist such as *areA*ᵈ and *nir*ᶜ (see Chapter 19).
[b] R denotes resistance, whilst S denotes sensitivity. The nitrogen source employed is crucial with respect to sensitivity or resistance status. Two further gene loci (designated *chlA* and *chlB*) have been identified and mutation within these leads to chlorate resistance but not to impaired growth on nitrate as a sole nitrogen source (see text and Cove 1976*a,b*, 1979).
[c] + denotes wild-type levels of growth, − denotes poor growth.

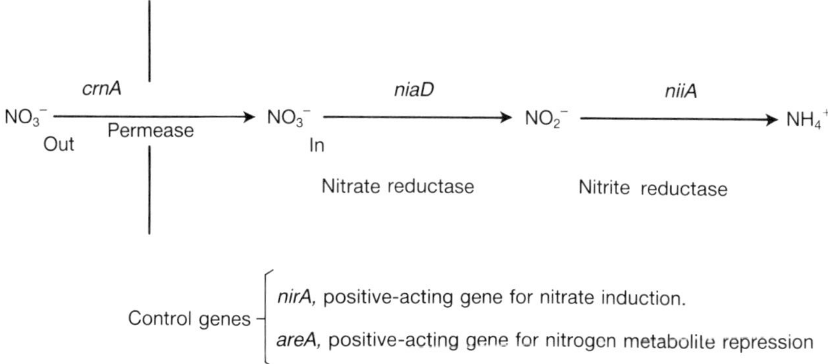

Fig. 6.1. The nitrate assimilation pathway in *A. nidulans*. The *nirA* gene maps on linkage group VIII whilst *areA* is located on linkage group III (Chapter 19).

genes were expressed from a single polycistronic mRNA species. Such modes of organization are common in certain prokaryotic organisms (for an interesting discussion on the existence or otherwise of filamentous fungal operons see Arst 1981). The genetic evidence against an operon type of organization in the case of the *A. nidulans* nitrate gene cluster can be summarized as follows. Firstly, no polypeptide chain termination mutations are known that exert polar effects on the downstream genes (as one sees in prokaryotic operons). For instance, two suppressible mutations (*niaD*500 and *niaD*501) have been identified that are likely to be chain termination mutations (Roberts *et al.* 1979; Martinelli *et al.* 1984). Significantly, these do not exhibit polarity on NiR activity (J. R. Kinghorn, unpublished). Secondly, certain alleles in the control gene *nirA*, (*vide infra*) have different effects on *niaD* and *niiA* expression. For example the regulatory mutation *nirA*ᶜ*1* results in complete constitutive expression of NR but only partial constitutive expression of NiR (Cove, 1979). A third line of genetic evidence against the operon possibility comes from the study of the *nis-5* mutant strain in which a large fragment from chromosome II has been inserted between the *niiA* and *niaD* genes, which reside on chromosome VIII (Rand and Arst 1977; Arst *et al.* 1979; Chapter 19). That the relocation of functionally unrelated genes between *niiA* and *niaD* does not abolish NiR or NR activities suggests that *niiA* and *niaD* expression cannot occur solely from a *niiAniaD* dicistronic (or tricistronic) transcript. A prediction arising from *nis-5* studies, namely that transcription of *niiA* initiates from the intergenic region between *niiA–niaD*, was made by Arst and colleagues (1979). These deductions, based on formal genetic analysis of the nitrate system, have been borne out by recent molecular genetic studies (*vide infra*).

An impressive fine structure genetic map of the *crnA–niiA–niaD* region (Fig. 6.3) has been constructed by Tomsett and Cove (1979). Their approach was essentially similar to that originally devised by Benzer for viral systems (Benzer and Champe 1961). A large number of *niaD* deletion strains were generated and

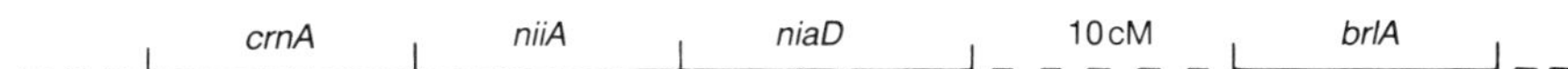

Fig. 6.2. Gene order of the nitrate assimilation cluster, orientation and distance (in genetic recombination terms) of the nitrate gene cluster with respect to the bristle locus (*brlA*). The three contiguous genes are located on linkage group VIII but are unlinked with *nirA* (cM = centimorgans).

while some mutant lesions were found to be contained entirely within *niaD* others extended into *niiA* or into *crnA*. Deletion mapping of the region has shown the existence of 9 deletion intervals within *niiA* and 26 within *niaD*. This contribution, which represents one of the more extensive analyses of the fine structure of any eukaryotic locus, has still to be exploited biochemically. Consequently the original goal—to relate the genetic map with function and structure of the gene products—remains unfulfilled since only a few mutants have been analysed biochemically (*vide infra*). What can be discerned so far are possible *niaD* protein-coding regions by comparing the position of mutant lesions possessing a heat-sensitive enzyme (Fig. 6.3, marked *) with the position of null mutations.

The pathway is under strong genetic regulation mediated by two unlinked control genes (Chapter 19 and references therein). *nirA* is a pathway-specific control gene involved in nitrate induction of NR and NiR activities. These activities, as well as the *crnA* permease, are regulated by ammonium or nitrogen metabolite repression via the product of *areA*, a 'global' control gene which regulates the expression of a number of other ammonium-repressible pathways. A feature additional to this interesting regulatory circuit is that NR *per se* plays a role in its own synthesis (autogenous control) as well as that of NiR (Cove, 1979).

Although chlorate is extensively and successfully employed to generate mutant strains in a range of organisms from prokaryotes to higher plants, the mechanisms involved in toxicity are still unclear. The generally accepted view is that chlorate is converted to chlorite by NR-associated catalytic activity and it is chlorite, not chlorate, that is toxic to wild-type cells. Although NR certainly does have chlorate reductase activity, Cove (1976*a*) points out that this cannot be the whole story since certain NR- (and chlorate reductase) deficient mutants of *A. nidulans* are sensitive to chlorate. Conversely, certain chlorate-resistant mutant strains (mutations within which map in two unlinked genes designated *chlA* and *chlB*) have appreciable levels of NR activity and consequently grow on nitrate as a sole nitrogen source. Another curious finding is that the level of toxicity exhibited by a given concentration of chlorate depends on the nitrogen source used (Cove 1976*b*) and, consequently, higher concentrations are required on some nitrogen sources than on others. Moreover, the choice of nitrogen source markedly affects the class and frequency of mutant isolated. For example *nirA⁻* mutant strains are rarely, if ever, generated when glutamate is the nitrogen source in selection experiments (Tomsett and Cove, 1979). A final noteworthy point, made by Cove (1979) is that nitrate itself may reduce the levels of enzyme systems required for

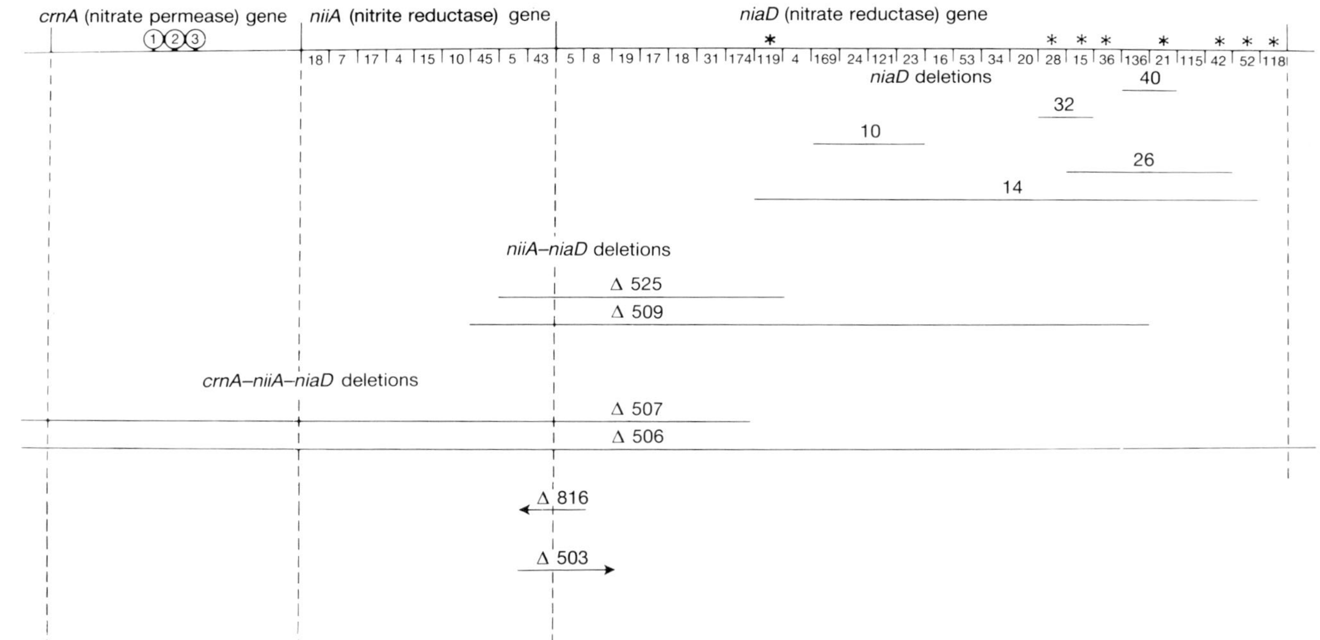

Fig. 6.3. Fine structure genetic map of the *crnA–niiA–niaD* region modified after Tomsett and Cove (1979). Deletion mutations extending from *niaD* into *niiA* or through *niiA* into *crnA* are prefixed Δ. Horizontal continuous lines are used to show the extent of each deletion. Although a large number of deletion mutations have been isolated and characterized, for clarity only a few typical representatives are shown. Deletion intervals lie between the short vertical lines. An arrowhead indicates that the extremity of the deletion is unknown. Many point mutations have been assigned to each region but for simplicity only one representative is included. An asterisk above a given region means that temperature-sensitive mutants exist in that region. Broken vertical lines denote the boundary between each gene. No fine structure map is available for the *crnA* gene; the positions of mutations (encircled) in *crnA* are unknown.

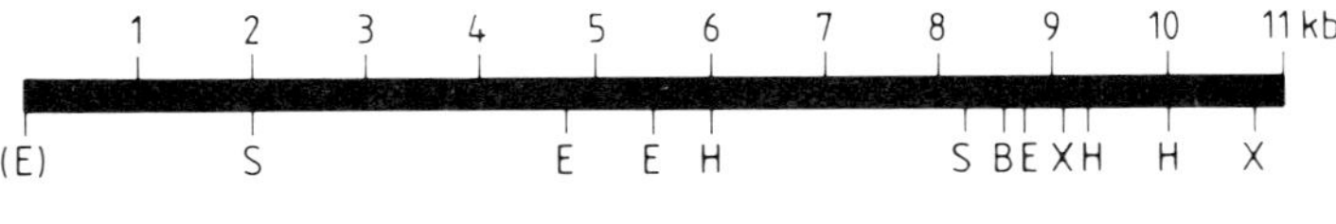

(a)

(b)

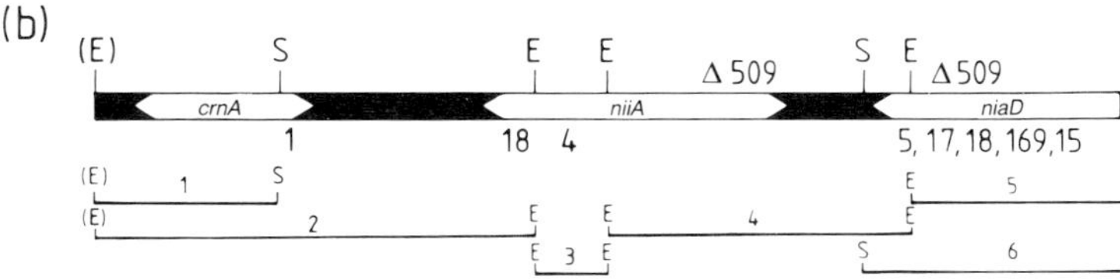

(c)

(d)

Fig. 6.4. Physical map of the cloned nitrate assimilation gene cluster of *A. nidulans*. Only those restriction endonuclease sites relevant to the discussion in the text are included. B = *Bam*H1, Bg = *Bgl*1, E = *Eco*R1, H = *Hind*III, S = *Sma*I, X = *Xba*I. (E) denotes a polylinker site. (a) The position of deleted regions in mutant strains *niiA–niaD*Δ509 and *crnA–niiA–niaD*Δ506 (as determined by DNA:DNA hybridization) are represented as horizontal lines. The locations of the extremities of the *niiA–niaD*Δ509 deletion have not been precisely located and hence are represented as broken lines. (b) The bar shows the position of the nitrate assimilation genes on the physical map as judged by the complementation of mutants. Numbers immediately below the bar represent the position of point mutations. Subclones generated for complementation studies with their corresponding number are shown by the horizontal lines. (c) Approximate position and size of *crnA*, *niiA* and *niaD* transcripts (as determined by DNA:RNA hybridization). Horizontal lines above the transcripts represent DNA fragments used as probes for hybridization studies. (d) Position of the *niiA* and *niaD* protein-coding sequences relative to the physical map. Arrows denote direction of transcription.

nitrogen catabolism and that chlorate can possibly mimic nitrate in this regulatory aspect although chlorate itself cannot act as a nitrogen source. This would result in the inability of wild-type cells to grow on nitrogen sources in the presence of chlorate due to nitrogen starvation rather than to chlorate toxicity *per se*. The complexity of chlorate resistance and possible reasons for bias frequency in favour of particular genotypes are discussed at length by Cove (1976*a,b*, 1979).

Biochemical organization

The information gained by study of the biochemistry of nitrate assimilation in *A. nidulans* has not, so far, matched the genetic contribution and, of the three steps in the pathway, data are available only on NADPH nitrate reductase. Early independent reports indicated a size of 190–197 kDa for the NR molecule (Downey, 1971; MacDonald and Coddington 1974), while a more recent study estimated 180 kDa (Minagawa and Yoshimoto 1982). These values are significantly smaller than those of 228 kDa (Garrett and Nason 1969; Pan and Nason 1978) or 290 kDa (Horner 1983) reported for the NR of *Neurospora crassa*. The subunit size of the *A. nidulans* enzyme has been reported to be of the order of 50–55 kDa (Downey and Focht 1974; Downey and Steiner 1979; Steiner and Downey 1982), raising the possibility of a tetrameric protein. Such a structure would be at variance with that reported for the NR of other fungi.

At least part of the variability in the estimations of the size of *A. nidulans* NR is undoubtedly due to the lengthy purification procedures, giving proteases ample opportunity to cleave the enzyme. More recently, Cooley and Tomsett (1985) attempted to determine the subunit size using an immunological approach which would be likely to minimize proteolytic degradation of the nitrate reductase subunit. They labelled *A. nidulans* by growth on [^{35}S]SO$_4$, ground the mycelia in the presence of the protease inhibitor PMSF and immunoprecipitated the labelled NR subunits with specific NR antibodies. Subsequent SDS–polyacrylamide gel electrophoresis of the immunoprecipitate showed the presence of a protein of 91 kDa in mycelia grown with nitrate, while mycelia grown without nitrate lacked this protein. This protein was also absent from immunoprecipitates of the extensive deletion mutant *crnA–niiA–niaD* Δ506 and from the internal deletion mutant *niaD*14 grown with nitrate as sole nitrogen source. These results suggest that the subunit size of *A. nidulans* NR is around 90 kDa, consistent with the holoenzyme (∼190 kDa) being a dimer of two such subunits and confirming the smaller size estimate of the *A. nidulans* NR when compared to *N. crassa*.

Like the NR of *N. crassa* (Garrett and Nason 1969; Chapter 20), the NR of *A. nidulans* contains FAD, haem (cytochrome b_{557}) and molybdenum (Cove and Coddington 1965; Pateman *et al.* 1967; MacDonald and Coddington, 1974), the latter carried as part of the MoCo (Chapter 14), encoded by the *cnx* genes (Chapter 19). Electron flow is from NADPH through enzyme-bound FAD, haem, and molybdenum to nitrate. In addition to this physiological reaction (NADPH–nitrate reductase), the enzyme also possesses NADPH cytochrome *c* reductase activity and reduced benzylviologen dye-NR activity, assayable *in vitro* but considered to be non-physiological. MacDonald and Coddington (1974) have described the enzyme as consisting of two functional regions, one of which is heat-labile, contains FAD and probably haem, and transfers electrons from NADPH to cytochrome *c*, whilst the other contains molybdenum and transfers electrons from reduced viologen dyes to nitrate.

The biochemical properties of only a few mutants have been described so far (MacDonald *et al.* 1974). The putative point mutations in the apoprotein carried by strain *niaD*8 and *niaD*21 result in a lack of all NR-associated activities whilst the mutations in strains *niaD*10 and *niaD*13 lead to loss of NADPH-nitrate reductase and NADPH-cytochrome *c* reductase activities with the retention of reduced benzylviologen-NR activity sedimenting at 7.6 S, the sedimentation coefficient of the wild-type dimer. Such results are difficult at present to interpret in terms of structure–function relationships within the NR molecule but do suggest that the mutations carried by *niaD*10 and *niaD*13 are perhaps located at sites in the proximal part of the protein associated with electron transfer from NADPH to cytochrome *c*. Distal regions of the protein molecule involved in binding MoCo, and in association between the subunits, appear not to be adversely affected. Strains carrying the point mutation *niaD*17 which is located towards the 5′ end of the *niaD* gene lack both NR activities but possess a 4.6 S (monomeric) NR-associated cytochrome *c* reductase activity, suggesting a defect in dimerization. A role for MoCo in dimerization has been suggested in *N. crassa* (Wahl *et al.* 1984) and this may indicate that the MoCo binding domain is located towards the 5′ end (N-terminal end) of the protein. This prediction is supported by amino acid sequence comparisons of two different MoCo-containing enzymes (Chapter 24). Finally the role of the *cnx* gene products in the dimerization process has also been studied and is discussed elsewhere (MacDonald *et al.* 1974; Cove 1979, Chapter 19).

Molecular organization

An 11 kb recombinant DNA fragment containing part of the *A. nidulans* nitrate gene cluster was isolated (Johnstone *et al.* 1989) by complementation of the *niiA*4 mutant phenotype using a self-cloning transformation system developed for *A. nidulans* (Johnstone *et al.* 1985). Positive identification and further characterization of the DNA clone was achieved by (a) comparing DNA hybridization profiles between wild-type and deletion mutant strains; (b) complementation studies with mutants; (c) messenger RNA analysis; (d) DNA sequencing; and (e) functional analysis of the *niiA–niaD* intergenic region. A brief summary of the results taken from Greaves (unpublished Thesis, 1989; Greaves *et al.* (1989) and Johnstone *et al.* (1989) is presented below and in Fig. 6.4.

Southern blot and DNA:DNA hybridization

Physical studies of genomic DNA from wild-type *A. nidulans* and from the mutants *crnA–niiA–niaD*Δ506, *niiA–niaD*Δ509 and *niaD*26, judged to be deletion mutations from fine structure genetic analyses (Tomsett and Cove 1979), were carried out. This approach provided positive identification of the recombinant DNA clone as well as yielding additional information on the extent and position of the deletion. As expected from the genetic map (Fig. 6.3), deletion mutant strain *crnA–niiA–niaD*Δ506, which completely lacks the 91 kDa NR protein, shows extensive loss of DNA from the predicted region of the physical map (Fig. 6.4a). The available physical data on *niiA–niaD*Δ509 are fairly consistent with the genetic map (Fig. 6.3) in that the mutational event has erased DNA from around the deduced 5′ region of *niiA*, the 5′ region of the *niaD* gene, and the intergenic region (Fig. 6.4a). In contrast, deletion strain *niaD26* which has a smaller (58 kDa and 55 kDa) than wild-type protein (Cooley and Tomsett, 1985) shows no detectable difference in fragment size as judged from DNA hybridization patterns. One explanation is that *niaD26* is a small deletion not obviously detectable at the DNA level (the extent of *niaD26* is relatively short according to the available genetic evidence; Fig. 6.3). The mutational event may have resulted in a frameshift giving rise to a termination codon and thus a protein of reduced subunit size.

Complementation of mutants

Plasmid pNIIA was transformed into *argB2* double mutants carrying one of *crnA1*, *niiA4*, *niiA18* or *niaD17* mutations to determine if all three genes were present on the recombinant clone. The results (summarized from Johnstone *et al.* 1989) are presented in Table 6.2. Approximately 50 per cent of *argB2*-derived transformants concomitantly complement *crnA* or *niiA* mutations, while only 2.2 per cent of *argB2* transformants complement *niaD17*. This frequency pattern is consistent with intact *crnA* and *niiA* genes but only part of the *niaD* gene being present on pNIIA.

In order to determine the approximate location of point mutants on the physical map, various stretches of pNIIA were subcloned into the *argB*-based vector pILJ16 (Johnstone *et al.* 1985) and used to complement certain point and deletion strains (Fig. 6.3; Greaves, unpublished Thesis, 1989; Johnstone *et al.* 1989). The probable location of mutations with respect to the physical map is indicated by the diagrammatic representation (Fig. 6.4a,b). As expected, the inferred position of the point mutations approximately corresponds to the order and to the orientation of the three genes determined on the basis of classical genetic analysis (Tomsett and Cove 1979). Subclone 2 but not subclone 1, was found to complement *crnA1*, showing that the *crnA1* mutation locates within subclone 2. This result also suggests the *crnA1* mutation *per se* does not reside within subclone 1 (although the remaining *crnA* sequences may extend into subclone 1). Complementation of *niiA18* also occurred using subclone 2, but not with subclone 1 indicating that both *crnA1* and *niiA8* mutations locate on the *Sma*I–*Eco*R1 2.8 kb stretch of subclone 2. That they are both represented on the same fragment is not inconsistent with the genetic fine structure analysis data,

Table 6.2. Complementation frequency of *crnA, niiA, niaD* nutations with recombinant plasmid pNIIA[a]

Mutation[b]	*argB2* selected transformants	Transformants screened for the following phenotype[c,d]		
		crnA[+]	*niiA*[+]	*niaD*[+]
*crnA*1	100	51 (51%)		
*niiA*4	4000		1600 (40%)	
*niiA*18	216		129 (57%)	
*niaD*17	506			44 (2.2%)

[a] Brief summary after Greaves, unpublished Thesis (1989), Johnstone *et al.* (1989).
[b] See Fig. 6.3 for genetic location of mutations.
[c] *crnA*, 10 mM nitrate and 10 mM methylammonium as sole sources of nitrogen.
  niiA, 10 mM nitrite as sole source of nitrogen.
  niaD, 10 mM nitrate as sole source of nitrogen.
[d] Only transformants showing wild-type growth are recorded here. A low level of arginine prototrophic transformants showing partial growth on nitrate or nitrite was observed.

which place the *niiA*18 mutation at the most *crnA*-proximal *niiA* deletion interval (Fig. 6.3). In contrast, none of the subclones appeared to complement the phenotype of the *niiA*4 mutation. One possible explanation for this is that the *niiA*4 mutation lies within the small 0.8 kb *Eco*R1 subclone 3 and that the fragment contains an internal segment of *niiA*. If this is so then integration of subclone 3 by a single cross-over into the *niiA* region would result in a tandem repeat of the mutant *niiA*4 allele with the 0.8 kb *Eco*R1 recombinant stretch of *niiA* (and hence lack of complementation of *niiA*4). That the left hand extremity of the *niiA–niaD*Δ509 deletion is within subclone 4 is also shown by the ability of subclone 4 to complement this deletion, allowing the mutant to grow on nitrite. These results also indicate that the *niiA* gene extends from fragment 2 through fragment 3 into fragment 4. Subclone 6 complements the phenotype of mutant strains *niaD*5, -17, -18, -169, and -15 indicating that these mutations are represented in this region of the physical map. Subclone 5, in contrast, will not complement these mutant phenotypes, suggesting that the small *Sma*I–*Eco*R1 600 bp stretch contains an internal segment of the *niaD* (*vis-à-vis* lack of complementation of *niiA*4 by subclone 3). Mutations *niaD*5, -17, -18, -189, and -15 cannot be ordered with respect to each other on fragment 6. Finally a low frequency of *argB*[+] transformants of *niiA–niaD*Δ509 (but not *crnA–niiA–niaD*Δ506) with pNIIA complement to give a nitrate (as well as a nitrite) utilization phenotype. This would confirm that the deletion *niiA–niaD*Δ509 terminates within pNIIA.

Table 6.3. Mitotic stability of the *niaD*$^+$ phenotype in transformed strains

Transformant number	Transformed by	Recipient strain	Cells screened	Chlorate-resistant strains	Number of chlorate-resistant strains which are also auxotrophic for arginine
Wild-type	–	–	10^9	290	–
T26	pNIIA	*niaD17*	5×10^6	269	194 (72%)
T36	pNIIA	*niiA–niaD*Δ509	5×10^6	338	129 (38%)

Northern blot and DNA:RNA hybridization

Analysis revealed that the *niaD* mRNA species of *A. nidulans* is approximately 2.8 kb (Johnstone *et al.* 1989; Fig. 6.4c). This messenger species could encode a subunit of some 93 kDa, which would agree with the biochemical evidence of 91 kDa (taking into account 5′ and 3′ untranslated sequences). This result supports the view that the *A. nidulans* NR subunit may be somewhat smaller than that of *N. crassa*, in which the messenger size is around 3.4 kb, sufficient to encode a polypeptide of 114 kDa (Chapter 20). Higher plant *niaD*-equivalent mRNAs also appear to be longer, *Cucurbita* (squash) and *Arabidopsis* messenger sizes being 3.2 kb (Chapter 21). The *A. nidulans* NiR (*niiA*) mRNA species was found to be around 3.4 kb (Johnstone *et al.* 1989), which is significantly larger than spinach species of around 2.3 kb (Chapter 18). This is broadly consistent with the indication that the *N. crassa* NiR is a dimer of monomer size 140 kDa (Lafferty and Garrett 1979; Prodouz and Garrett 1981) while the spinach NiR is a monomer of 63 kDa (Ida and Mikami 1986). This size difference between plant and fungal NiR is probably due to a FAD domain being present on the fungal enzyme (Chapter 24). Finally, the *A. nidulans* permease messenger was found to the approximately 1.6 kb. The fact that only individual transcripts are seen is in broad agreement with the genetic data, which suggest that an operon type of organization does not exist (*vide supra*). Similar conclusions are being drawn for the proline (C. Scazzocchio, personal communication), alcohol (Gwynne *et al.* 1987) and quinic acid (A. R. Hawkins, personal communication) gene clusters of *A. nidulans*. Regulation of the *niiA* and *niaD* genes appears to occur at the transcriptional level as detectable NiR and NR mRNAs are only seen in wild-type cells grown under induced but not repressed conditions (Johnstone *et al.* 1989). However, the possibility that mRNA stability is modulated by nitrate cannot be ruled out.

DNA sequence analysis

The *A. nidulans* NR and NiR genes have been sequenced (Johnstone *et al.* 1989) and analysis of the DNA sequence together with the inferred amino acid sequence reveals the basic structural architecture of *niiA* and *niaD* genes, the salient features of which are as follows. First, in accordance with most fungal genes both *niiA* and *niaD* have a number of short (< 100 bp) introns, seven in *niiA*, and six in *niaD*, with putative internal consensus sequences necessary for intron splicing (Gurr *et al.* 1987). Secondly, the inferred amino acid sequences have been compared with those from plants to identify putative functional domains (Chapter 24). Thirdly, the genes are divergently transcribed from an internal region (1272 bp intergenic region; Fig. 6.4d), an observation which confirms the earlier genetic evidence (*vide supra*). The sequencing of the *crn*A gene is in progress (E. I. Campbell and P. Montague, unpublished work) and this should provide interesting information on its architecture.

Functional analysis of the niiA–niaD control region

Functional tests have been used in an attempt to analyse the 1272 bp *niiA–niaD* internal region likely to contain *niiA* and *niaD* receptor sites for control proteins The *Eco*R1 fragment, encompassing the internal control region, was inserted in both orientations into the *lacZ* reporter gene of *Escherichia coli* encoding β-galactosidase (Greaves *et al.* 1989). Analysis of transformants containing these plasmid constructs, viz. 5′*niaD*::*lacZ*(pSTA20) and 5′*niiA*::*lacZ*(pSTA21), showed that β-galactosidase activity is subject to control phenomena normally exerted on *niiA* and *niaD* genes, such as nitrate induction and nitrogen metabolite repression. Furthermore β-galactosidase activity was found in the absence of nitrate (with a neutral nitrogen source) when constructs were transformed into a *niaD* mutant (but not found when transformed into the wild-type strain), supporting the hypothesis that NR regulates its own synthesis (Cove 1969).

In transformant strains designated Tnap3 and Tnap5 (<u>T</u>ransformant <u>n</u>it<u>r</u>ate reductase <u>p</u>romoter) with amplified copies (10–15) of the intergenic region, NR and NiR activities are reduced to around 50 per cent compared with the wild type when both are grown under inducing conditions (Greaves *et al.* 1989). This would support the notion that the *nirA* product is limiting (Cove 1969), although there are quantitative differences which might be explained by the very different approaches used by these two groups. In contrast, studies of acetamidase activities in these strains containing the multicopy intergenic region shows that the *areA* gene product is not limiting, a result which supports the work of Hynes using multicopy acetamidase *amdS* control sequence to titrate the *areA* product (Kelly and Hynes, 1987). However, it must be noted that only two multicopy transformant were available for these studies.

Integration of transformants

Combined classical and physical analyses of transformants of recipient *argB2 niiA4* showed that pNIIA had integrated overwhelmingly into or near the nitrate gene cluster rather than the normally more favoured *argB* site (or elsewhere) (Greaves, unpublished Thesis, 1989; Johnstone *et al.* 1989). This could be explained by the fact that the 11 kb fragment offers a greater region of homology, than the ~2 kb *argB* fragment on pNIIA, to aid integration. Surprisingly, even when strains carrying the deletion *niiA–niaDΔ509* were the recipients, integration took place preferentially at the nitrate assimilation locus. Further evidence comes from the study of the location of the integrates of the intergenic *niiA–niaD* 2.7 kb *Eco*R1/*lac* fusion plasmid constructs, pSTA20 and pSTA21 (*vide supra*). Again, the majority were found to locate at the *crnA–niiA–niaD* region (Greaves, unpublished Thesis, 1989; Greaves *et al.* 1989). There is an indication, therefore, that the nitrate assimilation gene cluster is favoured as a site of integration over the *argB* locus.

Mitotic stability of *niaD* transformants

Two transformants generated with pNIIA, namely T26, originating from *niaD*17 and T36 from the deletion strain *niiA, niaD*Δ509, were studied for mitotic stability of the *niaD* phenotype (Greaves, unpublished Thesis, 1989). The integrated sequences locate in the nitrate gene cluster. However, the stability of the integrated gene is lower than that of the wild type since the frequency of chlorate-resistant colonies arising from the transformants was found to be much higher (~200-fold) than the wild type (Table 6.3). On screening these for the arginine phenotype, a significant number showed concomitant loss of the *argB*$^+$ allele. Whether there is any significance in the difference in *argB*$^-$ auxotrophic numbers between the two transformants T26 and T36 is impossible to estimate from such a limited sample. Loss of *argB*$^+$ *niaD*$^+$ sequences is most likely due to excision of plasmid sequences—occurring at a frequency of 1 in 5000, at least in integrates at the resident locus. Presumably instability is due to a reversal of the integration process. No determination of *niaD* stability of integrates at ectopic sites has been determined. Studies of *niaD* stability have also been carried out in *A. niger* (Unkles *et al.* 1989) with similar results: genes transferred into recipient strains by gene-mediated transformation systems are much less stable than the wild-type gene. However, this phenomenon clearly requires further investigation.

The nitrate assimilation pathway as a transformation and selection system

Several fungal transformation and selection systems have been used, which have their various advantages and disadvantages (Chapter 22 and references therein). The nitrate assimilation pathway offers another possible selection system, particularly since *niaD* mutants are easy to isolate and characterize by simple growth tests (Table 6.1). Our studies with *A. nidulans* showed that the nitrate assimilation system could be of potential use since:

1. Growth of recipient mutants is very poor.
2. There is lack of abortive transformants.
3. Deletion mutants are available (Tomsett and Cove 1979) thus avoiding back mutations in recipient strains which could be confused with transformants (Chapter 22).
4. There is strong selection for integration at the resident gene locus.
5. Amplification is seen in only in a small number of transformants.

These latter two aspects might be a disadvantage in certain circumstances.

Future prospects

With the cloning of the *A. nidulans crnA–niiA–niaD* region, the *areA* and *nirA* control genes (Chapter 19), and counterpart genetic elements from other filamentous fungi (Chapter 20, 22), detailed questions regarding firstly the molecular nature of mechanisms that govern gene expression of the nitrate assimilation genes and secondly structure–function relationships of NR and NiR proteins *per se*, can now be realistically addressed. The isolation of *nirA* and *areA* control proteins in relatively large amounts may be within our grasp and consequently their sites of action within the *niiA–niaD* intergenic region can be determined by gel retardation assays coupled with deletion analyses of the *niiA* and *niaD* upstream receptor sites or upstream activator sequences (UAS). It would be particularly interesting to study the interactive relationship of *nirA* and *areA* proteins with respect to *niiA* and *niaD* receptor sites at the nucleotide and chromatin levels to determine how these control genes 'talk' to the structural genes that come within their regulatory domain. Finally, it should be relatively straightforward to determine if NR itself does indeed bind to the product of *nirA*, as postulated originally by Cove (1969).

DNA sequence analysis of point mutants ordered by fine structure genetic mapping (Fig. 6.3) should allow us to determine the nature of amino acid changes (inferred from the DNA sequence) and their positions on the *niiA* and *niaD* mutant proteins. Such studies coupled with biochemical studies of mutant proteins may reveal interesting alterations in domains which, *a priori*, would give us clearer structure function insights, for example, into the way in which nitrate reductase interacts with FAD, NADPH, haem, and its molybdenum cofactor. Preliminary information on the location of functional domains has been obtained by amino acid sequence comparisons between NR from fungal and plant sources and between NR and physiologically unrelated enzymes with shared functional domains (Chapter 24).

The 'coming-of-age' of *A. nidulans* molecular genetics, the enormous wealth of *A. nidulans* conventional or classical genetics, and the abundance of genetically well-defined mutants unable to utilize nitrate can now be used therefore to exploit the nitrate assimilation pathway to the very full.

Acknowledgements

I wish to acknowledge the support provided by the Science and Engineering Research Council for molecular research into the nitrate assimilation gene cluster of *Aspergillus nidulans*. I thank Professor H. Arst for making available the *crnA1* mutant strain, Dr S. Martinelli for *niaD*500 and *niaD*501 mutant strains and Dr B. Tomsett for his generosity in providing the author with his valuable collection of both point and deletion mutant strains. Receipt of unpublished data generated by postgraduate student Mr P. Greaves and the cooperation and collaboration of Dr I. Johnstone is acknowledged. Helpful suggestions made by Dr J. L. Wray were very much appreciated. I also acknowledge Shawlands Senior Secondary School, Glasgow, Scotland for Latin tuition.

References

Arst, H. N., Jr. (1981). Aspects of the control of gene expression in fungi. In *Genetics as a tool in Microbiology* (ed. S. W. Glover and D. A. Hopwood), Society for General Microbiology Symposium **31**, pp. 131–60. Cambridge University Press, Cambridge.

Arst, H. N., Jr. (1984). Regulation of gene expression in *Aspergillus nidulans. Microbiological Sciences* **1**, 137–40.

Arst, H. N., Jr. and Scazzocchio, C. (1985). Formal genetics and molecular biology of the control of gene expression in *Aspergillus nidulans.* In *Gene manipulations in Fungi* (ed. J. W. Bennett and L. L. Lasure), pp. 309–53. Academic Press, London.

Arst, H. N., Jr., Rand, K. N., and Bailey, R. (1979). Do the tightly linked structural genes for nitrate and nitrite reductases in *Aspergillus nidulans* form an operon? *Molecular and General Genetics* **174**, 89–100.

Benzer, S. and Champe, S. P. (1961). Ambivalent *rII* mutants of phage T4. *Proceedings of the National Academy to Science, USA* **47**, 1025–38.

Brownlee, A. G. and Arst, H. N., Jr. (1983). Nitrate uptake in *Aspergillus nidulans* and involvement of the third gene of the nitrate assimilation gene cluster. *Journal of Bacteriology* **115**, 1138–46.

Cooley, N. R. and Tomsett, B. A. (1985). Determination of the subunit size of NADPH-nitrate reductase from *Aspergillus nidulans. Biochimica et Biophysica Acta* **831**, 89–93.

Cove, D. J. (1969). Evidence for a near limiting intracellular concentration of a regulator substance. *Nature* **224**, 272–3.

Cove, D. J. (1970). Control of gene action in *Aspergillus nidulans. Proceedings of the Royal Society London, B* **176**, 267–75.

Cove, D. J. (1976a). Chlorate toxicity in *Aspergillus nidulans*: the selection and characterisation of chlorate resistant mutants. *Heredity* **36**, 191–203.

Cove, D. J. (1976b). Chlorate toxicity in *Aspergillus nidulans*: studies of mutants in nitrate assimilation. *Molecular and General Genetics* **146**, 147–59.

Cove, D. J. (1979). Genetics studies of nitrate assimilation in *Aspergillus nidulans. Biological Reviews* **54**, 291–327.

Cove, D. J. and Pateman, J. A. (1963). Independently segregating genetic loci concerned with nitrate reductase activity in *Aspergillus nidulans. Nature* **198**, 262–3.

Cove, D. J. and Coddington, A. (1965). Purification of nitrate reductase and cytochrome *c* reductase from *Aspergillus nidulans. Biochimica et Biophysica Acta* **110**, 312–18.

Downey, J. R. (1971). Characterisation of reduced nicotinamide adenine dinucleotide phosphate–nitrate reductase of *Aspergillus nidulans. Journal of Bacteriology* **105**, 759–63.

Downey, R. J. and Focht, W. J. (1974). The subunit size of *Aspergillus nidulans* nitrate reductase. *Microbios* **11**, 61–70.

Downey, R. J. and Steiner, F. X. (1979). Further characterisation of the reduced nicotinamide adenine dinucleotide phosphate:nitrate oxidoreductase in *Aspergillus nidulans. Journal of Bacteriology* **137**, 105–14.

Dunn-Coleman, N. S., Smarrelli, J., Jr., and Garrett, R. H. (1984). Nitrate assimilation in eukaryotic cells. *Review of Cytology* **82**, 1–47.

Garrett, R. H. and Nason, A. (1969). Further purification and properties of *Neurospora nitrate* reductase. *Journal of Biological Chemistry* **244**, 2870–82.

Grant, S., Roberts, C. F., Lamb, H. K., Stout, M., and Hawkins, A. R. (1988). Genetic regulation of the quinic acid utilisation (*qut*) gene cluster in *Aspergillus nidulans*. *Journal of General Microbiology* **134**, 347–58.

Greaves, P. (1989). Molecular genetical analysis of the nitrate utilisation gene cluster of Aspergillus nidulans. Ph.D. Thesis, University of St Andrews, United Kingdom.

Greaves, P. A., Montague, P., Grieve, C., and Kinghorn, J. R. (1989). Evidence from *lacZ* fusion studies that the internal region of the *Aspergillus nidulans niiA* and *niaD* genes for nitrate assimilation contain sequences which modulate the expression of both genes. *Molecular and General Genetics* (submitted).

Gurr, S. J., Unkles, S. E., and Kinghorn, J. R. (1987). The structure and organisation of nuclear genes of filamentous fungi. In *Gene structure in eukaryotic microbes, SGM Special Publications* **23**, (ed. J. R. Kinghorn), pp. 93–139. IRL Press, Oxford.

Gwynne, D. I., Buxton, F. P., Sibley, S., Davies, R. N., Lockington, R. A., Scazzocchio, C., and Sealey-Lewis, H. M. (1987). Comparison of the cis-acting control regions of two coordinately controlled genes involved in ethanol utilisation in *Aspergillus nidulans*. *Gene* **51**, 205–16.

Horner, R. D. (1983). Purification and comparison of *nit*-1 and wild-type nitrate reductase of *Neurospora crassa*. *Biochimica Biophysica Acta* **744**, 7–15.

Ida, S. and Mikami, B. (1986). Spinach ferredoxin-nitrite reductase: a purification procedure and characterisation of chemical properties. *Biochimica et Biophysica Acta* **871**, 167–76.

Johnstone, I. L. (1985). Transformation in *Aspergillus nidulans*. *Microbiological Sciences* **2**, 307–11.

Johnstone, I. L., Hughes, S. G., and Clutterbuck, A. J. (1985). Cloning of an *Aspergillus nidulans* developmental gene by transformation. *EMBO Journal* **4**, 1307–11.

Johnstone, I. L., Greaves, P., Kinghorn, J. R., Gurr, S. J., and Innes, M. (1989). Cloning and characterisation of the *Aspergillus nidulans* gene cluster for nitrate assimilation. *EMBO Journal* (submitted).

Kelly, J. M. and Hynes, M. J. (1987). Multiple copies of *amds* of *Aspergillus nidulans* cause titration of transacting regulatory proteins. *Current Genetics* **12**, 21–31.

Lafferty, M. A. and Garrett, R. H. (1974). Purification and properties of the *Neurospora crassa* assimilatory nitrate reductase. *Journal of Biological Chemistry* **249**, 7555–67.

MacDonald, D. W. and Coddington, A. (1974). Properties of the assimilatory nitrate reductase from *Aspergillus nidulans*. *European Journal of Biochemistry* **46**, 169–78.

MacDonald, D. W., Cove, D. J., and Coddington, A. (1974). Cytochrome-*c* Reductases from wild-type and mutant strain of *Aspergillus nidulans*. *Molecular and General Genetics* **128**, 187–99.

Martinelli, S. D., Roberts, T. J., Sealy-Lewis, H. M., and Scazzocchio, C. (1984). Evidence for a nonsense mutation in the *niaD* locus of *Aspergillus nidulans*. *Genetical Research* **43**, 241–2.

Minagawa, N. and Yoshimoto, A. (1982). Purification and characterisation of the assimilatory NADPH-nitrate reductase of *Aspergillus nidulans*. *Journal of Biochemistry* **91**, 761–79.

Pan, S.-S. and Nason, A. (1978). Purification and characterisation of homogeneous assimilatory reduced nicotinamide adenine dinucleotide phosphate-nitrate reductase from *Neurospora crassa*. *Biochimica et Biophysica Acta* **523**, 297–313.

Pateman, J. A. and Kinghorn, J. R. (1976). Nitrogen metabolism. In *The filamentous fungi* Vol II (eds. J. E. Smith and D. R. Berry), pp. 157–237. Academic Press, London.

Pateman, J. A., Rever, B. M., and Cove, D. J. (1967). Genetical and biochemical studies of nitrate reduction in *Aspergillus nidulans*. *Biochemical Journal* **104**, 103–11.

Pateman, J. A., Doy, C. H., Olsen, J. E., Kane, H. K., and Creasor, E. M. (1985). Molecular analysis of alcohol metabolism in *Aspergillus*. In *Molecular genetics of filamentous fungi* (ed. W. E. Timberlake), pp. 171–84. Alan R. Liss, New York.

Prodouz, K. N. and Garrett, R. H. (1981). *Neurospora crassa* NAD(P)H–nitrite reductase. Studies of its composition and structure. *Journal of Biological Chemistry* **256**, 1971–81.

Rand, K. N. and Arst, H. N., Jr. (1977). A mutation in *Aspergillus nidulans* which affects the regulation of nitrite reductase and is tightly linked to its structural gene. *Molecular and General Genetics* **155**, 67–75.

Roberts, T. J., Martinelli, S. D., and Scazzocchio, C. (1979). Allele specific, gene unspecific suppressors in *Aspergillus nidulans*. *Molecular and General Genetics* **177**, 57–64.

Scazzocchio, C. (1980). The genetics of the molybdenum containing enzymes. In *Molybdenum and molybdenum-containing enzymes* (ed. M. P. Coughlan), pp. 487–515. Pergamon Press, Oxford.

Steiner, F. X. and Downey, R. J. (1982). Isoelectric focussing and two-dimensional analysis of purified nitrate reductase. *Biochemica et Biophysica Acta* **706**, 203–11.

Tomsett, B. A. and Cove, D. J. (1979). Deletion mapping of the *niiA niaD* gene region of *Aspergillus nidulans*. *Genetic Research* **34**, 19–32.

Tomsett, B. A. and Garrett, R. H. (1980). The isolation and characterisation of mutants defective in nitrate assimilation in *Neurospora crassa*. *Genetics* **95**, 649–60.

Unkles, S. E., Campbell, E. I., Carrez, D., Contreras, R., Fiers, W., Grieve, C., van den Hondel, C. A. M. J. J., and Kinghorn, J. R. (1989). Transformation of *Aspergillus niger* using the homologous *niaD* gene. *Gene* (in press).

Wahl, R. C., Hageman, R. V., and Rajagopalan, K. V. (1984). The relationship of Mo, molybdopterin, and the cyanolyzable sulphur in the Mo cofactor. *Archives Biochemistry and Biophysics* **230**, 264–73.

Note added in proof: Recently Dr. Claudio Scazzocchio's group has also isolated DNA clones containing the *A. nidulans* nitrate gene cluster (C. Scazzocchio, personal communication).

7. Structure–function relationships of algal nitrate reductases

Larry P. Solomonson and Michael J. Barber

Introduction

Nitrate reductase (NR), the key enzyme in the process of nitrate assimilation, has been extensively studied in order to delineate properties related to its catalytic efficiency and regulation. Nitrate assimilation is an important process in a wide variety of prokaryotic and eukaryotic organisms including algae, yeast, fungi, and higher plants. Early studies suggested considerable variability in structural properties of assimilatory NR isolated from different sources. Later studies, however, indicate that the major structural properties of eukaryotic NR are very similar, though not identical. The most extensive studies have focused on the enzyme isolated from the eukaryotic, single-celled green alga, *Chlorella vulgaris*. This is a particularly good source of NR for structural studies because of the relatively high level of the enzyme, its stability in crude extracts, and ease of manipulation of culture conditions. Structural studies of NR from eukaryotic sources have been facilitated by the development of affinity chromatography procedures, based on the high affinity of the enzyme for the dye, cibacron blue, initially applied to the purification of *Chlorella* nitrate reductase (Solomonson 1975). Subsequently, NR from other sources, including higher plants, were also purified using this method (Guerrero *et al.* 1981). Similarly, ferredoxin–Sepharose was used to isolate the ferredoxin : nitrate reductase from the blue-green alga, *Anacystis nidulans*, in high yield (Manzano *et al.* 1978). Such rapid, high-yield procedures enabled researchers to purify the enzyme to homogeneity in quantities sufficient for extensive physical studies.

This review describes our current understanding of structure–function relationships which have arisen from these studies and is confined to assimilatory NR from eukaryotic sources with particular emphasis on the enzyme isolated from *Chlorella vulgaris*, since this system has been the most extensively studied. Assimilatory nitrate reductases from cyanobacteria have been partially characterized but appear to differ significantly from eukaryotic NR with respect to certain structure–function relationships (Guerro and Lara 1987). While the primary focus will be on algal NR, comparisons will be made with assimilatory NR from other eukaryotic sources where appropriate. Where results appear to be contradictory, we will offer possible explanations and judgements based on our experience with the *Chlorella* system. We hope that these opinions will stimulate further studies which will lead to a better understanding of the molecular complexities involved in the action of NR.

Prosthetic groups

NR is a complex enzyme containing several different redox-active prosthetic groups. Early work suggested that molybdenum is a component of assimilatory nitrate reductase (Nason and Takahashi 1958); later work confirmed this (Hewitt 1975). Nason and co-workers showed through reconstitution experiments that molybdenum may be present as a 'cofactor', common to several molybdenum-containing enzymes (Nason *et al.* 1971). Rajagopalan and co-workers subsequently identified this molybdenum cofactor (MoCo) as a molybdo-pterin derivative associated with several molybdoenzymes, including *Chlorella* NR (Johnson *et al.* 1980). Garrett and Nason (1967) first showed that *Neurospora* NR was associated with a *b*-type cytochrome: other purified NR were subsequently shown to exhibit the same spectral properties (Guerrero *et al.* 1981). NAD(P)H-NR activity from a number of sources is dependent on the presence of exogenous FAD (Nason and Takahashi 1958; Hewitt 1975; Guerro *et al.* 1981). However, *Chlorella* NR activity is not dependent on exogenous FAD, although chemical and spectral analysis of the purified enzyme indicated that FAD, haem and molybdenum are present in approximately a 1:1:1 ratio (Solomonson *et al.* 1975; Giri and Ramadoss 1979). No labile sulphide or non-haem iron was found to be associated with the enzyme. Subsequent prosthetic group analyses of NR from eukaryotic sources suggested that this stoichiometry may not hold for all NR from eukaryotic sources, which in some cases appear to contain one Mo per two haem molecules (Pan and Nason 1978; De la Rosa *et al.* 1981). Later studies, however, suggest the stoichiometry of prosthetic groups in nitrate reductases from eukaryotic sources to be 1:1:1 (Redinbaugh and Campbell 1985; Solomonson and McCreery 1986). Mo content is sometimes less than expected, presumably due to co-purification of non-functional enzyme or to loss of some Mo during purification.

Early studies showed that assimilatory NR exhibits a number of partial activities which may be classified as either NADH-dehydrogenase (diaphorase) activities, in which various artificial electron acceptors (ferricyanide, dichloro phenolindophenol, or cytochrome *c*) are reduced by NAD(P)H, or nitrate-reducing activities, in which the enzyme catalyses the reduction of nitrate by artificial electron donors such as reduced viologen dyes or reduced flavins (Guerrero *et al.* 1981). The latter activities are dependent on molybdenum but not flavin and are inhibited by cyanide but not by sulphydryl reagents such as *N*-ethylmaleimide. In contrast, the NADH-dehydrogenase partial activities are dependent on FAD but not molybdenum and are inhibited by sulphydryl reagents but not by cyanide (Garrett and Nason 1969; Garrett and Greenbaum 1973; Solomonson 1974; Solomonson and Spehar 1979; Guerrero *et al.* 1981).

The possible participation of sulphydryl groups in the transfer of reducing equivalents between NAD(P)H and FAD was proposed based on labelling studies of *Neurospora* nitrate reductase (Amy *et al.* 1977). More recent studies on *Chlorella* nitrate reductase indicated there is one essential sulphydryl group per haem which is required for NADH binding and activity but is probably not directly involved in electron transfer between prosthetic groups since it reacts with SH-specific reagents when the enzyme is in the fully oxidized state and is protected by the physiological electron donor, NADH (Barber and Solomonson 1986*a*). Furthermore, the results of microcoulometry experiments using *Chlorella* nitrate reductase indicated that a maximum of five electrons are taken up per haem group, consistent with FAD, haem, and Mo being the only redox-active prosthetic groups (Spence *et al.* 1988).

Functional size

There is a wide variation in reported molecular weights of assimilatory NR isolated from different sources, ranging from 200 to 500 kDa (Hewitt 1975; Guerrero *et al.* 1981). Similar variation was reported for the size and number of polypeptide subunits, leading to some rather complex structural models for NR (Notton and Hewitt 1979). However, recent work has clarified our understanding and enables us to account for some of the previously reported variations.

Through a combination of gel chromatography, sedimentation equilibrium, gel electrophoresis, cross-linking, and peptide mapping studies, Howard and Solomonson (1982) obtained rather conclusive evidence that *Chlorella* NR is a homotetramer with dihedral symmetry ('dimer of dimers'). The mode of association between subunits is isologous (head-to-head) rather than heterologous (head-to-tail). As pointed out by Monod *et al.* (1965), isologous associations can give rise only to even numbered oligomers and, furthermore, would tend to give rise to 'closed' or finite polymers since further polymerization could occur only by using 'new' binding sites. It was also shown that the active tetramer dissociates to active dimers at low enzyme concentrations without significant loss of activity, suggesting that association/dissociation plays no role in regulation of enzyme activity (Howard and Solomonson 1982). These results would rule out the trimeric model for *Chlorella* nitrate reductase which had been proposed previously (Solomonson *et al.* 1975; Giri and Ramadoss 1979). The nitrate reductase homotetramer exhibits a rather high partial specific volume at neutral pH which may have led to erroneous conclusions as to the molecular weight and quaternary structure of the native enzyme. No lipid or carbohydrate was found associated with the enzyme, suggesting that the high value for the partial specific volume may arise from preferential solvent interactions (Howard and Solomonson 1982). The observation that the enzyme can exist as an active dimer or tetramer may explain some of the reported variations in observed molecular weights of the native enzyme.

The subunit of *Chlorella* NR had a molecular weight, determined by SDS–polyacrylamide gel electrophoresis, of about 100 kDa and migrated as a single band (Solomonson *et al.* 1975). Similar results were obtained for *Neurospora* and barley NR (Pan and Nason 1978; Kuo *et al.* 1980). Reports of multiple, lower molecular weight subunits for NR from spinach and barley were subsequently shown to be due to proteolytic 'nicking' (Campbell and Wray 1983; Fido and Notton 1984). *In vitro* translation products of mRNA also appear to be in this molecular weight range (Cheng *et al.* 1986; Crawford *et al.* 1986; Commere *et al.* 1986). These results suggest a similar subunit structure for eukaryotic NR: a possible exception is the assimilatory NR of *Ankistrodesmus braunii* which has been purified in a high activity state and appears to be a homo-octomer with a subunit molecular weight of 59 kDa (De la Rosa *et al.* 1981).

Available evidence suggested that the 'functional unit' of NR is probably the homodimer. The technique of radiation inactivation analysis was utilized to determine the functional size of *Chlorella* NR at high and low concentrations of enzyme, where the principal species would be either tetrameric or dimeric, respectively. In both cases, the functional size obtained by this method was about 100 kDa, suggesting that each subunit in the tetramer or dimer can function independently (Solomonson and McCreery 1986). These results also confirmed earlier results which indicated that the subunits are identical and that each contains a full complement of prosthetic groups, consistent with the model shown in Fig. 7.1.

Functional domains

Radiation inactivation analysis of NADH-NR activity indicated that the 'functional unit' of the enzyme is the 100 kDa subunit. Of related interest are the topological arrangement and functional relationships of the prosthetic groups and binding sites of the native enzyme. Some information has been obtained using specific inhibitors which make possible the functional separation of certain partial activities of NR. Further information and characterization of functional domains of *Chlorella* NR has been achieved through a combination of limited proteolysis, specific molecular probes and radiation inactivation analysis of partial activities.

Functional sizes for the partial activities NADH-ferricyanide reductase, NADH-cytochrome *c* reductase, and nitrate-reducing activities, obtained by radiation inactivation analysis, were in each case different and significantly smaller than the minimum polypeptide mass, suggesting that these activities reside on functionally independent domains (Solomonson and McCreery 1986; Solomonson *et al.* 1987). The apparent functional size for NADH-cytochrome *c* reductase was significantly larger than for NADH-ferricyanide reductase (45 versus

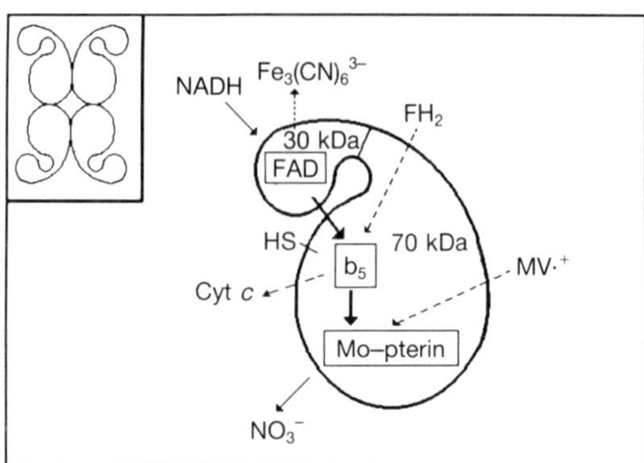

Fig. 7.1. Structure–function model of *Chlorella* nitrate reductase. Features of the model include four identical subunits with two different binding domains and the FAD/NADH and haem(b_5-type)/molybdenum domains connected by a protease-sensitive hinge region. Solid arrows indicate physiological electron transfer reactions, and dashed arrows indicate electron transfer reactions to or from artificial acceptors and donors (partial reactions).

34 kDa), suggesting that these alternate electron acceptors do not act at the same site (Solomonson *et al.* 1987). Since Mo is not required for either activity, this result may suggest that ferricyanide accepts electrons via the FAD centre while cytochrome *c* accepts electrons via the haem moiety of NR.

Incubation of native *Chlorella* NR with either trypsin, *Staphylococcus aureus* V8 protease, or a natural inactivator protease from maize results in loss of NADH-NR and NADH-cytochrome *c* reductase activities but no loss of nitrate-reducing activities. Incubation of NR with V8 protease or maize inactivator protease resulted in two different products, each of which retained a different partial activity (Solomonson *et al.* 1986). Nitrate-reducing activity was associated with a homotetrameric fragment of about 260 kDa which contained haem and Mo but no FAD whereas NADH-ferricyanide reductase was associated with a monomeric species of approximately 30 kDa which contained FAD and the NADH binding site. The haem associated with the nitrate-reducing fragment could no longer be reduced by the NADH, indicating that FAD is essential for the transfer of electrons between NADH and haem. CD, visible, and EPR spectral analysis of the limited proteolysis fragments compared to the native enzyme indicated no significant perturbation of the protein environment surrounding the prosthetic groups as a result of limited proteolysis (Solomonson *et al.* 1986; Kay *et al.* 1988). Combination of the results from radiation inactivation analysis and limited proteolysis studies suggest the presence of three major functional domains, shown in Fig. 7.2, corresponding to the various partial activities.

The role of the essential sulphydryl group of *Chlorella* NR was investigated by a combination of spin-labelling and inhibition experiments (Barber and Solomonson 1986*a*). Incubation of the enzyme with either NEM or a spin-labelled derivative of NEM resulted in a time-dependent inactivation of NADH-NR and

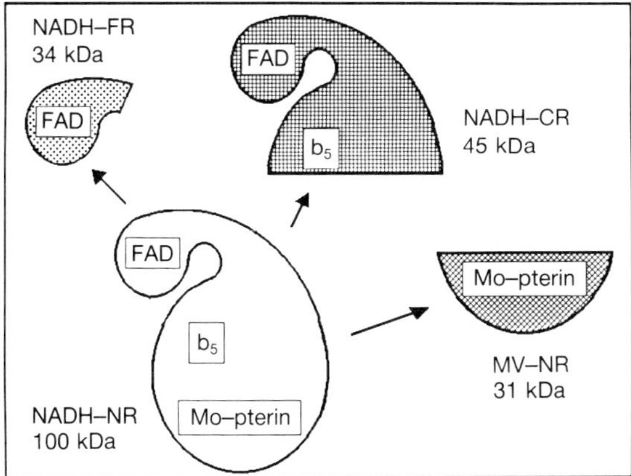

Fig. 7.2. Functional domain model of *Chlorella* nitrate reductase. Three major domains with various radiation inactivation target sizes are indicated together with their associated partial activities. NADH-NR, NADH-nitrate reductase; NADH–FR, NADH-ferricyanide reductase; NADH-CR, NADH-cytochrome *c* reductase; MV-NR, reduced methylviologen-nitrate reductase.

NADH-cytochrome *c* reductase activities with no effect on nitrate-reducing activity. This inactivation was prevented by NADH and involved the modification of a single sulphydryl group. The EPR spectrum of the spin-labelled enzyme revealed the presence of a single species with the nitroxide retaining substantial motional freedom. The absence of any change in the nitrogen hyperfine coupling constant of the bound label compared with that of the free nitroxide in buffer and its facile reduction by ascorbate indicated that the sulphydryl group is probably located on the external surface of the enzyme close to the NADH binding site. Cleavage of the spin-labelled enzyme using maize inactivating protease and separation into its FAD and haem/Mo domains followed by EPR spectroscopy revealed the modified sulphydryl group to be associated with the haem/Mo Fragment, suggesting a close interaction of these domains in the region of the nucleotide binding site.

Kinetic properties

Various steady-state kinetic mechanisms have been reported for assimilatory NR from different sources. Herrero *et al.* (1980) reported an iso ping-pong bi-bi kinetic mechanism for *Ankistrodesmus braunii* NR based on initial velocity and product inhibition patterns. Similarly, Campbell and Smarrelli (1978) proposed a two-site ping-pong mechanism for squash and maize NR. In contrast, McDonald and Coddington (1974) proposed a random order rapid-equilibrium mechanism for the assimilatory NR from *Aspergillus nidulans*. Howard and Solomonson (1981) also reported a rapid equilibrium random bi-bi kinetic mechanism for *Chlorella* NR based on initial velocity, product inhibition, and dead-end inhibitor patterns. These reports suggest fundamental differences in kinetic mechanisms for assimilatory NR. However, there are considerations which might explain these apparent differences. It is noteworthy that the product inhibition patterns for all eukaryotic assimilatory NR so far studied have been the same. The apparent differences have been in the initial velocity patterns, which have in some cases been parallel (Eaglesham and Hewitt 1975; Campbell and Smarrelli 1978; Herrero *et al.* 1980) suggesting ping-pong kinetics and in other cases have been intersecting (McDonald and Coddington 1974; Howard and Solomonson 1981) which indicates sequential kinetics. Parallel initial velocity patterns suggest, but are not sufficient evidence for, a ping-pong mechanism. The apparent patterns will be dictated by the relative magnitudes of the individual dissociation constants comprising the kinetic rate expression. Intersecting initial velocity patterns, on the other hand, are not consistent with a ping-pong mechanism but rather indicate a sequential mechanism. In addition, NR catalyses an essentially irreversible reaction ($K_{eq} = 10^{40}$) so the products (nitrite and NAD^+) should behave as dead-end inhibitors and product inhibition patterns established for reversible bi-substrate reactions would therefore not be applicable. Non-product dead-end inhibitors could be used to distinguish between possible kinetic mechanisms for NR but have been used only for the *Chlorella* enzyme (Howard and Solomonson 1981).

Kay and Barber (1986) have recently reported the kinetic parameters (Table 7.1) for the various partial activities of *Chlorella* NR and the effect of ionic strength on these parameters. Reduced flavin-NR and NADH-NR had the same V_{max} while all other partial activities had significantly higher V_{max} values. These activities (FH_2:NR and NADH:NR) were also the only activities that were stimulated by increased ionic strength. These results, together with the effect of ionic strength on the steady-state reduction of the haem prosthetic group of NR during turnover, suggested that the transfer of electrons from haem to Mo is the rate-limiting step in the transfer of electrons between NADH and nitrate.

Effectors

Various factors of possible physiological significance affect the activity of NR. Cyanide combines with the reduced enzyme to produce an inactive and stable enzyme–cyanide complex which can be rapidly and completely reversed by re-oxidation (Solomonson 1974; Lorimer *et al.* 1974). These interconversions have also been shown to occur *in vivo* (Lorimer *et al.* 1974; Solomonson and Spehar 1977). Cyanide combines with the enzyme in approximately a 1 : 1 stoichiometry (per subunit) and the dissociation constant of the reduced enzyme–cyanide complex is approximately 3×10^{-10} M, which is far below the levels of cyanide which would inhibit respiration.

Table 7.1. Redox potentials and K_m values for substrates and prosthetic groups of *Chlorella* nitrate reductase.

Substrate/prosthetic group	Midpoint potential (mV, pH 7)	K_m (μM)
Substrates		
Donors		
$NAD^+/NADH$	-320	4
$FAD/FADH_2$	-60	165
$MV/MV^{\cdot+}$	-440	5
Acceptors		
NO_3^-/NO_2^-	$+420$	110
$Fe_e^{3+}(CN)_6/Fe^{2+}(CN)_6$	$+360$	13
Cytochrome $c_{ox/red}$	$+250$	22
Prosthetic Groups		
$FAD/FAD^{\cdot-}$	-372	
$FAD^{\cdot-}/FADH_2$	-172	
$Haem_{ox/red}$	-164	
$Mo(VI)/Mo(V)$	$+19$	
$Mo(V)/Mo(IV)$	-31	

Cyanide does not interact with the haem prosthetic group of NR but rather appears to bind specifically to the Mo centre (Solomonson 1974; Solomonson *et al.* 1984). This conclusion is based on EPR evidence and the specificity of cyanide for the nitrate-reducing activity, which is known to be associated with the Mo centre of NR. It seems likely that cyanide stabilizes the Mo(IV) oxidation state thereby prohibiting the transfer of electrons from Mo to nitrate.

Phosphate has long been known to stimulate NR activity (Nason and Takahashi 1958), although a part of this stimulatory effect may have been due to ionic strength stimulation in some cases where ionic strength was not carefully controlled (Nicholas and Scawin 1956; Kinsky and McElroy 1958; Oji *et al.* 1987). This phosphate effect is also specific for nitrate-reducing activities and is probably due to interactions of phosphate with the Mo centre (Howard and Solomonson 1981; Solomonson *et al.* 1984; Kay and Barber 1986).

Thermodynamic properties

Oxidation–reduction midpoint potentials for the flavin, haem and molybdopterin prosthetic groups of *Chlorella* NR have been determined by visible, CD, and EPR potentiometry and microcoulometric titrations (Solomonson *et al.* 1984; Barber and Solomonson 1986b; Kay *et al.* 1986, 1988; Spence *et al.* 1988). The haem prosthetic group masks absorbance due to FAD or molybdopterin, precluding optical titration of these prosthetic groups in the holoenzyme. Optical titration of the FAD domain of proteolytically cleaved enzyme showed a reversible $n=2$ redox process with a midpoint potential of -288 mV which was shifted approximately 60 mV more positive in the presence of NAD^+ (Solomonson and Barber 1987). This value was confirmed by CD spectroscopy (-272 mV) and microcoulometry (-283 mV) (Kay *et al.* 1988; Spence *et al.* 1988). The values for the individual flavin couples, determined by CD titrations in the presence of dye mediators, were -372 mV for $FAD/FAD^{\cdot-}$ and -172 mV for $FAD^{\cdot-}/FADH_2$. Potentiometric titrations of the FAD centre of NR from other sources have not yet been done. CD titrations would be the method of choice in those cases where FAD is not tightly bound since free FAD would not interfere with CD spectral determinations of bound FAD.

The midpoint potential for the haem prosthetic group was determined to be -164 mV and showed a pH dependence of 20 mV/pH unit within the range of 5.5 to 7, suggesting the presence of a single, redox-associated, ionizable functional group in the protein with $pK_{ox}=5.8$ and $pK_{red}=6.2$. At pH7 and within the range 12–38°C, the midpoint potential of the haem decreased by approximately 1 mV per degree. Values for ΔS° and ΔH° were calculated to be -25.6 eu and -4.0 kcal/mol (Kay *et al.* 1986). Values for the midpoint potential of the haem prosthetic group of spinach NR and *Ankistrodesmus* NR, obtained previously, were -60 mV and -73 mV, respectively, indicating significant variation in the redox potential of the haem prosthetic group of NR from different sources (De la Rosa 1983; Barber *et al.* 1987). The midpoint potential of the haem prosthetic group of NR from the yeast, *Candida nitratophila*, has recently been determined to be -174 mV, similar to the value for the haem group of the *Chlorella* enzyme (Notton *et al.* 1987).

Low-temperature EPR titrations revealed that the midpoint potentials for both the Mo(VI)/Mo(V) and Mo(V)/Mo(IV) redox couples were pH-dependent, yielding values of -34 mV and -54 mV at pH 7 and -120 mV and -140 mV at pH 9. The shift in the potential of the titration curve with increasing pH indicated the involvement of a single proton in both redox steps (Barber and Solomonson 1986). Values for the Mo(VI)/Mo(V) and Mo(V)/Mo(IV) couples of spinach NR at pH 7, determined by low temperature EPR titrations were -8 and -42 mV respectively (Barber *et al.* 1987), comparable to the value for *Chlorella* NR determined under identical conditions and suggest minimal variation in the midpoint potentials of the Mo centre compared to the variation in midpoint potential of the haem centre.

Determinations of the Mo potentials in the frozen state may not accurately reflect the potentials at room temperature (Porras and Palmer 1982). The midpoint potentials for the Mo centre have recently been determined at room temperature by CD, microcoulometry, and room temperature EPR titrations. The respective midpoint potentials for the Mo(VI)/Mo(V) and Mo(V)/Mo(IV) redox couples of *Chlorella* nitrate reductase at room temperature and pH 7 were determined to be $+26$ and -40 mV by CD, $+15$ and -25 mV by room temperature EPR, and $+16$ and -27 mV by microcoulometric titrations (Kay *et al.* 1988; Spence *et al.* 1988). These values are all in good agreement but shifted to higher potential when compared to the values determined by low temperature EPR (Solomonson *et al.* 1984; Barber and Solomonson 1986). Thus, the freezing of samples to 100 K results in significant redistribution of electrons, particularly affecting the measured Mo(VI)/Mo(V) couple. The results of the room temperature redox titrations are summarized in Table 7.1.

Conclusions and future directions

Various structure–function relationships of NR have been described and are summarized in Figs 7.1 and 7.2. These models depict the sizes and relationships of the functional domains as well as the locations of the prosthetic groups.

Important information to be learned in the future includes the 3-dimensional structure of the enzyme in various physiological states determined by X-ray crystallography; the possible effect of physiological effectors on various thermodynamic parameters and the kinetic consequences of these changes; the rate constants for individual steps in electron transfer from NADH to nitrate, determined by transient-state kinetics; and the role of individual functional groups of the enzyme as determinants of overall catalytic efficiency. The latter area, involving site-directed mutagenesis, will be especially fruitful and may permit alterations in the enzyme which may be of potential benefit to agriculture and crop productivity.

Acknowledgements

We gratefully acknowledge the participation of current and past associates in work from the authors' laboratories described here, especially Drs W. D. Howard and C. J. Kay, and fruitful collaborations with other investigators, particularly Drs D. C. Eichler, M. J. McCreery, B. A. Notton, A. Oaks, K. V. Rajagopalan, and J. T. Spence. Research from the authors laboratories was supported by grants from NIH (GM 32696), USDA (GAM 8400528), NSF (DMB 8214001 and DMB 8615836), and NATO (86-0015).

References

Amy, N. K., Garrett, R. H., and Anderson, B. M. (1977). Reaction of the *Neurospora crassa* nitrate reductase with NAD(P) analogs. *Biochimica et Biophysica Acta* **480**, 83–95.

Barber, M. J. and Solomonson, L. P. (1986*a*). The role of the essential sulfhydryl group in assimilatory NADH : nitrate reductase of *Chlorella*. *Journal of Biological Chemistry* **261**, 4562–7.

Barber, M. J. and Solomonson, L. P. (1986*b*). Properties of the molybdenum domain of nitrate reductase. *Polyhedron* **5**, 577–80.

Barber, M. J., Notton, B. A., and Solomonson, L. P. (1987). Oxidation-reduction midpoint potentials of the molybdenum center in spinach NADH : nitrate reductase. *FEBS Letters* **213**, 372–4.

Campbell, W. H. and Smarrelli, I., Jr. (1978). Purification and kinetics of higher plant NADH : nitrate reductase. *Plant Physiology* **61**, 611–16.

Campbell, J. McA and Wray, J. L. (1983). Purification of barley nitrate reductase and demonstration of nicked subunits. *Phytochemistry* **22**, 2375–82.

Cheng, C.-L., Dewdney, J., Kleinhofs, A., and Goodman, H. M. (1986). Cloning and nitrate induction of nitrate reductase mRNA. *Proceedings of the National Academy of Sciences, USA* **83**, 6825–8.

Commere, B., Chirel, I., Kronenberger, J., Galangan, F., and Caboche, M. (1986). In vitro translation of nitrate reductase messenger RNA from maize and tobacco and detection with an antibody directed against the enzyme of maize. *Plant Science* **44**, 191–203.

Crawford, N. M., Campbell, W. H., and Davis, R. W. (1986). Nitrate reductase from squash : cDNA cloning and nitrate regulation. *Proceedings of the National Academy of Sciences, USA* **83**, 8073–6.

De la Rosa, M. A. (1983). Assimilatory nitrate reductase from the green alga *Ankistrodesmus braunii*. *Molecular and Cellular Biochemistry* **50**, 65–74.

De la Rosa, M. A., Vega, J. M., and Zumft, W. G. (1981). Composition and structure of assimilatory nitrate reductase from *Ankistrodesmus braunii*. *Journal of Biological Chemistry* **256**, 5814–19.

Eaglesham, A. R. J. and Hewitt, E. J. (1975). Inhibition of nitrate reductase from spinach leaf by adenosine nucleotides. *Plant Cell Physiology* **16**, 1137–49.

Fido, R. J. and Notton, B. A. (1984). Spinach nitrate reductase: further purification and removal of 'nicked' subunits by affinity chromatography. *Plant Science Letters* **37**, 87–91.

Garrett, R. H. and Nason, A. (1967). Involvement of a *b*-type cytochrome in the assimilatory nitrate reductase of *Neurospora crassa*. *Proceedings of the National Academy of Science, USA* **58**, 1603–10.

Garrett, R. H. and Nason, A. (1969). Further purification and properties of *Neurospora* nitrate reductase. *Journal of Biological Chemistry* **244**, 2870–82.

Garrett, R. H. and Greenbaum, P. (1973). The inhibition of the *Neurospora crassa* nitrate reductase complex by metal-binding agents. *Biochimica et Biophysica Acta* **302**, 24–32.

Giri, L. and Ramadoss, C. S. (1979). Physical studies on assimilatory nitrate reductase from *Chlorella vulgaris*. *Journal of Biological Chemistry* **254**, 11703–12.

Guerrero, M. G. and Lara, C. (1987). Assimilation of inorganic nitrogen. In *The Cyanobacteria* (eds P. Fay and C. VanBaalen), pp. 163–86. Elsevier, Amsterdam.

Guerrero, M. G., Vega, J. M., and Losada, M. (1981). The assimilatory nitrate-reducing system and its regulation. *Annual Review of Plant Physiology* **32**, 169–204.

Herrero, A. De la Rosa, M. A., Diez, J., and Vega, J. M. (1980). Catalytic properties of *Ankistrodesmus braunii* nitrate reductase. *Plant Science Letters* **17**, 409–15.

Hewitt, E. J. (1975). Assimilatory nitrate-nitrite reduction. *Annual Review of Plant Physiology* **26**, 73–100.

Howard, W. H. and Solomonson, L. P. (1981). Kinetic mechanism of assimilatory NADH : nitrate reductase from *Chlorella*. *Journal of Biological Chemistry* **256**, 12725–30.

Howard, W. D. and Solomonson, L. P. (1982). Quarternary structure of assimilatory NADH : nitrate reductase from *Chlorella*. *Journal of Biological Chemistry* **257**, 10243–50.

Kay, C. J. and Barber, M. J. (1986). Assimilatory nitrate reductase from *Chlorella* : effect of ionic strength and pH on catalytic activity. *Journal of Biological Chemistry* **261**, 14125–9.

Kay, C. J., Solomonson, L. P., and Barber, M. J. (1986). Thermodynamic properties of the heme prosthetic group in assimilatory nitrate reductase. *Journal of Biological Chemistry* **261**, 5799–802.

Kay, C. J., Barber, M. J., and Solomonson, L. P. (1988). CD and potentiometry of FAD, heme and Mopterin cofactors of assimilatory nitrate reductase. *Biochemistry* **27**, 6142–9.

Kinsky, S. C. and McElroy, W. D. (1958). *Neurospora* nitrate reductase: the role of phosphate flavin and cytochrome *c* reductase. *Archives of Biochemistry and Biophysics* **73**, 466–83.

Kuo, T., Kleinhofs, A. and Warner, R. L. (1980). Purification and partial characterization of nitrate reductase from barley leaves. *Plant Science Letters* **17**, 371–81.

Lorimer, G. H., Gewitz, H. S., Volker, W., Solomonson, L. P., and Vennesland, B. (1974). The presence of bound cyanide in the naturally inactivated form of nitrate reductase of *Chlorella*. *Journal of Biological Chemistry* **249**, 6074–9.

Manzano, C., Candau, P., and Guerrero, M. F. (1978). Affinity chromatography of *Anacyctis nidulans* ferredoxin–nitrate reductase on reduced ferredoxin–sepharose. *Analytical Biochemistry* **90**, 408–12.

McDonald, D. W. and Coddington, A. (1974). Properties of the assimilatory nitrate reductase from *Aspergillus nidulans*. *European Journal of Biochemistry* **46**, 169–78.

Monod, J., Wyman, J., and Changeux, J. P. (1965). On the nature of allosteric transitions: a plausible model. *Journal of Molecular Biology* **12**, 88–118.

Nason, A. and Takahashi, H. (1958). Inorganic nitrogen metabolism. *Annual Review of Microbiology* **12**, 203–46.

Nason, A., Lee, K.-Y., Pan, S.-S., Ketchum, P. A., Lamberti, A., and DeVries, J. (1971). *In vitro* formation of assimilatory reduced nicotinamide adenine dinucleotide phosphate: nitrate reductase from a *Neurospora* mutant and a component of molybdenum enzymes. *Proceedings of the National Academy of Sciences, USA* **68**, 3242–6.

Nicholas, D. J. D. and Scawin, J. H. (1956). A phosphate requirement for nitrate reductase from *Neurospora crassa*. *Nature* **178**, 1474–5.

Notton, B. A. and Hewitt, E. J. (1979). Structure and properties of higher plant nitrate reductase, especially *Spinacea oleracea*. In *Nitrogen assimilation of plants* (eds E. J. Hewitt and C. V. Cutting), pp. 227–44. Academic Press, London.

Notton, B. A., Kay, C. J., Barber, M. J., Solomonson, L. P., Kau, D., Cannons, A. C., and Hipkin, C. R. (1987). Heme Em and partial enzymic activities of nitrate reductase. *Federation Proceedings* **46**, 2098.

Oji, Y., Ryoma, Y., Wakinchi, N., and Okarnoto, S. (1987). Effect of inorganic orthophosphate on *in vitro* activity of NADH-nitrate reductase isolated from 2-row barley leaves. *Plant Physiology* **83**, 472–4.

Pan, S.-S. and Nason, A. (1978). Purification and characterization of homogeneous assimilatory reduced nicotinamide adenine dinucleotide phosphate–nitrate reductase from *Neurospora crassa*. *Biochimica et Biophysica Acta* **523**, 297–313.

Porras, A. G. and Palmer, G. (1982). The room temperature potentiometry of xanthine oxidase : pH-dependent redox behavior of the flavin, molybdenum and iron–sulfur centers. *Journal of Biological Chemistry* **257**, 11617–26.

Redinbaugh, M. G. and Campbell, W. H. (1985). Quaternary structure and composition of squash NADH : nitrate reductase. *Journal of Biological Chemistry* **260**, 3380–5.

Solomonson, L. P. (1974). Regulation of nitrate reductase activity by NADH and cyanide. *Biochimica et Biophysica Acta* **334**, 297–308.

Solomonson, L. P. (1975). Purification of NADH-nitrate reductase by affinity chromatography. *Plant Physiology* **56**, 853–5.

Solomonson, L. P. and Spehar, A. M. (1979). Model for the regulation of nitrate assimilation. *Nature* **265**, 373–5.

Solomonson, L. P. and McCreery, M. J. (1986). Radiation inactivation of assimilatory NADH:nitrate reductase from *Chlorella*: catalytic and physical sizes of functional units. *Journal of Biological Chemistry* **261**, 806–10.

Solomonson, L. P., Lorimer, G. H., Hall, R. H., Borchers, R., and Bailey, J. L. (1975). Reduced nicotinamide adenine dinucleotide-nitrate reductase of *Chlorella vulgaris*: purification, prosthetic groups, and molecular properties. *Journal of Biological Chemistry* **250**, 4120–7.

Solomonson, L. P. and Barber, M. J. (1987). Structure–function relationships of assimilatory nitrate reductase. *Inorganic nitrogen metabolism* (eds W. Ullrich, P. J. Aparicio, P. J. Syrett, and F. Castillo), pp. 71–75, Springer-Verlag, Berlin.

Solomonson, L. P., Barber, M. J., Robbins, A. P., and Oaks, A. (1986). Functional domains of assimilatory NADH:nitrate reductase from *Chlorella*. *Journal of Biological Chemistry* **261**, 11290–4.

Solomonson, L. P., McCreery, M. J., Kay, C. J., and Barber, M. J. (1987). Radiation inactivation analysis of assimilatory NADH:nitrate reductase: apparent functional size of partial activities associated with intact and proteolytically modified enzyme. *Journal of Biological Chemistry* **262**, 8934–9.

Spence, J. T., Barber, M. J., and Solomonson, L. P. (1988). Stoichiometry of electron uptake and oxidation–reduction potentials of NADH:nitrate reductase. *Biochemical Journal* **250**, 921–3.

8. Genetics and regulatory aspects of nitrate assimilation in algae

Emilio Fernández and Jacobo Cárdenas

The nitrate assimilatory pathway in algae

Most algae can use nitrate, nitrite, and ammonium as sole nitrogen sources. Nitrate and nitrite utilization begins with their transport into the cell, followed by reduction to ammonium, which is subsequently incorporated into glutamate by the glutamine synthetase/glutamate synthase pathway. Comprehensive reviews have been devoted to the different biochemical and regulatory aspects of nitrate assimilation in eukaryotes (Hewitt and Notton 1980; Srivastava 1980; Syrett 1981; Guerrero *et al.* 1981; Ullrich 1983; Dunn-Coleman *et al.* 1984). This review focuses on the genetics and regulation of nitrate assimilation in algae.

Nitrate reductase (NR) has been located in the pyrenoid of the chloroplast in green algae (López-Ruíz *et al.* 1985), which also seems to be the most probable location for nitrite reductase (NiR) (Vega *et al.* 1980; Guerrero *et al.* 1981). Thus, nitrate and nitrite have to cross at least two barrier membranes before their reduction takes place. In non-vacuolated green algae, nitrate and nitrite uptake and reduction are very tightly coupled processes which are difficult to study separately (Ullrich 1983): the mechanism and regulation of these uptake processes will be analysed in more detail below. Ammonium and methylammonium are taken up in green algae by the same carrier proteins in a process which shows saturation kinetics and inhibition by uncoupling agents (Pelley and Bannister 1979; Wheeler 1980; Wright and Syrett 1983; Florencio and Vega 1983*b*; Franco *et al.* 1987*a*).

NR from different algal species and strains uses NADH alone (EC 1.6.6.1) or either NADH or NADPH (EC 1.6.6.2) as electron donor to reduce nitrate to nitrite (Sosa and Cárdenas 1977; Hewitt and Notton 1980). The enzyme activity (overall NR activity) involves the same prosthetic groups present in other eukaryotic NR: FAD, cytochrome b_{557}, and molybdenum. Molybdenum is present as part of a pterin-based molybdenum cofactor (MoCo) which is common to all molybdoenzymes except dinitrogenase (Johnson 1980; Chapter 14). Algal NR also shows the two partial activities NAD(P)H-cytochrome c reductase (diaphorase) and $FADH_2$-NR or reduced viologen-NR (terminal NR), which are associated with different protein domains of the NR complex (Franco *et al.* 1984*a*; Solomonson *et al.* 1986).

In *Chlorella vulgaris* NR is a homotetramer with identical 100 kDa subunits containing a full complement of prosthetic groups (Howard and Solomonson 1982; Chapter 7). In contrast, the enzyme of *Monoraphidium (Ankistrodesmus) braunii* is an octamer, with subunits of about 59 kDa, containing four FAD, four haem groups, and two atoms of molybdenum. The enzyme molecule shows an 8-fold rotational symmetry with subunits alternately arranged in two planes (De la Rosa *et al.* 1981*a*). Thus it is difficult to reconcile the structure of NR from *Chlorella* with that of *Ankistrodesmus*. In *Chlamydomonas reinhardtii*, a heteromultimeric NR of 220 kDa with two kinds of subunits has been proposed (Franco *et al.* 1984*a*).

NiR in photosynthetic organisms uses reduced ferredoxin as electron donor to reduce nitrite to ammonium (Vega *et al.* 1980). The enzyme has been purified to homogeneity in *Chlorella fusca* (Zumft 1972) and *Porphyra yezoensis* (Ho *et al.* 1976) and has been shown to be a single polypeptide chain of about 63 kDa, most probably containing sirohaem as a prosthetic group. Very recently, NiR from *Chlamydomonas* has been purified, and shown to have a molecular weight of 86 kDa and to contain two kinds of subunits: one of 63 kDa with reduced methylviologen-NiR activity and containing one functional sirohaem, and another of 25 kDa which is easily lost during purification and has been correlated with the ferredoxin-dependent activity (Romero *et al.* 1987).

Genetics of nitrate assimilation in algae

Nitrate assimilation mutants

Chlorate resistance caused by defects in nitrate assimilation has made it possible to isolate many mutant strains deficient in this pathway. Three NR-deficient mutants of *Eudorina elegans*, designated *nar-1*, *nar-2*, and *nar-3*, were the first described in green algae (Toby and Kemp 1977). *nar-1* and *nar-2* are probably defective in either NR apoprotein or NR regulation, while *nar-3* appears to be deficient in the MoCo common to NR and xanthine dehydrogenase. Other mutant strains have been described in *Dunaliella tertiolecta* (Latorella *et al.* 1981) and *Chlorella sorokiniana* (Knobloch and Tischner 1986) but have not been genetically characterized.

Three genetic loci related to nitrate assimilation have been identified in the multicellular alga *Volvox*: two linked genes *nit A* and *nit B*, and a third unlinked locus *nit C* (Huskey *et al.* 1979). *nit A* and *nit C* mutants lack NR activity. The *nit B* locus has been implicated in the nitrate transport system. This conclusion is supported by the observation that the *nit B* mutant has NR activity but shows poor growth at low nitrate concentrations.

The mutant strains from *Chlamydomonas reinhardtii* isolated by Nichols and

Syrett (1978) and Sosa *et al.* (1978) are the most extensively characterized mutants in algae, both biochemically and genetically. Their genotypes and phenotypes are summarized in Table 8.1. We will refer to these mutants below.

Nichols and Syrett (1978) used u.v. irradiation to obtain about 200 chlorate-resistant mutants of *Chlamydomonas reinhardtii*. Most of the mutants grow well in minimal nitrate medium, although many of them are unable to use nitrate in the presence of acetate: the role of acetate in modifying the growth of these mutants is not understood. Three mutants, *nit A*, *nit B*, and *nit C*, identifying three unlinked Mendelian loci, were analysed and shown to be unable to grow with nitrate in either the presence or the absence of acetate (Nichols and Syrett 1978; Nichols *et al.* 1978). The mutant *nit A* lacks NR-diaphorase activity and has high levels of terminal NR activity. Nichols *et al.* (1978) have suggested that *nit B* is a regulatory mutant, which lacks any NR-related activity but is able to incorporate nitric nitrogen into insoluble nitrogenous compounds, unlike mutant *nit A*. The failure of the *nit B* mutant to grow with nitrate seems rather puzzling (Nichols *et al.* 1978). A possible explanation for this phenotype is that it may have very low levels of NR activity that allow incorporation of nitrate at a rate too slow to support the growth of the cells. It has been suggested that undetectably low levels of NR (< 5 per cent of wild type) are responsible for the slow growth observed in heterozygous diploids of allelic NR mutants and molybdate-repairable NR mutants of *Chlamydomonas* (Fernández and Matagne 1986; Fernández and Aguilar 1987). The *nit C* mutant cannot use either nitrate or hypoxanthine as sole nitrogen source and is thus similar to the MoCo-deficient mutants of *Aspergillus nidulans* and *Neurospora crassa* (Nichols and Syrett 1978; Cove 1979; Chapters 6, 19, 20).

Sosa *et al.* (1978) isolated 13 chlorate-resistant mutants from *Chlamydomonas renhardtii* using chemical mutagenesis, all of which were unable to utilize nitrate. Mutants similar to both *nit A* and *nit B* (above) were found. A third group of mutants had high levels of NR-diaphorase activity, but lacked overall and terminal NR. Two other mutants with normal enzyme levels were suggested to be affected in the nitrate transport system.

Most NR-deficient mutants in green algae were isolated by resistance to chlorate. This selection procedure, however, may not allow recovery of all different kinds of NR-deficient mutants since some remain sensitive to chlorate, at least in *A. nidulans* (Cove 1969).

In vitro complementation between NR mutants

The *in vitro* complementation system for NR reported in fungi (Nason *et al.* 1970; Garrett and Cove 1976) and higher plants (Mendel and Müller 1978) uses the ammonium-repressible NR-diaphorase subunits present in MoCo$^-$ mutants, which can assemble into an active NR complex when a functional MoCo is added.

Mutants 102, 104, and 307 from *Chlamydomonas* have been shown to be deficient in the MoCo common to the molybdoenzymes since they lack both

Table 8.1. Genotype and phenotype for NR of different wild and mutant strains of *Chlamydomonas reinhardtii*.

Strain	Assigned genotype	Phenotype	References[a]
6145c	+	Wild-type	1
305, *nit A*	*nit-1a*	Only terminal NR	1, 2, 3, 4
nit-1	*nit-1a*	No NR related activities, it has terminal NR-subunits	1,5
301	*nit-1b*	No terminal NR, it has NR-diaphorase and MoCo	1, 4, 6
203, *nit B*, *nit-2*	*nit-2*	No NR related activities, leaky growth in nitrate	1, 2, 3, 4
307	*nit-3*	No MoCo, NR subunits assembled into an inactive NR	1, 7
104	*nit-4*	No MoCo, repairable by high concentrations of molybdate	1, 7
102	*nit-5/nit-6*	No MoCo, repairable by high concentrations of molybdate	1, 7
21*gr*	*nit-5*	Wild-type	1
I$_3$	*nit-6*	Wild-type, tungstate-resistant	1, 7
137c	*nit-1/nit-2*	No NR related activities	1
2170	*ma-1*	Methylammonium-resistant, derepressed NR in ammonium and methylammonium media	8
2172	*ma-2*	Methylammonium-resistant, derepressed NR in methylammonium media	9

[a] References: 1, Fernández and Matagne 1984; 2, Nichols and Syrett 1978; 3, Nichols *et al.* 1978; 4, Sosa *et al.* 1978; 5, Fernández and Matagne 1986; 6, Fernández and Cárdenas 1982*a*; 7, Fernández and Aguilar 1987; 8, Franco *et al.* 1987*a*; 9, A. R. Franco, unpublished Thesis 1986.

xanthine dehydrogenase (XDH) and NR activities, and are incapable of restoring the NR activity of the mutant *nit-1* of *Neurospora crassa* in *in vitro* complementation experiments. In contrast, the wild type and mutants 203, 305 and 301 have XDH activity, and their extracts exhibit MoCo activity *in vitro* (Fernández and Cárdenas 1981*b*).

In common with the MoCo from many organisms (Nason *et al.* 1971; Johnson 1980; Kramer *et al.* 1984; Chapter 14), MoCo in *Chlamydomonas* is a heat-labile low molecular weight species constitutively present in the cells and easily inactivated in crude extracts. Inactive MoCo is produced in tungstate-grown cells and can be activated *in vitro* with 10 mM molybdate (Fernández and Cárdenas 1981*a*). In contrast to fungal and higher plant systems, the MoCo is incapable of restoring *in vitro* NR activity in the MoCo-deficient mutants 102, 104, and 307 of *Chlamydomonas* (Fernández and Cárdenas 1982*a*). However, overall NR activity is reconstituted *in vitro* between the ammonium-repressible NR-diaphorase proteins from these mutants and a protein of about 67 kDa containing the MoCo; this protein is present in the wild type and also in mutant 305, which has an NR complex with only terminal activity (Fernández and Cárdenas 1981*a*, 1983*c*; Franco *et al.* 1984*a*). The MoCo bound to this 67 kDa protein is not heat labile and, if it is isolated in inactive form from tungstate-treated cells, it cannot be reactivated by molybdate. The reconstituted NR is physicochemically and enzymatically indistinguishable from the wild-type NR complex (Fernández and Cárdenas 1981*a*).

Mutant 301 of *Chlamydomonas* possesses functional ammonium-repressible diaphorases and MoCo but lacks overall and terminal NR activities (Fernández and Cárdenas 1982*a*). The phenotype of mutant 301 has no parallel in other organisms and it seems to be a structural gene mutant because of its defective NR enzyme activity. In addition, its NR-diaphorase is able to reconstitute NR activity with the terminal NR from mutant 305. Mutants such as *nia D17* from *Aspergillus* have MoCo and NR-diaphorase activity but do not complement defective NR from other *nia D* strains in the same complementation group (Cove 1979). Very recently, *in vivo* interallelic complementation has been reported between some *nia* mutants of *Nicotiana plumbaginifolia*, although restored NR activity levels are lower than those found in *Chlamydomonas* (Gabard *et al.* 1987).

On the basis of the above results it has been proposed that the NR complex of *Chlamydomonas* consists of two kinds of subunit: diaphorase subunits, active *per se*, and terminal subunits, inactive unless assembled into the complex (Fernández and Cárdenas 1981*b*; Franco *et al.* 1984*a*). The octameric NR complex of *Ankistrodesmus braunii* has been suggested to have two types of subunit on the basis of the higher number of subunits than prosthetic groups (De la Rosa *et al.* 1981*a*).

The diaphorase subunit has been purified and characterized as a protein of about 45 kDa containing FAD and cytochrome b_{557}, whose kinetic parameters are similar to those reported for the diaphorase activity of the whole complex (Sosa and Cárdenas 1977; Fernández and Cárdenas 1983*a,b*).

The terminal NR activity of mutant 305 of *Chlamydomonas* shows the same enzyme properties as the corresponding activity in the wild-type complex, and has a lower estimated molecular weight, about 170 kDa *vs* 220 kDa (Fernández and Cárdenas 1981*a*, 1983*c*; Franco *et al.* 1984*a*). Mutant *nit A*, phenotypically very similar to the mutant 305, also has a smaller NR complex (Nichols *et al.* 1978); for both mutants, it has been suggested that the NADH-binding portion of the protein is absent. However, the wide variation in molecular size reported for wild-type NR and also for NR of mutants 305 and *nit A* (Barea and Cárdenas 1975; Fernández and Cárdenas 1981*b*; Hipkin *et al.* 1985) will require further estimations by more accurate molecular approaches. Hipkin *et al.* (1985) have proposed that a non-proteolytic dissociation of the native enzyme into smaller catalytically active proteins occurs *in vitro*, as previously reported for the enzyme of *Chlorella* (Howard and Solomonson 1982).

Genetic analysis

The genetic analysis of six of the *Chlamydomonas* mutants (305, 301, 203, 102, 104, and 307) isolated by Sosa *et al.* (1978) indicates that all except 102 are affected at a single Mendelian locus (see Table 8.1) (Fernández and Matagne 1984).

Crosses between mutants 305, lacking diaphorase activity, and 301, lacking terminal NR activity, do not give rise to wild-type recombinants. Similarly, neither of these mutants give rise to wild-type recombinants when crossed to mutant *nit-1* of *Chlamydomonas* which has been mapped to the linkage group IX (O'Brien 1986) and is allelic to the NR structural mutant *nit A* (Nichols and Syrett 1978). These data suggest that a single structural gene locus for NR, namely *nit-1*, exists in *Chlamydomonas*. *In vitro* complementation and biochemical data suggest that different NR subunits are defective in mutants 305 and 301, and their assignment to different complementation groups is further supported by *in vivo* complementation studies (Fernández and Matagne 1986). NR activities in heterozygous diploids formed between the various mutant strains support a heteromultimeric NR structure, in which subunits can exchange to form hybrid complexes. It has been proposed that the diaphorase and terminal subunits may be encoded by two different but tightly linked cistrons, *nit-1a* and *nit-1b*. Alternatively, NR in *Chlamydomonas* could be made from one gene, *nit-1*, whose product is post-translationally modified to produce the different subunits (Fernández and Matagne 1986). The complete clarification of this point will require the isolation and characterization of the NR structural gene.

Similar to mutant 305, mutant *nit-1*, isolated from strain 137c has defective diaphorase subunits and intact terminal subunits (Fernández and Matagne 1986). *In vivo*, interallelic complementation can take place between the defective NR of mutants 305 and *nit-1*, although very inefficiently, as reflected by the very poor growth of the diploids on nitrate medium and their very low NR levels.

The mutants 203 and *nit B* have the same phenotype (Nichols *et al.* 1978; Fernández and Cárdenas 1982a) and map at the same locus, *nit-2* (Nichols and Syrett 1978; Fernández and Matagne 1984), located in linkage group III (O'Brien 1986). These mutants lack NR activity, but grow on nitrite medium and have ammonium-repressible NiR. As discussed above, it seems probable that *nit-2* is a regulatory gene for NR. The fact that each *nit-2* mutation is recessive to its wild-type allele suggests positive control by the *nit-2* product on the expression of the ammonium-repressible structural gene(s) (Fernández and Matagne 1986). In addition, XDH, urate oxidase and the urate uptake system are represssed by ammonium in *Chlamydomonas* and XDH does not appear to be under the control of *nit-2* (Fernández and Cárdenas 1981b; Pineda *et al.* 1984, 1987). In this respect, Fernández and Matagne (1986) have suggested that there exists a major nitrogen regulatory gene, different from *nit-2*, mediating general ammonium repression in *Chlamydomonas*. Both pathway-specific regulatory loci and major nitrogen regulatory genes acting on general catabolite repression are known in fungi (Chapters 19 and 20).

MoCo-deficient mutants in *Chlamydomonas* occur at different unlinked loci: *nit-3* (mutant 307), *nit-4* (mutant 104), and *nit-5* and *nit-6* (mutant 102) (Fernández and Matagne 1984). *nit-5* and *nit-6* are cryptic genes since a double mutation, one at each cistron, is required for the appearance of the MoCo⁻ phenotype. Mutants resembling *nit-5* and *nit-6* have not been reported in other organisms. Not enough information is available to relate the mutant *nit C*, which is most probably MoCo⁻ (Nichols and Syrett 1978), to any of the mutants described above (see Table 8.1). The mutant alleles of either *nit-4*, and *nit-5/nit-6* are recessive in heterozygous diploids. *nit-3*, however, is co-dominant, suggesting that *nit-3* codes for a protein whose activity is limiting for the MoCo biosynthetic route (Fernández and Matagne 1986).

The MoCo-deficient mutants can be distinguished by their different responses to molybdate and tungstate when added to the culture media. Mutants at either of the cryptic genes, *nit-5* or *nit-6*, show a wild-type phenotype, i.e. they grow on nitrate medium. However, in contrast to other phenotypically wild-type strains of *Chlamydomonas* (see Table 8.1), mutant *nit-6* can grow on nitrate medium containing 2 mM tungstate (Fernández and Aguilar 1987). High concentrations of molybdate (15 mM) allow mutants 104 and 102 to grow on nitrate medium (Fernández and Aguilar 1987). These mutants resemble mutants *nit-9A* and *nit-9B* of *Neurospora* (Dunn-Coleman 1984). *cnxE* of *Aspergillus* (Cove 1979) and *cnxA* of *Nicotiana* (Mendel *et al.* 1986), which are thought to be affected in molybdate insertion or processing. Growth on nitrate of mutant 307 (*nit-3*) is not restored by molybdate (Fernández and Aguilar 1987). Thus, the *nit-4*, *nit-5*, and *nit-6* loci of *Chlamydomonas* appear to be involved in molybdate processing (uptake, storage, insertion in the pterin moiety of MoCo, etc.) but not in the synthesis of the modified pterin, in which *nit-3* probably participates. By using the *in vitro* complementation assay with NR-diaphorase from the *Neurospora* mutant *nit-1* (Fernández and Cárdenas 1981*a*), an 'empty MoCo' which is molybdate-activatable has been detected in mutants 104 and 102 but not in 307 (M. Aguilar, personal communication).

The ability to assemble the NR subunits into the enzyme complex has been related to the organic pterin moiety of the molybdenum cofactor (Johnson 1980; Mendel *et al.* 1981; Wahl *et al.* 1984). In this respect, mutants 104 (*nit-4*) and 102 (*nit-5/nit-6*) have the 45 kDa NR-diaphorase subunits assembled into a 220 kDa NR complex with only diaphorase activity. The mutant 307 (*nit-3*) is able to assemble NR subunits to a certain extent, and probably has a defective pterin cofactor (Fernández and Aguilar 1987). The mutant *cnxD* of *Nicotiana plumbaginifolia* has a phenotype similar to that of 307 (De Vries *et al.* 1986).

A number of studies on the role of ammonium on nitrate assimilation have used methylammonium, a non-metabolizable analogue of ammonium, to mimic the effects of ammonium (see below). *Chlamydomonas* cells cannot grow on nitrate medium containing methylammonium since nitrate uptake is inhibited and NR represssed (Franco *et al.* 1984*b*, 1987*a*). Spontaneous methylammonium-resistant mutants of *Chlamydomonas* showing an altered pattern for NR regulation have been isolated (A. R. Franco, unpublished Thesis 1986; Franco *et al.* 1987*a*). These are the first mutants of this kind isolated in phototrophic eukaryotes, none of which can use methylammonium as a nitrogen source. Two of these mutants have been biochemically and genetically analysed (Table 8.1). They are defective in two linked genes, *ma-1* and *ma-2*, which are separately responsible for the two components involved in the transport of both ammonium and methylammonium by *Chlamydomonas* cells (Franco *et al.* 1988). Mutant 2170 (*ma-1*) is defective in component 1 which has a high K_m and a high V_{max} for the transport of both cations; mutant 2172 (*ma-2*) lacks component 2 which has a low K_m and a low V_{max}. In mutant 2170, NR is derepressed on nitrate medium even in the presence of ammonium or methylammonium (Franco *et al.* 1987*a*). The characteristics of this mutant will be discussed below in comparison to the wild type.

Regulation of nitrate and nitrite assimilation in algae

Nitrate assimilation in algae appears to be regulated at three different levels: the uptake of nitrate and nitrite; the activity of the NR complex; the amount of NR and NiR enzymes present in the cells.

Control of nitrate and nitrite uptake and utilization

The control of the uptake of nitrate and nitrite depends on the presence of ammonium in the culture medium, the carbon and nitrogen status of the algal cells, and the availability of light.

Ammonium-grown cells of *Phaeodactylum tricornutum* lack the nitrate uptake system; its induction requires nitrogen deprivation, photosynthetic energy and *de novo* protein synthesis (Cresswell and Syrett 1981). Either nitrogen deprivation or nitrate (or chlorate) induce the subsequent nitrate uptake in *Chara corallina* (Deane-Drummond 1984*a*).

Nitrite inhibits nitrate uptake strongly in *Chlorella* (Syrett and Morris 1963) and *Chlamydomonas* (Thacker and Syrett 1972*a*), but only weakly in *Phaeodactylum* (Cresswell and Syrett 1982) and not at all in *Skeletonema costatum* (Serra *et al.* 1978*b*). The kinetic and regulatory similarities in the uptake of nitrate and nitrite (Cresswell and Syrett 1982; Florencio and Vega 1983*b*), as well as competitive inhibition of nitrite uptake by nitrate (Bilbao *et al.* 1981), suggest that both anions may be taken up by the same transport system. However, the differences in pH dependence (Ullrich 1974) and differential effects of monochromatic light (Calero *et al.* 1980) for nitrate and nitrite uptake have led to a proposal that the two anions are transported by two different systems (Ullrich 1983). Very recently, it has been shown that nitrate is a partially competitive inhibitor of nitrite uptake in *Chlamydomonas* even in the absence of a functional NR (Córdoba *et al.* 1986). This kinetic mechanism of inhibition is only compatible with the existence of two separate sites for the uptake of nitrate and nitrite in *Chlamydomonas*.

Ammonium blocks utilization of nitrate and nitrite, and this inhibition is relieved when ammonium disappears from the medium during its assimilation by the cells (Cárdenas, 1972; Syrett 1981; Guerrero *et al.* 1981). This inhibitory effect occurs at the level of uptake, as has been shown in diatoms (Serra *et al.* 1978*a,b*; Cresswell and Syrett 1979). Since ammonium does not inhibit nitrate uptake in carbon-deficient cells (Syrett and Morris 1963; Thacker and Syrett 1972*a*; Ullrich 1979), it has been suggested that an organic product of ammonium assimilation, and not ammonium itself, is the true inhibitor. Likewise, algae incubated with L-methionine-DL-sulphoximine (MSX), an inhibitor of glutamine synthetase, show no inhibition of nitrate uptake by ammonium (Rigano *et al.* 1979; Cullimore and Sims 1981; Di Martino Rigano 1982; Florencio and Vega 1983*b*), which reinforces the proposal that an ammonium derivative is the inhibitor of nitrate uptake. Some algae, such as *Phaeodactylum* (Cresswell and Syrett 1984), can metabolize MSX, suggesting that MSX effects have to be analysed with caution.

The effect of ammonium on nitrite uptake has been poorly investigated in algae. It has been suggested that an ammonium derivative is responsible for the ammonium inhibition of nitrite utilization (Florencio and Vega 1983*b*). Accordingly, either carbon starvation or MSX treatment partially relieves ammonium inhibition of nitrite uptake in *Chlamydomonas* (Córdoba *et al.* 1987). However, under both conditions, nitrite uptake is inhibited progressively by experimentally increasing the intracellular ammonium concentrations. Thus, both ammonium itself and an ammonium derivative appear to act together to completely block nitrite uptake by *Chlamydomonas* cells. This proposal is also supported by the fact that methylammonium, a non-metabolizable analogue of ammonium (Franco *et al.* 1984*b*), partially inhibits nitrite uptake (Córdoba *et al.* 1987). Methylammonium also partially inhibits nitrate and nitrite uptake in *Phaeodactylum*; the inhibition takes a long time to develop as a result of the slow rate of methylammonium accumulation into the cells (Cresswell and Syrett 1984).

Nitrate uptake might be also regulated cooperatively by ammonium itself and by ammonium derivative(s). Bagchi *et al.* (1985) have proposed this mechanism based on data with glutamine auxotrophs of *Anabaena cycadeae*.

The effect of ammonium and ammonium derivatives on nitrate uptake notwithstanding, algae growing under conditions of limiting nitrogen supply assimilate both ammonium and nitrate simultaneously (Syrett 1981 and references therein). This fact is not surprising since under these conditions intracellular levels of ammonium and its regulatory derivatives may be low, which would reduce ammonium inhibition of nitrate uptake. In fact, the mutant *ma-1* of *Chlamydomonas* is able to assimilate nitrate and ammonium simultaneously (A. R. Franco, unpublished Thesis 1986). This mutant lacks component 1 for ammonium uptake and has low intracellular ammonium concentrations on ammonium medium. The activities of ammonium-assimilating enzymes in this mutant are similar to those found in wild-type cells subjected to nitrogen deprivation, even if ammonium is present in the medium (Franco *et al.* 1987*a*).

Cellular nitrogen status is an important factor regulating inorganic nitrogen uptake. Nitrogen starvation increases both the ability of algal cells to take up nitrate, nitrite, ammonium and methylammonium and the enzyme levels of glutamine synthetase, glutamate synthase and glutamate dehydrogenase (Eppley and Renger 1974; Hipkin and Syrett 1977*a*; Syrett *et al.* 1986; Franco *et al.* 1987*a*). Control of the activities of ammonium-assimilating enzymes by means of the nitrogen status of the cell may be involved in the regulation of nitrate assimilation, as suggested for glutamine synthetase (Cullimore and Sims 1981).

CO_2 starvation under autotrophic conditions inhibits nitrate and nitrite utilization in many green algae (Grant 1968; Grant and Turner 1969). Under conditions of CO_2 deprivation, algal cells can reduce nitrate to nitrite and ammonium, which are excreted and accumulate in the medium in varying amounts (Larsson and Andersson 1981; Azuara and Aparicio 1983, 1984), in a process which is dependent on the pH (Eisele and Ullrich 1977; Di Martino Rigano *et al.* 1985).

Several authors have emphasized the light requirement for the uptake of nitrate, nitrite and ammonium by algae (Grant 1968; Grant and Turner 1969; Syrett 1981; Cresswell and Syrett 1981, 1982), and pointed out the close interaction between the effects of light and carbon metabolism (Thacker and Syrett 1972*b*). In *Chlamydomonas* both nitrate and ammonium can be assimilated in darkness provided that sufficient carbon reserves are available (Syrett 1981). Nitrate uptake appears to be a process which requires energy from either photosynthetic or oxidative origin (Tischner and Lorenzen 1979; Cresswell and Syrett 1981; Florencio and Vega 1982).

In *Cyanidium caldarium* two transport systems for nitrate uptake with different affinities for nitrate have been reported. The high-affinity system takes up nitrate through a mechanism in which two protons are transported across the plasmalemma for each nitrate anion, with the carrier first binding nitrate and then protons (Fuggi 1985). Similar co-transport systems with similar stoichiometries have been previously reported for nitrate uptake in the giant-celled alga *Hydrodictyon africanum* (Raven and De Michelis 1979) and in *Ankistrodesmus* (Eisele and Ullrich 1975).

In *Chara corallina* the net nitrate uptake rate has been shown to be the result of different influx and efflux rates of the ion. Nitrate influx is dependent on Rb^+, Na^+, K^+, and Ca^{2+} in the external medium and is inhibited by FCCP and by variations in pH and Cl^- concentrations (Deanne-Drummond 1984*b,c,d*). Net nitrate uptake into *Chara* is completely prevented by ammonium even at relatively low concentrations, so that the stimulation of nitrate efflux by ammonium is the mechanism responsible for the immediate effects of ammonium on the net uptake of nitrate into *Chara* (Deanne-Drummond 1984*c*).

Control of nitrate reductase activity

Short-term variation in NR activity levels has been described in eukaryotic algae. In many green algae NR can undergo a reversible inactivation process of redox nature both *in vivo* by addition of ammonium to the cells (Losada *et al.* 1970) and *in vitro* by incubation of enzyme preparations with NAD(P)H alone or NAD(P)H plus ADP (Losada 1976; Guerrero *et al.* 1981), superoxide radicals (Chaparro *et al.* 1979; De la Rosa *et al.* 1981*b*); or cyanide (Gewitz *et al.* 1974); in all cases, the inactive NR can be reactivated *in vitro* by oxidation with ferricyanide. Light seems to influence the redox state and thus the activity of NR, through a mechanism probably involving the flavin prosthetic group of the enzyme (Aparicio *et al.* 1976). In *Chlamydomonas* blue, but not red, light reversibly reactivates inactive NR *in vivo* (Maldonado and Aparicio 1987).

In order to explain NR redox inactivation, Losada and co-workers have proposed that the role of ammonium in reversibly inactivating NR results from its uncoupling action on photosynthetic photophosphorylation through an increase of intracellular levels of ADP and reducing power. The active form of NR in its oxidized state becomes reversibly inactive in response to changes in the energy charge and intracellular levels of reduced pyridine nucleotides (Guerrero *et al.* 1981 and references therein).

On the other hand, Vennesland's group found that the inactive form of NR *in vivo* is a cyanide complex of the reduced enzyme, generated through an indirect effect of ammonium on nitrate uptake (Vennesland and Guerrero 1979). This cyanide complex is the only major inactive form of NR in *Chlorella vulgaris* (Sherman *et al.* 1983). Nitrate reactivates this inactive NR enzyme in a process which may be in part dependent on *de novo* protein synthesis (Pistorius *et al.* 1976).

In *Ankistrodesmus*, inactivation of NR requires a one-electron reduction of the enzyme molecule and its further interaction with superoxide or cyanide, with the interconversion being centred in the molybdenum prosthetic group of the enzyme (De la Rosa *et al.* 1981*b*).

Using the *Chlamydomonas* mutant 305 (which lacks NAD(P)H-diaphorase activity), it has been reported that *in vitro* an active diaphorase moiety of NR is required for the enzyme inactivation by NAD(P)H, whereas *in vivo* inactivation of NR occurs in the absence of intracellular nitrate and can be performed by endogenous cellular reductants other than NAD(P)H (Córdoba *et al.* 1985).

When nitrate is removed from *Chlamydomonas* cultures, or its uptake blocked, NR undergoes *in vivo* redox inactivation in a process slower than the rapid inhibition of nitrate uptake (Florencio and Vega 1982,1983*b*). Thus it seems that nitrate transport system and not NR activity controls the quantities of nitrate nitrogen incorporated by the cells, and that NR inhibition is a consequence of the disappearance of intracellular nitrate because of an inhibited nitrate uptake. At present the physiological role for NR redox interconversion is not well understood. Very recently, it has been shown that the reversible inactivation of NR in *Chlamydomonas* is closely linked to the regulation of the intracellular enzyme levels (Franco *et al.* 1987*b*). In fact, *in vivo* inactivation of NR in wild-type and mutant strains leads to a rapid degradation of the enzyme through a 7- to 8-fold decrease in its half-life. In addition, the NR-diaphorase subunit in mutant 104, which is incapable of undergoing redox interconversion, is not degraded under any of the conditions tested (Franco *et al.* 1987*b*). It has been proposed that the reversibly inactivated form is the main target of an NR degradation system. This hypothesis may explain the existence of high NR levels in *Chlamydomonas* cells exposed to either cyanate (Florencio and Vega 1983*a*) or blue light (Azuara and Aparicio 1983), which are reactivating agents for the inactive NR. In *Cyanidium caldarium* the mechanism of the ammonium-promoted redox inactivation seems to include an inhibition site sensitive to heat and mild treatments (Rigano and Violante 1972; Rigano *et al.* 1980).

Another type of reversible inactivation has been described in *Chlorella* cultures synchronized by light/dark cycles (Hodler *et al.* 1972; Tischner 1976; Griffiths 1979). NR and NiR activities increase rapidly during the light period and decrease during the dark phase; this rapid increase of NR activity after illumination is due to the reactivation of an inactive enzyme form, rather than to *de novo* synthesis, by a mechanism different from the oxidation/activation discussed above (Tischner and Hütterman 1978; Tischner 1984). The NR of *Chlorella sorokiniana* is reversibly inactivated by L-cysteine through an inactivation factor that is thought to be a protein (Tischner and Schmidt 1984). NR from *Chlorella vulgaris* is sensitive *in vitro* to NR-inactivating proteins from corn roots and rice cell suspensions (Yamaya *et al.* 1980), which suggests that similar inactivating proteins may exist in algae.

Control of nitrate reductase and nitrite reductase enzyme levels

In eukaryotic algae, intracellular levels of either NR or NiR are high in the presence of nitrate or nitrite. These enzymes appear to be ammonium-repressible rather than nitrate- or nitrite-inducible since their synthesis can also take place under non-inducing conditions (Guerrero *et al.* 1981; Syrett 1981; Wheeler and Weidner 1983). Synthesis of NR in N-free medium has been reported in *Chlorella* (Morris and Syrett 1963; Vega *et al.* 1971; Funkhouser and Garay 1981), *Cyanidium* (Rigano and Violante 1973), *Ankistrodesmus* (Syrett and Hipkin 1973; Díez *et al.* 1977), and *Chlamydomonas* (Herrera *et al.* 1972; Hipkin *et al.* 1980; Florencio and Vega 1983*a*; Franco *et al.* 1987*b*). In *Chlorella vulgaris* (Di Martino Rigano *et al.* 1983) and *Chlamydomonas* (Hipkin *et al.* 1980, Fernández and Cárdenas 1982*b*), very low but measurable NR levels are induced in nitrogen-free medium, nitrate being required to produce high enzyme levels. In contrast, nitrogen starvation failed to induce NR of *Chlorella sorokiniana* cells (Tischner and

Lorenzen 1980). The synthesis of NR in N-free medium has been attributed to nitrate formation by cells (Kessler and Oesterheld 1970; Spiller *et al.* 1976; Funkhouser and Garay 1981), although this view is disputed by others (Syrett and Hipkin 1973; Hipkin *et al.* 1980). It seems most likely that in algae the major effect of nitrogen starvation is to eliminate the ammonium repression of NR. Synthesis of NiR in N-free medium has also been reported in *Chlorella* (Losada *et al.* 1970) and *Chlamydomonas* (Fernández and Cárdenas 1982*b*).

Light appears to be required for the synthesis of NR (Tischner 1976; Fischer and Simonis 1979; Griffiths 1979), although in *Chlamydomonas* a considerable level of this enzyme is present in cells grown in the dark under mixotrophic conditions (Thacker and Syrett 1972*a*; Florencio and Vega 1983*a*). The effect of light on NR synthesis can be related to the availability of suitable carbon and energy sources (Thacker and Syrett 1972*b*).

Ammonium represses NR in eukaryotic algae when present in culture media as the sole nitrogen source or in medium containing both ammonium and nitrate (Losada *et al.* 1970; Vega *et al.* 1971; Herrera *et al.* 1972; Syrett and Hipkin 1973; Solomonson *et al.* 1973; Morris 1974; Syrett and Leftley 1976; Pistorius *et al.* 1978; Rigano *et al.* 1979; Florencio and Vega 1983*a*). In addition, NR induction in algae is inhibited by the protein synthesis inhibitor cycloheximide (Losada *et al.* 1970; Vega *et al.* 1971; Hipkin and Syrett 1977*b*; Funkhouser and Ramadoss 1980; Funkhouser *et al.* 1980). However, a preformed mRNA for NR synthesis appears to be present in ammonium grown cells of *Chlorella, Ankistrodesmus,* and *Dunaliella* since the appearance of NR, after transfer to either nitrate- or N-deprived medium, is not inhibited by 6-methylpurine, an inhibitor of RNA synthesis (Hipkin and Syrett 1977*b*).

Ammonium-grown cells of *Chlorella vulgaris* contain a precursor protein immunologically related to NR (Funkhouser and Ramadoss 1980). Moreover, cycloheximide completely inhibits NR synthesis in cells transferred from ammonium to nitrate medium, but only if the inhibitor is added at the start of the induction period (Funkhouser *et al.* 1980). This suggests that in *Chlorella* NR synthesis involves first the synthesis of an inactive protein, which can occur in ammonium-grown cells, followed by the conversion of the inactive precursor to active enzyme, by means of another 'activator' protein. The synthesis of this putative 'maturase' takes place in the absence of ammonium, and is probably induced by added nitrate (Funkhouser and Ramadoss 1980).

In the presence of nitrate and in the absence of molybdate and tungstate, *Chlorella* cells develop high levels of a demolybdo-NR, with the full cytochrome *c*-reducing capacity of normal NR, but with very little capacity to reduce nitrate (Vega *et al.* 1971; Gewitz *et al.* 1981). Under defined conditions, the demolybdo-NR can be converted to active NR by inserting molybdate both *in vitro* (Ramadoss *et al.* 1981) and *in vivo* in the absence of *de novo* protein synthesis (Solomonson and Spehar 1977), which, however, does not preclude the possibility that another protein may participate in the NR activation by molybdate.

In *Chlamydomonas* cells grown in ammonium or incubated with methylammonium, 6-methylpurine prevents NR de-repression in nitrate medium indicating that, in contrast to *Chlorella*, *Chlamydomonas* cells do not contain preformed NR mRNA in ammonium medium; NR mRNA, however, is present in nitrogen-starved cells (A. R. Franco, unpublished Thesis 1986). In addition, cycloheximide completely inhibits NR synthesis when added at the beginning or at different times of the de-repression treatment (A. R. Franco, personal communication), which rules out the existence of a 'maturase' for possible NR precursors in *Chlamydomonas*.

In order to study ammonium repression of NR in *Chlamydomonas*, Franco *et al.* (1984*b*) used the ammonium analogue methylammonium, which is not metabolized by *Chlamydomonas* and mimics all ammonium effects on regulation of nitrate assimilation (Díez *et al.* 1977; Franco *et al.* 1984*b*, 1987*a*; A. R. Franco, unpublished Thesis 1986). Methylammonium is actively taken up by *Chlamydomonas* cells and transformed into a single product identified as γ-N-methylglutamine (Franco *et al.* 1984*b*). NR de-repression in methylammonium-treated cells is always accompanied by a decrease in the intracellular levels of methylammonium, because of its conversion to methylglutamine which accumulates in the cells. Thus, ammonium (methylammonium) itself has been proposed to be the co-repressor molecule of NR when its intracellular concentrations are above a threshold level. However, *Chlamydomonas* cells growing under non-saturating CO_2 tensions (Azuara and Aparicio 1983) show NR activity in the presence of ammonium in the medium. This apparent contradiction can be explained by considering that, in addition to ammonium, ammonium-derived metabolites are also required for NR repression under stressed conditions, i.e. with a limiting supply of nutrients; under such limiting conditions, the formation of these metabolites would be prevented.

NR mRNA is present in ammonium or methylammonium medium, in mutant *ma-1* of *Chlamydomonas*, but appearance of the NR activity requires the presence of nitrate (A. R. Franco, unpublished Thesis 1986). The low intracellular ammonium concentrations in the mutant can be increased experimentally by increasing either the pH of the medium or the extracellular ammonium concentration, which is accompanied by a progressive inhibition of expression of the constitutive NR mRNA in nitrate medium (A. R. Franco, unpublished thesis 1986). Inhibition of the rate of NR synthesis by ammonium had been previously suggested by Hipkin *et al.* (1980).

In ammonium-grown wild-type cells of *Chlamydomonas*, NR is actively synthesized after transfer to N-free medium but, after about 1 h, the enzyme undergoes spontaneously reversible inactivation, followed by degradation (Franco *et al.* 1987*b*). The structural mutant 305, like mutant *nit A* (Hipkin *et al.* 1980), also synthesizes NR in N-free medium, but the enzyme remains in the active form and its degradation does not take place, resulting in very high levels of the enzyme in the cells. Thus, nitrate does not seem to be required for the expression of NR mRNA but rather maintains the enzyme in its active form, preventing its degradation. Since the regulation of enzyme levels is different in the wild type *vs* mutants with altered enzyme structure, it was proposed that the integrity of the NR complex plays an important role in its own regulation (Fernández and Cárdenas 1982*b*). This autoregulation of *Chlamydomonas* NR levels appears to depend on the NR redox inactivation properties which are different in the wild type and mutant strains (Córdoba *et al.* 1985; Franco *et al.* 1987*b*).

Figure 8.1 shows a working model of the regulation of NR enzyme levels in *Chlamydomonas*. Ammonium itself is the co-repressor of NR under saturating conditions with respect to other cell nutrients. Once NR mRNA has been synthesized, intracellular ammonium levels seem to inhibit the expression of NR apoprotein needed to assemble the active enzyme. Nitrate stabilizes NR into the active form which, in the absence of nitrate, interconverts to inactive NR. This enzyme form has a very short half-life and degrades readily. Ammonium does not seem to activate the degradation of NR (Franco *et al.* 1987*b*).

It has been shown that nitrate stimulates the turnover rate of NR in *Chlorella*, and that after transferring the cells from ammonium to nitrate medium NR activity increases rapidly (Johnson 1979). This increase is caused by the activation of an inactive enzyme form present in ammonium medium, which is

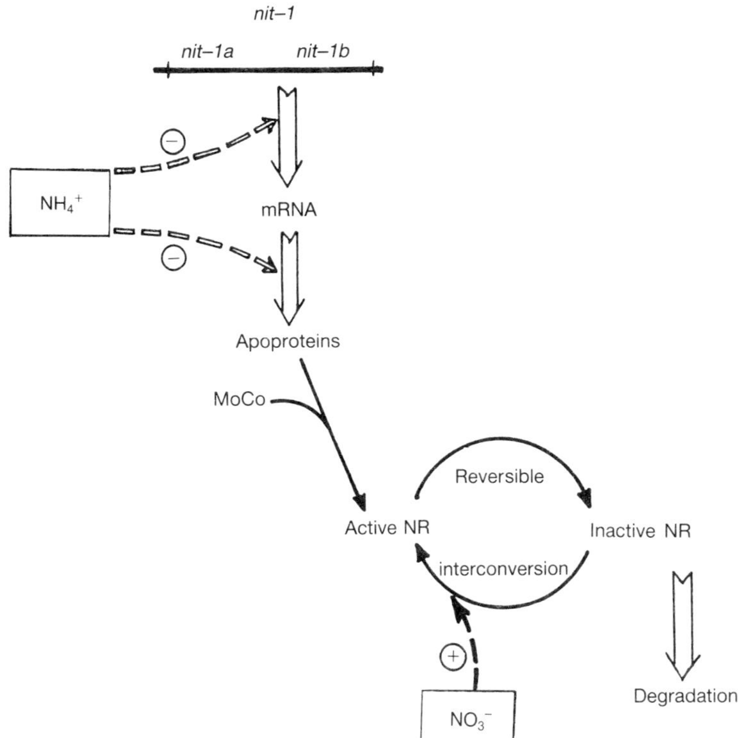

Fig. 8.1. Model for the regulation of NR levels in *Chlamydomonas reinhardtii.*

very stable *in vivo*. The clear differences in NR regulation in *Chlamydomonas* and *Chlorella* strongly suggest that at least two regulatory circuits are present in algae.

The regulation of NiR synthesis in algae has been less well studied. In *Chlorella*, NiR levels are high in cells grown phototrophically with nitrate or nitrite, as well as in ammonium-grown cells transferred to N-free medium (Losada *et al.* 1970; Cárdenas 1972). Like NR, NiR formation in diatoms (Eppley and Rogers 1970; Llama *et al.* 1979), *Platymonas striata* (Ricketts and Edge 1978), *Chlorella* (Losada *et al.* 1970) and *Chlamydomonas* (Herrera *et al.* 1972; Fernández and Cárdenas 1982*b*) is repressed in the presence of ammonium. In *Chlamydomonas* cells grown with ammonium a small amount of NiR is detected and thus the enzyme has been proposed to be semiconstitutive (F. Córdoba, unpublished Thesis 1985).

Acknowledgements

Authors wish to thank Dr Pete Lefebvre for encouragement and critical reading of the manuscript. The financial support from C.A.I.C.Y.T. (no. 1834), Junta de Andalucía (no. 111 and bolsa de viaje to J.C.) and F.I.S. (no. 86/624) (Spain) is gratefully acknowledged.

References

Aparicio, P. J., Roldán, J. M., and Calero, F. (1976). Blue light photoreactivation of nitrate reductase from green algae and higher plants. *Biochemical and Biophysical Research Communications* **70**, 1071–7.

Azuara, M. P. and Aparicio, P. J. (1983). *In vivo* blue-light activation of *Chlamydomonas reinhardtii* nitrate reductase. *Plant Physiology* **71**, 286–90.

Azuara, M. P. and Aparicio, P. J. (1984). Effects of quality of light, CO_2 tensions and NO_3^- concentrations on the inorganic nitrogen metabolism of *Chlamydomonas reinhardtii*. *Photosynthesis Research* **5**, 97–103.

Bagchi, S. N., Rai, U. N., Rai, A. N., and Singh, H. N. (1985). Nitrate metabolism in the cyanobacterium *Anabaena cycadeae*: regulation of nitrate uptake and reductase by ammonia. *Physiologia Plantarum* **63**, 322–6.

Barea, J. L. and Cárdenas, J. (1975). The nitrate-reducing system of *Chlamydomonas reinhardtii*. *Archives of Microbiology* **105**, 21–5.

Bilbao, M. M., Gabas, J. M., and Serra, J. L. (1981). Inhibition of nitrate uptake in the diatom *Phaeodactylum tricornutum* by nitrate, ammonium ions and some L-aminoacids. *Biochemical Society Transactions* **9**, 476–7.

Calero, F., Ullrich, W. R., and Aparicio, P. J. (1980). Regulation by monochromatic light of nitrate uptake in *Chlorella fusca*. In *The blue light syndrome* (ed. H. Senger), pp. 411–21. Springer-Verlag, Berlin.

Cárdenas, J. (1972). Nitrito reductasa de plantas superiores y algas. Ph.D. Thesis, University of Sevilla, Spain.

Chaparro, A., De la Rosa, M. A., and Vega, J. M. (1979). Involvement of oxygen in *Chlorella fusca* nitrate reductase inactivation by reduced nicotinamide adenine dinucleotide. *Zeitschrift für Pflanzenphysiologie* **95**, 77–85.

Córdoba, F. (1985). Regulación de la asimilación de nitrógeno inorgánico en *Chlamydomonas reinhardtii*. Ph.D. Thesis, University of Córdoba, Spain.

Córdoba, F., Cárdenas, J., and Fernández, E. (1985). Role of the diaphorase moiety on the reversible inactivation of the *Chlamydomonas reinhardtii* nitrate reductase complex. *Biochimica et Biophysica Acta* **827**, 8–13.

Córdoba, F., Cárdenas, J., and Fernández, E. (1986). Kinetic characterization of nitrite uptake and reduction by *Chlamydomonas reinhardtii*. *Plant Physiology* **82**, 904–8.

Córdoba, F., Cárdenas, J. and Fernández, E. (1987). Cooperative regulation by ammonium and ammonium derivatives of nitrite uptake in *Chlamydomonas reinhardtii*. *Biochimica et Biophysica Acta* **902**, 287–92.

Cove, D. J. (1976). Chlorate toxicity in *Aspergillus nidulans*. Studies with mutants altered in nitrate assimilation. *Molecular and General Genetics* **146**, 147–59.

Cove, D. J. (1979). Genetic studies of nitrate assimilation in *Aspergillus nidulans*. *Biological Review* **54**, 291–327.

Cresswell, R. C. and Syrett, P. J. (1979). Ammonium inhibition of nitrate uptake by the diatom *Phaeodactylum tricornutum*. *Plant Science Letters* **14**, 321–5.

Cresswell, R. C. and Syrett, P. J. (1981). Uptake of nitrate by the diatom *Phaeodactylum tricornutum*. *Journal of Experimental Botany* **32**, 19–25.

Cresswell, R. C. and Syrett, P. J. (1982). The uptake of nitrite by the diatom *Phaeodactylum*: interactions between nitrite and nitrate. *Journal of Experimental Botany* **33**, 1111–21.

Cresswell, R. C. and Syrett, P. J. (1984). Effects of methylammonium and of L-methionine-DL-sulfoximine on the growth and nitrogen metabolism of *Phaeodactylum tricornutum*. *Archives of Microbiology* **139**, 67–71.

Cullimore, J. V. and Sims, A. P. (1981). Glutamine synthetase of *Chlamydomonas*: its role in the control of nitrate assimilation. *Planta* **153**, 18–24.

De la Rosa, M. A., Vega, J. M., and Zumft, W. G. (1981*a*). Composition and structure of assimilatory nitrate reductase from *Ankistrodesmus braunii*. *Journal of Biological Chemistry* **256**, 5814–19.

De la Rosa, M. A., Gómez-Moreno, C., and Vega, J. M. (1981*b*). Interconversion of nitrate reductase from *Akistrodesmus braunii* related to redox changes. *Biochimica et Biophysica Acta* **662**, 77–85.

De Vries, S. E., Dirks, R., Mendel, R. R., Schaart, J. G., and Feenstra, W. J. (1986). Biochemical characterization of some nitrate reductase deficient mutants of *Nicotiana plumbaginifolia*. *Plant Science* **44**, 105–10.

Deane-Drummond, C. E. (1984*a*). The apparent induction of nitrate uptake by *Chara corallina* cells following pretreatment with or without nitrate and chlorate. *Journal of Experimental Botany* **35**, 1182–93.

Deane-Drummond, C. E. (1984*b*). Nitrate transport into *Chara corallina* cells using $^{36}ClO_3^-$ as an analogue for nitrate. I. Interaction between $^{36}ClO_3^-$ and NO_3^- and characterization of $^{36}ClO_3^-/NO_3^-$ influx. *Journal of Experimental Botany* **35**, 1289–98.

Deane-Drummond, C. E. (1984*c*). Nitrate transport into *Chara corallina* cells using $^{36}ClO_3^-$ as an analogue for nitrate. II. Comparison with ^{14}C methylamine fluxes at different pH and NH_4^+/NO_3^- interactions. *Journal of Experimental Botany* **35**, 1299–1308.

Deane-Drummond, C. E. (1984*d*). Nitrate transport into *Chara corallina* cells using $^{36}ClO_3^-$ as an analogue for nitrate. III. Interactions between chloride and nitrate transport processes. *Journal of Experimental Botany* **35**, 1733–43.

Di Martino Rigano, V., Vona, V., Fuggi, A., Di Martino, C., and Rigano, C. (1982). Effect of L-methionine-DL-sulfoximine, a specific inhibitor of glutamine synthetase, on ammonium and nitrate metabolism in the unicellular alga *Cyanidium caldarium*. *Physiologia Plantarum* **54**, 47–51.

Di Martino Rigano, V., Vona, V., Fuggi, A., Di Martino, C., and Rigano, C. (1983). Regulation of nitrate reductase in *Chlorella*. Nitrate requirement for the appearance of nitrate reductase activity. *Plant Science Letters* **28**, 265–72.

Di Martino Rigano, V., Martello, A., Di Martino, C., and Rigano, C. (1985). Effect of CO_2 and phosphate deprivation on the control of nitrate, nitrite and ammonium metabolism in *Chlorella*. *Physiologia Plantarum* **63**, 241–6.

Díez, J., Chaparro, A., Vega, J. M., and Relimpio, A. M. (1977). Studies on the regulation of assimilatory nitrate reductase in *Ankistrodesmus braunii*. *Planta* **137**, 231–4.

Dunn-Coleman, N. S. (1984). Biochemical characterization of the molybdenum cofactor mutants of *Neurospora crassa*: *in vivo* and *in vitro* reconstitution of NADPH-nitrate reductase activity. *Current Genetics* **8**, 581–8.

Dunn-Coleman, N. S., Smarelli, J., Jr., and Garrett, R. H. (1984). Nitrate assimilation in cukaryotic cells. *International Review of Cytology* **92**, 1–50.

Eisele, R. and Ullrich, W. R. (1975). Stoichiometry between photosynthetic nitrate reduction and alkalinization by *Ankistrodesmus braunii in vivo*. *Planta* **123**, 117–23.

Eisele, R. and Ullrich, W. R. (1977). Effect of glucose and CO_2 on nitrate uptake and coupled OH^- flux in *Ankistrodesmus braunii*. *Plant Physiology* **59**, 18–21.

Eppley, R. W. and Renger, E. H. (1974). Nitrogen assimilation of an oceanic diatom in nitrogen-limited continuous culture. *Journal of Phycology* **10**, 15–23.

Eppley, R. W. and Rogers, J. N. (1970). Inorganic nitrogen assimilation of *Ditylum brightwellii*. *Journal of Phycology* **6**, 344–51.

Fernández, E. and Aguilar, M. (1987). Molybdate repair of molybdopterin deficient mutants from *Chlamydomonas reinhardtii*. *Current Genetics* **12**, 349–55.

Fernández, E. and Cárdenas, J. (1981a). *In vitro* complementation of assimilatory NAD(P)H-nitrate reductase from mutants of *Chlamydomonas reinhardtii*. *Biochimica et Biophysica Acta* **657**, 1–12.

Fernández, E. and Cárdenas, J. (1981b). Occurrence of xanthine dehydrogenase in *Chlamydomonas reinhardtii*: a common cofactor shared by xanthine dehydrogenase and nitrate reductase. *Planta* **153**, 254–7.

Fernández, E. and Cárdenas, J. (1982a). Biochemical characterization of a singular mutant of nitrate reductase from *Chlamydomonas reinhardtii*. New evidence for a heteropolymeric enzyme structure. *Biochimica et Biophysica Acta* **681**, 530–7.

Fernández, E. and Cárdenas, J. (1982b). Regulation of the nitrate-reducing enzymes in wild and mutant strains of *Chlamydomonas reinhardtii*. *Molecular and General Genetics* **186**, 164–9.

Fernández, E. and Cárdenas, J. (1983a). Isolation and properties of the NAD(P)H-cytochrome *c* reductase subunit of *Chlamydomonas reinhardtii* NAD(P)H-nitrate reductase. *Biochimica et Biophysica Acta* **745**, 12–19.

Fernández, E. and Cárdenas, J. (1983b). Isoelectric focusing of the NAD(P)H-cytochrome *c* reductase subunit of *Chlamydomonas reinhardtii* nitrate reductase. *Zeitschrift für Naturforschung* **38c**, 35–8.

Fernández, E. and Cárdenas, J. (1983c). Nitrate reductase from a mutant strain of *Chlamydomonas reinhardii* incapable of nitrate assimilation. *Zeitschrift für Naturforschung* **38c**, 439–45.

Fernández, E. and Matagne, R. F. (1984). Genetic analysis of nitrate reductase-deficient mutants in *Chlamydomonas reinhardtii*. *Current Genetics* **8**, 635–40.

Fernández, E. and Matagne, R. F. (1986). *In vivo* complementation analysis of nitrate reductase-deficient mutants in *Chlamydomonas reinhardtii*. *Current Genetics* **10**, 397–403.

Fischer, S. and Simonis, W. (1979). Tagesperiodische Schwankungen und lichtinduzierte

Zunahme der Nitratreduktase-Aktivität bei Synchronkulturen von *Ankistrodesmus braunii*. *Zeitschrift für Pflanzenphysiologie* **92**, 143–52.

Florencio, F. J. and Vega, J. M. (1982). Regulation of the assimilation of nitrate in *Chlamydomonas reinhardtii*. *Phytochemistry* **21**, 1195–200.

Florencio, F. J. and Vega, J. M. (1983*a*). Regulation of the synthesis of the NAD(P)H-nitrate reductase complex in *Chlamydomonas reinhardtii*. *Zeitschrift für Pflanzenphysiologie* **111**, 223–32.

Florencio, F. J. and Vega, J. M. (1983*b*). Utilization of nitrate, nitrite and ammonium by *Chlamydomonas reinhardii*. Photoproduction of ammonium. *Planta* **158**, 288–93.

Franco, A. R. (1986). Estudio de mutantes del alga verde *Chlamydomonas reinhardtii* afectados en la asimilación del nitrógeno inorgánico. Ph.D. Thesis, University of Cordoba, Spain.

Franco, A. R., Cárdenas, J., and Fernández, E. (1984*a*). Heteromultimeric structure of the nitrate reductase complex of *Chlamydomonas reinhardtii*. *EMBO Journal* **3**, 1403–7.

Franco, A. R., Cárdenas, J., and Fernández, E. (1984*b*). Ammonium-(methylammonium) is the co-repressor of nitrate reductase in *Chlamydomonas reinhardtii*. *FEBS Letters* **176**, 453–6.

Franco, A. R., Cárdenas, J., and Fernández, E. (1987*a*). A mutant of *Chlamydomonas reinhardtii* altered in the transport of ammonium and methylammonium. *Molecular and General Genetics* **206**, 414–18.

Franco, A. R., Cárdenas, J. and Fernández, E. (1987*b*). Involvement of reversible inactivation in the regulation of nitrate reductase enzyme level in *Chlamydomonas reinhardtii*. *Plant Physiology* **84**, 665–9.

Franco, A. R., Cárdenas, J., and Fernández, E. (1988). Two different carriers transport both ammonium and methylammonium in *Chlamydomonas reinhardtii*. *Journal of Biological Chemistry* **263**, 14039–43.

Fuggi, A. (1985). Mechanism of proton-linked nitrate uptake in *Cyanidium caldarium*, an acidophilic non-vacuolated alga. *Biochimica et Biophysica Acta* **815**, 392–8.

Funkhouser, E. A. and Garay, A. S. (1981). Appearance of nitrate in soybean seedlings and *Chlorella* caused by nitrogen starvation. *Plant and Cell Physiology* **22**, 1279–86.

Funkhouser, E. A. and Ramadoss, C. S. (1980). Synthesis of nitrate reductase in *Chlorella*. II. Evidence for synthesis in ammonia-grown cells. *Plant Physiology* **65**, 944–8.

Funkhouser, E. A., Shen, T.-C., and Ackermann, R. (1980). Synthesis of nitrate reductase in *Chlorella*. I. Evidence for an inactive protein precursor. *Plant Physiology* **65**, 939–43.

Gabard, J., Marion-Poll, A., Cherel, I., Meyer, C., Muller, A., and Caboche, M. (1987). Isolation and characterization of *Nicotiana plumbaginifolia* nitrate reductase-deficient mutants: genetics and biochemical analysis of the NIA complementation group. *Molecular and General Genetics* **209**, 596–606.

Garrett, R. H. and Cove, D. J. (1976). Formation of NADPH-nitrate reductase activity *in vitro* from *Aspergillus nidulans nia D* and *cnx* mutants. *Molecular and General Genetics* **149**, 179–86.

Gewitz, H. S., Piefke, J, and Vennesland, B. (1981). Purification and characterization of demolybdo nitrate reductase (NADH-cytochrome *c* oxidoreductase) of *Chlorella vulgaris*. *Journal of Biological Chemistry* **256**, 11527–31.

Gewitz, H. S., Lorimer, G. H., Solomonson, L. P., and Vennesland, B. (1974). Presence of HCN in *Chlorella vulgaris* and its possible role in controlling the reduction of nitrate. *Nature* **249**, 79–81.

Grant, B. R. (1968). Effect of carbon dioxide concentration and buffer system on nitrate and nitrite assimilation in *Dunaliella tertiolecta*. *Journal of General Microbiology* **54**, 327–36.

Grant, B. R. and Turner, I. M. (1969). Light stimulated nitrate and nitrite assimilation in several species of algae. *Comparative Biochemistry and Physiology* **29**, 995–1004.

Griffiths, D. J. (1979). Factors affecting nitrate reductase activity in synchronous cultures of *Chlorella*. *New Phytologist* **82**, 427–37.

Guerrero, M. G., Vega, J. M., and Losada, M. (1981). The assimilatory nitrate-reducing system and its regulation. *Annual Review of Plant Physiology* **32**, 169–204.

Herrera, J., Paneque, A., Maldonado, J. M., Barea, J. L., and Losada, M. (1972). Regulation by ammonia of nitrate reductase synthesis and activity in *Chlamydomonas reinhardtii*. *Biochemical and Biophysical Research Communications* **48**, 996–1003.

Hewitt, E. J. and Notton, B. A. (1980). Nitrate reductase systems in eukaryotic and prokaryotic organisms. In *Molybdenum and molybdenum-containing enzymes* (ed. M. Coughlan), pp. 273–325. Pergamon Press, Oxford.

Hipkin, C. R. and Syrett, P. J. (1977a). Some effects of nitrogen-starvation on nitrogen and carbohydrate metabolism in *Ankistrodesmus braunii*. *Planta* **133**, 209–14.

Hipkin, C. R. and Syrett, P. J. (1977b). Post-transcriptional control of nitrate reductase formation in green algae. *Journal of Experimental Botany* **28**, 1270–7.

Hipkin, C. R., Al-Bassam, B. A., and Syrett, P. J. (1980). The roles of nitrate and ammonium in the regulation of the development of nitrate reductase in *Chlamydomonas reinhardtii*. *Planta* **150**, 13–18.

Hipkin, C. R., Hermann-Smith, J. A., and Syrett, P. J. (1985). The roles of nitrate, photosynthesis and protein turnover in the formation of nitrate reductase in the *nit A* mutant of *Chlamydomonas*. *Biochimica et Biophysica Acta* **838**, 191–6.

Ho, C.-H., Ikawa, T., and Nikizawa, K. (1976). Nitrite-reducing activity of modified cytochrome c_{553} from the red alga *Porphyra yeoyensis* Ueda. *Plant and Cell Physiology* **17**, 417–30.

Hodler, M., Morgenthaler, J. J., Eichenberger, W., and Grob, E. C. (1972). The influence of light on the activity of nitrate reductase in synchronous cultures of *Chlorella pyrenoidosa*. *FEBS Letters* **28**, 19–21.

Howard, D. W. and Solomonson, L. P. (1982). Quarternary structure of assimilatory NADH: nitrate reductase from *Chlorella*. *Journal of Biological Chemistry* **257**, 10243–50.

Huskey, R. J., Semenkovich, C. F., Griffin, B. E., Cecil, P. O., Callahan, A. M., Chace, K. V., and Kirk, D. L. (1979). Mutants of *Volvox carteri* affecting nitrogen assimilation. *Molecular and General Genetics* **169**, 157–61.

Johnson, C. B. (1979). Activation, synthesis and turnover of nitrate reductase controlled by nitrate and ammonium in *Chlorella vulgaris*. *Planta* **147**, 63–8.

Johnson, J. L. (1980). The molybdenum cofactor common to nitrate reductase, xanthine dehydrogenase and sulphite oxidase. In *Molybdenum and molybdenum-containing enzymes* (ed. M. Coughlan), pp. 345–383. Pergamon Press, Oxford.

Kessler, E. and Oesterheld, H. (1970). Nitrification and induction of nitrate reductase in nitrogen-deficient green algae. *Nature* **228**, 287–8.

Knobloch, O. and Tischner, R. (1986). Regulation of nitrate reductase synthesis investigated by using mutants of *Chlorella sorokiniana* partially NR-deficient. *Proceedings of the Advanced course on inorganic nitrogen metabolism* (eds. P. J. Aparicio, P. J. Syrett, W. R. Ullrich, and F. Castillo), p. 16. Jarandilla, Spain.

Kramer, S., Hageman, R. V., and Rajagopalan, K. V. (1984). *In vitro* reconstruction of nitrate reductase activity of the *Neurospora crassa* mutant *nit-1*: specific incorporation of molybdopterin, *Archives of Biochemistry and Biophysics* **233**, 821–9.

Larsson, C. M. and Andersson, M. (1981). Uptake and photoreduction of NO_3^- and NO_2^- in *Scenedesmus*: interactions with CO_2 fixation. In *Photosynthesis IV. Regulation of carbon metabolism* (ed. G. Akoyunoglou), pp. 741–50. Balaban, Philadelphia.

Latorella, A. H., Bromberg, S. K., Lieber, K., and Robinson, J. (1981). Isolation and partial characterization of nitrate assimilation mutants of *Dunaliella tertiolecta* (Chlorophyceae). *Journal of Phycology* **17**, 211–14.

Llama, M. J., Macarulla, J. M. and Serra, J. L. (1979). Characterization of the nitrate reductase activity in the diatom *Skeletonema costatum*. *Plant Science Letters* **14**, 169–75.

López-Ruíz, A., Verbelen, J. P., Roldán, J. M., and Díez, J. (1985). Nitrate reductase of green algae is located in the pyrenoid. *Plant Physiology* **79**, 1006–10.

Losada, M. (1976). Metalloenzymes of the nitrate-reducing system. *Journal of Molecular Catalysis* **1**, 245–64.

Losada, M., Paneque, A., Aparicio, P. J., Vega, J. M., Cárdenas, J., and Herrera, J. (1970). Inactivation and repression by ammonium of the nitrate reducing system in *Chlorella*. *Biochemical and Biophysical Research Communications* **38**, 1009–15.

Maldonado, J. M. and Aparicio, P. J. (1987). Photoregulation of nitrate assimilation in eukaryotic organisms. In *Inorganic nitrogen metabolism* (eds W. R. Ullrich, P. J. Aparicio, P. J. Syrett, and F. Castillo), pp. 76–81. Springer-Verlag, Berlin.

Mendel, R. R. and Müller, A. J. (1978). Reconstitution of NADH-nitrate reductase *in vitro* from nitrate reductase-deficient *Nicotiana tabacum* mutants. *Molecular and General Genetics* **161**, 77–80.

Mendel, R. R., Marton, L. and Müller, A. J. (1986). Comparative biochemical characterization of mutants at the nitrate reductase/molybdenum cofactor loci *cnx A*, *cnx B*, and *cnx C* of *Nicotiana plumbaginifolia*. *Plant Science* **43**, 125–9.

Mendel, R. R., Alikulov, Z. A., L'vov, P., and Müller, A. J. (1981). Presence of the molybdenum-cofactor in nitrate reductase-deficient mutant cell lines of *Nicotiana tabacum*. *Molecular and General Genetics* **181**, 395–9.

Morris, I. (1974). Nitrogen assimilation and protein synthesis. In *Algal physiology and biochemistry* (ed. W. D. P. Stewart), pp. 583–609. Blackwell, Oxford.

Morris, I. and Syrett, P. J. (1963). The development of nitrate reductase in *Chlorella* and its repression by ammonium. *Archives of Mikrobiology* **47**, 32–41.

Nason, A., Antoine, A. D., Ketchum, P. A., Frazier, W. A., III, and Lee, D. K. (1970). Formation of asssimilatory nitrate reductase by *in vitro* intercistronic complementation in *Neurospora crassa*. *Proceedings of the National Academy of Sciences, USA* **65**, 137–44.

Nason, A., Lee, K. Y., Pan, S. S., Ketchum, P. A., Lamberti, A., and De Vries, J. (1971). *In vitro* formation of assimilatory reduced nicotinamide adenine dinucleotide phosphate: nitrate reductase from a *Neurospora* mutant and a component of molybdenum-enzymes. *Proceedings of the National Academy of Sciences, USA* **68**, 3242–6.

Nichols, G. L. and Syrett, P. J. (1978). Nitrate reductase deficient mutants of *Chlamydomonas reinhardii*. Isolation and genetics. *Journal of General Microbiology* **108**, 71–7.

Nichols, G. L., Shehata, S. A. M., and Syrett, P. J. (1978). Nitrate reductase deficient mutants of *Chlamydomonas reinhardtii*. Biochemical characteristics. *Journal of General Microbiology* **108**, 79–88.

O'Brien, S. J. (ed.) (1986). *Genetic maps*. Cold Spring Harbor Laboratory Press, New York.

Pelley, J. L. and Bannister, T. T. (1979). Methylamine uptake in the green alga *Chlorella pyrenoidosa*. *Journal of Phycology* **15**, 110–12.

Pineda, M., Fernandez, E., and Cárdenas, J. (1984). Urate oxidase of *Chlamydomonas reinhardtii*. *Physiologia Plantarum* **62**, 453–7.

Pineda, M., Cabello, P., and Cárdenas, J. (1987). Ammonium regulation or urate uptake in *Chlamydomonas reinhartii*. *Planta* **171**, 496–500.

Pistorius, E. K., Gewitz, H. S., Voss, H., and Vennesland, B. (1976). Reversible inactivation of nitrate reductase in *Chlorella vulgaris in vivo*. *Planta* **128**, 73–80.

Pistorius, E. K., Funkhouser, E. A., and Voss, H. (1978). Effect of ammonium and ferricyanide on nitrate utilization by *Chlorella vulgaris*. *Planta* **141**, 279–82.

Ramadoss, C. S., Shen, T.-C., and Vennesland, B. (1981). Molybdenum insertion *in vitro* in demolybdo nitrate reductase of *Chlorella vulgaris*. *Journal of Biological Chemistry* **256**, 11532–7.

Raven, J. A. and De Michaelis, M. (1979). Acid base regulation during nitrate assimilation in *Hydrodictyon africanum*. *Plant Cell and Environment* **2**, 245–57.

Ricketts, T. R. and Edge, P. A. (1978). Nitrate and nitrite reductases in *Platymonas striata*, Butcher (Prasinophyceae). *British Phycological Journal* **13**, 167–76.

Rigano, C. and Violante, U. (1972). Effect of heat treatment on the activity *in vitro* of nitrate reductase from *Cyanidium caldarium*. *Biochimica et Biophysica Acta* **256**, 524–32.

Rigano, C. and Violante, U. (1973). Effect of nitrate, ammonia and nitrogen starvation on the regulation of nitrate reductase in *Cyanidium caldarium*. *Archives of Mikrobiology* **90**, 27–33.

Rigano, C., Di Martino Rigano, V., Vona, V., and Fuggi, A. (1979). Glutamine synthetase activity, ammonia assimilation and control of nitrate reduction in the unicellular red alga *Cyanidium caldarium*. *Archives of Microbiology* **121**, 117–20.

Rigano, C., Vona, V., Di Martino Rigano, V., and Fuggi, A. (1980). Active and inactive nitrate reductase. Effects of mild treatments with denaturing agents of protein. *Biochimica et Biophysica Acta* **613**, 26–33.

Romero, L. C., Galván, F., and Vega, J. M. (1987). Purification and properties of the siroheme-containing ferredoxin-nitrite reductase from *Chlamydomonas reinhardtii*. *Biochimica et Biophysica Acta* **914**, 55–63.

Serra, J. L., Llama, M. J., and Cadenas, E. (1978*a*). Nitrate utilization by the diatom *Skeletonema costatum*. I. Kinetics of nitrate uptake. *Plant Physiology* **62**, 987–90.

Serra, J. L, Llama, M. J., and Cadenas, E. (1978*b*). Nitrate utilization by the diatom *Skeletonema costatum*. II. Regulation of nitrate uptake. *Plant Physiology* **62**, 991–4.

Sherman, T. D., Erwin, M. A., and Funkhouser, E. A. (1983). Changes in nitrate reductase protein during loss and gain of nitrate reductase activity in *Chlorella vulgaris*. *Biochimica et Biophysica Acta* **749**, 265–9.

Solomonson, L. P. and Spehar, A. M. (1977). Model for the regulation of nitrate reductase. *Nature* **265**, 373–5.

Solomonson, L. P., Jetschmann, K., and Vennesland, B. (1973). Reversible inactivation of the nitrate reductase from *Chlorella vulgaris* Beijerinck. *Biochimica et Biophysica Acta* **267**, 544–57.

Solomonson, L. P., Barber, M. J., Robbins, A. P., and Oaks, A. (1986). Functional domains of assimilatory NADH: nitrate reductase from *Chlorella*. *Journal of Biological Chemistry* **261**, 11290–4.

Sosa, F. M. and Cárdenas, J. (1977). NADPH as electron donor for nitrate reduction in *Chlamydomonas reinhardtii*. *Zeitschrift für Pflanzenphysiologie* **85**, 171–5.

Sosa, F. M., Ortega, T., and Barea, J. L. (1978). Mutants from *Chlamydomonas reinhardtii* affected in their nitrate assimilation capability. *Plant Science Letters* **11**, 51–8.

Spiller, H., Dietsch, E., and Kessler, E. (1976). Intracellular appearance of nitrite and nitrate in nitrogen-starved cells of *Ankistrodesmus braunii*. *Planta* **129**, 175–81.

Srivastava, H. S. (1980). Regulation of nitrate reductase activity in higher plants. *Phytochemistry* **19**, 725–33.

Syrett, P. J. (1981). Nitrogen metabolism of microalgae. *Canadian Bulletin of Fisheries and Aquatic Sciences* **210**, 182–210.

Syrett, P. J. and Hipkin, C. R. (1973). The appearance of nitrate reductase activity in nitrogen-starved cells of *Ankistrodesmus braunii*. *Planta* **111**, 57–64.

Syrett, P. J. and Leftley, J. W. (1976). Nitrate and urea assimilation by algae. In *Perspectives in experimental biology*, vol. 2 (ed. N. Sunderland), pp. 221–34, Pergamon Press, Oxford and New York.

Syrett, P. J. and Morris, I. (1963). The inhibition of nitrate assimilation by ammonium in *Chlorella*. *Biochimica et Biophysica Acta* **67**, 566–75.

Syrett, P. J., Flynn, K. J., Molloy, C. J., Dixon, G. K., Peplinska, A. M., and Cresswell, R. C. (1986). Effects of nitrogen deprivation on rates of uptake of nitrogenous compounds by the diatom, *Phaeodactylum tricornutum* Bohlin. *New Phytologist* **102**, 39–44.

Thacker, A. and Syrett, P. J. (1972*a*). The assimilation of nitrate and ammonium by *Chlamydomonas reinhardtii*. *New Phytologist* **71**, 423–33.

Thacker, A. and Syrett, P. J. (1972*b*). Disappearance of nitrate reductase activity from *Chlamydomonas reinhardtii*. *New Phytologist* **71**, 435–41.

Tischner, R. (1976). Zur Induktion der Nitrat- und Nitritreduktase in vollsynchronen *Chlorella* Kulturen. *Planta* **132**, 285–90.

Tischner, R. (1984). A comparison of the high-active and low-active form of nitrate reductase in synchronous *Chlorella sorokiniana*. *Planta* **160**, 1–5.

Tischner, R. and Hüttermann, A. (1978). Light mediated activation of nitrate reductase in synchronous *Chlorella*. *Plant Physiology* **62**, 284–6.

Tischner, R. and Lorenzen, H. (1979). Nitrate uptake and nitrate reduction in synchronous *Chlorella*. *Planta* **146**, 287–92.

Tischner, R. and Lorenzen, H. (1980). Changes in the enzyme pattern in synchronous *Chlorella sorokiniana* caused by different nitrogen sources. *Zeitschrift für Pflanzenphysiologie* **100**, 333–41.

Tischner, R. and Schmidt, A. (1984). An L-cysteine dependent nitrate reductase inactivating factor in synchronous *Chlorella sorokiniana*. *Journal of Plant Physiology* **117**, 191–200.

Toby, A. L. and Kemp, C. L. (1977). Nitrate reductase mutants in *Eudorina elegans* (Chlorophyceae). *Journal of Phycology* **13**, 368–72.

Ullrich, W. R. (1974). Die nitrat- und nitritabhängige photosynthetische O_2-entwicklung in N_2 bei *Ankistrodesmus braunii*. *Planta* **116**, 143–52.

Ullrich, W. R. (1979). Die Nitritaufnahme bei Grünalgen und ihre Regulation durch äussere Faktoren. *Berlin Deutsche Botanische Gesellschaft* **92**, 273–84.

Ullrich, W. R. (1983). Uptake and reduction of nitrate: algae and fungi. In *Encyclopedia of plant physiology*, New series, Vol. 15A (eds A. Lauchli and R. L. Bieleski), pp. 376–97. Springer-Verlag, Berlin.

Vega, J. M., Herrera, J., Aparicio, P. J., Paneque, A., and Losada, M. (1971). Role of molybdenum in nitrate reduction by *Chlorella*. *Plant Physiology* **48**, 294–9.

Vega, J. M., Cárdenas, J., and Losada, M. (1980). Ferredoxin-nitrite reductase. *Methods in Enzymology* **48**, 294–9.

Vennesland, B. and Guerrero, M. G. (1979). Reduction of nitrate and nitrite. In *Encyclopedia of plant physiology*, New series, Vol. 6 (eds M. Gibbs and E. Latzko), pp. 425–44, Springer-Verlag, Berlin.

Wahl, R. C., Hageman, R. V., and Rajagopalan, K. V. (1984). The relationship of Mo, molybdopterin, and the cyanolyzable sulfur in the cofactor. *Archives of Biochemistry and Biophysics* **230**, 264–73.

Wheeler, P. A. (1980). Use of methylammonium as an ammonium analogue in nitrogen transport and assimilation studies with *Cyclotella cryptica* (Bacillariophyceae). *Journal of Phycology* **16**, 328–34.

Wheeler, W. N. and Weidner, M. (1983). Effect of external inorganic nitrogen concentration on metabolism, growth and activities of key carbon and nitrogen assimilatory enzymes of *Laminaria saccharina* (Phaeophyceae) in culture. *Journal of Phycology* **19**, 92–6.

Wright, S. A. and Syrett, P. J. (1983). The uptake of methylammonium and dimethylammonium by the diatom *Phaeodactylum tricornutum*. *New Phytologist* **95**, 189–202.

Yamaya, T., Solomonson, L. P., and Oaks, A. (1980). Action of corn and rice-inactivating proteins on a purified nitrate reductase from *Chlorella vulgaris*. *Plant Physiology* **65**, 146–50.

Zumft, W. G. (1972). Ferredoxin: nitrite oxidoreductase from *Chlorella*: purification and properties. *Biochimica et Biophysica Acta* **276**, 363–75.

9. Structure and regulation of nitrate reductase in higher plants

Wilbur H. Campbell

Introduction

The study of both the biochemistry and regulation of higher plant nitrate reductase (NR) have a long history. Although earlier evidence for reduction of nitrate by higher plants can be cited, the first report of a higher plant extract demonstrating nitrate reduction was by Evans and Nason (1953). Their purification and characterization of soybean pyridine nucleotide NR was a remarkable achievement in terms of the detail which they presented.

Although studies of the regulation of nitrate assimilation were published prior to the 1950s, the progress made by Hewitt and co-workers, studying the induction of NR by nitrate and the importance of molybdenum to this process is most significant (Afridi and Hewitt 1964; Hewitt and Notton 1980). In the 1960s and early 1970s, the study of the biochemistry of NR was often wedded to the study of the enzyme's regulation (Hageman 1979). In the studies of Schrader *et al.* (1968), Wray and Filner (1970), and Losada (1975), one finds the discovery of the partial activities of NR which have greatly aided in analysis of inhibitors and regulation of the enzyme. However, it was not until the late 1970s and early 1980s that the details of the structure of higher plant NR have become known and it is to this literature which I will devote much of this review.

In understanding the regulation of NR, the 1980s have produced a new type of analysis, in that the use of highly sensitive immunochemical methods has liberated the investigators from the limitations of activity assays for NR and its partial activities and permitted direct analysis of the enzyme's protein moiety (Campbell and Smarrelli 1986). Although the immunology of NR is covered in another chapter of this volume (Chapter 10), it would be impossible to review the regulation of NR without discussing the immunochemical approach and its implications in this field.

Finally, it is clear that we are entering a new era in the study of both the biochemistry and regulation of NR, which is being heralded by the isolation of genes for NR from a number of higher plants (Crawford *et al.* 1986; Cheng *et al.* 1986; Calza *et al.* 1987). Again this material forms the substance of several contributions to this volume (Chapters 12, 13 and 21). However, it would be negligent not to include these most recent findings with regard to their implications for the study of the biochemistry and regulation of NR.

Purification of higher plant nitrate reductase

Although the purification of an enzyme is often a rather trivial process and not pivotal to the understanding of it, the opposite is true for higher plant NRs. Without major advances in the purification of the enzyme, little new knowledge of NR biochemistry would have been gained in recent years. This seems to be true for two reasons: first, NR is present in only small amounts in leaves of plants and in even smaller amounts in roots, in general representing about 0.05 per cent of the total extractable protein (Campbell 1985; Campbell and Smarrelli 1986); second, NR is highly labile both in terms of easy proteolysis of its backbone but also in loss of cofactors, especially the molybdenum cofactor (Campbell and Wray 1983). Advances in purification are continuing today with the introduction of immunoaffinity chromatography, as will be described here and in Chapter 10.

The most significant advance in purification of NR was the introduction of Cibacron blue dye columns, which were first used by Solomonson (1975) for algal NR and by Campbell (1976) for higher plant NR. Blue Sepharose, as these blue dye columns have become known (Campbell and Smarrelli 1986), was the most significant advance ever introduced for purification of NR until the recent advances with immunoaffinity chromatography. In the direct application approach developed and routinely used by our laboratory (Campbell 1976, 1978, 1979; Campbell and Smarrelli 1978, 1986; Redinbaugh and Campbell 1985) crude extract of leaves or cotyledons is mixed in a batch mode with blue Sepharose and after the NR is bound, non-binding proteins are washed away on a filter funnel. The gel with bound NR is packed into an appropriate column and eluted by application of a buffered solution containing 0.1 mM NADH or NADPH depending on the form of enzyme which is being eluted and purified in the procedure (Redinbaugh and Campbell 1981; Robin *et al.* 1985). For squash, corn (maize), and soybean, we have been able to achieve 300- to 600-fold improvement in specific activity (i.e. μmol of nitrite produced/min/mg protein) using this protocol, which can be completed in 5 h or less, depending on the system (Campbell and Smarrelli 1986). In cases where the tissue contains more than one form of NR, such as corn roots and scutella and soybean leaves and cotyledons, it has been possible to separate two forms with little cross-contamination by elution first with NADPH and then with NADH, or the reverse depending on which enzyme form is more tightly held by the blue Sepharose (Campbell 1976, 1978; Orihuel-Iranzo and Campbell 1980; Redinbaugh and Campbell 1981; Robin *et al.* 1985).

Blue Sepharose has also proved useful in other laboratories and has been widely applied. For barley leaf NR, blue dextran Sepharose was used after the crude extract had been concentrated by ammonium sulphate fractionation and an overall purification of 570-fold was achieved (Kuo *et al.* 1980). In another laboratory, barley leaf NR was purified by ammonium sulphate fractionation and gel filtration before applying blue dextran Sepharose in order to achieve a total improvement in specific activity of 500- to 870-fold (Campbell and Wray 1983). In this case, the higher purity was achieved by including the protease inhibitor, leupeptin, and by modification of the extraction and purification buffers (Wray and Kirk 1981; Campbell and Wray 1983). Spinach leaf NR has also been purified by two different groups using blue Sepharose or a related material as a key step in the protocol (Notton *et al.* 1977; Fido and Notton 1984; Nakagawa *et al.* 1985). In addition to our studies of corn leaf NR, Oaks and coworkers have purified corn leaf NR using blue Sepharose among other steps (Nakagawa *et al.* 1984). Soybean NR have also been separated and purified using blue Sepharose, but this system is complicated by the fact that there are three forms of the enzyme in the leaves of this species (Streit *et al.* 1987). The soybeam NRs will be dealt with in a separate section of this review.

Although blue Sepharose provides a very significant improvement in the purity of higher plant NR preparations, gel electrophoresis generally shows that they are not homogeneous and further purification is necessary (Fido and Notton 1984; Redinbaugh and Campbell 1985). For squash NR, we prepared pure enzyme by metal chelate affinity chromatography, thus taking advantage of the heavy metal binding site we had shown to be present (Smarrelli and Campbell 1983; Redinbaugh and Campbell 1983). Fido and Notton (1984) used 5′(AMP Sepharose for spinach NR, while Nakagawa *et al.* (1985) used hydrophobic chromatography for further purification of this enzyme form. For soybean, high pressure liquid chromatography was used with an anionic exchanger to separate and further purify the two forms of NADH-NR after initial purification on blue Sepharose (Streit *et al.* 1987).

Most recently, we have used immunoaffinity chromatography to purify further squash and corn NRs after initial purification on blue Sepharose. The results of a purification of corn leaf NR on a monoclonal antibody column prepared with antibody from hybridoma cell line XV-11, which was raised against squash NR (Miller and Campbell 1986), are shown in Fig. 9.1. and the total purification is summarized in Table 9.1. Immunoaffinity chromatography can also be performed by applying the crude NR directly to the antibody column, as was recently described by Fido (1987). We have developed a similar approach for corn leaf NR. The success of this method depends on the availability of monoclonal antibodies, since we and others have found that polyclonal antibodies, even though specific for NR, do not work well as ligands for immunoaffinity chromatography (Campbell and Smarrelli 1986; Kleinhofs *et al.* 1986); this is probably due to the multiplicity of affinities for the enzyme's epitopes in a polyclonal antibody mixture, while monoclonal antibodies bind at only one site. Squash monoclonal antibody also has a high affinity for corn NR; although the monoclonal antibody binds to only one epitope on NR, this structural

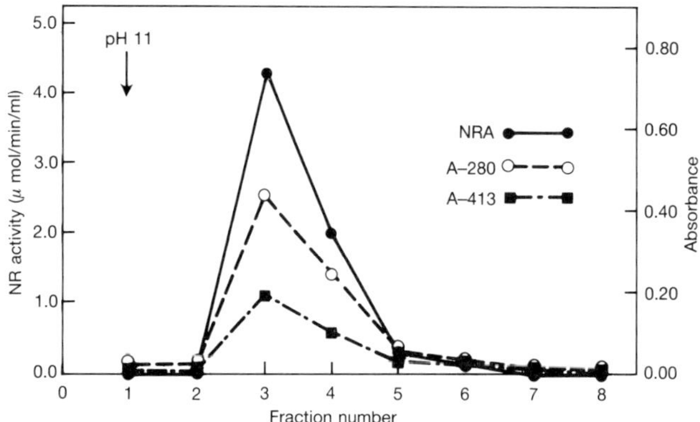

Fig. 9.1. Immunoaffinity chromatography using monoclonal antibody squash-8 (hybridoma XV-11) for purification of corn nitrate reductase. After purification on blue Sepharose as shown in Table 9.1, corn NR was applied to a 3 ml anti-NR column at a flow rate of about 1 ml/min. After loading and brief washing, the NR was eluted with 10 mM glycine, pH 11. One ml fractions were collected into tubes containing 0.1 ml of 2 M Tris-Cl, pH 7, in order to immediately adjust the pH after elution. For each fraction, the NR activity was assayed and absorbances at 280 and 413 nm were measured.

Table 9.1. Purification of maize leaf nitrate reductase

	Volume (ml)	Protein (mg)	Activity (units)*	Specific activity (units/mg)	Purity (-fold)
Crude extract	1200	2700	100	0.037	1
Blue Sepharose	100	3.0	60	20	540
Anti-NR Sepharose	2	0.15	7	47	1300

The experimental details of these purification procedures are given in the legend to Fig. 9.1. The crude extract and blue Sepharose purification were done as previously described (Campbell and Remmler 1986). This purification was done with 300 g of maize leaves, which were 5 days old and induced twice with 50 mM ammonium nitrate in Hoagland's solution.
* Units of activity = 1 μmol nitrite formed/min.

characteristic may be present in all higher plant forms of the enzyme, which results in monoclonal antibody columns being universal purification tools.

Structure of higher plant nitrate reductase

For some years, it has been known that NR was of large native molecular weight, but the lack of pure enzyme in sufficient quantities for detailed studies had hampered clear-cut conclusions and led to some misguided speculation about the enzyme's quaternary structure (Notton and Hewitt 1979; Hewitt and Notton 1980; Campbell and Smarrelli 1986). This was certainly complicated by the fact that the polypeptide backbone of higher plant NR is highly labile and quite susceptible to proteolytic degradation during the preparation of the enzyme (Small and Wray 1979; Wray *et al.* 1979). The use of protease inhibitors and methods for selecting between intact and degraded enzyme have led to preparation of holo-NR and to a clarification of the enzyme's polypeptide size (Wray and Kirk 1981; Campbell and Wray 1983; Fido and Notton 1984; Redinbaugh and Campbell 1985).

The most significant conclusion for further study of the enzyme is that all forms of higher plant NR contain a single polypeptide chain and are homodimers in their native forms, although some tendency for the homodimers to form a tetramer has been noted (Redinbaugh and Campbell 1985). The native molecular weight of squash NR has been determined to be 230 kDa by a single method involving only native gel electrophoresis and the relative size of the denatured polypeptide of squash enzyme is 115 kDa (Redinbaugh and Campbell 1985). Spinach NR has also been shown to be of similar dimensions (Fido and Notton 1984; Nakagawa *et al.* 1985). However, for monocotyledons the polypeptide has appeared to be smaller with a molecular weight of 110 kDa or less (Kuo *et al.* 1980; Kleinhofs *et al.* 1986). The controversy over whether the dicotyledon and monocotyledon NR have different size polypeptide chains will certainly be unequivocally resolved when the two genes have been cloned and sequenced, but we have gained further insight into this problem by comparing squash and corn NRs on the same denaturing gel (Fig. 9.2). Although the squash preparation contains a small amount of nicked polypeptide, of molecular weight of 10 kDa less than the intact polypeptide, the corn NR polypeptide is clearly slightly smaller than the squash form. A similar result was found by Western blotting when monocotyledon and dicotyledon NR were compared (Narayanan *et al.* 1983). Thus, the homodimer model for higher plant NR has gained wide acceptance and will be used as the working hypothesis in this review.

Prosthetic groups of higher plant nitrate reductase

Evans and Nason (1953) showed that soybean NR contained a flavin as a component of the enzyme. A few years later, molybdenum was recognized as a component of NR (Afridi and Hewitt 1964). Garrett and Nason (1967) showed *Neurospora crassa* NR contained a cytochrome based on its spectral properties: similar spectral properties for NR, from barley, spinach and squash, show that these proteins also contain a *b*-type cytochrome (Notton *et al.* 1977; Somers *et al.* 1982; Redinbaugh and Campbell 1985). A midpoint potential of -60 mV was determined for spinach NR (Fido *et al.* 1979). It was eventually shown that the molybdenum in NR and related oxygen-insensitive molybdenum-containing

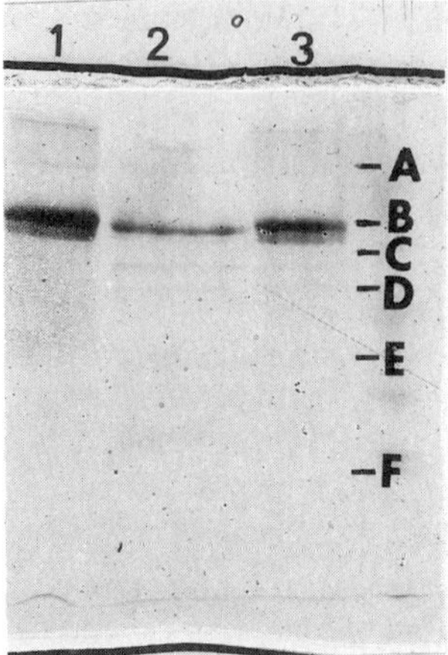

Fig. 9.2. Comparison of the molecular size of the polypeptide subunit of corn and squash NADH:nitrate reductase. NRs were purified by blue Sepharose followed by immunoaffinity chromatography as shown for corn in Table 9.1. After denaturing by boiling in 1% sodium dodecylsulfate and 1% 2-mercaptoethanol, the samples were electrophoresed for 30 min in a mini-slab gel containing 10% acrylamide. Lane 1: 4 μg squash NR. Lane 2: 2 μg corn NR. Lane 3: 3 μg squash NR. Molecular weight standards: A = 205 kDa; B = 116 kDa; C = 92.5 kDa; D = 66 kDa; E = 45 kDa; F = 29 kDa. The relative molecular mass of squash NR is estimated to be 115 kDa, while that of corn NR is 110 kDa. Squash NR contains a nicked form with a size of about 105 kDa, while corn contains no significant level of nicked forms.

enzymes was associated with an organic cofactor, which has now been identified as a pterin (Chapter 14). While these prosthetic groups were quantified in an algal NR some years ago, it was not until quite recently that we were able to obtain sufficient intact squash NR to establish the quantitative amounts of prosthetic groups in the higher plant enzyme (Redinbaugh and Campbell 1985). The results of our quantification of the prosthetic groups in squash NR are summarized in Table 9.2. It is now clear that each subunit of higher plant NR contains one equivalent of each prosthetic group and that each subunit appears able to catalyse nitrate reduction independent of the other. This was rather elegantly shown in the recent radiation inactivation study of *Chlorella* NR, which indicated that the reduction of nitrate by NADH required a protein the approximate size of the NR subunit and not the dimer or tetramer forms (Solomonson and McCreery 1986). Finally, it is important to state that NR models involving a single molybdenum shared between the two subunits of higher plant NR are not consistent with the data presented in Table 9.2 and should now be abandoned.

Domains of nitrate reductase

Most interestingly, NR contains a structural level which falls between the denatured polypeptide and the native dimeric structure. This level of structure is

Table 9.2. Prosthetic groups of squash nitrate reductase

Prosthetic group	Amount found	Stoichiometry
		(mole prosthetic group/mol enzyme subunit)
Flavin adenine dinucleotide	1.0	1
Haem	0.8	1
Iron	1.0	1
Molybdenum	0.7	1
Mo cofactor (Mo–pterin)	1.0	1

The data for this table were adapted from Redinbaugh and Campbell (1985).

associated with the prosthetic groups of the enzyme to some degree and is most widely known as the domain structure of the enzyme (Campbell and Smarrelli 1986). This is not a unique characteristic of NR but is also found among other redox enzymes similar to it such as xanthine oxidase, sulphite oxidase and yeast flavocytochrome *b-2* (Coughlan 1980; Xia *et al.* 1987). In fact, most complex enzymes with more than one functional characteristic have distinct structural domains.

Since NR has not been crystallized and studied by X-ray analysis (i.e. there is not a three-dimensional model for it), establishing that the enzyme has stable domains is most easily accomplished by degrading the enzyme with a protease and isolating the stable, functional domains. This was recently accomplished for *Chlorella* NR (Solomonson *et al.* 1986), but has not been shown for a higher plant NR. However, Poulle *et al.* (1987) recently showed that the protease isolated from corn roots, which has been termed a NR inactivase and characterized as such in several previous papers from that group as well as in work by Wallace (1977), degraded corn leaf NR into distinct molecular sizes as judged by denaturing gel electrophoresis. Since the pattern for corn leaf enzyme is very different from that obtained for *Chlorella* NR when it is treated with the same protease preparation (Solomonson *et al.* 1986; Poulle *et al.* 1987), it remains to be shown whether structurally stable and functionally active domains can be isolated for plant NR. However, sufficient evidence has been accumulated on monomeric forms of NR which retain their dehydrogenase functionality as well as isolated, degraded forms of NR, confidently to suggest that it is only a matter of time until it can be shown that higher plant NR functional domains can be isolated (Wallace and Johnson 1978; Small and Wray 1979; Wray *et al.* 1979; Kleinhofs *et al.* 1986; Campbell and Smarrelli 1986).

The following domain structure can be postulated based on studies of the partial activities of NR. NR must have a domain to bind NADH and it may perhaps have a separate domain to bind FAD. Together these domains will constitute the minimum needed for ferricyanide reduction, as was recently shown for *Chlorella* NR (Solomonson *et al.* 1986). If mammalian cytochrome *c* is to be reduced, then a further domain must be added in the form of the cytochrome b_{556} (Campbell and

Smarrelli 1986). The binding of Mo–pterin will be contained in a domain structure, but since this cofactor seems to bind only into the dimeric form of NR it may be difficult to isolate this domain by itself without what may be termed the dimer interface domain. A nitrate binding domain may exist, but this may be tightly linked to the Mo–pterin domain and may not be separable from it. To summarize, there are probably three domains for binding prosthetic groups, one each for the substrates and one more to account for the region of the polypeptide involved in binding to itself or in other words, six domains for the 115 kDa protein, which would give an average size of about 20 kDa for each domain.

Amino acid sequence of nitrate reductase

Since homogeneous NR protein has been difficult to obtain in sufficient amounts little progress has been made with amino acid sequencing, but amino acid compositions have been determined for barley, squash and corn NR (Kuo *et al.* 1982; Redinbaugh and Campbell 1985; E. R. Campbell, personal communication). Le and Lederer (1983) reported a partial amino acid sequence for *Neurospora crassa* NR, which showed the homology of the enzyme with mammalian cytochrome b_5 and other enzymes containing b_5-type cytochromes. With the recent advances in purification using immunoaffinity chromatography, sufficient amounts of squash and corn NRs are available for amino acid sequence analysis and these studies are currently being done. The initial results for both the corn and squash enzymes indicate that their amino terminal residues are blocked and not accessible by Edman degradation (E. R. Campbell, personal communication). Thus, at this time the most fruitful path to amino acid sequence data is by determining the nucleotide sequence of cDNA clones and genomic clones for NR, which has been done for tobacco (see Chapter 12). Since cDNA clones have been isolated for barley, squash, *Arabidopsis*, and corn as well as tobacco (Cheng *et al.* 1986; Crawford *et al.* 1986; Campbell and Gowri 1987, Chapters 12 and 21), a considerable amount of sequence data should be available in the near future, which will prove very useful for identification of conserved amino acid residues of higher plant NR.

Catalytic activity of higher plant nitrate reductase

Nitrate reductase is an oxidation/reduction catalyst, in that the reaction catalysed is a redox reaction and also in that the enzyme becomes reduced and oxidized cyclicly during catalytic turnover (Campbell and Smarrelli 1986). The reaction catalysed could be represented by the following half-reactions:

1) $\qquad$ NADH + NR (oxidized) → NAD$^+$ + H$^+$ + NR (reduced)

2) $\qquad$ nitrate + H$^+$ + NR (reduced) → nitrate + OH$^-$ + NR (oxidized)

$\qquad$ Net reaction: NADH + nitrate → NAD$^+$ + nitrite + OH$^-$

The overall reaction and each half-reaction are essentially irreversible (Hewitt and Notton 1980; Campbell and Smarrelli 1986). The first half-reaction appears to involve the FAD portion of NR and could informally be linked to the dehydrogenase partial activity of NR, which is dependent on the presence of FAD (Campbell and Smarrelli 1986). The stereospecificity of NR for removal of hydrogen from pyridine nucleotides is 'pro-R' (A-type) (Campbell and Smarrelli 1986). The species 'NR (reduced)' used above would represent NR with one pair of electrons distributed among FAD, cytochrome *b* and Mo of the enzyme. The second half-reaction appears to involve the Mo site of NR and could informally be linked to the reductase partial activity of NR, which is dependent on the presence of Mo (Hewitt and Notton 1980; Campbell and Smarrelli 1986).

Partial reactions catalysed by nitrate reductase

The partial reactions are of two major types: a NADH dehydrogenase activity and a dye-dependent NR activity (Hewitt and Notton 1980; Campbell and Smarrelli 1986). For the dehydrogenas activity, a variety of electron acceptors has been identified including ferricyanide, mammalian cytochrome *c*, dichlorophenol-indophenol, methylene blue, and water soluble quinones such as menadione and benzoquinone (Smarrelli and Campbell 1979; Campbell and Smarrelli 1986). It can now be unequivocally stated that the NADH ferricyanide reductase activity of NR requires only the flavin and that cytochrome *c* reduction requires both the flavin and the cytochrome *b* of NR (Solomonson *et al.* 1986). The dye-dependent NR activity has been measured with reduced flavins or viologen dyes until recently, when bromphenol blue was recognized as another electron donor for NR (Hoarau *et al.* 1986; Campbell 1986). Reduced bromphenol blue is the most effective electron donor characterized to date since it supports rates of nitrate reduction 5–10 times greater than NADH and six times greater than reduced methylviologen (Campbell 1986). It has been assumed that the reduced flavins may donate electrons to NR via cytochrome *b* while the reduced viologens acted directly at the Mo (Campbell and Smarrelli 1986). Reduced bromphenol blue appears to donate electrons to NR at either the cytochrome *b* or the Mo, but use of monoclonal antibodies which inhibit NR demonstrated that the methylviologen and bromphenol blue donate electrons at distinct sites on the enzyme (Campbell 1986).

Other multicentre redox enzymes are like NR in that a variety of electron donors and acceptors can be utilized (Coughlan 1980; Palmer and Olson 1980; Campbell and Smarrelli 1986). For example, xanthine oxidase can accept electrons from either xanthine or hypoxanthine and can reduce ferricyanide, cytochrome *c*, dichlorophenolindophenol, phenazine methosulphate, methylene blue, oxygen, nitroblue tetrazolium, and trinitrobenzene sulphonate (Coughlan 1980). These electron acceptors are reduced at different sites on the enzyme, as shown by studies using a variety of inhibitors, and it should be especially noted that cytochrome c and nitroblue tetrazolium cannot be directly reduced by the enzyme but that these reductions are mediated by oxygen via the superoxide radical in most cases (Coughlan 1980). In many of these enzymes, the effectiveness with which the enzyme reduces a substrate may depend on the state of the enzyme, with some enzyme modifications (i.e. either oxidation or reduction of sulphydryls) leading to increases in some activities and decreases in others (Palmer and Olson 1980).

Kinetic constants for higher plant nitrate reductase

Specific activities for NR range from about 90 to 120 μmol/min/mg protein for homogeneous preparations from all sources including higher plants, algae, fungi and bacteria (Hewitt and Notton 1980; Campbell and Smarelli 1986). Turnover numbers for NR vary from 140 to 190 s^{-1} (Campbell and Smarrelli 1986). Nitrate K_m values range from 0.03 to 0.2 mM, while NADH/NADPH K_m values range from 0.004 to 0.06 mM (Hewitt and Notton 1980; Campbell and Smarrelli 1986). Catalytic efficiency for NR depends on the substrate being considered: for NADH/NADPH, $k_{cat}/K_m = 1–4 \times 10^7\ s^{-1}\ M^{-1}$, and for nitrate the values are about one-tenth lower (Campbell and Smarrelli 1986). Thus, the reduction of NR by pyridine nucleotides tends to approach the diffusion limit, while the reduction of nitrate by reduced NR appears to be more rate limiting. The lower efficiency of nitrate reduction versus NADH oxidation, could be due to a rate limiting transfer of electrons between the FAD and the Mo and not necessarily due to a slower transfer of electrons from the Mo to nitrate. Most multicentre redox enzymes are rate limited by either product release or internal electron transfer and rapidly reduce electron acceptors (Palmer and Olson 1980).

Inhibitors of nitrate reductase

Sulphydryl-binding reagents block the enzyme at or near the pyridine nucleotide binding site, resulting in loss of both the complete and the dehydrogenase activities (Beevers and Hageman 1980; Hewitt and Notton 1980; Campbell and Smarrelli 1986). While substrate can protect against this inhibition, studies of NR inhibition with pyridine nucleotide analogues have indicated a catalytic role for sulphydryl groups (Hewitt and Notton 1980; Campbell and Smarrelli 1986). Metal-binding anions such as cyanide, azide and cyanate inhibit the complete and reductase activities of NR, apparently by blocking electron transfer from the terminal site where Mo is found (Hewitt and Notton 1980; Campbell and Smarrelli 1986). Cyanide is a competitive inhibitor of nitrate unless the NR is reduced in the absence of nitrate, in which case the cyanide forms a tight complex with the enzyme (Hewitt and Notton 1980; Guerrero *et al.* 1981; Howard and Solomonson 1982; Campbell and Smarrelli 1986).

Steady state kinetic mechanism

Our results (Campbell and Smarrelli 1978, 1086) and those of Renosto *et al.* (1981) indicate that the kinetic mechanism of NR is of the two-site ping-pong type. This is similar to the kinetic mechanism reported for other redox enzymes such as xanthine dehydrogenease and other Mo-containing hydroxylases (Coughlan 1980; Coughlan and Rajagopalan 1980), glutamate synthase (Rendina and Orme-Johnson 1978) and glutathione reductase (Mannervik 1976). The two-site ping-pong kinetic mechanism was originally described for transcarboxylase and other biotin-containing enzymes (Northrup 1969). However, a different kinetic mechanism was suggested for spinach NR (De la Rosa *et al.* 1980). Howard and Solomonson (1981) reported intersecting initial velocity patterns for *Chlorella* NR and suggested the kinetic mechanism was a novel ordered two-site mechanism. Renosto *et al.* (1982*a*) reported that fungal NR had similar kinetics at the suboptimum pH of 6.2 and attributed this to loose binding of FAD to the enzyme at this pH. Similar intersecting line initial velocity patterns had previously been observed for fungal NR (McDonald and Coddington 1974). Xanthine dehydrogenase was found to have intersecting line initial velocity patterns at some limiting concentrations of NAD (Coughlan 1980; Coughlan and Rajagopalan 1980). Although it cannot be considered to be fully resolved for NR until more detailed studies are done, it would appear that multicentre redox enzymes can display either parallel line double reciprocal plots, which reflect the presence of binary complexes between substrates and enzyme, or intersecting line double reciprocal plots, reflecting ternary complexes of substrates and enzyme (Palmer and Olson 1980). The type of kinetic plots obtained will depend on the substrate concentrations used and may change as the ratio of reductant to oxidant is varied at high and low substrate concentration.

Kinetics of partial reactions

Renosto *et al.* (1982*b*) have studied the kinetics of the reductase partial activity using $FADH_2$ as the electron donor and found that the kinetics conform to classical ping-pong mechanism. They suggest that $FADH_2$ donates electrons at a site other than the Mo centre, which is consistent with observations made by others on this partial reaction of NR (Campbell and Smarrelli 1986; Renosto *et al.* 1982*b*). De la Rosa and Palacian (1981) have reported a study of the cytochrome *c* reductase activity of spinach NR, which describes a hexa uni ping-pong mechanism for this partial reaction of NR. Again it was suggested that NADH and cytochrome *c* bind to different sites on the enzyme, and this observation is consistent with prior observations (Hewitt and Notton 1980; Campbell and Smarrelli 1986).

Internal electron transfer in nitrate reductase

Linear electron transfer from pyridine nucleotide to nitrate *via* the internal carriers FAD, cytochrome *b* and Mo was originally suggested for *N. crassa* NADPH-NR and has been confirmed by both visible spectral analysis and EPR (Hewitt and Notton 1980; Campbell and Smarrelli 1986; Jacob and Orme-Johnson 1980). Kay and Barber (1986) recently showed that ionic strength affected the rates of electron transfer within NR such that it appeared that internal electron transfer, especially the transfer of electrons from the cytochrome *b* to the molybdenum, was rate limiting. This observation is consistent with the rates of the partial reactions (i.e. cytochrome *c* reductase and methylviologen NR), which are both greater than the complete reaction, and this tends to indicate that internal electron transfer is the rate limiting step in NR catalysis (Campbell and Smarrelli 1986).

Soybean nitrate reductases

Higher plants in general contain NADH-NR with a pH optimum of 7.5 and no other forms of the enzyme (Campbell and Smarrelli 1986). Soybean has always been considered unusual in this regard since it contains an NAD(P)H-NR and did not appear to contain the common NADH-utilizing form of the enzyme (Jolly *et al.* 1976; Campbell 1976; Orihuel-Iranzo and Campbell 1980). Higher plant NR was first purified and characterized from soybean leaves and it was found to be a bi-specific enzyme using both NADH and NADPH about equally well with a pH optimum of 6.5 (Evans and Nason 1953). Subsequently, it was found that when soybean leaves were extracted in the presence of thiol-reducing agents, NADH-NR activity dramatically increased, such that it appeared that the bi-specific feature of soybean NR had been due to some artifact of the preparation, but the enzyme still had a pH optimum of 6.5 (Beevers *et al.* 1964; Jolly *et al.* 1976). Attempts to purify and separate soybean NR into different forms failed and a phosphatase converting NADPH to NADH was suggested as the explanation of the high level of NADPH-supported NR activity in soybean, although the presence of such a phosphatase was not shown (Wells and Hageman 1974). Eventually, Jolly *et al.* (1976) separated and purified NADH- and NAD(P)H-NRs from soybean leaf extracts and both of these enzyme forms had pH optima of 6.5. Campbell (1976) also separated these enzyme forms using blue dextran Sepharose affinity chromatography from soybean leaf extracts and later on we used blue Sepharose to separate these NRs from soybean cotyledons, showing that the enzyme forms peaked in activity at different developmental ages (Orihuel-Iranzo and Campbell 1980, 1981).

However, the complexity of the NR forms in soybean is greater than these studies had shown. An NR-deficient mutant of soybean lacking NAD(P)H-NR activity and having NADH-NR activity has been generated (Ryan *et al.* 1983; Nelson *et al.* 1983). Using this mutant and various N growth conditions for wild-type plants, NAD(P)H-NR was shown to be a constitutive form (i.e. not requiring nitrate for expression). The leaves of the mutant had an NADH-NR with pH optimum of 7.5 under nitrate-inducing conditions and it was concluded that this was obscured in wild-type leaves due to the high levels of constitutive NAD(P)H-NR (Nelson *et al.* 1984*a*). These results showed that soybean leaves contained the form of NR found most commonly in higher plants. It was also found that the only form of the enzyme expressed in soybean tissue cultures was NADH-NR (Nelson *et al.* 1984*a*). The soybean NR-deficient mutant was further studied with an improved purification for separating the soybean leaf NR using blue Sepharose and a mouse polyclonal antibody raised against NAD(P)H-NR (Robin *et al.* 1985). Using the combination of these tools, Robin *et al.* (1985) showed that the soybean nr$_1$ mutant was devoid of NAD(P)H-NR, which was clearly present in Western blots of the partially purified enzyme from the wild type. Subsequently, Streit *et al.* (1985) showed that soybean leaves contained three forms of nitrate reductase: NAD(P)H-NR (6.5), NADH-NR (7.5) and NADH-NR (6.5). The demonstration of the third form (i.e. NADH-NR pH 6.5 optimum) with a preference for NADH as electron donor and low pH optimum, explained why NADH-NR (7.5) was not found in the earlier studies (Streit and Harper 1986; Streit *et al.* 1987). Since the soybean nr$_1$ mutant is deficient in both NAD(P)H-NR (6.5) and NADH-NR (6.5), the mutant was the key to demonstrating the presence of NADH-NR (7.5) in soybean leaves (Nelson *et al.* 1984*b*, 1986; Streit *et al.* 1985). Since the mutant grows normally on nitrate and soybean tissue cultures utilize nitrate (Nelson *et al.* 1983, 1984*a*), it is now clear that the NADH-NR (7.5) is the main assimilatory NR of soybean, as it is in all other higher plants. Recently, Smarrelli *et al.* (1987) showed that squash NR cDNA hybridized to the soybean NADH-NR mRNA and could be used to study the expression of the enzyme's mRNA. Thus, it can be expected that the soybean NADH-NR gene will soon be cloned and its sequences compared to that of other higher plant NR which have been cloned and sequenced.

Soybean NAD(P)H-NR has two unusual biochemical properties when compared to any other NR: a pH optimum of 6.5 and a nitrate K_m of 7.5 mM (Jolly *et al.* 1976; Campbell 1976). Studies of the pH optima for the complete and partial reactions catalysed by soybean NAD(P)H-NR showed that the dehydrogenase activity had a normal pH optimum of 7.5, like most NADH-NR of higher plants, but the complete reaction and reductase partial activities had the lower pH optima of 6.5 (Campbell 1976). Thus, it was clear that the change in the NAD(P)H-NR which gave it this low pH optimum was at the nitrate binding and reaction site and since the reactions that take place at this site are rate-limiting for reduction of nitrate, the complete reaction also has this same low pH optimum (Campbell 1976). Detailed studies of the kinetic mechanism of the NAD(P)H-NR showed that it operated with a two-site ping-pong mechanism and that NADH and NADPH compete for the same electron donor site on the enzyme (Orihuel-Iranzo and Campbell 1981). The K_m values for NADH and NADPH were 1–3 μmol, which were in the range reported for NADH-NR and were not unusual; the K_m values for nitrate were 7.5 mM with NADPH as electron donor at pH 6.5 and 9.5 mM with NADPH at pH 7.5, while 6.7 mM with NADH at pH 6.5 and increasing to 16.5 mM with NADH at pH 7.5 (Orihuel-Iranzo and Campbell 1981). These K_m values were similar to apparent K_m values previously reported for NAD(P)H-NR (Jolly *et al.* 1976), who also found that NAD(P)H–NR was relatively insensitive to inhibition by cyanide and azide, which are very effective inhibitors of NADH-NR. Immunological data had established that NAD(P)H-NR was a member of the homologous family of higher plant NR both by demonstration of cross reactivity with anti-NR prepared against squash and corn NADH-NR, but also by antibodies against NAD(P)H-NR reacting with squash and corn NADH-NR (Smarrelli and Campbell 1981; P. Robin and W. H. Campbell, personal communication).

Does NAD(P)H-NR have the same prosthetic groups as other NRs? In the original study of NAD(P)H-NR by Evans and Nason (1953), they had shown that this NR was a flavo-enzyme and probably contained FAD, which is consistent with its dehydrogenase properties (Jolly *et al.* 1976; Campbell 1976). We found that NAD(P)H-NR had a visible spectrum identical with squash NADH-NR, which indicated that haem-Fe was present in the form of cytochrome b_{556} (P. Robin and W. H. Campbell, personal communication). Recently, Streit *et al.* (1987) showed that all three soybean NR forms have spectral properties indicative of the presence of a cytochrome b_{556} as a component of their structures. Nelson *et al.* (1986) showed that purified NAD(P)H-NR could be used to reconstitute NR activity of a barley Mo cofactor (MoCo)-deficient mutant, which tends to establish that soybean NAD(P)H-NR contains the same MoCo as other NR. Thus, NAD(P)H-NR appears to have a normal complement of NR prosthetic groups.

The NADH-NR (6.5) appears to share properties with both other forms of soybean NR: it resembles NAD(P)H-NR in pH optimum, gel electrophoretic mobility and sedimentation coefficient (Robin *et al.* 1985; Streit *et al.* 1985), while it resembles NADH-NR in having a greater preference for NADH over NADPH, low K_m for nitrate (0.2 mM), and is eluted from blue Sepharose by NADH and not by NADPH (Robin *et al.* 1985; Streit *et al.* 1985). Molecular weights of soybean NR forms have not yet been determined but both gel filtration and sedimentation data indicate that NAD(P)H-NR is smaller than the common form of NADH-NR (Streit *et al.* 1985). We estimated the native molecular weight using soybean anti-NR and Western blots and found the NAD(P)H-NR had a molecular weight of 150 kDa, while NADH-NR, which was a mixture of pH 6.5 and 7.5 forms, had a molecular weight of 230 kDa (P. Robin and W. H. Campbell, personal communication). However, after the three forms of soybean NR were separated and purified, their subunit size appeared to be the same when estimated by denaturing gel electrophoresis, which is inconsistent with the different sizes found for the native molecular sizes of the NAD(P)H- and NADH-NR forms (Streit *et al.* 1987).

Harper (1981) found that nitrogen oxides were evolved during the *in vivo* NR assay with soybean leaf discs. Mulvaney and Hageman (1984) investigated this gas evolution and determined that the evolution of nitrogen oxides were not as large as the evolution of acetaldehyde oxime. They demonstrated that these gaseous compounds were made from [^{15}N]nitrate fed to the soybean leaf discs and that only very small amounts of such oxidized nitrogen compounds were evolved from corn leaves in a similar assay. More recently, the evolution of the oxime compound was disputed (J. E. Harper, personal communication). Since the unusual difference between soybean leaves and those of other plants is the soybean NAD(P)H-NR, we reasoned that this enzyme might be responsible for the evolution of nitrogen oxides and related compounds. We had earlier observed that nitrite could act as a substrate for soybean NAD(P)H-NR, but could not determine the nature of the products of nitrite reduction (Orihuel-Iranzo and Campbell 1981).

To test whether soybean NAD(P)H-NR could generate gaseous products, we set up a simple system where NADH or NADPH and nitrite or nitrate was mixed with purified enzyme in a closed vessel which had been evacuated and then capped with a trap containing 0.2 M KOH (P. Robin and W. H. Campbell, personal communication). The results of these experiments are presented in Table 9.3. It is

Table 9.3. Nitrogen oxide evolved by soybean NAD(P)H:nitrate reductase

Assay mixture*	NO_x evolved and trapped (nmol NO_2^-)
NO_2^-	0.6
$NO_2^- + NADH$	0.7
$NO_2^- + NADH + NR$	24.7
$NO_2^- + NADPH + NR$	23.6
$NO_3^- + NADH + NR$	9.6
$NO_2^- + NADH + NR + azide$	0.7

* Soybean NAD(P)H:NR was extracted and purified as described in Robin *et al.* (1985). The assay mixture contained 25 mM potassium phosphate, pH 6.5, 0.25 mM EDTA, 5 mM potassium nitrite or 10 mM potassium nitrate, and 0.5 mM NADPH or NADH with an appropriate amount of purified NR added to each cuvette. The cuvette was closed with a side-arm trap containing 1.2 ml of 0.2 N KOH. After 8 to 12 h of standing as a closed system, the nitrite content of the KOH trap was determined using a standard diazo-dye method. One mM sodium azide was used to inhibit the reaction (Robin, unpublished).

clear that nitrogen oxides were evolved during the assay by the action of NAD(P)H-NR with either NADH or NADPH as electron donor and nitrite or nitrate as electron acceptor. The process could be inhibited by azide, which indicates that the action takes place at the Mo site of the NAD(P)H-NR. Since

NAD(P)H-NR seems to have MoCo at this active site, its ability to reduce nitrate to nitrite and to further reduce nitrite to nitrogen oxides appears to be related to the changed amino acid sequence of this enzyme relative to other NR, which is reflected in its low pH optimum and high K_m for nitrate (Campbell 1976, Orihuel-Iranzo and Campbell 1981). It is worth noting that the K_m for nitrite for soybean NAD(P)H-NR, determined by monitoring disappearance of NADH or NADPH, was found to be quite low (0.1 mM) (Orihuel-Iranzo and Campbell 1981). An additional confirmation that NAD(P)H-NR accounts for the evolution of nitrogen oxides from soybean leave is that the mutant soybean nr_1, which lacks NAD(P)H-NR (Robin *et al.* 1985), does not evolve nitrogen oxides during the *in vivo* NR assay (Nelson *et al.* 1984*a*), and soybean tissue cultures, which do not express NAD(P)H-NR, also lack nitrogen oxide gas evolution (Nelson *et al.* 1984*a*). Thus, it is clear that NAD(P)H-NR is the catalyst for formation of nitrogen oxides in soybean leaves. But what purpose can such a process serve in soybean metabolism? Further study of the NAD(P)H-NR may reveal more about this unusual enzyme and what the true products of its *in vivo* catalysis are.

NAD(P)H-Nitrate reductase of monocotyledons

Various observations had suggested that corn leaf and scutellar NR activity would be supported by both NADH and NADPH, but all attempts to isolate an NADPH-NR from corn had failed (Wells and Hageman 1974; Campbell and Smarelli 1986). After soybean NAD(P)H-NR had been purified using blue dextran Sepharose affinity chromatography, this method was applied to corn scutella and resulted in isolation of an NAD(P)H-NR with a pH optimum of 7.5 and a nitrate K_m of 0.6 mM, which indicates that this enzyme differs significantly from soybean NAD(P)H-NR (Campbell 1978). An NAD(P)H-NR from rice seedlings, which had a pH optimum of 7.4 and a nitrate K_m of 1.1 mM, had also been isolated (Shen *et al.* 1979). Among the NR-deficient mutants of barley, some were found to contain an NAD(P)H-NR with a pH optimum of 7.7 and a nitrate K_m of 1.2 mM (Dailey *et al.* 1982; Harker *et al.* 1986). We characterized an NAD(P)H-NR from the roots of corn which had a pH optimum of 7.5 and a nitrate K_m of 0.3 mM (Redinbaugh and Campbell, 1981). Recently, electrophoretic evidence has been obtained for two forms of NR in the leaf extracts of several monocotyledons: one form was clearly the NADH-NR which predominates in all monocotyledons, but the second form was correlated with the NAD(P)H-NR of the barley NR-deficient mutants (Heath-Pagliuso *et al.* 1984). This second NR form appeared to be constitutive in oats and other monocotyledons in that it did not require nitrate for expression and it also had a low turnover rate (Heath-Pagliuso *et al.* 1984). We observed a similar NR with high electrophoretic mobility which was expressed in corn leaves in the absence of nitrate (Remmler and Campbell 1986). A recent study showed that NAD(P)H-NR of corn was genetically distinct from the major NADH-NR (Sorger *et al.* 1986). Thus, it is now clear that monocotyledons contain two forms of NR.

The assimilatory NR is the NADH-NR in monocotyledons, except in barley mutants deficient in this enzyme where the NAD(P)H-NR seems capable of enhanced expression and can act as the assimilatory NR (Dailey *et al.* 1982). Why monocotyledons have NAD(P)H-NR and how its expression is regulated in relation to the regulation of NADH-NR in the same tissue are facinating questions. However, it is clear that monocotyledon NAD(P)H-NR (7.5) is not like soybean NAD(P)H-NR (6.5) in any rcspect other than the bispecific pyridine nucleotide electron donation site.

Regulation of higher plant nitrate reductase

There are several aspects to the question of regulating NR, which is considered the key regulatory point in nitrate assimilation in plants (Beevers and Hageman 1980; Losada *et al.* 1981; Guerrero *et al.* 1981; Campbell and Smarrelli 1986). Regulation can be effected by changing the level or amount of NR; this may be achieved by altering the balance between biosynthesis and degradation such that different steady-state concentrations of NR are present in the plant cell (Zielke and Filner 1971). At another level, regulation can involve altering the catalytic effectiveness of the existing NR or, in other words, altering the activity level of the enzyme present in the cell. In addition, overall control of nitrate assimilation rests with the availability of nitrate and energy to drive its reduction. Since nitrate levels dramatically affect NR levels, it has been difficult to separate availability of nitrate as a direct regulator of the level of activity of NR from its influence on the steady-state concentration of NR. Thus, the cells appears to tune its nitrate assimilation level to nitrate availability in an integrated manner. This also appears to be true in relation to the energy level of the cell, but this has not been as thoroughly studied as substrate availability. Nitrate reductase activity levels are clearly very responsive to cellular conditions in general and have been used as an indicator by many investigators concerned with various effectors which alter cellular metabolism (Beevers and Hageman 1972; Vennesland and Guerrero 1979; Srivastava 1980; Naik *et al.* 1982; Abrol *et al.* 1983). In order to gain an in-depth understanding of the regulation of NR, the underlying molecular mechanisms must be studied. This requires the development of tools not dependent on the activity of the enzyme for measuring various aspects of the enzyme's regulation such as amounts of enzyme protein. Ultimately, the unravelling of the molecular biology of NR and an understanding of the regulation of its expression must be achieved.

Appearance of nitrate reductase activity in plants

In studying corn (maize) plants, Hageman and Flesher (1960) used aetiolated plants for study of 'induction' of NR activity in plants either pretreated with nitrate or not pretreated. In no-nitrate plants given nitrate and light at the same time, the induction of NR activity was initially slow but became rapid after some hours. For nitrate-treated plants, there was no lag in appearance of NR activity when light was given. They concluded that both nitrate and light influenced NR activity of corn seedlings (Hageman and Flesher 1960). Many experiments have been done since to address the importance of light in the induction of NR activity relative to nitrate (Beevers and Hageman 1972; Srivastava 1980; Abrol *et al.* 1983). In general, it has been concluded that nitrate is required for light induction to occur in that the influence of light can only be observed if the plants are preconditioned with nitrate (Hageman and Flesher 1960; Travis *et al.* 1970; Beevers and Hageman 1972; Abrol *et al.* 1983). Red light can be substituted for white light in triggering the induction of NR activity in several different higher plants, which suggests that phytochrome is probably responsible for the light induction of NR activity (Beevers and Hageman 1972; Srivastava 1980; Abrol *et al.* 1983). However, a number of experiments have shown that NR activity can be induced in the dark by applying nitrate if the plants have adequate energy reserves to synthesize proteins (Travis *et al.* 1970; Huffaker 1982; Abrol *et al.* 1983). Thus, it appears nitrate by itself is a sufficient stimulus to trigger NR induction and that light is not an obligatory requirement. On the other hand, the level of NR activity in the dark is low and light appears to be a significant factor in raising the enzyme levels (Abrol *et al.* 1983).

Appearance of nitrate reductase protein in plants

Analysis of the appearance or induction of NR activity leaves open the question of whether new enzyme was made *de novo* when the stimulus was given or whether existing enzyme in some latent form was activated by the stimulus. To answer this question, a very sensitive method for detecting the enzyme's protein component is necessary. While biochemical methods such as gel electrophoresis can be used to detect proteins such as ribulose bisphosphate carboxylase, present in plant cells in large amounts, these methods lack the sensitivity to detect the small amount of NR protein present in leaves which is of the order of 1 μg/g leaf (Campbell and Smarrelli 1986). Immunochemical methods are sensitive enough to detect these low levels of NR protein (Kleinhofs *et al.* 1986; Campbell and Smarrelli 1986). Rocket immunoelectrophoresis was the first method used to analyse for NR protein in conjunction with studies of the enzyme's appearance in relation to application of nitrate (Somers *et al.* 1983*a,b*). Since immunochemical methods involving immunoprecipitation (such as rocket immunoelectrophoresis) cannot detect less than about 100 ng antigen (Campbell and Ripp 1984), it was necessary to look at large changes in NR levels and to concentrate the plant extracts prior to analysis (Somers *et al.* 1983*b*). Nevertheless, large increases in detectable NR protein were found which appeared to follow the increases in NR activity, and these results were confirmed by Western blotting which is a more sensitive immunochemical method (Somers *et al.* 1983*b*; Kleinhofs *et al.* 1986).

A more sensitive method for detecting the enzyme's protein, resulting in easier analysis of large numbers of samples, is the enzyme-linked immunosorbent assay (ELISA) for NR (Campbell and Ripp 1984; Campbell and Remmler 1986; Campbell and Smarrelli 1986). Several other groups have also developed ELISA for NR some of which use monoclonal antibodies (Notton *et al.* 1985; Cherel *et al.* 1985, 1986; Maki *et al.* 1986; Whitford *et al.* 1987). The ELISA can detect NR protein at the ng level in crude extracts of plant tissue without significant interference (Campbell and Remmler 1986; Whitford *et al.* 1987). Thus, the lowest levels of NR activity that can be detected with standard nitrate appearance assays can also be detected with the ELISA (Campbell and Remmler 1986). The most difficult aspect of the ELISA for NR is establishing a reliable standard for calibration of the assay; in general we have done this by comparing homogeneous samples of NR to partially purified enzyme and showing that the same slope is obtained in relation to both activity and protein content of the pure samples (Campbell and Remmler 1986). Thus, in general, we calibrate the ELISA with blue Sepharose-purified NR, which appears to contain only active enzyme, and use the specific activity of pure NR to convert activity to enzyme protein (Campbell and Remmler 1986).

We modelled our experimental approach on that of Hageman and Flesher (1960), comparing NR protein levels in N-starved, etiolated maize leaves when the plants were either given no nitrate or pretreated with nitrate for 4 h prior to giving light and nitrate together (Remmler and Campbell 1986). We observed the same lag as Hageman and Flesher (1960) in appearance of both NR activity and protein in non-pretreated plants and rapid appearance of both activity and protein in plants pretreated with nitrate (Remmler and Campbell 1986). However, the importance of performing these experiments with aetiolated tissue was that we observed no background of NR protein using either the ELISA or Western blot analysis (Remmler and Campbell 1986). Thus it was clear that NR was synthesized *de novo* when nitrate was applied either in the dark or the light and even in nitrate-pretreated plants no large pool of enzyme protein was built up prior to the light treatment, but rather light appeared to simulate rapid synthesis of NR (Remmler and Campbell 1986; Campbell 1987). We also tested the effects of ammonium on nitrate induction of NR and found that ammonium did not repress enzyme synthesis but was, in fact, stimulatory (Remmler and Campbell 1986). Since the ELISA is equally sensitive to the activity assay and as easy to do, our studies included an hourly time course comparing appearance of NR activity and protein for both pretreated and non-pretreated plants (Remmler and Campbell 1986). These analyses led to the observation that NR protein appeared earlier and more rapidly than the enzyme's activity, as if some process other than protein synthesis limited the appearance of activity (Remmler and Campbell 1986; Campbell 1987). This suggests that either the assembly of NR subunits into the active enzyme is slower than synthesis of the enzyme's protein or that availability of one of the enzyme's prosthetic groups is limiting.

Appearance of molybdenum cofactor in plants

Extensive analyses of MoCo in higher plant extracts have shown that nitrate application stimulates the appearance of Mo cofactor in N-starved barley plants (Mendel *et al.* 1982*a,b*; 1985; Mendel 1983). Ammonium was also shown not to substitute for nitrate in stimulating the appearance of MoCo (Mendel *et al.* 1985). This group has tried to make a distinction between total and free Mo cofactor based on how the extracts were treated prior to analysis, althought it is unclear whether this is a totally valid distinction (Mendel *et al.* 1985). We have limited our analysis of MoCo in maize leaves to total available cofactor and not tried to assess 'free' versus 'total' (Campbell *et al.* 1987). Because MoCo analyses are somewhat cumbersome, due to the nature of the bio-assay, our analyses were limited to investigations of steady-state conditions and we have not yet looked at the kinetic aspect of this question (Campbell *et al.* 1987). However, we found that only nitrate stimulated increases in total MoCo availability and that light had no significant influence (Campbell *et al.* 1987). We have extended these analyses to include a comparison of NR activity and protein as well as availability of MoCo under steady-state conditions of presence and absence of nitrate in light and dark after 24 h of treatment and the results are summarized in Table 9.4. We conclude

Table 9.4. Relative changes in nitrate reductase and molybdenum cofactor levels as influenced by light and nitrogen

		NR activity	NR protein	MoCo
No nitrogen	Dark	1	1	1
	Light	9	14	1
50 mM nitrate	Dark	8	10	2
	Light	55	58	2

The plants for these experiments were grown as described in Campbell *et al.* (1987) and the Mo cofactor data were adapted from that paper. The leaves were extracted and assayed for nitrate reductase activity and protein as described by Remmler and Campbell (1986). Nitrate reductase protein was determined using an ELISA method.

that under conditions of nitrate limitation and in the absence of light, NR protein is synthesized in excess relative to NR activity, perhaps reflecting lack of Mo cofactor availability (Table 9.4). Under conditions when nitrate is sufficient but light is not provided, NR protein levels exceed the levels of activity only slightly, probably because sufficient MoCo is available. More sensitive and facile assays for

MoCo will allow us to analyse the kinetics of NR activity and protein appearance in relation to the availability of MoCo.

Reversible inactivation of nitrate reductase

Since algal NRs appcar to have reversible mechanisms for inactivation and reactivation of enzyme activity (Vennesland and Guerrero 1979; Guerrero *et al.* 1981; Losada *et al.* 1981), several types of experiments have been done in attempts to show that similar mechansms exist for higher plants (Beevers and Hageman 1980; Campbell and Smarrelli 1986). Much of this effort has focused on cyanide as a key regulatory factor in controlling NR activity, which has yet to be proven (Vennesland and Guerrero 1979; Guerrero *et al.* 1981; Aryan *et al.* 1983). Somers *et al.* (1983b) showed that rocket immunoelectrophoresis could detect both the active form of NR and its cyanide-inactivated form. They also noted that when NR activity fell after nitrate was removed from the plants the NR protein decreased more slowly, as if an inactive form were present (Somers *et al.* 1983b). In our studies (Remmler and Campbell 1986; Campbell 1987), when nitrate-sufficient corn plants were transferred to darkness after 24 h light periods, NR activity decreased by 30% in about 1 h while NR protein remained constant, based on ELISA analysis. Furthermore, if the lights were turned back on after 1 h of darkness, the activity of the enzyme was rapidly recovered, with little change in the ELISA-detectable protein suggesting that enzyme was reactivated without new synthesis (Remmler and Campbell 1986; Campbell 1987). Thus, it appears that a mechanism for reversible inactivation of NR does exist in higher plants, but the molecular basis of it is yet to be shown.

We recently found that some of the monoclonal antibodies for corn NR can increase the activity of the enzyme. The data for one of these monoclonals are summarized in Table 9.5. The stimulation of activity is not dramatic, yielding an increase of only about 50%. However, the effect of the antibody appears to be global in that not only is NADH nitrate reduction stimulated but also both types of partial activities. The only other known stimulator of NR activity is phosphate (Nicholas and Scawin 1956), which was recently studied for barley NR (Oji *et al.* 1987). The phosphate stimulation of corn NR was evaluated in comparison to the antibody stimulation and the two were found to be additive (Table 9.5). In addition, phosphate stimulation is localized on the nitrate-reducing portion of the enzyme and has no influence on dehydrogenase activities (Howard and Solomonson 1981). Thus, the influence of phosphate and the monoclonal antibody on NR appear to be *via* different mechanisms and do not interfere with one another, the antibody being a global effector while the phosphate effect is localized at the nitrate site and may actually affect the nitrate K_m. While the ability of an antibody to stimulate NR activity suggests that the enzyme has a flexible conformation which can change—leading to an increase in activity—it is not at all clear that this can explain the differences observed in dark inactivation

Table 9.5. Activation of corn nitrate reductase by monoclonal antibody
Co-1 as compared to activation by phosphate[a]

Enzyme activity	Antibody	Buffer system	
		HEPES	Phosphate
NADH-NR	−	100	217
	+	154	275 [127]
NADH-FeCNR	−	100	104
	+	118	129 [124]
NADH-Cyt *c* R	−	100	118
	+	133	153 [130]
MVr-NR	−	100	209
	+	151	280 [134]
BPBr-NR	−	100	227
	+	125	271 [119]
FADH$_2$-NR	−	100	221
	+	176	316 [143]

[a] The percentage activation is expressed for each activity as compared to the base activity in the absence of antibody and phosphate. The activation by antibody as compared to the base activity in phosphate is indicated in square brackets. Two buffers were used: 50 mM HEPES, pH 7.5, and 30 mM K-phosphate, pH 7.5, with each containing 40 mM NaCl and 1 mM EDTA, except for MVr:NR where 50 mM Tris-Cl, pH 7.5, containing the additives was substituted for HEPES. Abbreviations: NR, nitrate reductase; FeCN R, ferricyanide reductase; Cyt *c* R, cytochrome *c* reductase; MVr, reduced methylviologen; and BPBr, reduced bromphenol blue.

experiments. In fact, experiments showing that dark-inactivated NR can be reactivated by treatment with ferricyanide suggest that NR is chemically altered when inactivated and that this chemical change can be reversed (Aryan *et al.* 1983; Wallace 1987).

Irreversible inactivation of nitrate reductase

It was clear in the experiments of Somers *et al.* (1983) that when nitrate was removed from barley plants decreases in NR activity were accompanied by losses in enzyme protein. This was also true in our experiments involving dark inactivation of NR activity (Remmler and Campbell 1986). After the initial drop in activity where no protein was lost, if darkness was continued, the activity continued to fall to very low levels over the next 12–24 h and protein also decreased, reaching low levels (Remmler and Campbell 1986). These results can best be explained by proteolytic degradation of NR (Sherrard and Dalling 1979; Huffaker 1982; Wallace and Oaks 1986). Thus, it appears that any reversible inactivation of NR is probably only a short-term effect and that enzyme is rapidly degraded if either nitrate or light is removed. In Western blot analysis of the degradation process, neither Somers *et al.* (1983) nor Remmler and Campbell (1986) observed a build-up of partially degraded forms of NR suggesting that nicked forms of the enzyme have a short half-life and that the enzyme protein is rapidly converted to amino acids. The proteolytic enzymes responsible for this rapid turnover of NR may be specific as has been suggested for some proteases which have been isolated from higher plants (Wallace and Oaks 1986). However, more general proteases may degrade NR when it becomes susceptible under the non-inducing conditions. This area of NR regulation is one where more study is required to clarify molecular mechanisms and enzymes involved.

Regulation of nitrate reductase at the level of transcription

The most recent advance in the study of NR involves the isolation of cDNA clones for barley, squash, tobacco, and corn (Cheng *et al.* 1986; Crawford *et al.* 1986; Calza *et al.* 1987; Campbell and Gowri 1987). These tools and their applications are described elsewhere in this volume (Chapters 12, 13, and 21). It is in this area of study that the next major advance in our understanding of NR and its role in regulation of nitrate assimilation will come. Several of the first experiments done in demonstrating the authenticity of the NR cDNA clones have focused on the steady state levels of poly(A)+RNA in nitrate-treated and non-treated plants (Cheng *et al.* 1986; Crawford *et al.* 1986; Calza *et al.* 1987; Campbell and Gowri 1987). In general, as might be expected from earlier studies (Beevers and Hageman 1980), the non-nitrate-treated plants have little NR mRNA while those exposed to nitrate contain higher level of this mRNA as shown by *in vitro* translation analysis (Commere *et al.* 1986). Recently, it was shown that squash NR cDNA could be used to study the regulation of mRNA for the NADH-NR form in soybean leaves and that glutamine had a modulating effect on nitrate control of the level of this message (Smarrelli *et al.* 1987). It is clear that these studies can be extended to investigate the effects of various potential regulators such as red light and to determine if these effects are acting at the level of transcription. In addition, cloning and sequencing of genomic DNA for NR along with its 5′- and 3′-flanking regions will permit study of the importance of regulatory nucleotide sequences and perhaps identification of those involved in 'nitrate induction'. This might subsequently lead to the isolation of *trans*-acting proteins which bind to these sequences. In addition, the presence of light-activating sequences similar to those found for various chloroplast genes may be found and the interesting relationship of the family of light-activating sequences might then be studied (Campbell and Gowri 1987). The cDNA clones may be useful in studying mutants of NR in higher plants in order to identify the specific site of mutation. From this variety of new approaches, it can be expected that more specific details will become known about the molecular mechanism for nitrate 'induction' of NR and the relationship of light activation to the nitrate process.

Acknowledgements

I thank the National Science Foundation for grant DMB 85-02672, which currently supports our research on the biochemistry of nitrate reductase, and the U.S. Department of Agriculture, Competitive Grants Office for grant 86-CRCR-1-2071, which supports our studies on the regulation of nitrate reductase. I also thank my students and colleagues who have provided unpublished data described in this review.

References

Abrol, Y. P., Sawhney, S. K., and Naik, M. S. (1983). Light and dark assimilation of nitrate in plants. *Plant, Cell and Environment* **6**, 595–9.

Afridi, M. M. R. K. and Hewitt, E. J. (1964). The inducible formation and stability of nitrate reductase in higher plants. I. Effects of nitrate and molybdenum on enzyme activity in cauliflower (Brassica oleracea var. botrytis). *Journal of Experimental Botany* **15**, 251–71.

Aryan, A. P., Batt, R. G., and Wallace, W. (1983). Reversible inactivation of nitrate reductase by NADH and the occurrence of partially inactive enzyme in the wheat leaf. *Plant Physiology* **71**, 582–7.

Beevers, L. and Hageman, R. H. (1972). The role of light in nitrate metabolism in higher plants. In *Current topics in photobiology and photochemistry*, Vol. VII (ed. A. C. Giese), pp. 85–113. Academic Press, New York.

Beevers, L. and Hageman, R. H. (980). Nitrate and nitrite reduction. In *The biochemistry of plants*, Vol. 5 (ed. B. J. Miflin), pp. 115–68. Academic Press, New York.

Beevers, L., Flesher, D., and Hageman, R. H. (1964). Studies on the pyridine nucleotide specificity of nitrate reductase in higher plants and its relationship to sulfhydryl level. *Biochimica et Biophysica Acta* **89**, 453–64.

Calza, R., Huttner, E., Vincentz, M., Rouze, P., Galangau, F., Vaucheret, H., Cherel, I., Meyer, C., Kronenberger, J., and Caboche, M. (1987). Cloning of DNA fragments complementary to tobacco nitrate reductase mRNA and encoding for epitopes common to the nitrate reductases from higher plants. *Molecular and General Genetics* **209**, 552–67.

Campbell, W. H. (1976). Separation of soybean leaf nitrate reductases by affinity chromatography. *Plant Science Letters* **7**, 239–44.

Campbell, W. H. (1978). Isolation of NAD(P)H:nitrate reductase from the scutellum of maize. *Zeitschrift für Pflanzenphysiologie* **88**, 357–61.

Campbell, W. H. (1979). Affinity purification and kinetic studies of squash cotyledon NADH nitrate reductase. In *Nitrogen assimilation of plants* (eds. E. J. Hewitt and C. V. Cutting), pp. 321–30. Academic Press, New York.

Campbell, W. H. (1985). The biochemistry of higher plant nitrate reductase. In *Nitrogen fixation and carbon dioxide metabolism* (ed. P. W. Ludden and J. E. Burris), pp. 143–51. Elsevier Science Publishing Co, New York.

Campbell, W. H. (1986). Properties of bromophenol blue as an electron donor for higher plant NADH: nitrate reductase. *Plant Physiology* **82**, 729–32.

Campbell, W. H. (1987). Regulation of nitrate reductase in maize: An immunochemical approach. In *Inorganic nitrogen metabolism* (eds. W. R. Ullrich, P. J. Aparicio, P. J. Syrett and F. Castillo), pp. 99–103. Springer-Verlag, Berlin.

Campbell, W. H. and Gowri, G. (1987). Regulation of maize leaf NADH:nitrate reductase m-RNA by nitrate and light. *Journal of Cellular Biochemistry* **11B**, 49.

Campbell, W. H. and Remmler, J. L. (1986). Regulation of corn leaf nitrate reductase. I. Immunochemical methods for analysis of the enzyme's protein component. *Plant Physiology* **80**, 435–41.

Campbell, W. H. and Ripp, K. G. (1984). An ELISA for higher plant nitrate reductase. *Annals of the New York Academy of Sciences* **435**, 123–5.

Campbell, W. H. and Smarrelli, J. (1978). Purification and kinetics of higher plant NADH, nitrate reductase. *Plant Physiology* **61**, 611–16.

Campbell, W. H. and Smarrelli, J. (1986). Nitrate reductase: biochemistry and regulation. In *Biochemical basis of plant breeding*, Vol. II (ed. C. Neyra), pp. 1–45. CRC Press, Boca Ratan, Florida.

Campbell, J. M. and Wray, J. L. (1983). Purification of barley nitrate reductase and demonstration of nicked subunits. *Phytochemistry* **22**, 2375–82.

Campbell, W. H., DeGarcia, D. J., and Campbell, E. R. (1987). Regulation of molybdenum cofactor of maize leaf. *Phytochemistry* **27**, 2149–50.

Cheng, C.-L., Dewdney, J., Kleinhofs, A., and Goodman, H. M. (1986). Cloning and nitrate induction of nitrate reductase mRNA. *Proceedings of the National Academy of Sciences, USA* **83**, 6825–8.

Cherel, I., Grosclaude, J., and Rouze, P. (1985). Monoclonal antibodies identify multiple epitopes on maize leaf nitrate reductase. *Biochemical and Biophysical Research Communications* **129**, 685–91.

Cherel, I., Marion-Poll, A., Meyer, C., and Rouze, P. (1986). Immunological comparisons of nitrate reductases of different plant species using monoclonal antibodies. *Plant Physiology* **81**, 376–8.

Commere, B., Cherel, I., Kronenberger, J., Galangau, F., and Caboche, M. (1986). *In vitro* translation of nitrate reductase messenger RNA from maize and tobacco and detection with an antibody directed against the enzyme of maize. *Plant Science* **44**, 191–203.

Coughlan, M. P. (1980). Aldehyde oxidase, xanthine oxidase and xanthine dehydrogenase: Hydroxylases containing molybdenum, iron–sulfur and flavin. In *Molybdenum and molybdenum-containing enzymes* (ed. M. P. Coughlan), pp. 119–85. Pergamon Press, Oxford.

Coughlan, M. P. and Rajagopalan, K. V. (1980). The kinetic mechanism of xanthine dehydrogenase and related enzymes, *European Journal of Biochemistry* **105**, 81–4.

Crawford, N. M., Campbell, W. H., and Davis, R. W. (1986). Nitrate reductase from squash: cDNA cloning and nitrate regulation. *Proceedings of the National Academy of Sciences, USA* **83**, 8073–6.

Dailey, F. A., Warner, R. L., Somers, D. A., and Kleinhofs, A. (1982). Characteristics of a nitrate reductase barley mutant deficient in NADH, nitrate reductase. *Plant Physiology* **69**, 1200–4.

De la Rosa, F. F. and Palacian, E. (1981). Spinach nitrate reductase. Kinetic studies of NADH-diaphorase. *Plant Science Letters* **21**, 1–8.

De la Rosa, F. F., Palacian, E., and Castillo, F. (1980) Studies on the kinetic mechanism of nitrate reductase from spinach (Spinacea oleracea), *Revista Expanola de Fisiologia* **36**, 279–83.

Evans, H. J. and Nason, A. (1953). Pyridine nucleotide-nitrate reductase from extracts of higher plants. *Plant Physiology* **28**, 233–54.

Fido, R. J. (1987). Purification of nitrate reductase from spinach (*Spinacea oleracea* L.) by immunoaffinity chromatography using a monoclonal antibody. *Plant Science* **50**, 111–6.

Fido, R. J. and Notton, B. A. (1984). Spinach nitrate reductase: further purification and removal of 'nicked' subunits by affinity chromatography. *Plant Science Letters* **37**, 87–91.

Fido, R. J., Hewitt, E. J., Notton, B. A., Jones, O. T. G., and Nasrulhaq-Boyce, A. (1979). Haem of spinach nitrate reductase, low temperature spectrum and mid-point potential. *FEBS Letters* **99**, 180–5.

Garrett, R. H. and Nason, A. (1967) Involvement of a *b*-type cytochrome in the assimilatory nitrate reductase of *Neurospora crassa*. *Proceedings of the National Academy of Sciences, USA* **58**, 1603–8.

Guerrero, M. G., Vega, J. M., and Losada, M. (1981). The assimilatory nitrate-reducing system and its regulation. *Annual Review of Plant Physiology* **32**, 169–204.

Hageman, R. H. (1979). Integration of nitrogen assimilation in relation to yield. In *Nitrogen assimilation of plants* (eds. E. J. Hewitt and C. V. Cutting), pp. 591–611. Academic Press, New York.

Hageman, R. H. and Flesher, D. (1960). Nitrate reductase activity in corn seedlings as affected by light and nitrate content of nutrient media. *Plant Physiology* **35**, 700–8.

Harker, A. R., Narayanan, K. R., Warner, R. L., and Kleinhofs, A. (1986). NAD(P)H bispecific nitrate reductase in barley leaves: partial purification and characterization. *Phytochemistry* **25**, 1275–9.

Harper, J. E. (1981). Evolution of nitrogen oxide(s) during *in vivo* nitrate reductase assay of soybean leaves. *Plant Physiology* **68**, 1488–93.

Heath-Pagliuso, S., Huffaker, R. C., and Allard, R. W. (1984). Inheritance of nitrite reductase and regulation of nitrate reductase, nitrite reductase and glutamine synthetase isozymes. *Plant Physiology* **76**, 353–8.

Hewitt, E. J. and Notton, B. A. (1980). Nitrate reductase systems in eukaryotic and prokaryotic organisms. In *Molybdenum and molybdenum-containing enzymes* (ed. M. P. Coughlan), pp. 275–325. Pergamon Press, Oxford.

Hoarau, J., Hirel, B., and Nato, A. (1986). New artificial electron donors for *in vitro* assay of nitrate reductase isolated from cultured tobacco cells and other organisms. *Plant Physiology* **80**, 946–9.

Howard, W. D. and Solomonson, L. P. (1981). Kinetic mechanism of assimilatory NADH, nitrate reductase from Chlorella. *Journal of Biological Chemistry* **256**, 12725–30.

Howard, W. D. and Solomonson, L. P. (1982). Quaternary structure of assimilatory NADH, nitrate reductase from Chlorella. *Journal of Biological Chemistry* **257**, 10243–50.

Huffaker, R. C. (1982). Biochemistry and physiology of leaf proteins. In *Encyclopedia of plant physiology*, New series Vol. 14A, *Nucleic acids and proteins in plants. I. Structure, biochemistry and physiology of proteins* (eds. D. Boulter and B. Parthier), pp. 370–400. Springer-Verlag, Berlin.

Jacob, G. S. and Orme-Johnson, W. H. (1980). Prosthetic groups and mechanism of action of nitrate reductase from *Neurospora crassa*. In *molybdenum and molybdenum-containing enzymes* (ed. M. P. Coughlan), pp. 327–44. Pergamon Press, Oxford.

Jolly, S. O., Campbell, W. H., and Tolbert, N. E. (1976). NADPH- and NADH-nitrate reductases from soybean leaves. *Archives of Biochemistry and Biophysics* **174**, 431–9.

Kay, C. J. and Barber, M. J. (1986). Assimilatory nitrate reductase from *Chlorella*. Effect of ionic strength and pH on catalytic activity. *Journal of Biological Chemistry* **261**, 14125–9.

Kleinhofs, A., Narayanan, K. R., Somers, D. A., Kuo, T. M., and Warner, R. L. (1986). Immunochemical methods for higher plant nitrate reductase. In *Immunology in plant sciences* (eds. H. F. Linskens and J. F. Jackson), pp. 191–211. Springer-Verlag, Berlin.

Kuo, T., Kleinhofs, A., and Warner, R. L. (1980). Purification and partial characterization of nitrate reductase from barley leaves. *Plant Science Letters* **17**, 371–81.

Kuo, T. M., Somers, D. A., Kleinhofs, A., and Warner, R. L. (1982). NADH-nitrate reductase in barley leaves. Identification and amino acid composition of subunit protein. *Biochimica et Biophysica Acta* **708**, 75–81.

Le, K. H. D. and Lederer, F. (1983). On the presence of a heme-binding domain homologous to cytochrome *b*-5 in *Neurospora crassa* assimilatory nitrate reductase. *EMBO Journal* **2**, 1909–14.

Losada, M. (1975). Metalloenzymes of the nitrate-reducing system. *Journal of Molecular Catalysis* **1**, 245–64.

Losada, M., Guerrero, M. G., and Vega, J. M. (1981). The assimilatory reduction of nitrate. In *Biology of inorganic nitrogen and sulfur* (eds. H. Bothe and A. Trebst), pp. 30–63, Springer-Verlag, Berlin.

Maki, H., Yamagishi, K., Sato, T., Ogura, N., and Nakagawa, H. (1986). Regulation of nitrate reductase activity in cultured spinach cells as studied by an enzyme-linked immunosorbent assay. *Plant Physiology* **82**, 739–41.

Mannervik, B. (1976). The kinetic mechanism of glutathione reductase: A branching reaction scheme applicable to many flavoproteins. In *Flavins and flavoproteins* (ed. T. P. Singer), pp. 485–91. Elsevier, Amsterdam.

McDonald, D. W. and Coddington, A. (1974). Properties of the assimilatory nitrate reductase from *Aspergillus nidulans*. *European Journal of Biochemistry* **46**, 169–78.

Mendel, R. R. (1983). Release of molybdenum cofactor from nitrate reductase and xanthine oxidase by heat treatment. *Phytochemistry* **22**, 817–19.

Mendel, R. R., Alikulov, Z. A., and Muller, A. J. (1982*a*). Molybdenum cofactor in nitrate reductase-deficient tobacco mutants. II. Release of cofactor by heat treatment. *Plant Science Letters* **25**, 67–72.

Mendel, R. R., Alikulov, Z. A., and Muller, A. J. (1982*b*). Molybdenum cofactor in nitrate reductase-deficient tobacco mutants. III. Induction of cofactor synthesis by nitrate. *Plant Science Letters* **27**, 95–101.

Mendel, R. R., Kirk, D. W., and Wray, J. L. (1985). Assay of molybdenum cofactor of barley. *Phytochemistry* **24**, 1631–4.

Miller, A. and Campbell, W. H. (1986). Monoclonal antibodies for higher plant nitrate reductase. *Current Topics in Plant Biochemistry and Physiology* **5**, 192.

Mulvaney, C. S. and Hageman, R. H. (1984). Acetaldehyde oxime, a product formed during the *in vivo* nitrate reductase assay of soybean leaves. *Plant Physiology* **76**, 118–24.

Naik, M. S., Arrol, Y. P., Nair, T. V. R., and Ramarao, C. S. (1982). Nitrate assimilation—its regulation and relationship to reduced nitrogen in higher plants. *Phytochemistry* **21**, 495–504.

Nakagawa, H., Poulle, M., and Oaks, A. (1984). Characterization of nitrate reductase from corn leaves (*Zea mays* cv W64A × W182E). Two molecular forms of the enzyme. *Plant Physiology* **75**, 285–9.

Nakagawa, H., Yonemura, Y., Yamamoto, H., Sato, T., Ogura, N., and Sato, R. (1985). Spinach nitrate reductase. Purification, molecular weight, and subunit composition. *Plant Physiology* **77**, 124–8.

Narayanan, K. R., Somers, D. A., Kleinhofs, A., and Warner, R. L. (1983). Nature of cytochrome *c* reductase in nitrate reductase-deficient mutants of barley. *Molecular and General Genetics* **190**, 222–6.

Nelson, R. S., Ryan, S. A., and Harper, J. E. (1983). Soybean mutants lacking constitutive nitrate reductase activity. I. Selection and initial plant characterization. *Plant Physiology* **72**, 503–9.

Nelson, R. S., Horn, M. E., Harper, J. E., and Widholm, J. M. (1984*a*). Nitrate reductase activity and nitrogenous gas evolution from heterotropic, photomixotrophic and photoautotrophic soybean suspension cultures. *Plant Science Letters* **34**, 145–52.

Nelson, R. S., Streit, L., and Harper, J. E. (1984*b*). Biochemical characterization of nitrate and nitrite reduction in wild-type and a nitrate reductase mutant of soybean. *Physiologia Plantarum* **61**, 384–90.

Nelson, R. S., Streit, L., and Harper, J. E. (1986). Nitrate reductase from wild-type and *nrl*-mutant soybean (*Glycine max* [L.] Merr.) leaves. II. Partial activity, inhibitor and complementation analyses. *Plant Physiology* **80**, 72–6.

Nicholas, D. J. D. and Scawin, J. H. (1956). A phosphate requirement for nitrate reductase from *Neurospora crassa*. *Nature* **178**, 1474–5.

Northrup, D. B. (1969). Transcarboxylase. VI. Kinetic analysis of the reaction mechanism. *Journal of Biological Chemistry* **244**, 5808–19.

Notton, B. A. and Hewitt, E. J. (1979). Structure and properties of higher plant nitrate reductase, especially Spinacea oleracea. In *Nitrogen assimilation of plants* (eds. E. J. Hewitt and C. V. Cutting), pp. 227–44. Academic Press, New York.

Notton, B. A., Fido, R. J., and Hewitt, E. J. (1977). The presence of functional haem in a higher plant nitrate reductase. *Plant Science Letters* **8**, 165–70.

Notton, B. A., Fido, R. J., and Galfre, G. (1985). Monoclonal antibodies to a higher plant nitrate reductase, Differential inhibition of enzyme activities. *Plants* **165**, 114–9.

Oji, Y., Ryoma, Y., Wakiuchi, N., and Okamoto, S. (1987). Effect of inorganic orthophosphate on *in vitro* activity of NADH-nitrate reductase from 2-row barley leaves. *Plant Physiology* **83**, 472–4.

Orihuel-Iranzo, B. and Campbell, W. H. (1980). Development of NAD(P)H: and NADH:nitrate reductase activities in soybean. *Plant Physiology* **65**, 595–9.

Orihuel-Iranzo, B. and Campbell, W. H. (1981). Pyridine nucleotide specificity and kinetics of soybean leaf NAD(P)H:nitrate reductase. In *Photosynthesis*, Vol. IV (ed. G. Akoyunoglou), pp. 715–20. Balaban Int. Science Services, Philadelphia.

Palmer, G. and Olson, J. S. (1980). Concepts and approaches to the understanding of electron transfer processes in enzymes containing multiple redox centers. In *Molybdenum and molybdenum-containing enzymes* (ed. M. P. Coughlan), pp. 187–220. Pergamon Press, Oxford.

Poulle, M., Oaks, A., Bzonek, P., Goodfellow, V. J., and Solomonson, L. P. (1987). Characterization of nitrate reductases from corn leaves (*Zea mays* cv W64A × W182E) and *Chlorella vulgaris*. *Plant Physiology* **85**, 375–8.

Redinbaugh, M. G. and Campbell, W. H. (1981). Purification and characterization of NAD(P)H:nitrate reductase and NADH:nitrate reductase from corn roots. *Plant Physiology* **68**, 115–20.

Redinbaugh, M. G. and Campbell, W. H. (1983). The purification of squash NADH, nitrate reductase by zinc chelate affinity chromatography. *Plant Physiology* **71**, 205–7.

Redinbaugh, M. G. and Campbell, W. H. (1985). Quaternary structure and composition of squash NADH, nitrate reductase. *Journal of Biological Chemistry* **260**, 3380–5.

Remmler, J. L. and Campbell, W. H. (1986). Regulation of corn leaf nitrate reductase. I. Synthesis and turnover of the enzyme's activity and protein. *Plant Physiology* **80**, 442–7.

Rendina, A. R. and Orme-Johnson, W. H. (1978). Glutamate synthase, on the kinetic mechanism of the enzyme from *Escherichia coli* W. *Biochemistry* **17**, 5388–95.

Renosto, F., Ornitz, D., Peterson, D., and Segal, I. H. (1981). Nitrate reductase from Penicillium chrysogenum, Purification and kinetic mechanism. *Journal of Biological Chemistry* **256**, 8616–25.

Renosto, F., Schmidt, N. D., and Segel, I. H. (1982*a*). Nitrate reductase from Penicillium chrysogenum: kinetic mechanism at sub-optimum pH. *Biochemical and Biophysical Research Communications* **107**, 12–18.

Renosto, F., Schmidt, N. D., and Segel, I. H. (1982*b*). Nitrate reductase from *Penicillium chrysogenum*: the reduced flavinadenine dinucleotide-dependent reaction. *Archives of Biochemistry and Biophysics* **219**, 12–20.

Robin, P., Streit, L., Campbell, W. H., and Harper, J. E. (1985). Immunochemical characterization of nitrate reductase forms from wild-type (cv. Williams) and *nrl* mutant soybean. *Plant Physiology* **77**, 232–6.

Ryan, S. A., Nelson, R. S., and Harper, J. E. (1983). Soybean mutants lacking constitutive nitrate reductase activity. II. Nitrogen assimilation, chlorate resistance and inheritance. *Plant Physiology* **72**, 510–14.

Schrader, L. E., Ritenour, G. L., Eilrich, G. L., and Hageman, R. H. (1968). Some characteristics of nitrate reductase from higher plants. *Plant Physiology* **43**, 930–40.

Shen, T., Funkhouser, E. A., and Guerrero, M. G. (1979). NADH- and NADPH-anitrate reductases in rice seedlings. *Plant Physiology* **58**, 292–4.

Sherrard, J. H. and Dalling, M. J. (1979). *In vitro* stability of nitrate reductase from wheat leaves. *Plant Physiology* **63**, 346–53.

Small, I. and Wray, J. L. (1979). Breakdown of barley NADH-nitrate reductase to functional NADH-cytochrome *c* reductase species. *Biochemical Society Transaction* **7**, 737–9.

Smarrelli, J. and Campbell, W. H. (1979). NADH dehydrogenase activity of higher plant nitrate reductase (NADH). *Plant Science Letters* **16**, 139–47.

Smarrelli, J. and Campbell, W. H. (1981). Immunological approaches to structural comparisons of assimilatory nitrate reductases. *Plant Physiology* **68**, 1226–30.

Smarrelli, J. and Campbell, W. H. (1983). Heavy metal inactivation and chelator stimulation of higher plant nitrate reductase. *Biochimica et Biophysica Acta* **742**, 435–45.

Smarrelli, J., Malone, M. J., Watters, M. T., and Curtis, L. T. (1987). Transcriptional control of the inducible nitrate reductase isoform from soybean. *Biochemical and Biophysical Research Communications* **146**, 1160–5.

Solomonson, L. P. (1975). Purification of NADH-nitrate reductase by affinity chromatography. *Plant Physiology* **56**, 853–7.

Solomonson, L. P. and McCreery, M. J. (1986). Radiation inactivation of assimilatory NADH:nitrate reductase from *Chlorella*. Catalytic and physical sizes of functional units. *Journal of Biological Chemistry* **261**, 806–10.

Solomonson, L. P., Barber, M. J., Robbins, A. P., and Oaks, A. (1986). Functional domains of assimilatory NADH:nitrate reductase from *Chlorella*. *Journal of Biological Chemistry* **261**, 11290–4.

Somers, D. A., Kuo, T., Kleinhofs, A., and Warner, R. L. (1982). Barley nitrate reductase contains a functional cytochrome *b*-557. *Plant Science Letters* **24**, 261–4.

Somers, D. A., Kuo, T., Kleinhofs, A., and Warner, R. L. (1983*a*). Nitrate reductase-deficient mutants in barley. Immunoelectrophoretic characterization. *Plant Physiology* **71**, 145–9.

Somers, D. A., Kuo, T., Kleinhofs, A., and Warner, R. L., and Oaks, A. (1983*b*). Synthesis and degradation of barley nitrate reductase. *Plant Physiology* **72**, 949–52.

Sorger, G., Gooden, D. O., Earle, E. D., and McKinnon, J. (1986). NADH nitrate reductase and NAD(P)H nitrate reductase in genetic variants and regenerating callus of maize. *Plant Physiology* **82**, 473–8.

Srivastava, H. S. (1980). Regulation of nitrate reductase activity in higher plants.

Phytochemistry **19**, 725–33.

Streit, L. and Harper, J. E. (1986). Biochemical characterization of soybean mutants lacking constitutive NADH:nitrate reductase. *Plant Physiology* **81**, 593–6.

Streit, L., Nelson, R. S., and Harper, J. E. (1985). Nitrate reductases from wild-type and *nrl*-mutant soybean (*Glycine max* [L.] Merr.) leaves. I. Purification, kinetics and physical properties. *Plant Physiology* **78**, 80–4.

Streit, L., Martin, B. A., and Harper, J. E. (1987). A method for the separation and partial purification of the three forms of nitrate reductase present in wild-type soybean leaves. *Plant Physiology* **84**, 654–7.

Travis, R. L., Jordan, W. R., and Huffacker, R. C. (1970). Light and nitrate requirements for induction of nitrate reductase activity in *Hordeum vulgare*. *Physiologia Plantarum* **23**, 678–85.

Vennesland, B. and Guerrero, M. G. (1979). Reduction of nitrate and nitrite. In *Encyclopedia of plant physiology*, New series, Vol. 6, Photosynthesis II. Photosynthetic carbon metabolism and related processes (eds. M. Gibbs and E. Latzko), pp. 425–44. Springer-Verlag, Berlin.

Wallace, W. (1977). Proteolytic inactivation of enzymes. In *Regulation of enzyme synthesis and activity in higher plants* (ed. H. Smith), pp. 177–95. Academic Press, London.

Wallace, W. (1987). Regulation of nitrate utilization in higher plants. In *Inorganic nitrogen metabolism* (eds. W. R. Ullrich, P. J. Aparicio, P. J. Syrett and F. Castillo), pp. 223–30. Springer-Verlag, Berlin.

Wallace, W. and Johnson, C. B. (1978). Nitrate reductase and soluble cytochrome *c* reductases in higher plants, *Plant Physiology* **61**, 748–52.

Wallace, W. and Oaks, A. (1986). Role of proteinases in the regulation of nitrate reductase. In *Plant proteolytic enzymes*, Vol. II (ed. M. J. Dalling), pp. 81–9. CRC Press, Boca Ratan, Florida.

Wells, G. and Hageman, R. H. (1974). Specificity for nicotinamide adenine dinucleotide by nitrate reductase. *Plant Physiology* **54**, 136–41.

Whitford, P. N., Fido, R. J., and Notton, B. A. (1987). An enzyme linked immunosorbent assay for nitrate reductase using monoclonal antibodies. *Phytochemistry* **26**, 2467–70.

Wray, J. L. and Filner, P. (1970). Structural and functional relationship of enzyme activities induced by nitrate in barley. *Biochemical Journal* **119**, 715–25.

Wray, J. L. and Kirk, D. W. (1981). Inhibition of NADH-nitrate reductase degradation in barley leaf extracts by leupeptin. *Plant Science Letters* **23**, 207–13.

Wray, J. L., Small, I. S., and Brown, J. (1979) A model for the subunit composition of higher plant NADH-nitrate reductase, *Biochemical Society Transactions* **7**, 739–41.

Xia, Z., Shamala, N., Bethge, P. H., Lim, L. W., Bellamy, H. D., Xuong, N., Lederer, F., and Mathews, F. S. (1987). Three-dimensional structure of flavocytochrome *b*-2 from baker's yeast at 3.0 Å resolution. *Proceedings of the National Academy of Sciences, USA* **84**, 2629–33.

Zielke, H. R. and Filner, P. (1971). Synthesis and turnover of nitrate reductase induced by nitrate in cultured tobacco cells. *Journal of Biological Chemistry* **246**, 1772–80.

10. Immunology of nitrate reductase with special reference to higher plants

Brian A. Notton

Twenty-five years ago Murray and Sandwal (1963) used an antiserum prepared against a semi-purified respiratory nitrate reductase (NR) from *Escherichia coli* to precipitate assimilatory NR from the same organism. The precipitate was shown to have benzylviologen NR activity. At about the same time, Pateman *et al.* (1964) used antibodies raised against crude extracts of *Aspergillus nidulans*, which were shown to inhibit NADH NR activity, to examine extracts of NR-deficient mutants. They used cross-reacting material present in some of the extracts to reduce the amount of antibody available to inhibit added active enzyme. The effect of antiserum raised against a semi-purified spinach NR on the overall (NADH-nitrate reductase) and partial (dehydrogenase and terminal nitrate-reducing) activities of the enzyme was examined by Graf *et al.* (1975*a*). All three activities were inhibited (dehydrogenase incompletely) at high concentrations of both the whole immune serum and its γ-globulin fraction: at lower concentrations the enzyme activities were stimulated. This stimulation applied particularly to the terminal nitrate reducing activity. A 'standard' enzyme was used to measure the relative inhibitory capability of the serum and its γ-globulin fraction against the three activities. The demonstrated stimulation led to speculation about the presence of stimulatory antibodies, which had been previously reported for anti-penicillinase (Pollock 1964). Graf *et al.* (1975*b*) attempted to quantify the amount of antigenic, enzymically inactive protein present in a solution by measuring the increase in the amount of antiserum required to cause 50 per cent inactivation of a standardized enzyme preparation by the competing protein. This technique was then used (Notton *et al.* 1974) to show that molybdenum-deficent spinach plants, grown with an ammonium supplement, contained enzymically inactive cross-reacting material to the antiserum, whereas plants grown with only nitrate-nitrogen did not. This cross-reacting material, which was concentrated in fractions associated with the holoenzyme during purification, was thought to be aponitrate reductase. The supposed apoenzyme was also shown to be much less stable than the equivalent holonitrate reductase and it was concluded that molybdenum cofactor (MoCo) conferred stability to the enzyme complex.

These experiments, carried out with crude extracts or semi-purified nitrate reductase as immunogens, were early attempts to utilize the resultant anitsera as analytical tools. The following selection of reported research will show the progress that has been made recently in the production and use of antibodies to probe the macromolecular structure, formation, function and phylogenicity of nitrate reductase.

Nitrate reductase is difficult to obtain in a pure, undegraded state. The word 'pure' has different meanings, immunologically and analytically, in that a trace contaminant may be extremely immunogenic, resulting in the production of a disproportionate amount of antibodies directed against that contaminant. Fortunately NR is a good immunogen and a few hundred μg are sufficient to induce an immune response in an animal. However, to obtain mono-specific antiserum a number of different techniques have had to be used.

The term 'mono-specific' has been used loosely in the plant literature to imply a solution either containing no antibodies directed against other proteins present in the immunogen source material, but still containing antibodies endogenous to the animal, or, more correctly, containing only antibodies directed against the original immunogen.

Amy and Garrett (1979) used two different methods to purify their antiserum to *Neurospora crassa* nitrate reductase. They either precipitated out contaminating antibodies with an extract from ammonium-grown *N. crassa*, which contained contaminating antigens but not NR, or they localized, by staining briefly for enzyme activity, a NR–antinitrate reductase complex formed by immunoelectrophoresis, and used that complex as a fresh immunogen. The resultant, purified, antibodies were then used to examine extracts of *N. crassa* NR-deficient mutants for the presence of NR-related proteins by three methods; protection of inhibition of added active enzyme, rocket immunoelectrophoresis and crossed immunoelectrophoresis.

Nakagawa *et al.* (1986) purified antiserum to spinach NR by chromatography of the antiserum against contaminating proteins present in the original immunogen covalently coupled to Sepharose. The unbound fraction contained antibodies with greatly increased specificity.

The antiserum obtained to *Chlorella vulgaris* NR (Funkhouser and Ramadoss 1980) was purified by passing the antiserum over a column of the purified enzyme covalently coupled to Sepharose, followed by elution of the bound antibodies under acid conditions.

NR from squash (Smarrelli and Campbell 1981) and barley (Somers *et al.* 1983) was identified on gels after polyacrylamide electrophoresis, excised from the gels and pulverized in buffer. This purified enzyme was used as an immunogen.

A number of workers have used antiserum to NR from one plant species for immunological comparisons of structures of NR obtained from different eukaryotic sources. Smarrelli and Campbell (1981) used antiserum to NR extracted from squash to compare the enzyme from four higher plants (squash, spinach, corn and soybean), *C. vulgaris* and *N. crassa*. Immunodiffusion studies revealed a high degree of similarity between the NADH-enzyme from squash and spinach, while the NADH-enzyme from corn and soybean and the NAD(P)H-enzyme from soybean were less closely related to the enzyme from squash. The algal and fungal enzymes were very low in cross-reactivity and antigenically different to the enzyme from squash. Although the antiserum was able to inhibit the overall activity of all species tested, the amount of antiserum required varied considerably. A similar experiment was performed using antiserum to barley nitrate reductase (Snapp *et al.* 1984) against the enzyme obtained from nine higher plant species. Rocket immunoelectrophoresis and antiserum titration showed that the enzymes from barley and wheat were indistinguishable, that from rye was slightly different, while oat and maize nitrate reductases were easily distinguished from the barley enzyme. Extracts of dicotyledonous species, tobacco, soybean and pea, failed to form rockets on immunoelectrophoresis against the barley antiserum. Enzyme activity of all species was inhibited by the barley antiserum but the amounts required for 50 per cent inhibition, particularly by the enzyme from dicotyledonous species, were much higher (2- to 10-fold). This suggests that although structural differences exist between the eukaryotic NR examined, including algae and fungi, there is sufficient conservation of antigenic sites to cause inhibition of enzyme activity without necessarily causing precipitation.

The presence and quantification of cross-reacting material in extracts of barley NR-deficient mutants was examined by their ability to protect an active barley NR from inactivation by mono-specific antiserum to the barley enzyme (Kuo *et al.* 1981). The same extracts were found to form rockets on immunoelectrophoresis against the antiserum only in the presence of native enzyme. This was considered to be due to the formation of an immunoprecipitate by the native enzyme, which allowed the product of the mutation to interact with the complex and therefore increase the height of the rockets (Somers *et al.* 1983).

Mono-specific antibodies, prepared either by chromatography of the IgG fraction of an immune serum against NR linked to Sepharose or the IgG fraction of antiserum obtained by Protein A–Sepharose chromatography, have been used in a sandwich type ELISA for quantification of antigenic NR in crude extracts of spinach (Maki *et al.* 1986), squash (Campbell and Ripp 1984) and corn (Campbell and Remmler 1986). Peroxidase-labelled antibodies were used to detect antigenic enzyme bound to unlabelled antibodies fixed either to polystyrene beads or the wells of a microtitre plate. The methods were extremely sensitive, amounts of NR in the range 0.5–4 ng (spinach) and 0.5–10 ng (squash and corn) could be determined. The same anti-spinach NR as above was used to immuno-gold localize the enzyme in spinach leaves (Kamachi *et al.* 1987). The results obtained showed that the enzyme was specifically located in the chloroplast.

A major advance in the immunological study of NR came with the possibility of producing monoclonal antibodies (Kohler and Milstein 1975) which are, by definition, mono-specific. Provided a suitable assay is available to distinguish the protein of interest from contaminating immunogens, a pure immunogen is not initially essential for the production of the monoclonal antibodies, althought it is obviously desirable, thus minimizing the number of cell lines producing monoclonal antibodies to impurities. The paper by Galfre and Butcher (1986) is a useful introduction to the techniques involved and includes a section entitled 'Diary of a successful experiment' which describes, in some detail, the production of monoclonal antibodies to spinach NR. Some of the early results on the characterization and properties of these monoclonal antibodies were reported by Notton *et al.* (1985).

NR is inherently unstable. Our attempts to purify the enzyme from several other higher plants, particularly cereals, using the methodology previously used for spinach NR (Fido and Notton, 1984) in sufficient quantities to measure midpoint potentials and perform the same physiochemical analysis as we have done with the spinach enzyme (Notton *et al.* 1977; Fido *et al.* 1979; Gutteridge *et al.* 1983; Barber *et al.* 1987) were unsuccessful. It is also well known that estimation of the enzyme content of tissues by measurement of enzyme activity is subject to a number of variables which influence apparent activity. These include ionic strength of the assay medium, the presence of endogenous FAD, a sulphydryl reagent, and EDTA, as well as protease inhibitors.

Notton *et al.* (1985) therefore set out to prepare a set of monoclonal antibodies that would be useful in the large scale purification of the enzyme from a number of different species and in the quantification of the NR protein rather than enzyme activity. Our objectives were based on previous observations that antiserum to NR raised against the enzyme from a wide range of eukaryotic sources had the ability to inhibit the activity of the enzyme from a large number of different species, without necessarily causing an immunoprecipitate. This indicated a conservation, albeit limited, of antigenic sites. These sites should be identifiable using monoclonal antibodies.

The initial immunogen (spinach NR) was purified 3400-fold (Fido and Notton 1984), which made it more likely that supernatants from confluent hybrid myelomas (Galfre and Milstein 1981) which gave positive direct ELISA against the same purified enzyme would contain monoclonal antibodies directed against NR and not against a contaminant. Twenty-four positive cell lines were obtained. Confirmation of NR as the antigen was initially obtained by testing for inhibition of total enzyme activity and six cell lines were selected for further cloning, screening and production of ascites fluid (Galfre and Milstein 1981). These cell lines were designated AFRC MAC 74–79.

Examination of the ability of the various monoclonal antibodies to inhibit the partial activities of spinach NR showed that MACs 74, 75, 76, and 77 inhibited terminal activities ($FMNH_2$ and MV-nitrate reductase) whereas MACs 78 and 79 partially inhibited dehydrogenase activity (NADH-cytochrome *c* reductase). The uniqueness of the binding site was determined by competition between a radiolabelled monoclonal antibody and an unlabelled monoclonal antibody for binding to the antigen. Only MACs 76 and 77 were competitive in a way which suggested some overlap of antigenic site. The remainder were unique.

The monoclonal antibodies were also tested for their ability to inhibit the NADH-NR activity of other plant species. MAC 74 was found to be as effective against the enzyme in extracts of marrow, pea, cucumber, oil seed rape, barley, wheat, oats, and maize leaves as against spinach, whereas MACs 75, 76, and 77 were effective to varying degrees, dependent on the species. MACs 78 and 79 inhibited only marrow NR while stimulating the enzyme from the other species, particularly cereals. This stimulation could have been due to a stabilization of existing activity or, less likely, a change in conformation of the enzyme due to antibody binding.

Treatment of the spinach NR with SDS followed by eletrophoresis in the presence of SDS (Laemmli 1970) and 'Western' blotting onto nitrocellulose before incubating with each of the monoclonal antibodies (Tobin *et al.* 1979) showed that the binding sites of MACs 74–77 were conformationally dependent and only MAC 78 and 79 recognized the SDS-treated enzyme.

When the IgG subclass of each of the monoclonal antibodies was determined, it was found that only MAC 75 was of a subclass suitable for purification using Protein A–Sepharose affinity chromatography. This relatively small number of IgG which bind to Protein A is apparently a common feature of monoclonal antibodies raised in rats as opposed to mice. Our monoclonal antibodies were therefore purified by the more traditional technique of $(NH_4)_2SO_4$ precipitation and ion-exchange chromatography. Fractions from the columns were examined by direct ELISA and by their ability to inhibit nitrate reductase activity and fractions were found which contained the monoclonal antibodies and were positive by both criteria. With two of the monoclonal antibodies, ELISA-positive fractions which did not have enzyme inhibiting properties were also found. This effect was subsequently shown to be due to protein in the ascites fluid which caused non-specific binding in the ELISA. These could be diluted out to zero ELISA signal while inhibitory fractions remained ELISA-positive at the same dilution.

Using MAC 74, because of its ability to recognize all higher plant species so far tested, an indirect sandwich ELISA was developed which used antiserum obtained from rabbits as the first layer (Whitford *et al.* 1987). The amount of antigenic NR subsequently bound was indirectly detected by measuring the amount of MAC 74 bound to the enzyme, using a peroxidase-labelled sheep anti-rat antibody. Conditions were optimized with respect to concentrations of antibodies and incubation times to reflect economy of both time and materials. The assay can be completed in one day and the final ELISA signal, being a colourimetric measure of the peroxidase activity, was found to be linearly related to the amount of NR up to ~ 1.5 ng and to the log of the amount of NR from 20 to 1200 ng. All of the other monoclonal antibodies gave a dose–response signal which varied in intensity, dependent on the avidity of the antibody for the enzyme.

Provided a polyclonal antiserum can be obtained which will bind the nitrate reductase of the species under examination, it should be possible to use MAC 74 to measure the antigenic enzyme.

Using this ELISA, Whitford *et al.* (1987) were able to show the distribution of antigenic NR in roots, petiole and individual leaves of a young spinach plant. Antigenic NR reflected enzyme activity and most of the enzyme was shown to be present in the leaves. The inducible nature of NR was utilized to follow the decay and reappearance of the enzyme on removal and resupply of nitrate to spinach plants in water culture. The ELISA signal and enzyme activity fell in the same proportion on nitrate starvation indicating decay of the enzyme but, during resupply of nitrate, the ELISA signal exceeded the enzyme activity perhaps indicating that, as well as resynthesis of the antigenic enzyme, some conformational change takes place before the enzyme is capable of expressing its activity. A similar result was reported by Maki *et al.* (1986) on induction of the enzyme with nitrate in spinach cell suspension culture.

Attempts to purify active NR by immunoaffinity chromatography using polyclonal antibodies have resulted in very poor yields (Kleinhofs *et al.* 1986, Chapter 9) and my own experience has been similar. The multi-site binding of the enzyme to the polyclonal antibody is such that extreme conditions of pH etc. are required to release it, resulting in degradative loss of activity.

Because of their limited points of attachment to their antigen, monoclonal antibodies are ideally suited for immunoaffinity chromatography of active enzyme, provided they have sufficient avidity to hold the antigen during loading and washing and will release it by mild changes in conditions.

Using MAC 74, chosen because of its potential for application to other plant species, Fido (1987) purified NR from spinach. The enzyme, contained in a crude extract, was concentrated by $(NH_4)_2SO_4$ precipitation and re-solution, before being passed slowly through a column containing non-immune rat globulin linked to Sepharose. This served the dual purpose of filtering out particulate matter and removing non-specifically absorbed proteins. The eluate passed through the immunocolumn and nitrate reductase was bound. After washing the column until the eluate was free of 280 nm-absorbing material the enzyme was eluted. Preliminary experiments showed that neither strong salt solutions, such as 2 M $(NH_4)_2SO_4$ or 1 M phosphate at pH 7.5, nor 0.2 M glycine–HCl at pH 2.8 followed by immediate neutralization, released significant amounts of active enzyme. However, 1 M KNO_3 eluted active enzyme, particularly if the column was allowed to equilibrate in the presence of the nitrate for 1 h before elution. Stronger nitrate solutions decreased the amount of active enzyme recovered, although probably not the amount of antigenic NR. Recoveries were of the order of 60 per cent of the applied activity, with a specific activity purification of 1500-fold. Examination of the product using SDS–PAGE revealed that a number of fragments had also co-purified. The SDS–PAGE pattern was similar to that obtained from conventionally purified spinach NR (Fido and Notton 1984) and showed fragments which were proteolytically 'nicked' subunits. These had been reported for the barley enzyme (Campbell and Wray 1983) and the squash enzyme (Redinbaugh and Campbell 1985). The spinach enzyme fragments, which remained associated with the enzyme during either native PAGE or molecular sieving on Biogel A1.5 m, were removed by affinity chromatography on 5′-AMP–Sepharose as previously described by Fido and Notton (1984). Only a fraction (∼50 per cent) of the active NR was recovered from the affinity chromatography column using NADH as an eluant. SDS–PAGE of this solution showed only a doublet with molecular weights of 115 and 110 kDa, with the smaller component predominating. Active enzyme was located on the column after NADH elution and some enzyme was removed in an inactive form by SDS in dilute Tris buffer. Subsequent SDS–PAGE of this fraction also showed a doublet in which the 115 kDa fraction predominated.

Using MAC 74 for immunoaffinity chromatography, NR from several other plant species have also been purified.

As part of a structural comparison between NR from higher plants and *C. vulgaris* [a joint project between myself and Professors M. J. Barber and L. P. Solomonson (University of South Florida)], monoclonal antibodies 74–79 and supernatants from our previously uncloned cell lines were screened by enzyme inhibition and direct ELISA for their ability to recognize the *Chlorella* enzyme. Although polyclonal antisera against both spinach and *Chlorella* NR was found to inhibit all activities of the enzyme, none of the monoclonal antibodies or the supernatants from the uncloned cell lines inhibited enzyme activity of the *Chlorella*. However, three of the monoclonal antibodies and six of the supernatants recognized the *Chlorella* enzyme by direct ELISA. Two of the supernatants in particular had high avidity and these were subcloned to produce monoclonal antibodies, designated AFRC MAC 231 and 232. The order of avidity of the monoclonal antibodies against the *Chlorella* enzyme was $232 \gg 231 > 77 > 79 > 75$ while MAC 78 and interestingly, MAC 74, did not recognize the enzyme at all. The corresponding order for the spinach enzyme was $232 \gg 231 > 75 > 79 > 77 > 74 > 78$. MAC 79 also recognized the SDS-treated *Chlorella* enzyme, as it does the spinach enzyme. MAC 231 and 232 recognized both the spinach and *Chlorella* enzyme after SDS treatment showing that the antigenic sites were not conformationally dependent.

Monoclonal antibodies have also been obtained against maize leaf NR. Cherel *et al.* (1985) started with a semi-purified immunogen and screened the monoclonal antibodies by their ability to inhibit enzyme activity and by Western blotting of the enzyme from a gel run under denaturing conditions. The position of the antibody–antigen complex was aligned with the position of enzyme activity revealed within the gel. The eight monoclonal antibodies obtained were classified into six groups, four of which recognized the native enzyme strongly (groups 1, 2, 4, and 5) and two weakly (groups 3 and 6). Groups 4, 5, and 6 also strongly recognized the denatured enzyme. Group 1 inhibited overall enzyme activity, terminal activity and dehydrogenase activity while group 2 inhibited only total activity and terminal activity. Six of these monoclonal antibodies, one from each group, were used in an immunological comparison of the enzyme from nine species (Cherel *et al.* 1986). Two of the monoclonal antibodies inhibited activity of the four other monocotyledonous species tested (sugar cane, pearl millet, asparagus and barley) while one of these two also inhibited the enzymes from *Nicotiana tabacum* and *N. plumbaginifolia*, sunflower and also the two NR [NADH and NAD(P)H] of soybean, but at much higher concentrations than those required for the moncotyledons. Neither of these two monoclonal antibodies was able to inhibit the activity of the enzyme from *Chlorella pyrenoides* or *N. crassa*. Using a two-site ELISA with a monoclonal antibody as the first coating to bind the enzyme from the different plant species and with polyclonal antiserum to the maize enzyme as the second antibody layer, being detected with an alkaline phosphatase-labelled antispecies, four of the monoclonal antibodies recognized barley enzyme, three of these four recognized pearl millet and sugar cane enzymes but only the two inhibitory monoclonal antibodies recognized NR from asparagus.

Monoclonal antibodies raised against squash and corn NR (Miller and Campbell 1986) have been used to differentiate between the binding sites for NADH, methylviologen and bromophenol blue (Hoarau *et al.* 1986) as alternative electron donors for nitrate reduction (Campbell 1986). One monoclonal antibody obtained against the squash enzyme inhibited all three activities of both species, while other monoclonal antibodies were able to differentiate between species and donor binding sites.

Both native and SDS-treated subunits of NR from barley have been used as immunogens to raise monoclonal antibodies in mice (Kudrna *et al.* 1987). Of the twenty-one positive clones obtained some were specific to the native form of the enzyme and were therefore conformation-dependent, while others were found to be conformation-independent, recognizing both immunogens. Only one clone recognized only the SDS-treated subunit specifically. The use of mice enabled the production of many monoclonal antibodies of the correct subclass of globulin to bind to protein A and this allowed further characterization.

With the emphasis of work on NR moving rapidly towards the molecular and genetic level, monoclonal antibodies which are either specific in binding only to the enzyme from the eukaryotic species used to produce the immunogen, or species-independent and will bind to an invariant site on the enzyme irrespective of its origin, will become increasingly important probes. Identification of differences and similarities in the amino acid composition and folding of the protein obtained from plants, fungi, algae and yeasts is now possible using monoclonal antibodies. This will help in understanding how the enzyme obtained from different eukaryotic sources varies in its kinetic efficiency to reduce nitrate and, therefore, what steps have to be taken to make plants perform this important process in the most efficient manner.

Acknowledgements

Long Ashton Research Station is funded through the Agricultural and Food Research Council. The author would like to thank his colleagues Roger Fido,

Peter Whitford and Edgar Watson for their contribution to some of the work described.

References

Amy, N. K. and Garrett, R. H. (1979). Immunoelectrophoretic determination of nitrate reductase in *Neurospora crassa. Analytical Chemistry* **95**, 97–107.

Barber, M. J., Notton, B. A., and Solomonson, L. P. (1987). Oxidation reduction mid-point potentials of the molybdenum centre in spinach NADH-nitrate reductase. *FEBS Letters* **213**, 372–4.

Campbell, J. McA. and Wray, J. L. (1983). Purification of barley nitrate reductase and demonstration of nicked subunits. *Phytochemistry* **22**, 2375–82.

Campbell, W. H. (1986). Properties of bromophenol blue as an electron donor for higher plant NADH-nitrate reductase. *Plant Physiology* **82**, 729–32.

Campbell, W. H. and Remmler, J. L. (1986). Regulation of corn leaf nitrate reductase. I. Immunochemical methods for analysis of the enzyme's protein component. *Plant Physiology* **80**, 435–41.

Campbell, W. H. and Ripp, K. G. (1984). An ELISA for higher plant nitrate reductase. *Annals of the New York Academy of Science* **435**, 123–5.

Cherel, I., Grosclaude, J., and Rouze, P. (1985). Monoclonal antibodies identify multiple epitopes on maize leaf nitrate reductase. *Biochemical Biophysical Research Communications* **129**, 686–93.

Cherel, I., Marion-Poll, A., Mayer, C., and Rouze, P. (1986). Immunological comparisons of nitrate reductases of different plant species using monoclonal antibodies. *Plant Physiology* **81**, 376–8.

Fido, R. J. (1987). Purification of nitrate reductase from spinach (*Spinacea oleracea* L.) by immunoaffinity chromatography using a monoclonal antibody. *Plant Science* **50**, 111–15.

Fido, R. J. and Notton, B. A. (1984). Spinach nitrate reductase: further purification and removal of 'nicked' subunits by affinity chromatography. *Plant Science Letters* **37**, 87–91.

Fido, R. J., Hewitt, E. J., Notton, B. A., Jones, O. T. G., and Nasrulhaq-Boyce, A. (1979). Haem of spinach nitrate reductase: low temperature spectrum and mid-point potential. *FEBS Letters* **99**, 180–2.

Funkhouser, E. A. and Ramadoss, C. S. (1980). Synthesis of nitrate reductase in *Chlorella*. II. Evidence for synthesis in ammonium-grown cells. *Plant Physiology* **65**, 944–8.

Galfre, G. and Butcher, G. W. (1986). Making antibodies. In *Immunology in plant science* (ed. T. L. Wang), Society for Experimental Biology. Seminar series 29. pp. 1–26. Cambridge University Press, Cambridge.

Galfre, G. and Milstein, C. (1981). Preparation of monoclonal antibodies: strategies and procedures. *Methods in Enzymology* **73**, 3–46.

Graf, L., Notton, B. A., and Hewitt, E. J. (1975*a*). The effects of immune and non-immune rabbit sera on the various activities of spinach (*Spinacea oleracea* L.) nitrate reductase. *Plant Science Letters* **4**, 69–75.

Graf, L., Notton, B. A., and Hewitt, E. J. (1975*b*). Serological estimation of spinach nitrate reductase. *Phytochemistry* **14**, 1241–3.

Gutteridge, S., Bray, R. C., Notton, B. A., Fido, R. J., and Hewitt, E. J. (1983). Studies by electron paramagnetic resonance spectroscopy of the molybdenum centre of spinach (*Spinacea oleracea* L.) nitrate reductase. *Biochemical Journal* **213**, 137–42.

Hoarau, J., Hirel, B., and Nato, A. (1986). New artificial electron donors for *in vitro* assay of nitrate reductase isolated form cultured tobacco cells and other organisms. *Plant Physiology* **80**, 946–9.

Kamachi, K., Ameniya, Y., Ogura, N., and Nakagawa, H. (1987). Immuno-gold localization of nitrate reductase in spinach (*Spinacea oleracea*) leaves. *Plant Cell Physiology* **28**, 333–8.

Kleinhofs, A., Narayanan, K. R., Somers, D. A., Kuo, T. M., and Warner, R. L. (1986). Immunochemical methods for higher plant nitrate reductase. In *Immunology in plant sciences* (eds. H. F. Linskens and J. F. Jackson), pp. 190–211. Springer-Verlag, Berlin.

Kohler, G. and Milstein, C. (1975). Continuous culture of fused cells secreting antibody of predefined specificity. *Nature* **256**, 595–7.

Kuo, T. M., Kleinhofs, A., Somers, D. A., and Warner, R. L. (1981). Antigenicity of nitrate reductase-deficient mutants in *Hordeum vulgare* L. *Molecular and General Genetics* **181**, 20–3.

Kudrna, D. A., Kakefuda, G., Warner, R. L., and Kleinhofs, A. (1987) Production and partial characterization of monoclonal antibodies specific toward barley nitrate reductase. In *Abstracts of the Second International Symposium on Nitrate Assimilation— Molecular and Genetic Aspects*, St Andrews, UK. B1.

Laemmli, U. K. (1970). Cleavage of structural proteins during the assembly of the head of bacteriophage T4. *Nature* **227**, 680–5.

Maki, H., Yanagishi, K., Sato, K., Ogura, N., and Nakagawa, H. (1986). Regulation of nitrate reductase activity in cultured spinach cells as studied by an enzyme-linked immunosorbent assay. *Plant Physiology* **82**, 739–41.

Miller, A. and Campbell, W. H. (1986). Monoclonal antibodies for higher plant nitrate reductase. In *Proceedings of the fifth annual symposium in plant biochemistry and physiology*, Vol 5 (ed. D. D. Randall), p. 192. University of Missouri, Columbia.

Murray, E. D. and Sandwal, B. D. (1963). An immunological enquiry into the identity of assimilatory and dissimilatory nitrate reductase from *Escherichia coli*. *Canadian Journal of Microbiology* **9**, 781–90.

Nakagawa, H., Yamagishi, K., Yamagishi, N., Sato, T., Ogura, N., and Oaks, A. (1986) Immunological characterisation of nitrate reductase in different tissues of spinach seedlings. *Plant Cell Physiology* **27**, 627–33.

Notton, B. A., Graf, L., Hewitt, E. J., and Povey, R. C. (1974). The role of molybdenum in the synthesis of nitrate reductase in cauliflower (*Brassica oleracea* L var. *Botrytis* L.) and spinach (*Spinacea oleracea* L.). *Biochemica et Biophysica Acta* **364**, 45–58.

Notton, B. A., Fido, R. J., and Hewitt, E. J. (1977). The presence of functional haem in a higher plant nitrate reductase. *Plant Science Letters* **8**, 165–70.

Notton, B. A., Fido, R. J., and Galfre, G. (1985). Monoclonal antibodies to a higher plant nitrate reductase: differential inhibition of enzyme activities. *Planta* **165**, 114–19.

Pateman, J. A., Cove, D. J., Rever, B. M., and Roberts, D. B. (1964). A common cofactor for nitrate reductase and xanthine dehydrogenase which also regulates the synthesis of nitrate reductase. *Nature* **201**, 58–60.

Pollock, M. R. (1964). Stimulating and inhibiting antibodies for bacterial penicillinase. *Immunology* **7**, 707–23.

Redinbaugh, M. G. and Campbell, W. H. (1985). Quaternary structure and composition of squash NADH:nitrate reductase. *Journal of Biological Chemistry* **260**, 3380–5.

Smarrelli, J., Jr. and Campbell, W. H. (1981). Immunological approach to structural comparisons of assimilatory nitrate reductases. *Plant Physiology* **68**, 1226–30.

Snapp, S., Somers, D. A., Warner, R. L., and Kleinhofs, A. (1984). Immunological comparisons of higher plant nitrate reductases. *Plant Science Letters* **36**, 13–18.

Somers, D. A., Kuo, T. M., Kleinhofs, A., and Warner, R. L. (1983). Nitrate reductase-deficient mutants in barley. Immunoelectrophoretic characterisation. *Plant Physiology* **71**, 145–9.

Tobin, H., Staehelin, T., and Gordon, J. (1979). Electrophorestic transfer of proteins from polyacrylamide gels to nitrocellulose sheets: procedure and some applications. *Proceedings of the National Academy of Science, USA* **76**, 4350–4.

Whitford, P. N., Fido, R. J., and Notton, B. A. (1987). An enzyme linked immunosorbent assay for nitrate reductase using monoclonal antibodies. *Phytochemistry* **26**, 2467–70.

11. Biochemical and somatic cell genetics of nitrate reduction in *Nicotiana*

Andreas J. Müller and Ralf R. Mendel

Introduction

Nicotiana tabacum and N. *plumbaginifolia* were chosen for genetic studies of nitrate assimilation because they can be manipulated in cell culture and regenerated to plants more easily than most other higher plant species. This offered the possibility of utilizing cell culture techniques for the isolation and genetic analysis of mutants. Due to this *in vitro* approach, more nitrate reductase (NR) deficient mutants and especially many more non-leaky mutants have been isolated and characterized in *Nicotiana* than in any other higher plant. In this review we will summarize what we believe to be the main results of the mutational analysis of nitrate reduction carried out in these two *Nicotiana* species.

The two species differ in their genome structure. As a true diploid, the wild species, N. *plumbaginifolia* ($2n = 20$), is better suited to the mutational analysis of gene functions than the amphiploid (allotetraploid) species N. *tabacum* ($2n = 48$). Both the isolation and the analysis of recessive mutants are complicated by the amphiploid genome structure of the latter; nevertheless, many NR-deficient mutants have been isolated in N. *tabacum*, and the characterization of these mutants has contributed substantially to our understanding of the genetics of nitrate assimilation in higher plants. Moreover, the NR-deficient mutants of N. *tabacum* provide a unique opportunity to study special problems of gene interaction in amphiploid plant species. Finally, these mutants have proved to be useful marker systems for somatic hybridization and gene transfer experiments.

Isolation and genetic analysis of NR-deficient mutants

All of the NR-deficient mutants described in these two *Nicotiana* species have been isolated by selection for chlorate resistance in cell cultures derived from haploid plants. This approach was first successful in N. *tabacum*, although, as we now know, in this amphiploid species haploidization does not lead to single copies of the NR-specific genes. However at the time these *N. tabacum* mutants were isolated cell culture techniques for *N. plumbaginifolia* and other higher plant species from which monohaploids could be obtained were still in their infancy.

N. tabacum

Selection for chlorate resistance in amphihaploid ($n = 24$) cell suspension cultures of *N. tabacum* led to the isolation of cell lines that lacked NR activity and had an absolute requirement for reduced nitrogen (Müller and Grafe 1975, 1978; Müller 1978; Buchanan and Wray 1982; Evola 1983*a*). These mutant cell lines were tested for simultaneous loss of xanthine dehydrogenase activity and by this criterion classified into NR-specific mutants (Nia⁻ phenotype) and mutants lacking functional molybdenum cofactor (MoCo) (Cnx⁻ phenotype). In the selection experiments carried out in our laboratory, a total of 36 Nia⁻ and four Cnx⁻ lines were isolated from about 10^9 mutagenized cells. The last of these experiments included regeneration of mutant plants on medium containing ammonium succinate as sole nitrogen source, grafting the regenerated plants onto wild-type stocks, searching for plants that were fertile as a consequence of spontaneous diploidization, and genetic analysis through crosses (Müller 1978, 1983).

Stable transmission of NR deficiency was observed in all the 16 mutant lines (15 Nia⁻ and one Cnx⁻) that produced fertile plants. NR-deficient progeny seedlings could easily be identified by growth tests for both nitrate utilization and chlorate resistance. The ability to distinguish mutant and wild-type seedlings in simple growth tests allowed extensive segregational analyses of progeny from crosses between mutant and wild-type plants and revealed that the NR deficiency was conferred by two recessive nuclear mutations at unlinked loci in all the lines studied. This showed that amphihaploid cells of *N. tabacum* contain duplicate forms of the NR-specific genes and, therefore, two independent mutations are needed to abolish NR activity.

Nicotiana plumbaginifolia

The successful isolation of rare double mutants in *N. tabacum* demonstrated the efficiency of the selection procedures employed and suggested that NR-deficient mutants could be recovered at much higher frequencies from monoploid cells, which contain only one copy of each gene. Several experiments in which monoploid ($n = 10$) protoplast cultures of *N. plumbaginifolia* were screened for chlorate resistance have been carried out during the last 6 years. Marton *et al.* (1982*a*) isolated and characterized seven Nia⁻ and four Cnx⁻ cell lines. Negrutiu *et al.* (1983) isolated 26 Nia⁻ and four Cnx⁻ lines, regenerated them to fertile plants and showed that, as expected, the NR deficiency was transmitted to the progeny as a single recessive nuclear gene. Recent experiments using improved procedures for the selection of NR-deficient cells (Grafe *et al.* 1986) and for the regeneration of mutant plants led to the isolation of 141 Nia⁻ and 70 Cnx⁻ lines (Gabard *et al.* 1987). Of these 211 mutant lines, 147 produced seed progeny and were shown to be homozygous for a recessive mutation conferring NR deficiency. The frequency of spontaneous NR-deficient mutants in monoploid protoplast cultures of *N. plumbaginifolia* was found to range from 5×10^{-5} to 10^{-6}.

Complementation analysis

Both sexual crosses and somatic hybridization have been used to determine the allelism of the NR-deficient mutants. In *N. plumbaginifolia* all of the 100 Nia⁻ mutants tested fall into one complementation group, and the 62 Cnx⁻ mutants tested fall into six complementation groups (Marton *et al.* 1982*b*; Dirks *et al.* 1985; Gabard *et al.* 1987. Overlapping of complementation groups has not been found. Thus, the mutants tested define seven genes (*nia, cnxA, cnxB, cnxC, cnxD, cnxE, cnxF*). The linkage relationships of these nuclear genes are not yet known.

The results show that *N. plumbaginifolia* has at least as many *cnx*-type genes as *Aspergillus nidulans*. However, it remains to be determined whether all of the *cnx* genes described in *Aspergillus* have their counterpart among the *N. plumbaginifolia* genes identified so far. The possibility has to be considered that in a higher plant more genes are involved with the MoCo function than in a fungus. It is noteworthy that all of the 100 Nia⁻ mutants tested have been assigned to a single complementation group and are therefore all alleles of the *nia* gene which, by other evidence, has been identified as the NR structural gene. Theoretically, the Nia⁻ phenotype could also be produced by mutations in genes that control the expression of the NR structural gene or the synthesis of prosthetic groups other than the MoCo.

In *N. tabacum*, 19 Nia⁻ mutants have been tested for complementation and found to be allelic to each other (Müller 1983, and unpublished). The four Cnx⁻ lines isolated in our laboratory as independent mutations fall into a single complementation group which is homologous to *cnxA* of *N. plumbaginifolia* (Grafe and Müller 1983). The four Cnx⁻ cell lines described by Buchanan and Wray (1982) have been shown to be homologous to *cnxB* of *N. plumbaginifolia* (Xuan *et al.* 1983).

The NR-deficient mutants are not equally distributed over the complementation groups. In *N. plumbaginifolia*, about 60 per cent of the recovered mutants occurred in the *nia* gene, about 20 per cent in the *cnxA* gene and the remaining 20 per cent in the other five (or six) *cnx* genes. These differences are difficult to explain by selection bias since the rare types of *cnx* mutants grew in cell culture as well as the *cnxA* and *nia* mutants and were not less resistant to chlorate. Moreover, *cnxA* mutants were as difficult to regenerate to plants as the other types of *cnx* mutants. It is therefore most likely that the *nia* and *cnxA* genes mutate more frequently than the other *cnx* genes. The same conclusion can be drawn from the results obtained in *N. tabacum*. The experiments carried out in our laboratory gave 36 Nia⁻, four *cnxA* and no other types of *cnx* mutants. This ratio is consistent with that seen in

N. plumbaginifolia, if one considers that for a certain gene the frequency of double mutants should equal the squared value of the frequency of single mutants. The low frequency of *cnx* mutants in *N. tabacum* is evidence that not only the *nia* gene but also all the *cnx* genes are present in duplicate.

The use of somatic hybridization for complementation analysis of NR-deficient mutants merits some discussion. The parasexual test for genetic complementation was introduced by Glimelius *et al.* (1978) who showed that fusion between protoplasts of the Nia63 and the Cnx68 mutants of *N. tabacum* resulted in hybrid cells with restored NR activity and that these hybrids could be detected by their ability to grow on nitrate. This approach has since been used to classify several NR-deficient mutants of *N. tabacum* (Grafe and Müller 1983, Evola 1983*b*; Xuan *et al.* 1983) and *N. plumbaginifolia* (Marton *et al.* 1982*b*; Dirks *et al.* 1985). Biasini and Marton (1985) developed from it a rapid assay which uses *in vivo* NR activity as the criterion of complementation. Recently, the complementation behaviour of 28 Cnx⁻ mutants of *N. plumbaginifolia* was studied in 187 fusion combinations (Gabard *et al.* 1988).

Compared to sexual crosses, somatic hybridization of NR-deficient mutants has the advantage that the complementation test can be performed shortly after isolation of the mutant cell lines and long before mature plants have been regenerated; mutants that are sterile can be tested for complementation; and the homology of mutants in different species can be determined. Interspecific combinations of NR-deficient mutants have been studied in the case of *N. tabacum* × *N. plumbaginifolia* (Xuan *et al.* 1983) and of *Hyoscyamus muticus* × *N. tabacum* (Lazar 1983). This approach offers the possibility of classifying NR-deficient mutants of many plant species by fusing them to the well characterized *N. plumbaginifolia* mutants.

Chlorate resistance and NR deficiency

The successful isolation of NR-deficient mutants by selection for chlorate resistance and the observation that the level of chlorate toxicity in cultured cells and seedlings was closely correlated with the level of NR activity (Grafe *et al.* 1986; Müller 1983) indicate that the toxicity of chlorate to *Nicotiana* cells is almost completely dependent on NR activity, presumably because NR catalyses the reduction of chlorate to the toxic chlorite. Moreover, all NR-deficient mutants, irrespective of whether they were Nia⁻ or Cnx⁻ and whether or not they retained the wild-type pattern of regulation (see below) were highly resistant to chlorate. Obviously, the chlorate toxicity observed in *Nicotiana* is different from that in *Aspergillus* (Cove 1979). It should be noted that all *Nicotiana* mutants were selected for loss of the constitutive NR activity. To avoid competition between chlorate and nitrate, selection for chlorate resistance was carried out with nitrate-free selection media which contained ammonium succinate or amino acids as sole nitrogen source. Mutants that were impaired in the induction of NR by nitrate while retaining the constitutive NR level could, therefore, not be isolated.

Stably chlorate-resistant cell lines that exhibited intermediate or normal levels of NR activity have repeatedly been isolated in *Nicotiana* (Müller and Grafe 1978; Buchanan and Wray 1982; Marton *et al.* 1982a). Such lines have not been characterized in detail and no insights into the mechanism of their resistance have been gained. Whether or not mutants impaired in the uptake of chlorate and nitrate can be selected as chlorate-resistant cell lines remains to be investigated.

Mutations in the structural gene for nitrate reductase

The function of the nia gene

The gene defined by the *nia* mutants of *N. tabacum* and *N. plumbaginifolia* has been identified as the NR structural gene, based on the observation that several *nia* mutants contain modified forms of the NR protein (see below). Mutations in the *nia* gene simultaneously affect both the constitutive and the nitrate-induced NR activities, and many of them abolish NR activity completely in all tissues examined (cotyledons, leaves and roots at various stages of plant development, cultured cells) (Müller 1983). In addition, most of the *nia* mutants isolated in *N. plumbaginifolia* (Negrutiu *et al.* 1983, Gabard *et al.* 1987) and *N. tabacum* (Müller 1983) were incapable of utilizing nitrate for growth. These observations imply that the enzyme encoded by the *nia* gene is responsible for all detectable NR activities and is essential for nitrate utilization. In other words, *Nicotiana* has only one NR.

The NR of *N. tabacum* uses NADH (but not NADPH) as physiological electron donor. It resembles other higher plant and fungal NR in that it has a molecular weight of about 200 kDa, contains FAD, haem of the cytochrome *b* type and molybdopterin (MoCo) as prosthetic groups and displays two types of partial activities: the molybdenum-independent diaphorase activity (usually measured as NADH-cytochrome *c* reductase activity) and the molybdenum-dependent terminal NR activities (i.e. nitrate reduction with artificial electron donors such as FMNH, FADH, and reduced viologen dyes) (Mendel and Müller 1979, 1980).The tobacco NR is composed of two subunits of equal size (Somers *et al.* 1986; Horau *et al.* 1986). In *N. plumbaginifolia*, the subunits are expected to be identical since this species has only one *nia* locus.

The nia gene of N. tabacum is present in duplicate

The genetic analysis of Nia⁻ mutants has shown that *N. tabacum* contains duplicate forms of the NR structural gene at unlinked loci (*nia1*, *nia2*). All of the Nia⁻ mutants analysed have proved to be double mutants at these loci (Müller 1983). Since the amphiploid species *N. tabacum* is believed to have arisen by hybridization between *N. sylvestris* and *N. tomentosiformis*, presumably one of the homeologous *nia* loci is from *N. sylvestris* and the other of *N. tomentosiformis* origin. Homozygous single mutants at either the *nia1* or the *nia2* locus have been selected from the progeny of crosses between Nia⁻ double mutants and wild-type plants. The characterization of these single mutants revealed that either of the two *nia* loci is able to produce fully active NADH-NR (see Table 11.4). The two types of single mutants did not differ in their regulation of *in vivo* NR activity and grew as well as wild-type plants with nitrate as sole nitrogen source. The possibility must be considered therefore that such single mutants occur spontaneously and are occasionally maintained as a tobacco variety.

The polypeptides encoded by the two *nia* loci are 904 amino acids long, have a predicted molecular weight of 102 kDa, and differ from each other only at a few positions (see Chapter 12). The complementation studies reported below show that both *nia* loci are functional simultaneously and that the subunits produced by these two loci can associate as a hybrid molecule (or homeodimer). It is to be expected therefore that wild-type plants of *N. tabacum* contain three different forms of NADH-NR: two types of homodimers and a homeodimer. It should be noted that techniques that could distinguish very similar forms of NR (e.g. isoelectrofocusing) have not yet applied to the tobacco NR. For this reason it is also not known how similar the NR of *N. tabacum* and *N. plumbaginifolia* are and whether there is any polymorphism for NR structure within these species.

Biochemical diversity of nia mutants

Evidence for the presence of inactive NR protein in the *nia* mutants of the two *Nicotiana* species has been obtained from tests for the partial activities of NR, by immunological tests for NR cross-reacting material and by studying genetic complementation between different *nia* mutants. The results of these studies show that most of the *nia* mutants contain modified NR protein and thus meet the expectation for structural gene mutants that have occurred spontaneously or after treatment with mutagens known to induce preferentially base-pair substitutions. Three main types of NR proteins with impaired NADH-NR activity could be distinguished:

1. NR proteins that retained the ability to catalyse reduction of nitrate with artificial electron donors, i.e. the terminal NR activity.
2. NR proteins that retained the diaphorase activity of NR.
3. NR proteins that lacked both types of partial activities and were only detectable as CRM (or indirectly by tests for genetic complementation).

In *N. tabacum*, 36 Nia⁻ mutants were assayed for NR partial activities (Mendel and Müller 1979; Müller *et al.* in preparation). Three of these mutants exhibited terminal NR activity (Table 11.1) and all were impaired in the diaphorase activity of NR. However, the possibility that some mutants retained low levels of diaphorase activity, could not be excluded. The CRM level in cultured cells or in leaves of nitrate-induced plants (Table 11.1) has been determined in 16 *nia* mutants of *N. tabacum* (Schiemann and Müller 1985). Only one of these mutants did not contain CRM; this might be considered an unusually high proportion of CRM-positive mutants. However, since the CRM present in a given double mutant may have been produced by both *nia* loci or by only one of them, the proportion of CRM-positive mutant alleles may be lower.

Table 11.1. NR activity and CRM level in *nia1, nia2* double mutants of *Nicotiana tabacum* (Müller *et al.*, in preparation)

Mutant line	Seeding growth on nitrate	NR activity*				CRM*	
		in vivo		*NADH*		*BVH*	
		C	L	C	L	C	L
Nia20	0	0	0	0	0	0	32
Nia21	0	0	0	0	0	0	29
Nia23	0	0	0	0	0	0	47
Nia25	0	0	0	0	0	0	23
Nia26	0	0	0	0	0	12	35
Nia27	0	0	0	0	0	0	(6)
Nia28	0	0	0	0	0	18	24
Nia33	0	0	0	0	0	0	61
Nia36	0	0	0	0	0	0	24
Nia40	0	0	0	0	0	0	33
Nia30	6	0	0	0	0	0	94
Nia22	12	0	5	0	0	0	156
Nia31	10	0	4	0	0	0	179
Nia29	38	0	36	0	2	0	51
Nia34	54	2	48	0	4	0	77
Nia46	100	35	68	nd	12	nd	nd
Wild-type	100	100	100	100	100	100	100

C = determined in cell cultures, maximum values after addition of nitrate.
L = determined in young leaves of plants grafted on to wild-type stocks.
* = % of wild-type activity.

Only a few of the more than 100 *nia* mutants identified in N. *plumbaginifolia* have been assayed for the presence of NR protein so far. One of these mutants was shown to possess a dimeric NR molecule that has retained its diaphorase activity while lacking terminal NR activity (De Vries *et al.* 1986). A recent study describes three *nia* mutants that have retained terminal NR activity, two mutants with inactive NR protein identified as CRM, and one further mutant in which no CRM could be detected with the monoclonal antibody used (Gabard *et al.* 1987).

Some of the *nia* mutants studied in N. *tabacum* exhibited residual NR activity

(Table 11.1). A detailed examination of these mutants revealed that the growth response of seedlings to nitrate is a more sensitive indicator of residual NR activity than any NR assay. This could most convincingly be demonstrated with the Nia30 mutant, which showed a significant growth response to nitrate although NR activity was not detectable in any of the tissues examined, even by the very sensitive *in vivo* assay (Müller 1983). Studies performed with other low-activity mutants showed that the *in vivo* assay regularly gave higher relative values for the mutant level of NR activity than the *in vitro* assay (Table 11.1). These differences might be explained by assuming that the mutant NR was more unstable upon extraction than the wild-type NR. However, the differences can also be explained by the observation that the NR activities measured in the *in vivo* and *in vitro* assays are differently affected by regulatory mechanisms (see below).

Whether or not residual NR activity can be detected depends not only on the assay used but also on physiological factors. It has repeatedly been observed for several low-activity *nia* mutants that cell cultures initiated from mutant plants and grown on the usual auxin-rich media did not exhibit NR activity and were incapable of growth with nitrate as sole nitrogen source (Müller and Mendel 1982). The capability for growth on nitrate is only regained after transfer to media with a higher cytokinin/auxin ratio. It is obvious that mutant cell lines described as fully deficient in NR activity may contain a low-activity allele of the *nia* gene.

Genetic complementation between nia *mutants*

The assignment of mutants to the *nia* complementation group is based on the observation that crosses between these mutants resulted in NR deficient hybrid progeny. However, some combinations of *nia* mutants gave hybrids that exhibited NR activity, although at a lower than wild-type level. These cases of partial complementation for NR activity could easily be identified by the seedling test for growth on nitrate. In *N. plumbaginifolia*, 10 complementing combinations of *nia* mutants have been found so far (Gabard *et al.* 1987). Ten complementing combinations could also be identified in *N. tabacum* by intercrossing 14 *nia* mutants (including four mutants with residual NR activity) in all possible combinations (Müller 1983, and unpublished). The levels of *in vitro* NR activity in the complementing hybrids were considerably lower than the wild-type level (Table 11.2).

Complementation between different mutant alleles is common among structural genes that code for homomultimeric enzymes and results from the formation of hybrid enzymes composed of two different mutant subunits. In the case of the *N. tabacum* hybrids, conformational complementation seems to be the essential mechanism leading to the partial restoration of the observed NR activity. The enhanced level of NR activity would thus indicate that the conformation of the hybrid dimer is less abnormal and less unstable than that of the dimers produced by either mutant alone.

Table 11.2. Partial complementation in
hybrids between different *nia1*,*nia2* double
mutants of *N. tabacum* (Müller *et al.*, in
preparation)

Hybrid	NR activity*	
	in vivo	*in vitro* (NADH)
Nia22 × Nia23	51	4
Nia22 × Nia25	58	4
Nia22 × Nia28	46	3
Nia22 × Nia21	24	1
Nia31 × Nia23	51	3
Nia31 × Nia25	56	4
Nia31 × Nia28	62	5
Nia21 × Nia23	16	0.5
Nia21 × Nia25	14	0.5
Wild-type	100	100

NR activity in leaves of plants grown in soil or as grafts on
wild-type stocks.
* = % wild-type activity.

The complementation observed in *N. plumbaginifolia* is intragenic (or interallelic) in the strict sense. In contrast, the complementation observed in *N. tabacum*
may result not only from interaction between the subunits produced by different
alleles at the same locus but also from interaction between the product of a *nia1*
allele and that of a *nia2* allele. Evidence for the second type of interaction has been
obtained by genetic analysis, which showed that in at least three of the 10
complementing combinations studied the enhanced NR activity was due to
interaction between the two *nia* loci and, consequently, could be transmitted to
homozygous progeny plants. Selfing of the heterozygous hybrid *nia1-23/31*;
nia2-23/31 for instance resulted in a homozygous hybrid line *nia1-23/23*; *nia2-
31/31*, which stably maintained the enhanced NR level of its heterozygous
parent. The plants of this line can grow to maturity on nitrate as sole nitrogen
source whereas the Nia23 mutant is completely incapable of growth on nitrate
and the Nia31 mutant usually dies after forming small, yellow-green plantlets
(Müller, in preparation).

Reversion and interallelic recombination

Four Nia⁻ double mutants of *N. tabacum* (Grafe and Müller 1982, Müller and
Grafe, in preparation) and two Nia⁻ mutants of *N. plumbaginifolia* (Dirks *et al.*
1986) have been tested for the occurrence of reversion to Nia⁺. All of the mutants
tested produced revertants which were recovered at frequencies of about 10^{-6} in

protoplast cultures by their ability to grow with nitrate as the sole nitrogen source. Several of these revertants were regenerated to fertile Nia$^+$ plants. Selfed progeny of the revertant plants segregated approximately 3 : 1 for the ability of seedlings to grow on nitrate. This indicated that, as expected, the revertants were heterozygous for the mutation that restored NR activity.

Müller and Grafe (in preparation) subjected four revertants of the *N. tabacum* double mutants Nia30, Nia40, and Nia20 to a more detailed crossing analysis. The results of these studies showed that the reversion was not due to a trans-acting suppressor mutation, demonstrated at which of the two *nia* loci the back-mutation had occurred, and showed that the back-mutation restored NR activity only partially. It is interesting to note that the reversion frequencies seen in *N. tabacum* were of the same order of magnitude as those seen in *N. plumbaginifolia.* This shows that recombinational interactions between the homeologous *nia* loci occur no more frequently than reversions, if at all.

Meiotic interallelic recombination was studied by screening the selfed progeny obtained from heteroallelic hybrids between Nia$^-$ double mutants for nitrate-utilizing seedlings (Müller and Grafe, in preparation). For instance, Nia$^+$ recombinants occurred among selfed progeny from *nia1-30/27;nia2-30/27* plants at a frequency of 11/5400. No revertants were found when about the same number of progeny seedlings obtained from the homoallelic parental mutants (Nia30 and Nia27) were screened for nitrate utilization. These results showed that interallelic meiotic recombination occurs much more frequently than meiotic recombination between mutant alleles at different *nia* loci.

Physiological consequences of abolishing NR activity

The *nia* mutants of *Nicotiana* provided the opportunity to study the physiological consequences of a complete (and specific) loss of NR activity for the first time in higher plants. NR-deficient mutants previously identified in other higher plant species were only of limited value for this purpose because they either retained residual NR activity or were simultaneously impaired in other functions (as in the case of the *cnx* mutants).

A null-activity *nia* mutant (Nia28) was shown not to differ significantly in growth rate or morphology from wild-type plants when grown with ammonium as sole nitrogen source (Müller 1983). Similar results have since been obtained with several other *nia* mutants (e.g. Gabard *et al.* 1987). Ammonium-grown wild-type and Nia$^-$ plants suffered to the same extend from insufficient pH control in the medium (Müller 1983; Hamill and Cocking 1986). These observations indicate that NR has no other important function than to catalyse the reduction of nitrate. For instance, the normal growth of Nia$^-$ plants shows that the iron-reducing ability of NR plays no essential physiological role.

The uptake of nitrate seems not to be affected by *nia* mutations: Nia$^-$ cell cultures of *N. tabacum* take up nitrate at the same rate as wild-type cell cultures; in both types of cells the uptake showed the same multi-phase relationship to nitrate concentration (Saalbach and Müller 1982). The accumulation of nitrate in Nia⁻ plants may lead to bleaching and/or wilting of the leaves.

Mutations abolishing NR activity are lethal under field conditions, mainly because the soil is insufficiently buffered against the acidification caused by ammonium utilization in the absence of nitrate utilization but also because the Nia⁻ plants accumulate nitrate. The lethality of Nia⁻ plants implies that under natural conditions nitrate is not only the preferred but the essential nitrogen source for tobacco (and most other higher plants).

Mutations in genes controlling the synthesis of MoCo

In *N. plumbaginifolia* at least six genes (*cnxA* to *cnxF*) are involved in the synthesis of active MoCo. A mutation in any of these genes leads to the simultaneous loss of NR and XDH activities due to a defect in the common MoCo. The phenotype of the Cnx⁻ mutants allows conclusions to be drawn about the physiological role of XDH and other enzymes possibly containing MoCo. Very early it was shown that active XDH is not essential for cell cultures of tobacco (Mendel and Müller 1976). Since then large numbers of Cnx⁻ mutants of *N. plumbaginifolia* belonging to six complementation groups have become available in the form of regenerated plants and their progeny. Some of the Cnx⁻ mutants differ from each other in their morphology. However, it is not yet clear whether these differences represent pleiotropic effects of the *cnx* mutation or are due to independent genetic changes. Generally, a mutation in any of the six *cnx* genes leads to a similar phenotype, deviating from that of wild-type and Nia⁻ plants: weak growth, and narrow, flimsy leaves (Negrutiu 1983; Gabard *et al.* 1987). This phenotype is probably due to the defect of XDH and other as yet undefined enzymes which require MoCo.

A detailed biochemical analysis of the Cnx⁻ mutants would allow further insight into the localization of the defect and hence contribute to a better understanding of the synthesis of MoCo in plants. It was, therefore, necessary to find parameters and establish methods that would allow a more precise description of the impairment caused by a given cnx mutation. The NR apoprotein is not affected by a *cnx* mutation since such mutants retain their diaphorase activity. The obvious complementary nature of Cnx⁻ and Nia⁻ mutants has not only been demonstrated genetically but was also confirmed biochemically by reconstitution of NADH-NR in mixed extracts prepared by co-homogenization of Cnx⁻ and Nia⁻ cells (Mendel and Müller 1978; Marton *et al.* 1982*a*). In these experiments the inactive NR apoprotein of Cnx⁻ mutants could be reactivated *in vitro* by the active MoCo present in Nia⁻ cells. All attempts to demonstrate *in vitro* restoration of NADH-NR activity by homogenizing together cells mutated in different *cnx* genes remained unsuccessful (Bache *et al.* in preparation). Since *cnx* mutants complement each other genetically but not *in vitro* after extract mixing, it may be concluded that either the products of the *cnx* genes are very unstable under *in vitro* conditions or that the synthesis of MoCo is linked to processes and/or structures that are rendered non-functional once the cell has been extracted. There is one exception: the function of *cnxA* can be circumvented under *in vitro* conditions (see below).

Assay for MoCo

The MoCo was shown by Rajagopalan's group (Chapter 14) to be a low molecular weight molybdopterin exhibiting no catalytic activities on its own but becoming biologically active on association with an appropriate apoprotein. In order to assay active MoCo *in vitro* we used extracts of the *nit-1* mutant of *Neurospora crassa* which completely lacks molybdopterin and its precursors (Kramer *et al.* 1984) and used the activity of restored *Neurospora* NADPH-NR as a measure of the amount of active MoCo. Wild-type tobacco extract had only a low *nit-1* complementing activity but this could be increased several-fold by subjecting the extract to acid treatment. Thus the tobacco MoCo occurs in a bound form which is not readily accessible to the *nit-1* system and it must be released by pre-treatment with acid (Mendel *et al.* 1981, 1982*a*) or, more efficiently, heat (Mendel 1983). MoCo has to be protected by performing pre-treatments under anaerobic conditions and adding appropriate thiol reagents such as reduced glutathione (Mendel and Alikulov 1983; Alikulov and Mendel 1984) and high concentrations of molybdate (Mendel *et al.* 1981; Mendel 1983).

Complementation of nit-1

When the *nit-1* assay was performed in the presence of a high concentration of molybdate, *cnxA* mutants were shown to contain a cofactor moiety which could complement *nit-1* to high activity (Mendel *et al.* 1981, 1986). Omitting molybdate from the assay prevented *nit-1* complementation, whereas Nia$^-$ and wild-type cells do complement *nit-1* in the absence of molybdate (Mendel *et al.* 1981). It can be assumed, therefore, that the defect in the *cnxA* mutants resides in a lack of molybdenum as the catalytically active ligand metal for the cofactor, while the structural pteridine moiety of the cofactor does not seem to be impaired by the mutation. Extracts of *cnxB* and *cnxC* (Mendel *et al.* 1986), *cnxD* (De Vries *et al.* 1986), *cnxE* and *cnxF* (Bache *et al.*, in preparation) mutants do not complement *nit-1* (Table 11.3).

Mo content

To rule out the possibility that the *cnxA* mutants are uptake mutants for Mo the cellular Mo content was determined. Mutants in the *cnxA* gene and in all of the other *cnx* genes contained Mo in amounts similar to wild-type and Nia$^-$ cells (Mendel *et al.* 1984, 1986; Bache *et al.*, in preparation) and hence are unlikely to be uptake mutants for Mo (Table 11.3).

Table 11.3. Characteristics of *cnx* mutants of *N. plumbaginifolia* (Mendel *et al.* 1986; De Vries *et al.* 1986; Bache *et al.*, in preparation)

	Complementation groups (no. of mutants examined)					
	A (18)	B (2)	C (3)	D (6)	E (2)	F (2)
Complementation of *Neurospora nit-1*	+	−	−	−	−	−
Molybdenum content	+	+	+	+	+	+
Repair by molybdate *in vivo*	+	−	−	−	−	−
Repair by molybdate *in vitro*	+	−	−	−	−	−
Repair *in vitro* by heterologous MoCo	+	+	+	+	+	+
Dimerization of NR apoprotein	+	−	±	+	±	nt
	(3)	(1)	(2)	(2)	(1)	

Repair by molybdate in vivo

The *cnxA* mutants were shown to contain a potentially active molybdopterin moiety, presumably lacking Mo *in situ*. Subsequently, the cells were grown on media containing high levels of molybdate in attempts to circumvent the role of the gene product impaired by the *cnx* mutations. Table 11.3 shows that of all the *cnx* mutants tested only *cnxA* lines were repairable under these conditions (Mendel *et al.* 1981, 1986; Gabard *et al.* 1988; Bache *et al.*, in preparation).

Repair by molybdate in vitro

When *cnxA* cells were extracted in the presence of high concentrations of molybdate together with thiol reagents (Mendel and Müller 1985), NADH-NR activity was regained (Table 11.3). Thus, in the presence of high levels of molybdate the function controlled by the *cnxA* gene is neither required *in vivo* nor *in vitro* for the formation of active NADH-NR. No other *cnx* mutant could be repaired *in vitro* by molybdate (Mendel *et al.* 1986; De Vries *et al.* 1986; Bache *et al.*, in preparation).

Repair in vitro by heterologous MoCo

The addition of heterologous MoCo prepared from bovine milk xanthine oxidase to cell extracts of *cnx* mutants efficiently restores NADH-NR in all types of *cnx* mutants (Mendel and Müller 1985; De Vries *et al.* 1986; Bache *et al.*, in preparation) thereby demonstrating both the functional integrity of the *nia*-coded NR apoprotein and the universal function of the MoCo. As well as repair by heterologous MoCo of animal, bacterial (*E. coli*) and fungal (*N. crassa*) origin, repair by homologous MoCo purified from extracts of wild-type tobacco and *nia* mutants was also possible (Mendel and Müller 1985). Furthermore, inactive MoCo derived from *cnxA* cells reactivated *in vitro*, and added to extracts of *cnxA* mutants reconstituted NADH-NR (Mendel and Müller 1985). The ability of NR apoprotein to act as an acceptor for MoCo of different origins seems to be a general characteristic property of *cnx* mutants in *Nicotiana*.

Dimerization of NR monomers

MoCo not only forms part of the catalytically active centre of NR but is also essential for dimerizing the monomeric subunits of NR. By performing a sucrose density gradient centrifugation of extracts from cnx mutants and measuring the diaphorase activity, one can check the dimerization state of the NR apoprotein (Table 11.3). Only *cnxA* (Mendel and Müller 1979; Mendel *et al.* 1986) and one of the *cnxD* mutants (De Vries *et al.* 1986) showed dimeric NR apoproteins: *cnxB* and *cnxC* mutants were monomeric (Mendel *et al.* 1986), whilst *cnxE* and other *cnxD* mutants gave intermediate results (Bache *et al.*, in preparation).

Functions of the cnx gene products

cnxA and *cnxD*. The results reviewed here are interpreted as showing that *cnxA* and *cnxD* cells contain a MoCo that, although defective in its catalytic properties, is still able to mediate dimerization of NR monomers. The defect caused by the *cnxD* mutation is obviously quite different from and considerably more 'severe' than that caused by *cnxA* mutations since unlike *cnxA* $^-$, *cnxD* $^-$ cells cannot be repaired by molybdate *in vivo* or *in vitro* and do not complement *nit-1*. Mutation in *cnxD* therefore causes a partial structural defect in the molybdopterin moiety of MoCo, but leaves the site(s) of molybdopterin involved in dimerization unaffected. A mutation in *cnxA*, however, leaves the whole molybdopterin structurally unaffected. The MoCo of *cnxA* mutants can be repaired by different approaches. The NR apoprotein of *cnxA* mutants is dimeric, can be purified by affinity chromatography (Mendel 1980) and shows properties similar to wild-type NR (Mendel and Müller 1980). Thus it seems reasonable to suggest that the *cnxA* gene product is not involved in the synthesis of molybdopterin but is essential for inserting Mo into the pteridine moiety of MoCo. Mo insertion appears to be the final step in the biosynthesis of active MoCo, and is thus the link between two metabolic pathways, one of which involves the biosynthesis of the pteridine moiety of MoCo, the other pathway involving uptake, transport and processing (chelation) of the molybdate anion (Mendel and Müller 1985).

In *N. tabacum* four allelic isolates of *cnxB* (Buchanan and Wray 1982; Xuan *et al.* 1983) have phenotypic properties identical to *cnxA* mutants, except for their inability to be repaired *in vivo* by growth on high molybdate (Mendel *et al.* 1984). This suggests that they are defective in the Mo processing pathway. The

N. tabacum cnxB isolates differ considerably from the two independent *cnxB* mutants of *N. plumbaginifolia* (Marton *et al.* 1982*b*; Gabard *et al.* 1988) to which we refer below.

cnxB, cnxC, cnxE. There are no currently measurable biochemical differences between *cnxB, cnxC*, and *cnxE* mutants, therefore, further classification parameters will have to be developed in order to discriminate biochemically between these loci. Mutants in these genes presumably lack molybdopterin or possess a form so heavily damaged that it cannot be detected by the approaches used, suggesting that the products of genes *cnxB, cnxC, cnxE*, and also *cnxD* are involved in the synthesis of the molybdopterin moiety of MoCo. Since the pterin ring system also forms part of other compounds, some of which are essential for cellular metabolism (e.g. folic acid, flavins), a mutational block in its synthesis would be lethal for the cell. Hence the products of these *cnx* genes are more likely to be involved in the synthesis of the molybdopterin-specific alkyl side chain and the insertion of the functional groups, in particular the important sulphur atoms. The existence of MoCo carrier/storage proteins should also be taken into consideration since free pteridines are generally of low solubility in aqueous media.

MoCo is a unique structure which occurs in all organisms so far examined. This book demonstrates that mutants defective in MoCo have been described in many species ranging from prokaryotes to lower and higher eukaryotes. Perhaps the pathway for MoCo biosynthesis is similar in all of these organisms? The isolation and molecular comparison of *cnx* genes of organisms from different phylogenetic origin will answer this question.

Regulation of NR synthesis and activity

The synthesis of NR in *Nicotiana* is partially constitutive. Both cell cultures (Müller and Grafe 1978) and plants (Müller 1983) of *N. tabacum* exhibit considerable levels of NR activity when grown with ammonium or amino acids as sole nitrogen source. The increase in NR activity that occurs after addition of nitrate is due to *de novo* synthesis of NR protein (Schiemann and Müller 1985). Several workers have shown that ammonium succinate, glutamine and other amino acids act as repressors of NR activity in cultured tobacco cells (e.g. Marion-Poll *et al.* 1984). Repression of NR activity by ammonium succinate requires active glutamine synthetase (Müller *et al.* 1982).

The multitude of studies on NR regulation that has been performed in wild-type *N. tabacum* (only a few of which have been cited above) create a need for the identification of regulatory gene mutants in this organism. Unfortunately, as in other higher plants, such mutants are still lacking in *Nicotiana*. Information contributing to a better understanding of the mechanisms that regulate NR activity has however been obtained with the help of Nia$^-$ and Cnx$^-$ mutants.

Gene dosage effects

The influence of *nia*[+] gene dosage on NR activity was studied using stepwise substitution of null activity alleles for the four *nia*[+] gene copies present in wild-type *N. tabacum*. Progeny from back-crosses of Nia[−] double mutants to wild-type plants and plants containing only one or two active *nia*[+] gene copies were examined. All of these genotypes grew as well as wild-type plants with nitrate as sole nitrogen source, showing that one *nia*[+] gene copy is sufficient for nitrate reduction to occur at optimum rates.

Nevertheless, both the constitutive and the nitrate-induced levels of *in vivo* NR activity in seedlings were found to depend on *nia*[+] gene dosage. Sensitivity to chlorate toxicity was also closely correlated with the number of *nia*[+] gene copies. During further development of the plants, however, the differences between the genotypes diminished. When leaves of 6- to 14-week-old plants were examined, all the genotypes exhibited about the same level of *in vivo* NR activity, which suggested stringent feedback regulation (Müller 1983). In contrast, the extracted (*in vitro*) NR activity was linearly related to the number of *nia*[+] gene copies at all developmental stages (Müller and Saalbach, in preparation) as shown in the example given in Table 11.4. These observations led to the following conclusions:

1. Tobacco is obviously an 'overproducer' of NR. It synthesizes considerable amounts of NR even in the absence of the substrate, and when wild-type plants are grown with nitrate, they maintain a level of NR that is much higher than needed for optimum nitrate utilization.
2. Both the constitutive and the nitrate-induced level of NR are dependent on the number of *nia*[+] gene copies; under comparable conditions one-gene plants contain about four times less NR than wild-type plants.
3. The *in vivo* NR activity in leaves of nitrate grown plants (but not of young seedlings) is stringently regulated by feedback mechanisms and therefore does not depend on *nia*[+] gene dosage. The results can (at least formally) be explained by postulating that the feedback control leads to some kind of reversible inactivation of NR molecules and that the inactive state of the enzyme is maintained during the *in vivo* assay but reversed upon extraction.

Regulation of the synthesis of inactive NR protein

NR in *N. tabacum* appears to be autogenously regulated in a way similar to that proposed for the NR of *Aspergillus nidulans* (Cove 1979), since some NR-deficient mutants do not need nitrate for maximum induction of the synthesis of the inactive NR. Cell cultures of the two *cnxA* mutants of *N. tabacum* examined in this respect (Cnx68, Cnx101) exhibited constitutive levels of inactive NR that were higher than the nitrate-induced wild-type level and could not generally be further increased by nitrate. The level of inactive NR was estimated on the basis of both the diaphorase activity of NR (Mendel and Müller 1979) and the NR-CRM level

Table 11.4. Effect of *nia*[+] gene dosage on NR activity in leaves of 40-day-old *N. tabacum* plants grown in soil (Müller and Saalbach, in preparation)

Genotype (*nia*1; *nia*2)	*in vivo* NR (nmol/g fresh wt/min)	NADH-NR (nmol/g fresh wt/min)
+/+; +/+	17.2	34.8
+/+; 28/28	15.9	14.4
28/28; +/+	17.8	18.7
+/28; 28/28	16.8	9.3
28/28; +/28	16.1	6.4

(Schiemann and Müller 1985). The two *cnxA* mutants also showed an increased constitutive level of nitrite reductase.

In contrast, the terminal NR activity in three *nia* mutants (Nia93, Nia26, Nia28) (Mendel and Müller 1979; Müller *et al.*, in preparation) and the NR-CRM level in the Nia28 and Nia63 mutants (Schiemann and Müller 1985) were inducible by nitrate. Other NR-deficient mutants of *N. tabacum* and *N. plumbagini-folia* have not yet been tested for inducibility of the synthesis of NR protein. It is only known that some *nia* mutants have a higher than wild-type level of NR-CRM when grown on nitrate (Table 11.1). Recently, a *nia* mutant of *N. plumbaginifolia* (D51) was described that exhibited 5- to 10-fold higher levels of NR-CRM and terminal NR activity than wild-type plants (Gabard *et al.* 1987). Whether this overproduction of NR protein can be related to the autoregulatory role of NR is not clear, since the constitutive level of NR protein in this mutant has not yet been determined. It is also not clear whether feedback regulation is involved.

Regulation of MoCo synthesis

The level of active MoCo measured by the *nit-1* assay in cell cultures of *N. tabacum* was found to be inducible by nitrate and to change during passage in a similar way to the changes of NR activity (Mendel *et al.* 1982b). These observations suggested a common regulation of NR and MoCo synthesis. Further support for this idea was gained from results obtained with *cnx* and *nia* mutants: *cnxA* mutants that exhibited a high constitutive level of NR protein also maintained a high constitutive MoCo level; loss of NR activity in one group of *nia* mutants was accompanied by a drastic reduction of the MoCo level (no maximum at the end of the log phase, no induction by nitrate); other *nia* mutants, especially those that had retained inducible terminal NR activity, exhibited a nitrate-inducible MoCo level. Thus at least one step in MoCo synthesis is linked to NR synthesis. This may simply mean that the *nit-1* assay in heat-treated plant extracts measures only that MoCo which has been released from NR and other molybdoenzymes and ignores the precursors of the enzyme-bound MoCo.

The work reviewed above indicates that NR-deficient mutants may be of considerable help in elucidating the mechanisms that regulate the assimilation of nitrate in higher plants. Interesting results are to be expected when the influence of defined exogenous factors on the levels of NR activity (both *in vivo* and *in vitro*), NR protein and NR mRNA are studied not only in wild-type plants, but also in certain mutants and in plants with reduced *nia* $^+$ gene dosage. We hope that in the future more workers will make use of the possibilities offered by the *Nicotiana* mutants for an analysis of the regulatory network.

References

Alikulov, Z. A. and Mendel, R. R. (1984). Molybdenum cofactor from tobacco cell cultures and milk xanthine oxidase: involvement of sulfhydryl groups in dimerization activity of cofactor. *Biochemie und Physiologie der Pflanzen* **179**, 693–705.

Biasini, G. and Marton, L. (1985). A rapid assay for genetic complementation of nitrate reductase deficiency via bulk protoplast fusion. *Molecular and General Genetics* **198**, 353–5.

Buchanan, R. J. and Wray, J. L. (1982). Isolation of molybdenum cofactor-defective lines of *Nicotiana tabacum*. *Molecular and General Genetics* **188**, 228–34.

Cove, D. J. (1979). Genetic studies of nitrate assimilation in *Aspergillus nidulans*. *Biological Reviews* **54**, 291–327.

De Vries, S. E., Dirks, R., Mendel, R. R., Schaart, J. G., and Feenstra, W. J. (1986). Biochemical characterization of nitrate reductase deficient mutants of *Nicotiana tabacum*. *Plant Science* **44**, 105–10.

Dirks, R., Negrutiu, I., Sidorov, V., and Jacobs, M. (1985). Complementation analysis by somatic hybridization and genetic crosses of nitrate reductase-deficient mutants of *Nicotiana plumbaginifolia*. *Molecular and General Genetics* **201**, 339–43.

Dirks, R., Negrutiu, I., Heinderycks, M., and Jacobs, M. (1986). Genetic analysis of revertants for the nitrate reductase function of *Nicotiana plumbaginifolia*. *Molecular and General Genetics* **202**, 309–11.

Evola, S. V. (1983*a*) Chlorate-resistant variants of *Nicotiana tabacum* L. I. Selection *in vitro* and phenotypic characterization of cell lines and regenerated plants. *Molecular and General Genetics* **189**, 447–54.

Evola, S. V. (1983*b*) Chlorate-resistant variants of *Nicotiana tabacum* L. II. *Parasexual genetic characterization*. *Molecular and General Genetics* **189**, 455–7.

Gabard, J., Marion-Poll, A., Cherell, I., Meyer, C., Müller, A. J., and Caboche, M. (1987). Isolation and characterization of *N. plumbaginifolia* nitrate reductase-deficient mutants: genetical and biochemical analysis of the nia complementation group. *Molecular and General Genetics* **209**, 596–606.

Gabard, J., Pelsy, F., Marion-Poll, A., Caboche, M., Saalbach, I., Grafe, R., and Müller, A.J. (1988). Genetic analysis of nitrate reductase-deficient mutants of *Nicotiana plumbaginifolia*: evidence for six complementation groups among 70 classified molybdenum cofactor-deficient mutants. *Molecular and General Genetics* **213**, 206–13.

Glimelius, K., Eriksson, T., Grafe, R., and Müller, A. J. (1978). Somatic hybridization of nitrate reductase-deficient mutants of Nicotiana tabacum by protoplast fusion. *Physiologia Plantarum* **44**, 273–7.

Grafe, R. and Müller, A. J. (1982). Revertants of the nitrate reductase structural gene mutant nia-130 of Nicotiana tabacum. In *Nitrate assimilation—molecular and genetic aspects*. International Symposium, Abstracts, p. 26. Gatersleben, GDR.

Grafe, R. and Müller, A. J. (1983). Complementation analysis of nitrate reductase-deficient mutants of Nicotiana tabacum by somatic hybridization. *Theoretical and Applied Genetics* **66**, 127–30.

Grafe, R., Marion-Poll, A., and Caboche, M. (1986). Improved *in vitro* selection of nitrate reductase-deficient mutants of *Nicotiana plumbaginifolia*. *Theoretical and Applied Genetics* **73**, 299–304.

Hamill, J. D. and Cocking, E. (1986). Control of pH and its effect on ammonium utilization by nitrate reductase deficient plants, cells, and protoplasts of *Nicotiana tabacum*. *Journal of Plant Physiology* **123**, 289–98.

Hoarau, J., Hirel, B., and Nato, A. (1986). New artificial electron donors for *in vitro* assay of nitrate reductase isolated from cultured tobacco cells and other organisms. *Plant Physiology* **80**, 946–9.

Kramer, S., Hageman, R. V., and Rajagopalan, K. V. (1984). *In vitro* reconstitution of nitrate reductase activity of the *Neurospora crassa* mutant *nit1*: specific incorporation of molybdopterin. *Archives of Biochemistry and Biophysics* **233**, 821–9.

Lazar, G. B., Fankhauser, H., and Potrykus, I. (1983). Complementation analysis of a nitrate reductase-deficient *Hyoscyamus muticus* cell line by somatic hybridization. *Molecular and General Genetics* **189**, 359–64.

Marion-Poll, A., Huet, J. C., and Caboche, M. (1984). Regulation of nitrate reductase in protoplast-derived cells: influence of exogeneously supplied nitrate, ammonium and amino acids. *Plant Science Letters* **34**, 61–72.

Marton, L., Dung, T. M., Mendel, R. R., and Maliga, P. (1982*a*). Nitrate reductase-deficient cell lines from haploid protoplast cultures of *Nicotiana plumbaginifolia*. *Molecular and General Genetics* **186**, 301–4.

Marton, L., Sidorov, V., Biasini, G., and Maliga, P. (1982*b*). Complementation in somatic hybrids indicates four types of nitrate reductase-deficient lines in *Nicotiana plumbaginifolia*. *Molecular and General Genetics* **187**, 1–3.

Mendel, R. R. (1980). Comparative affinity chromatography of nitrate reductase from wild-type and molybdenum cofactor-defective cell cultures of *Nicotiana tabacum*. *Biochemie und Physiologie der Pflanzen* **175**, 216–27.

Mendel, R. R. (1983). Release of molybdenum co-factor from nitrate reductase and xanthine oxidase by heat treatment. *Phytochemistry* **22**, 817–19.

Mendel, R. R. and Alikulov, Z. A. (1983). Reversible immobilization of molybdenum cofactor on a gel matrix via sulphydryl groups. *Journal of Chromatography* **267**, 409–13.

Mendel, R. R. and Müller, A. J. (1976). A common genetic determinant of xanthine dehydrogenase and nitrate reductase in *Nicotiana tabacum*. *Biochemie und Physiologie der Pflanzen* **170**, 538–41.

Mendel, R. R. and Müller, A. J. (1978). Reconstitution of NADH-nitrate reductase *in vitro* from nitrate reductase-deficient *Nicotiana tabacum* mutants. *Molecular and General Genetics* **161**, 78–80.

Mendel, R. R. and Müller, A. J. (1979). Nitrate reductase-deficient mutant cell lines of *Nicotiana tabacum*. Further biochemical characterization. *Molecular and General Genetics* **177**, 145–53.

Mendel, R. R. and Müller, A. J. (1980) Comparative characterization of nitrate reductase from wild-type and molybdenum cofactor-defective cell cultures of *Nicotiana tabacum*. *Plant Science Letters* **18**, 277–88.

Mendel, R. R. and Müller, A. J. (1985). Repair *in vitro* of nitrate reductase-deficient tobacco mutants (cnxA) by molybdate and by molybdenum cofactor. *Planta* **163**, 370–5.

Mendel, R. R., Alikulov, Z. A., Lvov, N. P., and Müller, A. J. (1981). Presence of the molybdenum-cofactor in nitrate reductase-deficient mutant cell lines of *Nicotiana tabacum*. *Molecular and General Genetics* **181**, 395–9.

Mendel, R. R., Alikulov, Z. A., and Müller, A. J. (1982*a*). Molybdenum cofactor in nitrate reductase-deficient tobacco mutants. II. Release of cofactor by heat treatment. *Plant Science Letters* **26**, 67–72.

Mendel, R. R., Alikulov, Z. A., and Müller, A. J. (1982*b*). Molybdenum cofactor in nitrate reductase-deficient tobacco mutants. III. *Plant Science Letters* **27**, 95–101.

Mendel, R. R., Buchanan, R. J., and Wray, J. L. (1984). Characterization of a new type of molybdenum cofactor-mutant in cell cultures of *Nicotiana tabacum*. *Molecular and General Genetics* **195**, 186–9.

Mendel, R. R., Marton, L., and Müller, A. J. (1986). Comparative biochemical characterization of mutants at the nitrate reductase/molybdenum cofactor loci *cnxA*, *cnxB*, and *cnxC* of *Nicotiana plumbaginifolia*. *Plant Science* **43**, 125–9.

Müller, A. J. (1978). Nitrate reductase mutants and the cell culture approach to higher plant genetics. In *Problems in general genetics*, Proceedings XIV International Congress of Genetics, vol. II, 1, p. 229. Mir Publishers, Moscow.

Müller, A. J. (1983). Genetic analysis of nitrate reductase-deficient tobacco plants regenerated from mutant cells. Evidence for duplicate structural genes. *Molecular and General Genetics* **192**, 275–81.

Müller, A. J. and Grafe, P. (1975). Mutant cell lines of *Nicotiana tabacum* deficient in nitrate reductase. In *XII International Botanical Congress*, Abstracts, p. 304. Nauka Publishers, Leningrad.

Müller, A. J., and Grafe, R. (1978). Isolation and characterization of cell lines of *Nicotiana tabacum* lacking nitrate reductase. *Molecular and General Genetics* **161**, 67–76.

Müller, A. J. and Mendel, R. R. (1982). Nitrate reductase-deficient tobacco mutants and the regulation of nitrate assimilation. In *Plant tissue culture* 1982, p. 233. Maruzen Press, Tokyo.

Müller, A. J., Saalbach, I., Mendel, R. R., and Schiemann, J. (1982). Regulation of nitrate reductase activity in cultured tobacco cells. In *Nitrate assimilation—molecular and genetic aspects*, International Symposium, Abstracts, p. 28, Gatersleben, GDR.

Negrutiu, I., Dirks, R., and Jacobs, M. L. (1983). Regeneration of fully nitrate reductase-deficient mutants from protoplast culture of *Nicotiana plumbaginifolia* (Viviani). *Theoretical and Applied Genetics* **66**, 341–7.

Saalbach, T. and Müller, A. J. (1982). The kinetics of nitrate uptake by cultured cells of nitrate reductase-deficient tobacco mutants. In *Nitrate assimilation—molecular and genetic aspects*, International Symposium, Abstracts, p. 27, Gatersleben, GDR.

Schiemann, J. and Müller, A. J. (1985). Detection of nitrate reductase cross-reacting material in wild-type and mutant cells of *Nicotiana tabacum*. *Biochemie und Physiologie der Pflanzen* **180**, 63–74.

Somers, D. A., Narayanan, K. R., Kleinhofs, A., Cooper-Bland, S., and Cocking, E. C. (1986). Immunological evidence for transfer of the barley nitrate reductase structural gene to *Nicotiana tabacum* by protoplast fusion. *Molecular and General Genetics* **204**, 296–301.

Xuan, L. T., Grafe, R., and Müller, A. J. (1983). Complementation of nitrate reductase-deficient mutants in somatic hybrids between *Nicotiana* species. In *Protoplasts 1983*, Poster Proceedings, 6th International Protoplast Symposium, p. 76–7. Birkhäuser Verlag, Basel.

12. Molecular genetics of nitrate reduction in *Nicotiana*

Michel Caboche, Isabelle Cherel, Fabienne Galangau, Marie-Angele Grandbastien, Christian Meyer, Therese Moureaux, Frederique Pelsy, Pierre Rouze, Herve Vaucheret, Francoise Vedele and Michel Vincentz

Introduction

Nitrate utilization plays a key function in the physiology of higher plants. Apart from its function in the regulation of osmotic pressure in plant tissues, nitrate is the main source of mineral nitrogen used to synthesize amino acids and proteins. Nitrate reduction is the first step of this assimilation process, and the expression of nitrate reductase (NR) activity has been extensively studied under various physiological and genetic conditions (Kleinhofs *et al.* 1985; Wray 1986). Surprisingly, it is only recently that molecular tools such as monoclonal antibodies and cDNA probes have been obtained to study this enzyme. In this report we summarize the studies performed in our laboratory on the NR from corn, tomato and *Nicotiana*.

Most of our efforts have been concentrated on *N. tabacum* and *N. plumbaginifolia* since these species allow a combination of different approaches, including plant cell genetics, biochemistry, and molecular biology. Our interest in studying the tomato and corn enzymes are obviously related to the future prospect of evaluating the impact of nitrate reduction on crop production. Four areas of research will be presented: immunochemistry, genetics, molecular biology and physiology of nitrate reductase expression.

Immunochemical analysis of nitrate reductase

Mouse monoclonal antibodies were raised against partially purified preparations of corn and tobacco NR obtained by blue Sepharose column chromatography and PAGE electrophoresis. Seven epitopic classes were obtained by classification of these monoclonal antibodies on the basis of competitive binding to NR as measured by ELISA testing (Cherel *et al.* 1985). Three of these classes are

Table 12.1. Epitopic groups among monoclonal antibodies raised against corn and tobacco nitrate reductases

Epitopic group clones	1 30(6) 96(9)25	2 28(2)	3 25(15)	4 8(23) 7(10)12	5 42(22)	6 15(21)	7 17NP 7(6)
Western blotting							
Native form	+ + +	+ + + +		+ + +	+ + +	+	+
Denatured	(+)	−	−	+ + +	+ + +	+ + +	−
Enzymatic activity inhibition							
NADH-NR	+	+	−	−	−	−	(+)
MV-NR	+	+	−	−	−	−	(+)
NADH-CR	+	−	−	−	−	−	−
BOB-NR	−	−	−	−	−	−	+

Western blottings (Cherel *et al.* 1985) and enzymatic activity inhibition studies were performed on corn nitrate reductase.
Western blottings: + + + strong recognition by the antibody; + weak but significant recognition by the antibody; − no recognition by the antibody.
Enzymatic activity inhibition: NADH-NR = NADH nitrate reductase; MV-NR = methylviologen nitrate reductase; NADH-CR = NADH cytochrome c reductase; BOB-NR = bromophenol blue nitrate reductase. + more than 90% inhibition of the enzymatic activity; (+) weak inhibition of the enzymatic activity, may not be significant; − no significant inhibition of the enzymatic activity.

inhibitory to the catalytic activity of corn NR (Table 12.1). A monoclonal antibody raised against the tobacco NR was found to inhibit the recently discovered bromophenol blue NR activity carried by the NR protein (Meyer *et al.* 1987). A comparison of the ability of these different classes of antibodies to recognize NR extracted from other plant species showed that most of the epitopes recognized on the corn enzyme and other monocotyledon enzymes were not detected on the enzymes extracted from dicotyledons. Only one class, represented by the monoclonal antibody 96(9)25, was able to recognize NR from all higher plants tested. However, this monoclonal antibody was not able to recognize the *Aspergillus nidulans* or *Escherichia coli* nitrate reductases. These observations may suggest that this ubiquitous enzyme evolves rapidly.

Genetic analysis of the nitrate reductase structural gene

N. plumbaginifolia is a convenient plant species on which to perform cell genetics. Haploid protoplasts of *N. plumbaginifolia* were used to select for chlorate-resistant clones: from ~ 200 chlorate-resistant clones selected (Grafe *et al.* 1986; Marion-Poll *et al.* 1984) and regenerated (Caboche 1987), 60 per cent were specifically deficient for NR activity. These clones still expressed normal levels of xanthine dehydrogenase activity and were therefore able to synthesize the molybdenum cofactor (MoCo). They were presumed to be affected in the biosynthesis of the apoenzyme of NR. All of these plants displayed a chlorotic morphology when transferred to the greenhouse (Saux *et al.* 1986), and when grafted onto wild-type Wisconsin tobacco a large proportion was able to flower and set seed. These seeds were unable to grow on a culture medium containing nitrate as sole nitrogen source. Genetic analysis of these mutant clones showed that they generally did not complement each other and all were homozygous for a recessive nuclear mutation (Gabard *et al.* 1987). This complementation group was classified as *nia*, the structural gene for NR (Müller 1983) on the basis of the following observations. An ELISA assay of the amount of NR protein found in the different mutants showed that some of the clones were still producing an inactive apoenzyme whereas others contained no NR protein as determined by ELISA (Gabard *et al.* 1987). A methylviologen NR activity and a bromophenol blue NR activity were still detectable in some of the *nia* clones. This complementation group therefore appears heterogeneous for the structure and expression of an inactive NR activity, as predicted for a collection of mutants affected in the structural gene encoding for NR (Table 12.2). Some of the *nia* mutants over-expressed the apoenzyme polypeptide chain, suggesting that the catalytic activity of NR, or the enzyme itself, may play a regulatory function in its expression as previously observed for *A. nidulans* (Chapters 6 and 19) and *Neurospora crassa* (Chapter 20). Intragenic complementation between several *nia* mutants was observed (Gabard *et al.* 1987) and may be related to the dimeric structure of the enzyme, in which two non-functional subunits affected in different steps of the catalytic activity of NR may produce a functional heterodimeric enzyme. The reconstitution of a functional NR activity by mixing extracts of two complementing *nia* mutants favours this hypothesis.

Molecular cloning of a cDNA encoding for nitrate reductase

In order to characterize further the regulation of NR a strategy for the cDNA cloning of the apoenzyme of NR was developed as described previously (Cheng *et al.* 1986; Crawford *et al.* 1986).

Total polyadenylated mRNA extracted from tobacco leaves was size-fractionated on a sucrose gradient (Commere *et al.* 1986) and the 3–5 kb area of the gradient was collected and used for the biosynthesis of cDNA according to the procedure of Gubler and Hoffman (1983). After linker addition the cDNA preparation was used to build a library in the expression vector λgt11, which was screened with a polyclonal rabbit antibody raised against purified corn NR. Twelve immunoreactive recombinants were isolated and further characterized from the screening of $\sim$ 350 000 recombinant phages; eight contained inserts sharing sequence homologies. They gave consistently different signals upon immunological analysis, suggesting that they were independent clones.

Table 12.2. Biochemical characteristics of various *nia* mutants

Line	Apoenzyme amount (μg/g fresh wt.)	NADH-NR (nmol/min/g fresh wt.)	Methylviologen-NR (nmol/min/g fresh wt.)	FMNH2-NR (nmol/min/g fresh wt.)
Wild-type	0.20	21.0	17.8	40.6
E87	0.11	NS	NS	NS
F59	0.07	0.1	6.9	18.8
D51	1.95	0.05	87.6	242.0
E56	NS	NS	3.6	2.5
12	NS	NS	NS	NS
15	0.13	NS	NS	NS

Apoenzyme concentrations were deduced from ELISA measurements (Cherel *et al.* 1986), assuming an unmodified affinity of the antibodies for the antigen.
NS: not detectable.

The structure and orientation of cDNAs in the cloning site of the vector were studied using the restriction enzymes *Eco*R1, *Kpn*1, and *Sst*1. Inserts were found in both orientations in the vector, confirming the expression of inserted sequences under the control of the β-galactosidase promoter as expected, and also under the control of another phage promoter. One of the recombinant phages, λ13-29 was further characterized (Calza *et al.* 1987). This clone expressed a β-galactosidase fusion protein of molecular weight 170 kDa which was used to immunoselect antibodies from the NR antiserum (Fig. 12.1). These antibodies were able to specifically inhibit the catalytic activity of tobacco NR. In a set of monoclonal antibodies directed against NR from tobacco, two were found to recognize the fusion protein expressed by clone 13-29.

The 1.6 kb cDNA insert of clone 13-29 was subcloned in pEMBL plasmids and sequenced according to the method of Sanger (1977) after ordered deletion by the *Exo*III–S1 method. The analysis of the sequence of this 1.6 kb fragment revealed an open reading frame (ORF) covering the entire length of the cDNA in frame with the β-galactosidase-coding sequence of the vector. This ORF shared homologies with the porphyrin-binding site of various b_5 cytochromes (Nobrega *et al.* 1971), (Fig. 12.2), and a strong sequence homology with cytochrome b_5 reductase was also detected. This latter enzyme is a flavoprotein able to reduce cytochrome b_5 with the reducing power of NADH (Yabisui *et al.* 1984). Sequence data analysis of the cDNA therefore confirm the identity of this clone and suggest that NR may have evolved by the fusion of the coding sequences of two independent gene families (Calza *et al.* 1987).

The 1.6 kb insert of clone 13-29 was used as a probe in Northern blot experiments (Calza *et al.* 1987). Hybridization signals were detected in tobacco mRNA preparations with a 3.7 kb mRNA species. Under low stringency conditions weak cross hybridization was detected with tomato but not with corn

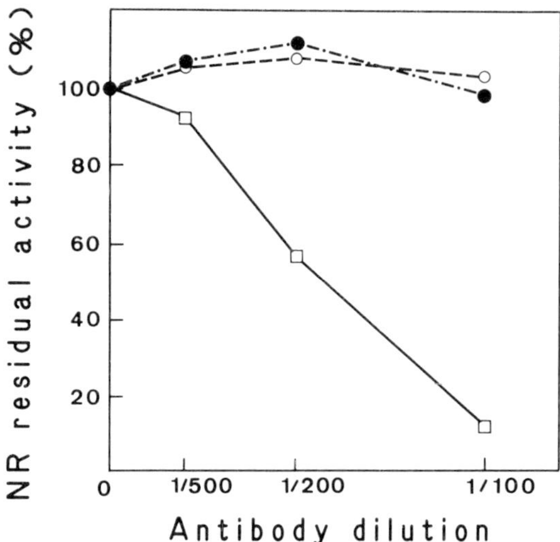

Fig. 12.1. Inhibition of nitrate reductase catalytic activity by antibodies immunoselected on the fusion protein expressed by recombinant clone 13-29. (□) NR polyclonal antibodies immunoselected on plaques from clone 13-29; (○) Non immune antiserum innumose-lected on plaques from clone 13-29; (●) NR polyclonal antibodies immunoselected on plaques from λgt11.

mRNA preparations. This suggests a rapid rate of evolution of NR in higher plants.

Expression of the structural gene of nitrate reductase

The parallel use of Northern blot, ELISA, and NR assay allows a better characterization of the expression of nitrate reductase activity in tissues. Tobacco plants starved for nitrogen for 1 week expressed approximately the same level of NR mRNA as plants continuously grown on nitrate. In contrast, nitrogen-starved plants expressed a 10-fold decrease of NR activity and NR protein compared to nitrate-grown plants, suggesting a post-transcriptional control of NR expression. Upon nitrate feeding, nitrogen-starved tobacco plants rapidly accumulated NR mRNA as previously described for barley and *Arabidopsis* (Cheng *et al.* 1986; Crawford *et al.* 1986) suggesting that NR expression is also regulated at the transcription level and is inducible by nitrate (Fig. 12.3).

The level of NR mRNA was followed over a 24 h period. An accumulation of NR mRNA was correlated with the night period, while during the day NR mRNA decreased significantly, become virtually undetectable by the end of the day. This dramatic change in the level of NR mRNA was not reflected at the protein and activity levels since these were found to decrease by a factor of two at most

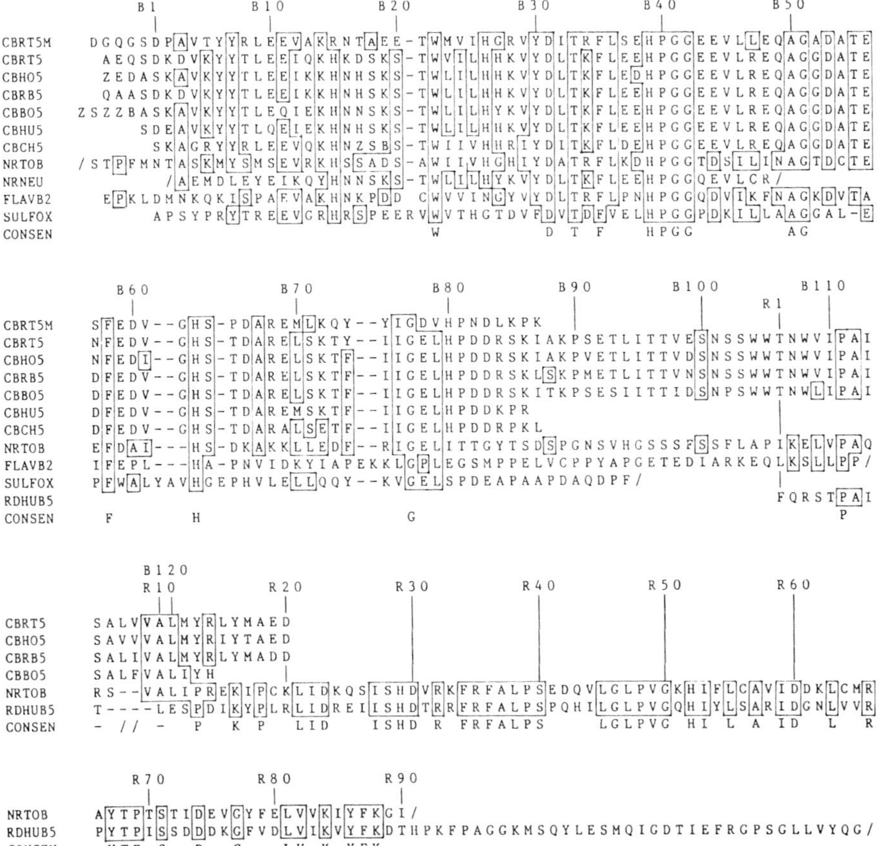

Fig. 12.2. Sequence homologies between nitrate reductase, cytochrome b_5 and cytochrome b_5 reductase. The predicted amino acid sequence of tobacco NR has been aligned in the haem-binding domain (B1–B120) with protein sequences of the cytochrome b_5 superfamily and in the FAD/NADH domain (R1–R90) with the protein sequence of human cytochrome b_5 reductase. CBRT5, CBHO5, CBRB5, CBBO5, CBHU5, CBCH5: Microsomal cytochrome b_5 from rat, horse, rabbit, bovine, human and chicken (Nobrega and Ozols 1971); CBRT5M: Mitochondrial cytochrome b_5 from rat (Lederer *et al* 1983); NRTOB: Nitrate reductase from tobacco; NRNEU: Nitrate reductase from *Neurospora crassa* (Le and Lederer 1983); FLAVB2: Flavocytochrome b_2 from *S. cerevisiae* (Lederer *et al.* 1985); SULFOX: Sulfite oxidase from *S. cerevisiae* (Guiard and Lederer 1979); RDHUB5: Cytochrome b_5 reductase from human (Yabisui *et al.* 1984).

between the beginning and the end of the light period (Fig. 12.3). A similar effect of light was observed when parallel studies were performed on tomato, using a tomato-specific probe, suggesting that although light is considered to be required to promote nitrate assimilation, it also seems to antagonize the expression of NR mRNA. When NR activity was followed during a 24 h period a small fluctuation in the level of the enzyme activity was observed, barely reflecting the change of

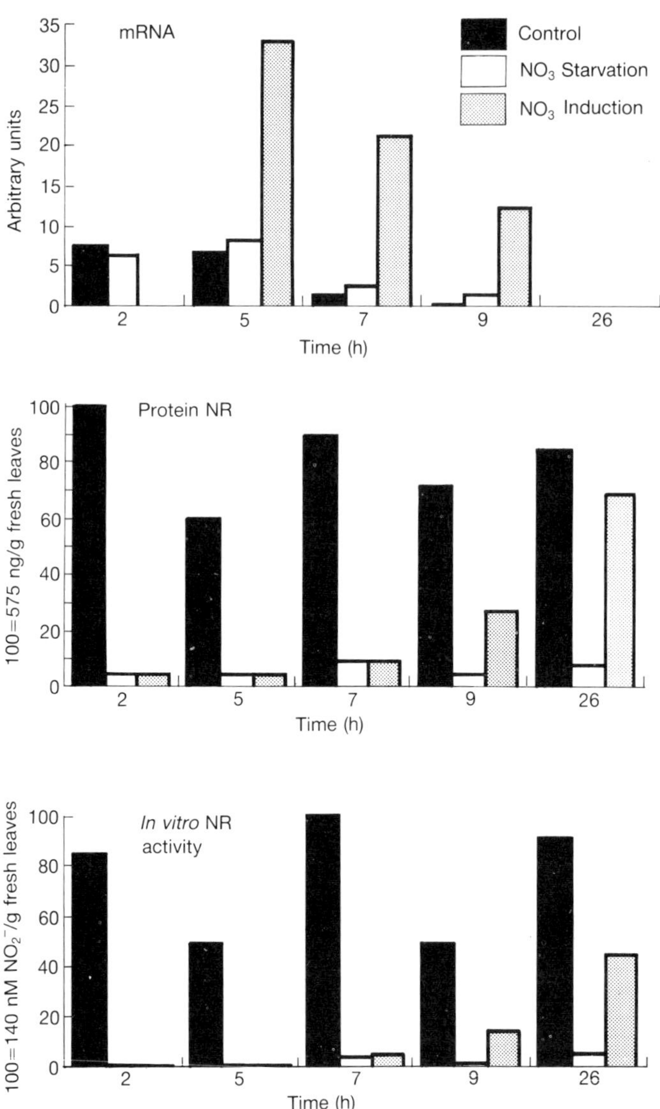

Fig. 12.3. Histogram of nitrate reductase expression under various physiological conditions. Plants were either grown permanently in the presence of nitrate (control), nitrate-starved for 7 days or nitrate-starved for the same period and then supplied with nitrate 2 h after the beginning of the light period (nitrate induction).

NR mRNA levels. Whether these variations reflect stimulation of transcription or stabilization of the mRNA transcript by nitrate is under study.

Diurnal changes of NR mRNA levels were also found to occur in roots. In roots the maximal level of NR mRNA was approximately decreased by a factor of 10 compared to leaves. However, NR mRNA levels in roots increased during the night period and decreased during the day period. This suggests that NR mRNA levels may be controlled by a circadian rhythm rather than directly regulated by light itself.

Cloning, characterization and functionality of the NR structural genes from tobacco

The 1.6 kb cDNA insert of clone 13-29 was used to screen genomic libraries of tobacco constructed in phages EMBL3B and 4. Two genomic clones were isolated and further characterized. Since the cDNA used for this screening contained only approximately 60 per cent of the sequence of the mRNA encoding for nitrate reductase the genomic clones were first characterized by subcloning the sequences located 5′ and 3′ to the area of homology with the cDNA. These subclones were used as probes in Northern blot experiments to identify genomic fragments containing exon-coding sequences. A physical map suggests that each of these genomic clones contains two introns of different size but located in the same position in the coding sequence (Fig. 12.4). Each of these clones has been attributed to one or the other of the ancestors of tobacco, *N. sylvestris* and *N. tomentosiformis*. These results are in agreement with the genetics (Müller 1983) and phylogeny of *N. tabacum*.

Using electroporation, these genomic clones have been cotransferred with the plasmid pABD1 which confers kanamycin and also paromomycin resistance to the recipient-transformed cells. Electroporation was performed in protoplasts isolated from the NIA 130 nitrate reductase-deficient mutant isolated by Müller (1983). Paromomycin-resistant colonies were regenerated from transfected protoplast cultures and regenerated on an ammonium-containing medium in the absence of selective pressure for nitrate utilization. Regenerated shoot cultures from several transformants were able to grow and root on a medium containing nitrate as sole nitrogen source. These results confirm the identity and integrity of the isolated genomic clones by genetic complementation. The sequence analysis of these clones is under way.

Discussion

NR provides an interesting model for the study of gene expression in higher plants. The availability of mutants deficient in NR allows the characterization of the expression of the corresponding cloned functional gene in deficient plants. NR expression can be efficiently selected for or against in tissue culture and can therefore be considered as a genetic tool comparable to the HAT system used for mammalian cell genetics. Classical techniques involving the use of reporter genes

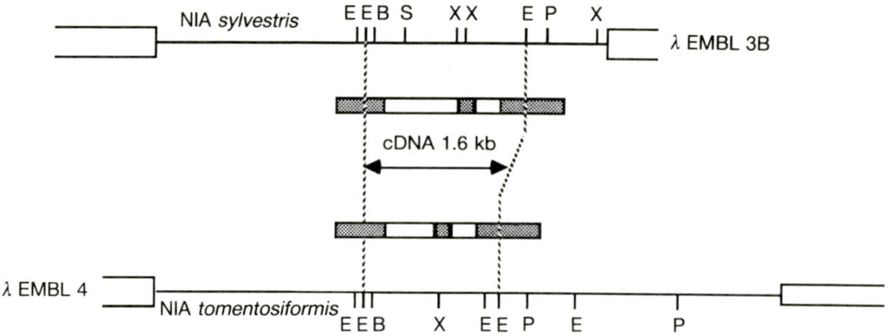

Fig. 12.4. Physical maps of nitrate reductase homologous tobacco genes. The physical map of two genomic clones isolated by screening genomic libraries with the NR cDNA clone 13-29 were compared. Significant homologies were found between the two clones. The sequencing of part of these clones has allowed a more precise description of these genes. Black boxes refer to coding sequences and open boxes to introns. *Eco*R1 sequences at the borders of the cDNA clone 13-29 are also found in the two genomic clones and are represented by dashed lines.

to study the regulatory sequences of NR structural genes or the use of NR coding sequences to study the expression of other plant gene promoters are now achievable goals.

NR expression appears to be regulated at different levels. Clear evidence has been obtained showing that NR mRNA pools are significantly increased by providing nitrate to nitrogen-starved plants (Cheng *et al.* 1986; Crawford *et al.* 1986). Whether this mainly reflects an increased rate of transcription or a decreased rate of breakdown of the mRNA remains to be studied. Nitrogen depletion does not result in a significant decrease of NR mRNA pools, suggesting that nitrate is not stringently required for the expression of NR. One of the questions raised by the preliminary data obtained in tobacco and tomato concerns the integration of nitrate reduction into the physiology of a photosynthetic organism. Nitrate reduction is connected with photosynthesis *via* ferredoxin-mediated nitrite reduction, a light-dependent process. Nitrite accumulation into chloroplasts is generally considered as deleterious to chloroplast integrity and it is expected that during a dark period nitrate reduction should be shut down to prevent this accumulation of nitrite. It is therefore surprising that NR mRNA induction takes place during a dark period. Experiments are under way to study factors that may affect NR mRNA levels such as the availability of reduced nitrogen.

The MoCo of NR plays a regulatory function in the biosynthesis of the NR apoenzyme of *E. coli* (Pascal *et al.* 1982). *A. nidulans* mutants deficient for the biosynthesis of the MoCo synthesize a defective NR holoenzyme, resulting in an altered regulation of the expression of this holoenzyme (Cove and Pateman 1969). A collection of *N. plumbaginifolia* mutants affected in various steps of the biosynthesis of MoCo will be used to study an eventual regulatory function of this cofactor in higher plants. The availability of mutants affected in the catalytic activity of NR will allow us to test for a possible autoregulatory role for this enzyme.

References

Caboche, M. (1987). Nitrogen, carbohydrate and zinc requirements for the efficient induction of shoot morphogenesis from protoplast-derived colonies of *Nicotiana plumbaginifolia*. *Plant Cell, Tissue and Organ Culture* **8**, 197–206.

Calza, R., Huttner, E., Vincentz, M., Rouze, P., Galangau, F., Vaucheret, H., Cherel, I., Meyer, C., Kronenberger, J., and Caboche, M. (1987). Cloning of DNA fragments complementary to tobacco nitrate reductase mRNA and encoding for epitopes common to the nitrate reductases from higher plants. *Molecular and General Genetics* **209**, 552–62.

Cheng, C. L., Dewdney, J., Kleinhofs, A., and Goodman, H. M. (1986). Cloning and nitrate induction of nitrate reductase mRNA. *Proceedings of the National Academy of Sciences, USA* **83**, 6825–8.

Cherel, I., Grosclaude, J., and Rouze, P. (1985). Monoclonal antibodies identify multiple epitopes on maize leaf nitrate reductase. *Biochemical and Biophysical Research Communications* **129**, 686–93.

Cherel, I., Marion-Poll, A., Meyer, C., and Rouze, P. (1986). Immunological comparisons of nitrate reductases of different plant species using monoclonal antibodies. *Plant Physiology* **81**, 376–8.

Commere, B., Cherel, I., Kronenberger, J., Galangau, F., and Caboche, M. (1986). *In vitro* translation of nitrate reductase messenger RNA from maize and tobacco and detection with an antibody directed against the enzyme of maize. *Plant Science* **44**, 191–203.

Cove, D. J. and Pateman, J. A. (1969). Autoregulation of the synthesis of nitrate reductase in *Aspergillus nidulans*. *Journal of Bacteriology* **97**, 1374–8.

Crawford, N. M., Campbell, W. H., and Davis, R. W. (1986). Nitrate reductase from squash: cDNA cloning and nitrate regulation. *Proceedings of the National Academy of Sciences, USA* **85**, 5006–10.

Gabard, J., Marion-Poll, A., Cherel, I., Meyer, C., Müller, A. J., and Caboche, M. (1987). Isolation and characterization of *N. plumbaginifolia* nitrate-reductase defficient mutants: genetic and biochemical analysis of the *nia* complementation group. *Molecular and General Genetics* **209**, 596–606.

Grafe, R., Marion-Poll, A., and Caboche, M. (1986). Improved *in vitro* selection of nitrate reductase-deficient mutants of *Nicotiana plumbaginifolia*. *Theoretical and Applied Genetics* **73**, 299–304.

Gubler, U. and Hoffmann, B. J. (1983). A simple and very efficient method for generating cDNA libraries. *Gene* **25**, 263–9.

Guiard, B. and Lederer, F. (1979). Aminoacid sequence of the b5-like heme-binding domain from chicken sulfite oxydase. *European Journal of Biochemistry* **100**, 441–53.

Kleinhofs, A., Warner, R. L., and Narayanan, K. R. (1985). Current progress towards an understanding of the genetics and molecular biology of nitrate reductase in higher plants. *Oxford Surveys of Plant Molecular and Cell Biology* **2**, 91–121.

Le, K. H. D. and Lederer, F. (1983). On the presence of a heme-binding domain homologous to cytochrome b_5 in *Neurospora crassa* assimilatory nitrate reductase. *EMBO Journal* **2**, 1909–14.

Lederer, F., Ghrir, R., Guiard, B., Cortial, S., and Ito, A. (1983). Two homologous cytochrome b_5 in a single cell. *European Journal of Biochemistry* **132**, 95–102.

Lederer, F., Cortial, S., Becam, A. M., Haumont, P. Y., and Perez, L. (1985). Complete amino acid sequence of flavocytochrome b_2 from baker's yeast. *European Journal of Biochemistry* **152**, 419–23.

Marion-Poll, A., Huet, J. C., and Caboche, M. (1984). Regulation of nitrate reductase in protoplast-derived cells: influence of exogenously supplied nitrate, ammonium and amino acids. *Plant Science Letters* **34**, 61–72.

Meyer, C., Cherel, I., Moureaux, T., Hoarau, J., Gabard, J., and Rouze, P. (1987). Bromphenol blue:nitrate reductase activity in *Nicotiana plumbaginifolia*. An immuno-chemical and genetic approach. *Biochimie* **69**, 735–42.

Müller, A. J. (1983). Genetic analysis of nitrate reductase-deficient tobacco plants regenerated from mutant cells. Evidence for duplicate structural genes. *Molecular and General Genetics* **192**, 275–81.

Nobrega, F. G., Ozols, J. (1971). Amino acid sequences of tryptic peptides of cytochromes b_5 from microsomes of human, monkey, porcine and chicken liver. *Journal of Biological Chemistry* **246**, 1706–17.

Pascal, M. C., Burini, J. F., Ratouchniak, J., and Chippaux, M. (1982). Regulation of the nitrate reductase operon: Effect of mutations in *chlA, B, D* and *E* genes. *Molecular and General Genetics* **188**, 103–6.

Sanger, F., Nicklen, S., and Coulson, A. R. (1977) DNA sequencing with chain termination inhibitors. *Proceedings of the National Academy of Sciences, USA.* **74**, 5463–7.

Saux, C., Gabard, J., and Morot-Gaudry, J. F. (1986). Modifications métaboliques liées à la perte de l'activité nitrate réductase chez un mutant de *Nicotiana plumbaginifolia*. *Comptes Rendus de l'Academie des Sciences, Paris, Série III* **302**, 391–4.

Wray, J. L. (1986). The molecular genetics of higher plant nitrate assimilation. In *A genetic approach to plant biochemistry* (eds A. D. Blonstein and P. J. King), pp. 101–57, Springer, New York.

Yabisui, T., Miyata, T., Iwanaga, S., Tamura, M., Yoshida, S., Takeshita, M., and Nakajima, H. (1984). Amino acid sequence of NADH-cytochrome b_5 reductase of human erythrocytes. *Journal of Biochemistry* **96**, 579–82.

13. Molecular genetics of nitrate reductase in barley

A. Kleinhofs, R. L. Warner, J. M. Lawrence, J. M. Melzer, J. M. Jeter, and D. A. Kudrna

Introduction

Nitrate assimilation is the primary pathway by which plants obtain reduced nitrogen. On a global basis, nitrate assimilation results in the reduction of $\sim 2 \times 10^4$ megatons of nitrogen per year (Guerrero *et al.* 1981). This is approximately 100-fold greater than that accounted for by nitrogen fixation (Burns and Hardy 1975).

Nitrate assimilation first requires reduction to ammonium which is carried out by two enzymes, nitrate reductase (NR) and nitrite reductase (NiR) (Beevers and Hageman 1969, 1972, 1983; Hewitt 1975; Hewitt *et al.* 1976; Garrett and Amy 1978; Cove 1979; Losada and Guerrero 1979; Vennesland and Guerrero 1979; Guerrero *et al.* 1981; Wray 1986). The rate-limiting step is reduction of nitrate to nitrite by NR (Beevers and Hageman 1969; Hewitt 1975). Literature on the genetics, biochemistry, and physiology of NR has been recently reviewed (Kleinhofs *et al.* 1985; Wray 1986).

Studies of NR-deficient mutants in barley have identified five molybdenum co-factor (MoCo) function genes and one gene each for the two NR apoproteins (NADH-NR and NAD(P)H-NR). This is similar to the intensively studied *Aspergillus nidulans* system where six different loci encode the MoCo functions (Cove 1979; Arst *et al.* 1982; Chapters 6, 19), and a single locus codes for the NR apoprotein (Pateman *et al.* 1967; Chapter 6). Unlike *Aspergillus nidulans* (Cove 1979; Chapter 19), regulatory mutants have not been identified in higher plants.

The study of higher plant NR regulation was facilitated by the cloning of the NADH-NR gene from barley (Cheng *et al.* 1986). Poly(A)$^+$ RNA from nitrate-induced barley leaves was enriched 5- to 10-fold for NR mRNA by size fractionation and cloned into the expression vector λgt11. Positive clones were identified by screening with NR-specific polyclonal antibody. Hybrid selection translation was used to confirm the identification of one plasmid clone (bNRp10) as NR-specific. This clone was used to study the nature of the NR-deficient mutants described in this paper.

Mutant description

Thirty-four NR-deficient mutants have been partially characterized by genetic, biochemical and molecular techniques. Genetic complementation analysis placed these mutants into seven complementation groups designated *nar-1* through *nar-7* (Table 13.1). An additional 12 mutants are currently being characterized.

Table 13.1. Summary of barley nitrate reductase genes and their function

Gene designation	Alleles (no.)	Function
nar-1	22	NADH–NR structural gene
nar-2	3	Molybdenum cofactor
nar-3	3	Molybdenum cofactor
nar-4	1	Molybdenum cofactor
nar-5	3	Molybdenum cofactor
nar-6	1	Molybdenum cofactor
nar-7	1	NAD(P)H-NR structural gene

All mutants were assayed for NADH-, NADPH- and $FMNH_2$-NR as well as cytochrome *c* reductase, NiR, and xanthine dehydrogenase activities. All mutants with approximately normal xanthine dehydrogenase activity belong to the *nar-1* complementation group, except mutant Az 70 which is designated *nar-7w*. All NR-deficient mutants with pleiotropic very low or absent xanthine dehydrogenase activity are presumed to be MoCo mutants and belong to complementation groups *nar-2–nar-6*.

nar-1 locus

The *nar-1* locus is represented by 22 mutant alleles. Greatly reduced but significant NADH- and NADPH-NR activities were observed in young seedling shoots of all *nar-1* alleles (Table 13.2). Wild-type (cvs. Steptoe and Winer) seedling shoots have high NADH-NR activity but no NADPH-NR activity. However, both NADH- and NADPH-NR activities were present in roots of wild-type seedlings (Table 13.3). The NADPH-NR activity in wild-type roots and in mutant shoots is due to a bi-specific NAD(P)H-NR with a preference for the NADPH electron donor (Dailey *et al.* 1982; Harker *et al.* 1986). Ratios of NADPH- to NADH-NR activity of ~1.3 were observed for most of the *nar-1* mutants, indicating that both activities were due to the bi-specific NAD(P)H-NR. In several mutant shoots, i.e., Az 32, 57, 64, EMS 31, and Xno 29, the NADPH-NR/NADH-NR ratio was less than 1, indicating that the activities were a result of both the

Table 13.2. Wild type and *nar-1* mutant enzyme activities and cross-reacting material

Selection number	Gene and allele	Nitrate reductase (μmol/g fresh wt/h)					Shoots*	
		Shoots			Roots		Cytochrome *c* Reductase	CRM**
		NADH	NADPH	FMNH$_2$	NADH	NADPH		
Wild type								
cv. Steptoe	–	50.8	<0.1	26.1	11.7	6.1	237	+++
cv. Winer	–	55.6	<0.1	26.0	7.4	3.5	234	+++
nar-1								
Az 12	*nar-1a*	4.4	5.6	0	4.7	6.3	15	0
Az 13	*nar-1b*	3.5	4.2	0	4.7	6.1	110	++
Az 23	*nar-1c*	3.7	4.5	0	5.9	7.7	103	+
Az 28	*nar-1d*	2.2	2.4	0	4.8	6.3	298	++++
Az 29	*nar-1e*	5.0	7.0	0	8.7	12.2	24	0
Az 30	*nar-1f*	6.0	7.5	0	5.6	7.4	22	0
Az 31	*nar-1g*	4.4	5.2	0	4.3	5.6	136	+++
Az 32	*nar-1h*	4.4	2.7	43.2	6.6	8.8	25	++++
Az 33	*nar-1i*	3.9	5.1	0	6.7	9.5	114	++
Az 56	*nar-1m*	4.8	6.3	0	4.4	5.7	16	0
Az 57	*nar-1n*	3.3	1.9	0	4.9	6.6	197	++
Az 63	*nar-1p*	5.3	7.2	0	5.3	6.7	20	0
Az 64	*nar-1q*	3.4	1.4	0	5.1	6.3	165	++++
Az 65	*nar-1r*	5.6	7.4	0	5.6	7.4	0	0
Az 76	*nar-1ab*	3.1	3.8	25.6	4.8	6.8	14	+++
Az 77	*nar-1ac*	4.0	5.4	0	6.0	8.4	16	0
Az 79	*nar-1ai*	4.2	5.6	0	5.2	7.4	5	0
Az 80	*nar-1aj*	5.4	7.1	0	6.9	9.1	18	0
BSMV1	*nar-1ao*	5.6	7.6	0	7.0	9.6	113	++
EMS29	*nar-1k*	3.8	5.1	0	4.5	6.3	20	0
EMS31	*nar-1l*	7.5	2.2	0	5.1	6.8	121	++
Xno29	*nar-1j*	7.2	3.8	0	3.8	4.5	57	++

* 24 h induction, all others were induced for 48 h.
** 0, +, ++, +++, ++++ indicate increasing concentrations of NR cross-reacting material.

Table 13.3. Effect of the nitrate reductase structural genes *nar-1* and *nar-7* on enzyme activities and cross-reacting material

Selection number	Gene and allele	Nitrate reductase (μmol/g fresh wt/h)					Shoots*		CRM*
		Shoots			Roots		Cytochrome c reductase	Xanthine dehydro-genase*	
		NADH	NADPH	FMNH$_2$	NADH	NADPH			
cv. Steptoe	*Nar-1 Nar-1; Nar-7 Nar-7*	38.6	<0.1	26.1	6.2	4.2	237	+++	+++
Az 12	*nar-1a nar-1a; Nar-7 Nar-7*	0.9	2.8	0	0.3	1.1	15	+++	0
Az 70	*Nar-1 Nar-1; nar-7w*								
	nar-7w	39.4	<0.1	28.1	4.2	<0.1	235	+++	++++
Az 12; Az 70	*nar-1a nar-1a; nar-7w*								
	nar-7w	0.1	0.1	0	<0.1	<0.1	13	++	0

*0, +, ++, +++, ++++ indicate increasing concentrations of NR cross-reacting material or enzyme activity.

NADH-specific and the NAD(P)H-bi-specific NRs. In these cases the mutation in the NADH-NR is presumably leaky, permitting some NADH-NR activity.

The NADH-NR activities were reduced in the roots of all *nar-1* mutant seedlings compared with the wild type (Table 13.2). In all mutants the root NADPH-NR activity was equal to or higher than the wild type from which the mutant was derived. Wild-type roots have both NADH-specific and NAD(P)H-bi-specific NRs, thus the NADH-NR activities are higher than the NADPH-NR activities. Inactivation of the NADH-NR by mutation should, therefore, result in reduction of NADH-NR activity without affecting the NADPH-NR activity. This is observed with most of the *nar-1* alleles; the exceptions have increased root NADPH-NR activities which remain to be explained.

All mutants were deficient in shoot $FMNH_2$-NR activity except Az 32 (*nar-1h*) and Az 76 (*nar-1ab*) which had activity equal to or greater than the wild type (Table 13.2). $FMNH_2$-NR activities were not measured in roots.

The NR-associated cytochrome *c* reductase (CR) activities in the mutants varied from nearly zero to greater than the wild type (Table 13.2). These were closely correlated with the presence or absence of nitrate reductase-specific protein as measured by cross-reaction with NADH-NR-specific antiserum (CRM) (Table 13.2). The two exceptions are mutants Az 32 and Az 76 which lack NADH-NR but are $FMNH_2$-NR positive. The NR protein in these mutants is unable to transfer electrons from NADH to cytochrome *c* or nitrate but retains the capacity to transfer electrons from $FMNH_2$ to nitrate.

All of the *nar-1* alleles are viable in the field under our conditions, with the possible exception of mutant Xno 29 (*nar-1j*). This mutant did not grow well under our field conditions, but it is not clear whether this is due to its NR deficiency. Mutant Xno 29 has both NADH- and NAD(P)H-NR with a total NR activity higher than most other *nar-1* mutants (Table 13.2). When crossed with cv. Steptoe, the NR-deficient phenotype did not segregate as a single gene. Therefore, other mutations besides the one at the *nar-1* locus may be present in Xno 29.

nar-7 locus

Mutant Az 12 (*nar-1a*) seeds were remutagenized with sodium azide and subjected to a second round of selection for mutants totally deficient in NR activity. Two mutants, designated Az 69 and Az 70, with very low or no nitrate reductase activity were selected. In one of these (Az 69) the second mutation was in a MoCo function (to be discussed later). The other mutant (Az 70) was MoCo-positive, but NADPH-NR-deficient (Warner *et al.* 1987). In order to separate the new mutation from the *nar-1a* (Az 12) background, the double mutant was crossed to the parent cultivar, Steptoe, and the F_2 progeny were grown to maturity, allowed to self-pollinate and produce F_3 seed. Progeny of F_2 individuals homozygous for shoot NADH-NR were analysed for root NADPH-NR activity. Individuals lacking root NADPH–NR activity and homozygous NADH-NR positive (*Nar-1 Nar-1*; *nar-7 nar-7*) were transplanted. This line was deficient in root NADPH-NR activity while all other NR and associated activities (FMNH$_2$-NR and CR) were normal (Table 13.3). The xanthine dehydrogenase and NiR activities were also normal (data not shown).

These results show that the *nar-7* locus controls NAD(P)H-NR, but not the NADH-NR. Since all genotypes were xanthine dehydrogenase positive, *nar-7* is not involved in MoCo functions. No other NR-associated activities were affected in the Az 70 mutant. NAD(P)H-NR structural gene function is proposed for *nar-7*, although some regulatory function specific to the NAD(P)H-NR cannot be ruled out at this time. The segregation data also show that the NADH-NR gene (*nar-1*) is not linked to the NAD(P)H-NR gene (*nar-7*).

The physiological function of the NAD(P)H-NR is not clear. We have not detected the NAD(P)H-NR activity in cv. Steptoe shoots in either growth chamber or field-grown material (although not all possible physiological states have been systematically sampled). The *Nar-1 Nar-1*; *nar-7w nar-7w* (Az 70) plants appear phenotypically normal under field conditions. The *nar-1a nar-1a*; *Nar-7 Nar-7* (Az 12) plants are slightly lighter green in colour, show delayed growth in early spring and are later maturing than the wild type; nevertheless the grain yield is only slightly reduced (Oh *et al.* 1980).

Molybdenum cofactor loci (nar-2–6)

All other mutants analysed so far fall into five complementation groups (*nar-2–6*). These mutants are characterized by pleiotropic loss of NR and xanthine dehydrogenase activities and are presumed to be MoCo function deficient. The NADH-NR activities were low, but variable (Table 13.4). The NADPH-NR activities, while present in both shoots and roots, tended to be very low and never exceeded the NADH-NR activity (Table 13.4). It appears that low levels of both NR are present in the shoots of most MoCo mutants. The root NADH- and NADPH-NR activities were lower than wild type in all mutants except Az 66 and 68, which had root NADH- and NADPH-NR activities comparable to the wild type (Table 13.4). All MoCo (*nar-2–6*) mutants had significant NR-associated CR activity and NR-CRM reflecting the presence of NR apoprotein (Table 13.4).

Gene mapping

The *nar-1* locus was mapped to chromosome 6 using wheat–barley addition line DNA probed with the barley NR cDNA clone bNRp10 PstI insert (Melzer *et al.* 1988). Further mapping is in progress and preliminary evidence suggests that *nar-1* is located at 6S, the short arm of chromosome 6.

The *nar-2* locus was mapped to the centromeric region of chromosome 7 by linkage to the marker genes *nld* and *mt2* (Melzer *et al.* 1988). Analysis of F$_3$ head

Table 13.4. Wild-type and MoCo mutant enzyme activities and NR cross-reacting material

Selection number	Gene and allele	Nitrate reductase (µmol/g fresh wt/h)				Shoots*			Field status†
		Shoots		Roots		Cytochrome c Reductase	CRM**	Xanthine dehydro-genase**	
		NADH	NADPH	NADH	NADPH				
Wild type									
cv. Steptoe	–	52.4	<0.1	10.8	6.4				V
cv. Steptoe*		32.2	<0.1	5.9	4.8	237	+++	+++	V
cv. Winer	–	52.8	<0.1	8.3	3.5	234	+++	+++	V
Az 34	*nar-2a*	2.6	0.9	1.8	1.8	128	++	0	V
R9401*	*nar-2ad*	<0.1	<0.1	0.1	0.1	94	+++	0	L
R9201*	*nar-2ag*	<0.1	<0.1	<0.1	<0.1	107	+++	0	L
Xno18	*nar-3a*	2.5	0.7	2.9	2.7	183	++	0	L
Xno19	*nar-3b*	1.4	0.4	2.2	2.0	163	++	0	L
Az 71	*nar-3x*	2.6	0.8	2.2	1.5	121	++	0	V
Az 72*	*nar-4y*	0.2	0.1	0.4	0.3	148	++++	0	L
Az 62	*nar-5o*	3.2	0.9	2.7	2.5	81	++	0	V
Az 66	*nar-5s*	2.9	1.0	6.6	5.2	88	++++	0	V
Az 68	*nar-5u*	2.5	1.0	5.3	4.3	95	++++	0	V
Az 69*	*nar-6v*	0.6	0.3	1.0	0.4	121	++	0	L

* 24 h induction with nitrate, all others were induced for 48 h.
** 0, +, ++, +++, ++++ indicate increasing concentration of NR cross-reacting material (CRM) or enzyme activity.
† V = viable, L = lethal.

rows located *nar-2* 8.4 ± 2.1 cM from *nld* and 23.0 ± 4.6 cM from *mt2*. This locates *nar-2* at 54.7 ± 3.1 cM from *s*, near the centromere of chromosome 7. Close linkage of *nar-2* with *ddt* and *lys3* was detected but could not be quantitated due to deviations from the expected $1:2:1$ segregations for the *ddt* and *lys3* genes.

Mapping of the *nar-3, -4, -5, -6* and *-7* loci is in progress. The *nar-7* locus is not closely linked to the *nar-1* locus, as indicated by separation of the two loci among limited number of F_2 individuals examined from the cross of *nar-1 nar-1*; *nar-7 nar-7* by the $++$; $++$ wild-type (Warner *et al.* 1987).

Southern and Northern analyses

Eighteen alleles of the *nar-1* gene were analysed by Southern and Northern analyses in order to determine the nature of the mutation at the molecular level. The restriction fragment length polymorphism (RFLP) analysis was somewhat complicated by the discovery that the parent cultivar (Steptoe) population was genetically heterogeneous at or near the *nar-1* locus. Individual Steptoe plants had one of two RFLP patterns at the *nar-1* locus which were traced to the parent cultivars of Steptoe, Unitan and WA3564 (Jeter, unpublished Thesis, 1987). The different *nar-1* mutant alleles (induced by sodium azide or ethyl methane-sulphonate treatments) did not show any major insertions or deletions. Insertions or deletions of less than 100 bp would probably not have been detected in this analysis, thus, it appears that these mutants are probably due to point mutations. Sixteen of the mutants analysed by Southern blots were also subjected to Northern analysis of the RNA samples isolated from nitrate-induced seedlings and probed with the bNRp10 PstI insert (Melzer, unpublished Thesis, 1987). These data showed that all of these mutants produced a NR-specific mRNA of approximately the same size as the wild-type NR mRNA, supporting the conclusion that these mutants are due to point mutations. Substantial differences were observed in the quantity of the NR mRNA accumulated by the different mutants (Table 13.5). These data indicate that some of the mutations in the *nar-1* locus affect the regulation of NR mRNA accumulation. If this effect is at the mRNA transcription, transport or stability level cannot be determined from the available data.

Discussion

The first higher plant NR-deficient mutant was isolated in *Arabidopsis thaliana* (Oostindier-Braaksma and Feenstra 1973). Since then large numbers of additional mutants have been isolated in *Arabidopsis* (Braaksma and Feenstra 1982*a,b,c*; Scholten *et al.* 1985; Wang *et al.* 1986), barley (Tokarev and Shumny 1977; Warner *et al.* 1977; Kleinhofs *et al.* 1980, 1985; Bright *et al.* 1983), *Datura*

Table 13.5. Nitrate reductase mRNA isolated from leaves of NR-deficient mutants after induction with nitrate

Selection number	Allele designation	Experiment 1		Experiment 2	
		Time* (h)	NR mRNA (% of Steptoe)	Time* (h)	NR mRNA (% of Steptoe)
Az 12	*nar-1a*	4	34.9	18	87.9
Az 13	*nar-1b*	4	146.0	18	339.4
Az 23	*nar-1c*	4	273.0	18	218.2
Az 28	*nar-1d*	4	276.2		
Az 29	*nar-1e*	4	52.4	18	130.3
Az 30	*nar-1f*	4	31.7		
Az 31	*nar-1g*	4	200.0		
Az 32	*nar-1h*	4	211.1		
Az 33	*nar-1i*	4	200.0		
Az 34	*nar-2a*	4	150.8		
Az 65	*nar-1r*	4	23.8	18	54.5
Az 68	*nar-5u*	4	120.6		
Az 71	*nar-3x*	4	125.4		
Az 77	*nar-1ac*	4	41.3		
Az 56	*nar-1m*	16	42.1	18	42.4
Az 57	*nar-1n*	16	294.7		
Az 63	*nar-1p*	16	313.2		
Az 64	*nar-1q*	16	331.6		
Xno 18	*nar-3a*	16	471.1	18	518.2
Xno 29	*nar-1j*	16	226.3		
EMS 29	*nar-1k*	16	147.4	18	133.3
EMS 31	*nar-1l*	16	371.1		
BSMV 1	*nar-lao*	16	102.6	18	97.0
Winer	wild-type	16	181.6		

* Time after induction by nitrate.

(King and Khanna 1980), *Hyoscyamus* (Gebhardt *et al.* 1981; Lazar *et al.* 1983; Strauss *et al.* 1981), *Nicotiana* (Mendel and Müller 1976, 1978, 1979; Muller and Grafe 1978; Buchanan and Wray 1982; Evola 1983; Negrutiu *et al.* 1983), pea (Kleinhofs *et al.* 1978; Feenstra and Jacobsen 1980; Warner *et al.* 1982), petunia (Steffen and Schieder 1983, 1984), and soybean (Nelson *et al.* 1983; Carroll and Gresshoff 1986; Streit and Harper 1986). Most of the mutants were selected based on their reduced sensitivity to chlorate. In barley, pea, and soybean an additional screening technique based on a semiquantitative *in vivo* NR assay on individual M_2 seedlings was used. This technique, although not as rapid as chlorate screening, permits selection of mutants in species where multiple NR are present.

In barley, chlorate selection results in the isolation of MoCo function mutants in most cases (Tokarev and Shumny 1977; Shumny and Tokarev 1981; Bright *et al.* 1983). The one exception is mutant Xno 29, isolated by Tokarev and Shumny (1977), which is a structural gene *nar-1* allele. The reason for the selection of MoCo mutants in barley by chlorate screening is clear since we now know that barley has both NADH-NR and NAD(P)H-NR. We assume that both enzymes share the same MoCo, thus explaining why MoCo-defective mutants can be selected by chlorate screening.

All of the mutants with the Az or EMS designation described herein were selected by the rapid *in vivo* assay technique. This procedure yields numerous structural gene (*nar-1*) mutants and some MoCo function mutants. The preponderance of structural gene *versus* MoCo mutants is also observed with chlorate selection in systems where only a single NR gene is found (Müller and Grafe 1978; Negrutiu *et al.* 1983).

Since some of the MoCo function genes in barley are represented by a single allele, we suspect that not all of the possible loci have been uncovered. Five MoCo genes have been observed in *Escherichia coli* (*chlA, B, D, E,* and *G*) (Reiss *et al.* 1987; Chapter 3), and six in *Aspergillus nidulans* (*cnx ABC, E, F, G, H, J*) (Cove 1979; Arst *et al.* 1982; Chapters 6, 19), two organisms in which NR genetics has been extensively investigated. In the case of *E. coli*, however, the *chlA* and *E* loci are believed to be complex (Reiss *et al.* 1987) and the *A. nidulans cnx ABC* locus is complex (Cove 1979). If these complex loci are not maintained as such in higher plants, it is possible that additional MoCo function loci will be uncovered.

The lack of representatives with true zero NR activity among our structural gene (*nar-1*) mutants presented an interesting problem. The discovery of the NAD(P)H-NR in our mutants (Dailey *et al.* 1982) suggested that the measured NADPH-NR activities are really due to the NAD(P)H-bi-specific enzyme. If this were the case, we would expect that the NADPH-NR/NADH-NR activity ratio would be constant. This is based on the observation that the partially purified NAD(P)H-NR has a NADPH-NR/NADH-NR activity ratio of approximately 1.3. This NADPH-NR/NADH-NR activity ratio is observed in the majority of the *nar-1* alleles (Table 13.2). Notable exceptions are mutants Az 32, 57, 64, EMS 31, and Xno 29. We propose that these alleles are due to leaky NADH-NR mutations and contain both NADH-specific and NAD(P)H-bi-specific NR activities. We have preliminary evidence that this is the case for mutant Xno 29. Direct evidence will be obtained when monoclonal antibodies positive for the NAD(P)H-NR but negative for the NADH-NR become available.

The root NR activities are more difficult to interpret. The wild-type barley roots, unlike the shoots, contain both the NADH- and NAD(P)H-NR. If the NADH-NR was eliminated by the mutation, we would expect a reduction in NADH-NR activity, no change in NADPH-NR activity, and a NADPH-NR/NADH-NR activity ratio of approximately 1.3. Within fairly broad experimental error, which is to be expected considering the difficulties when working with roots, this expectation is realized for most of the mutants. Several mutants, however, show NADPH-NR activities which appear to be higher than the wild-type root NADPH-NR activity. This observation, although unexplained at this time, illustrates the complex regulatory relationships among the genes controlling the NADH-NR and NAD(P)H-NR. For example, in the wild-type barley shoots, no NADPH-NR activity can be detected until the NADH-NR activity is removed by mutation. Consequently, the NAD(P)H-NR-deficient mutants can only be selected in NADH-NR-deficient background. To date, we only have one mutant of the presumed NAD(P)H-NR structural gene (designated *nar-7*), therefore, broad conclusions are not warranted. This mutant, Az 70, was selected in Az 12 (*nar-1* structural gene) background and separated from it by crossing to the wild-type cultivar Steptoe (Warner *et al.* 1987). The Az 12, Az 70 double mutant has essentially no NADH- or NADPH-NR activities (Table 13.3). The Az 70 mutant in the absence of the Az 12 mutation is normal with respect to NR activity except for the absence of NADPH-NR activity in the roots (Table 13.3). Thus, these two genes (*nar-1* and *nar-7*) appear to be independent and together control essentially all of the NR activity in barley.

Work with micro-organisms suggests that the mutagens azide and EMS cause base substitutions and not gross chromosomal alterations (Kleinhofs *et al.* 1974; Drake and Baltz 1976). The *nar-1* mutants analysed in this investigation appear to be due to base substitutions or small deletions and/or insertions. The Southern hybridizations did not detect size variation within the NR hybridizing fragments that could be attributed to mutations. Similarly, Northern analysis did not detect size variation in the mutant NR mRNA. However, small deletions or insertions (< 100 bp) would probably not have been detected.

Quantifying NR mRNA using dot-blot hybridizations revealed significant differences among the *nar* mutants. Several *nar-1* NADH-NR structural gene mutants accumulated 2–3 times more NR mRNA than the wild-type control. Six *nar-1* mutants contain less NR mRNA than the wild-type control at 4 and 16 or 18 h after nitrate induction. These mutants, Az 12, 29, 30, 56, 65, and 77, are deficient in cytochrome *c* reductase activity and NR cross-reacting material. Thus, they probably represent mutants with reduced mRNA or protein stability. Ribosome binding site or stop codon mutations would also result in such effects.

Mutants that accumulate more than wild-type levels of NR mRNA suggest that some form of feedback regulation present in the wild type is absent in these mutants. Autoregulation by NR has been proposed in fungi (Pateman and Cove 1967; Cove 1970, 1979; Tomsett and Garrett 1981; Chapters 6, 19, 20, and 22) and in *Escherichia coli* (Bonnefoy *et al.* 1986; Chapter 3). Products of nitrate assimilation, such as glutamine, have also been implicated in the feedback regulation of fungal NR (Dunn-Coleman *et al.* 1979; Premakumar *et al.* 1979). The possible role of any of these mechanisms in regulating barley NR remains to be investigated. The mutants described should provide powerful tools for those investigations.

Acknowledgements

This investigation was supported in part by U.S. Department of Agriculture Grant No. 86-CRCR-1-2004 and National Science Foundation Grant DMB-8505095.

References

Arst, H. N. Jr., Tollervey, D. W., and Sealy-Lewis, H. M. (1982). A possible regulatory gene for the molybdenum containing cofactor in *Aspergillus nidulans*. *Journal of General Microbiology* **128**, 1083–93.

Beevers, L. and Hageman, R. H. (1969). Nitrate reduction in higher plants. *Annual Review of Plant Physiology* **20**, 495–522.

Beevers, L. and Hageman, R. H. (1972). The role of light in nitrate metabolism in higher plants. In *Photophysiology* (ed. A. C. Giese), pp. 85–113. Academic Press, New York.

Beevers, L. and Hageman, R. H. (1983). Uptake and reduction of nitrate: Bacteria and higher plants. In *Inorganic plant nutrition* (eds. A. Lanchi and R. L. Bielesk), pp. 351–75. Springer-Verlag, Vienna.

Bonnefoy, V., Pascal, M.-C., Ratouchniak, J., and Chippaux, M. (1986). Autoregulation of the *nar* operon encoding nitrate reductase in *Escherichia coli*. *Molecular and General Genetics* **204**, 180–4.

Braaksma, F. J. and Feenstra, W. J. (1982*a*). Reverse mutants of the nitrate reductase-deficient mutant B25 of *Arabidopsis thaliana*. *Theoretical and Applied Genetics* **61**, 263–71.

Braaksma, F. J. and Feenstra, W. J. (1982*b*). Isolation and characterization of nitrate reductase-deficient mutants of *Arabidopsis thaliana*. *Theoretical and Applied Genetics* **64**, 83–90.

Braaksma, F. J. and Feenstra, W. J. (1982*c*). Nitrate reduction in the wild type and a nitrate reductase deficient mutant of *Arabidopsis thaliana*. *Physiologia Plantarum* **54**, 351–60.

Bright, S. W. J., Norbury, P. B., Franklin, J., Kirk, D. W., and Wray, J. L. (1983). A conditional-lethal *cnx*-type nitrate reductase-deficient barley mutant. *Molecular and General Genetics* **189**, 240–4.

Buchanan, R. J. and Wray, J. L. (1982). Isolation of molybdenum cofactor defective cell lines of *Nicotiana tabacum*. *Molecular and General Genetics* **188**, 228–34.

Burns, R. C. and Hardy, R. W. F. (1975). *Nitrogen fixation in bacteria and higher plants*. Springer, Berlin.

Carroll, B. J. and Gresshoff, P. M. (1986). Isolation and initial characterization of constitutive nitrate reductase-deficient mutants NR328 and NR345 of soybean (*Glycine max*). *Plant Physiology* **81**, 572–6.

Cheng, C.-L., Dewdney, J., Kleinhofs, A., and Goodman, H. M. (1986). Cloning and nitrate induction of nitrate reductase mRNA. *Proceedings of the National Academy of Sciences, USA* **83**, 6825–8.

Cove, D. J. (1970). Control of gene action in *Aspergillus nidulans*. *Proceedings of the Royal Society, London, Series B*, **176**, 267–75.

Cove, D. J. (1979). Genetic studies of nitrate assimilation in *Aspergillus nidulans*. *Biological Reviews of the Cambridge Philosophical Society* **54**, 291–327.

Dailey, F. A., Warner, R. L., Somers, D. A., and Kleinhofs, A. (1982). Characteristics of a nitrate reductase in a barley mutant deficient in NADH nitrate reductase. *Plant Physiology* **69**, 1200–4.

Drake, J. W. and Baltz, R. H. (1976). The biochemistry of mutagenesis. *Annual Review of Biochemistry* **45**, 11–38.

Dunn-Coleman, N. S., Tomsett, A. B., and Garrett, R. H. (1979). Nitrogen metabolite repression of nitrate reductase in *Neurospora crassa*: Effect of the *gln--*1a locus. *Journal of Bacteriology* **139**, 697–700.

Evola, S. V. (1983). Chlorate-resistant variants of *Nicotiana tabacum* L. I. Selection *in vitro* and phenotypic characterization of cell lines and regenerated plants. *Molecular and General Genetics* **189**, 447–54.

Feenstra, W. J. and Jacobsen, E. (1980). Isolation of a nitrate reductase deficient mutant of *Pisum sativum* by means of selection for chlorate resistance. *Theoretical and Applied Genetics* **58**, 39–42.

Garrett, R. H. and Amy, N. K. (1978). Nitrate assimilation in fungi. *Advances in Microbiology* **18**, 1–65.

Gebhardt, C., Schnebli, V. and King, P. J. (1981). Isolation of biochemical mutants using haploid mesophyll protoplasts of *Hyoscyamus muticus*. II. Auxotropic and temperature-sensitive clones. *Planta* **153**, 81–9.

Guerrero, M. G., Vega, J. M., and Losada, M. (1981). The assimilatory nitrate-reducing system and its regulation. *Annual Review of Plant Physiology* **8**, 169–204.

Harker, A. R., Narayanan, K. R., Warner, R. L., and Kleinhofs, A. (1986). NAD(P)H bispecific nitrate reductase in barley leaves: partial purification and characterization. *Phytochemistry* **25**, 1275–9.

Hewitt, E. J. (1975). Assimilatory nitrate-nitrite reduction. *Annual Review of Plant Physiology* **26**, 73–100.

Hewitt, E. J., Hucklesby, D. P., and Notton, B. A. (1976). Nitrate assimilation. In *Plant biochemistry* (ed. J. Bonner and J. E. Varner), pp. 633–81. Academic Press, New York.

Jeter, J. M. (1987). Molecular analysis of the *nar1* locus in barley. Unpublished M.S. thesis. Washington State University.

King, J. and Khanna, V. (1980). A nitrate reductase-less variant isolated from suspension cultures of *Datura innoxia* (Mill.). *Plant Physiology* **66**, 632–6.

Kleinhofs, A., Sander, C., Nilan, R. A., and Konzak, C. F. (1974). Azide mutagenicity – mechanism and nature of mutants produced. In *Polyploidy and induced mutations in plant breeding*, pp. 195–9. International Atomic Energy Agency, Vienna.

Kleinhofs, A., Warner, R. L., Muehlbauer, F. J., and Nilan, R. A. (1978). Induction and selection of specific gene mutations in *Hordeum* and *Pisum*. *Mutation Research* **51**, 29–35.

Kleinhofs, A., Kuo, T. M., and Warner, R. L. (1980). Characterization of nitrate reductase-deficient barley mutants. *Molecular and General Genetics* **177**, 421–5.

Kleinhofs, A., Warner, R. L., and Narayanan, K. R. (1985). Current progress towards an understanding of the genetics and molecular biology of nitrate reductase in higher plants. In *Oxford surveys of plant molecular and cell biology*, Vol. 2 (ed. B. J. Miflin), pp. 91–121. Oxford University Press, Oxford.

Lazar, G. B., Fankhauser, H., and Potrykus, I. (1983). Complementation analysis of a nitrate reductase deficient *Hyoscyamus muticus* cell line by somatic hybridization. *Molecular and General Genetics* **189**, 359–64.

Losada, M. and Guerrero, M. G. (1979). The protosynthetic reduction of nitrate and its regulation. In *Photosynthesis in relation to model systems* (ed. J. Barber), pp. 365–408. Elsevier, Amsterdam.

Melzer, J. M. (1987). Effect of nitrate, light and mutation on nitrate reductase messenger RNA in barley. Unpublished D.Phil. Thesis. Washington State University.

Melzer, J. M., Kleinhofs, A., Kudrna, D. A., Warner, R. L., and Blake, T. K. (1988). Genetic mapping of the barley nitrate reductase-deficient *nar1* and *nar2* loci. *Theoretical and Applied Genetics*, **75**, 767–71.

Mendel, R. R. and Müller, A. J. (1976). A common genetic determinant of xanthine dehydrogenase and nitrate reductase in *Nicotiana tabacum*. *Biochemie und Physiologie der Pflanzen* **170**, 538–41.

Mendel, R. R. and Müller, A. J. (1978). Reconstitution of NADH-nitrate reductase *in vitro* from nitrate reductase-deficient *Nicotiana tabacum* mutants. *Molecular and General Genetics* **161**, 77–80.

Mendel, R. R. and Müller, A. J. (1979). Nitrate reductase deficient mutant cell lines of *Nicotiana tabacum*. *Molecular and General Genetics* **177**, 145–53.

Müller, A. J. and Grafe, R. (1978). Isolation and characterization of cell lines of *Nicotiana tabacum* lacking nitrate reductase. *Molecular and General Genetics* **161**, 67–76.

Negrutiu, I., Dirks, R., and Jacobs, M. (1983). Regeneration of fully nitrate reductase-deficient mutants from protoplast culture of *Nicotiana plumbaginifolia* (Viviani). *Theoretical and Applied Genetics* **66**, 341–7.

Nelson, R. S., Ryan, S. A., and Harper, J. E. (1983). Soybean mutants lacking constitutive nitrate reductase activity. I. Selection and initial plant characterization. *Plant Physiology* **72**, 503–9.

Oh, J. Y., Warner, R. L., and Kleinhofs, A. (1980). Effect of nitrate reductase deficiency upon growth, yield and protein in barley. *Crop Science* **20**, 487–90.

Oostindier-Braaksma, F. J. and Feenstra, W. J. (1973). Isolation and characterization of chlorate-resistant mutants of *Arabidopsis thaliana*. *Mutation Research* **19**, 175–85.

Pateman, J. A. and Cove, D. J. (1967). Regulation of nitrate reduction in *Aspergillus nidulans*. *Nature* **215**, 1234–7.

Pateman, J. A., Rever, B. M., and Cove, D. J. (1967). Genetic and biochemical studies of nitrate reduction in *Aspergillus nidulans*. *Biochemical Journal* **104**, 103–11.

Premakumar, R., Sorger, G. J., and Gooden, D. (1979). Nitrogen metabolite repression of nitrate reductase in *Neurospora crassa*. *Journal of Bacteriology* **137**, 119–26.

Reiss, J., Kleinhofs, A., and Klingmüller, W. (1987). Cloning of seven differently complementing DNA fragments with *chl* functions from *Escherichia coli* K12. *Molecular and General Genetics* **206**, 352–5.

Scholten, H. J., de Vries, S. E., Nijdam, H., and Feenstra, W. J. (1985). Nitrate reductase deficient cell lines from diploid cell cultures and lethal mutant M_2 plants of *Arabidopsis thaliana*. *Theoretical and Applied Genetics* **71**, 263–71.

Shumny, V. K. and Tokarev, B. I. (1981). Genetic control of nitrate reductase activity in barley. In *Barley genetics IV, Fourth International Barley Genetics Symposium* (ed. M. J. C. Asher, R. P. Ellis, H. M. Hayter, and R. N. H. Whitehouse), pp. 881–5. Edinburgh University Press, Edinburgh.

Steffen, A. and Schieder, O. (1983). Selection and characterization of nitrate reductase deficient mutants of *Petunia*. In *Protoplasts 1983* (ed. I. Potrykus, C. T. Harms, A. Hinnen, R. Hutter, P. J. King, and R. D. Shillito), pp. 162–3. Birkhauser Verlag.

Steffen, A. and Schieder, O. (1984). Biochemical and genetical characterization of nitrate reductase deficient mutants of *Petunia*. *Plant Cell Report* **3**, 134–7.

Strauss, A., Bucher, F., and King, P. J. (1981). Isolation of biochemical mutants using haploid mesophyll protoplasts of *Hyoscyamus muticus*. I. A NO_3^- non-utilizing clone. *Planta* **153**, 75–80.

Streit, L. and Harper, J. E. (1986). Biochemical characterization of soybean mutants lacking constitutive NADH:nitrate reductase. *Plant Physiology* **81**, 593–6.

Tokarev, B. I. and Shumny, V. K. (1977). Detection of barley mutants with low level of nitrate reductase activity after seed treatment with ethylmethane-sulphonate. *Genetika*

(Moskva) **13**, 2097–103.

Tomsett, A. B. and Garrett, R. H. (1981). Biochemical analysis of mutants defective in nitrate assimilation in *Neurospora crassa*: evidence for autogenous control by nitrate reductase. *Molecular and General Genetics* **184**, 183–90.

Vennesland, B. and Guerrero, M. G. (1979). Reduction of nitrate and nitrite. In *Encyclopedia of plant physiology*, Vol. 6 (ed. M. Gibbs and E. Latzko), pp. 424-44. Springer, Berlin.

Wang, X. M., Scholl, R. L., and Feldman, K. A. (1986). Characterization of a chlorate-hypersensitive, high nitrate reductase *Arabidopsis thaliana* mutant. *Theoretical and Applied Genetics* **72**, 328–36.

Warner, R. L., Lin, C. J., and Kleinhofs, A. (1977). Nitrate reductase-deficient mutants in barley. *Nature* **269**, 406–7.

Warner, R. L., Kleinhofs, A., and Muehlbauer, F. J. (1982). Characterization of nitrate reductase-deficient mutants in pea. *Crop Science* **22**, 389–93.

Warner, R. L., Narayanan, K. R., and Kleinhofs, A. (1987). Inheritance and expression of NAD(P)H nitrate reductase in barley. *Theoretical and Applied Genetics* **74**, 714–17.

Wray, J. L. (1986). The molecular genetics of higher plant nitrate assimilation. In *A genetic approach to plant biochemistry*, Vol. 3 (ed. A. Blonstein and P. King), pp. 101–57. Springer-Verlag, Vienna.

14. Chemistry and biology of the molybdenum cofactor

K. V. Rajagopalan

The presence of tightly-bound molybdenum in rat liver xanthine oxidase was discovered in 1953 (Richert and Westerfeld 1953). Since then a number of molybdenum-containing enzymes have been identified in diverse organisms. All of the purified molybdenzymes so far characterized are complex proteins containing multiple prosthetic groups. The fact that the three molybdoenzymes of animal origin, xanthine oxidase, aldehyde oxidase, and sulphite oxidase, can be purified in relatively large amounts has led to detailed spectroscopic characterization of their prosthetic groups by techniques such as electronic absorption spectroscopy (Massey 1973), electron paramagnetic resonance (Bray 1975), and extended X-ray absorbance fine structure analysis (Garner and Bristow 1985). The results of such studies have provided strong evidence for the involvement of thiolate ligands in the coordination field of Mo in these enzymes, but have not given any indication of the presence of a non-protein moiety in association with the Mo in these proteins. Even in the case of rat liver sulphite oxidase, which can be proteolytically cleaved to generate a domain containing only the Mo center, the electronic absorption spectrum displayed by the domain is mainly due to the Mo–S linkages at the active site (Johnson and Rajagopalan 1977). Ironically, the fungal enzyme nitrate reductase, which is difficult to purify and thus has not been studied in any great detail, has played a vital role in generating the evidence for the presence, quantitation and ultimate characterization of the Mo cofactor (MoCo).

The first and most important contribution in relation to the MoCo was the role played by microbial genetics, initially with *Aspergillus nidulans* and later with a variety of other organisms. Genetic studies on the expression of xanthine dehydrogenase and nitrate reductase (NR) in *A. nidulans* (Cove and Pateman 1963; Pateman *et al.* 1964) led to the isolation of pleiotropic mutants, termed *cnx*, incapable of expressing either enzyme activity. The fact that mutants in five distinct genetic loci displayed the same phenotype led to the proposal of the existence of a cofactor common to the two molybdoproteins (Pateman *et al.* 1964).

Other putative MoCo-deficient mutants, lacking, as a result of a single mutation, all Mo enzymes usually present in the particular organism studied, have been identified in *Escherichia coli* (Haddock and Jones 1977; Bachmann 1983; Miller and Amy 1983), *Salmonella typhimurium* (Stouthamer 1969; Sanderson and Hartman 1978), *Citrobacter freundii* (de Graaff *et al.* 1973),

Klebsiella aerogenes (Stouthamer 1970), *Pseudomonas aeruginosa* (Jeter *et al.* 1984), *Proteus vulgaris* (Claassen *et al.* 1981), and *Nostoc muscorum* (Bagchi and Singh 1984); *Drosophila melanogaster* (Scazzocchio 1980; Warner and Finnerty 1981); in the higher plants *Nicotiana plumbaginifolia* (Márton *et al.* 1982), *N. tabacum* (Buchanan and Wray 1982), *Hordeum vulgare* (Bright *et al.* 1983), and *Hyoscyamus muticus* (Fankhauser *et al.* 1984); and in humans (Johnson *et al.* 1980*b*; Ogier *et al.* 1982; Wadman *et al.* 1983; Munnich *et al.* 1983).

The second major contribution to the evolving story of the MoCo came from the studies of Nason and co-workers (Nason *et al.* 1970, 1971; Ketchum *et al.* 1970) using the pleiotropic *Neurospora crassa* mutant termed *nit-1* (equivalent to one of the *A. nidulans cnx* genes), which also is incapable of growth on nitrate or xanthine. These studies provided corroboration of the common cofactor theory but, more importantly, provided evidence for the transferability of the cofactor and thereby made it possible to develop a biological assay for the cofactor. Ketchum *et al.* (1970) showed that the normally inactive *nit-1* apo-NR could be reconstituted *in vitro* by exogenously derived MoCo from a number of plant, bacterial, and animal sources, the one exception being the enzyme nitrogenase (Pienkos *et al.* 1977). The protocol developed by Nason and co-workers for the assay of MoCo is shown in Fig. 14.1.

Preparation of *N. crassa nit–1* extract

1. *Nit–1* cells grown on ammonia, and induced for nitrate reductase by incubation with nitrate.
2. Mycelia homogenized in 0.1 M phosphate. 1 mM EDTA, 1 mM DTT.
3. Supernatant prepared by centrifugation at 15 000 rpm.

Reconstitution assay		Nitrate reductase assay	
10 mM phosphate, pH7 4		0.1 M phosphate, pH7 4	200 μl
20 mM Na molybdate	100 μl	0.1 M nitrate	100 μl
0.1 mM EDTA		0.1 mM FAD	50 μl
Nit–1 Extract	100 μl	50 mM Sulfite	50 μl
Co-factor source	50 μl	2 mM NADPH	50 μl
Incubated 10 min at 25°C		Reconstitution mixture	50 μl

Apo–NR+Mo cofactor $\longrightarrow$ Holo–NR

Incubated 10 min at 25°C

$$NADPH + NO_3^- \longrightarrow NADP+ + NO_2^- + H_2O$$

Fig. 14.1. The activity assay for MoCo. Conditions shown here are for the short assay. Quantitative assay involves incubation of the reconstitution assay at 4°C–15°C for 5–12 h.

The availability of such an assay was a prerequisite for the isolation of the MoCo. However, its usefulness has been blunted by the rapidity with which the cofactor loses its activity during attempted purification (Johnson 1980). The assay is still useful, qualitatively for detecting the presence or absence of cofactor activity in mutant cell lines and quantitatively for assessing the relative efficiencies of cofactor preparations from purified molybdoenzymes. Hageman and Rajagopalan (1986) have assessed various procedures used for the extraction of cofactor from biological materials for use in the *nit-1* assay.

As shown in Fig. 14.1, the assay procedure involves several steps, each of which needs to be performed optimally in order to derive quantitative information. Hageman and Rajagopalan (1986) have summarized the optimum conditions for release of cofactor from biological samples, growth and induction of *nit-1* cells and the reconstitution assay. Most studies of *nit-1* reconstitution have been qualitative, performed at room temperature for 10–60 min, testing for the presence or absence of cofactor activity. It has recently been shown that when reconstitution is performed at 4-15°C over a period of several hours, all of the cofactor added to the reconstitution mixture is transferred to the apo-NR and can be quantitated by the known turnover number of *N. crassa* NR (Hawkes and Bray 1984; Wahl *et al.* 1984; Hageman and Rajagopalan 1985). Reconstitution of *nit-1* detects both Mo-free and Mo-containing cofactor when molybdate (~ 10 mM) is included in the reconstitution mixture. Even when the cofactor source is initially present as the Mo complex, the addition of Mo gives higher and more accurate measures of total MPT because of the lability of the intact complex when it is separated from protein and in aqueous solution (Wahl *et al.* 1984). Other assays for the cofactor have been proposed, based on reconstitution of apoproteins from other pleiotropic mutants, or from W-grown cells (Johnson 1980), but these have not found wide acceptance.

A major problem with the *nit-1* reconstitution assay, when used for identifying cofactor-deficient mutant cells, is the fact that it involves the use of crude extracts which may convert cofactor precursors in one extract to active cofactor, bypassing a defective biosynthetic step in one or both cell lines. This problem has led to confusion concerning the presence or absence of MoCo in various systems. For example, cofactor activity was found in the pleiotropic, molybdoprotein-deficient mutations *chlA1* and *chlE5* of *E. coli* by reconstitution of the *N. crassa nit-1* NR (Amy 1981), but not in other *chlA* and *chlE* strains (Miller and Amy 1983). It was later shown that not only are *chlA* and *chlE* tightly linked to the complementation groups *chlM* and *chlN*, respectively, but *chlA1* contains an activity capable of converting a precursor in *nit-1* extracts to active molybdopterin (Johnson and Rajagopalan 1987a). Additional contradictions arose when reconstitution of pleiotropic mutants from *N. tabacum* (*cnx68*) and *H. muticus* (*MA-2*) were used to measure the cofactor activities of *E. coli chl* mutants. In this case it was claimed that cofactor activity was present in *chlA1* but not in *chlE5* (Taylor *et al.* 1983), although the assay system itself was of questionable validity since the strains *cnx68* and *MA-2* are not true cofactor-deficient mutants but instead are unable to incorporate molybdenum into MoCo unless grown on high levels of molybdate (Mendel *et al.* 1981; Fankhauser *et al.* 1984).

The first proof for *in vitro* complementation was obtained when it was shown that the excluded fractions (from a Sephadex G-25 column) of extracts from *chlA* strains of *E. coli* were capable of converting a precursor present in the included fractions from an extract of *nit-1 N. crassa*, and without this component of the *nit-1* extract the *chlA* extracts had no molybdopterin activity (Johnson and Rajagopalan 1987*b*). These results indicated that *chlA* is a true molybdopterin-deficient strain and that it produces a macromolecular factor capable of converting a molybdopterin precursor in *nit-1* strains to active molybdopterin. It is quite possible that other cases of precursor conversion have unknowingly been reported as the presence of MoCo, and caution should be used when interpreting the results of mixing experiments with unfractionated extracts.

Although numerous studies have been directed at the structure of the MoCo no detailed characterization of the active molecule has been possible due to its extreme lability, even in the absence of oxygen. Early studies had shown that the cofactor was of low molecular weight by virtue of its dialysability (Ketchum and Swarin 1973). Other biochemical studies indicated both an aromatic component and amino acid residues due to its interaction with Sephadex and reaction with ninhydrin, respectively (Alikulov *et al.* 1980). The presence of amino acid residues fit well with the interpretation stated above, of an earlier analysis of pleiotropic, temperature-sensitive mutants of xanthine oxidase and NR in *A. nidulans* (MacDonald and Cove 1974) and with the finding that similar small dialysable peptides could be obtained under conditions where cofactor activity was also dialysable from both xanthine oxidase and sulphite oxidase (Johnson 1980). However, it was later shown that MoCo preparations of high activity lacked stoichiometric amounts of amino acids. Thus, unless the apo-NR contains additional cofactor components, such as a small peptide required as a noncovalently bound part of the cofactor, it may be concluded that the MoCo does not contain multiple organic constituents.

The first insight into the chemical nature of MoCo was derived from the observation that upon aerobic denaturation of molybdoproteins fluorescence developed which resembled that of oxidized pterins. This led to the proposal that the organic moiety of the cofactor contained the pteridine ring system (Johnson *et al.* 1980*a*). Procedures were then developed to isolate stable oxidized derivatives of the cofactor, termed Form A and Form B (Fig. 14.2). The details of the chemical characterization of Form A and Form B, which in themselves define many of the features of the active species, have been published (Johnson *et al.* 1984). The structure of Form A revealed that the cofactor pterin contains a 6-alkyl sidechain containing four carbons, while the structure of Form B showed, in addition, the presence of sulphur in the pterin. Despite the apparent marked difference between the two compounds, it was apparent that the fused thiophene ring of Form B could have arisen from a linear C-6 side chain, by the reaction of a sulphhydryl group attached to a side chain carbon at the C-7 of the pterin ring.

A third, inactive derivative of the MoCo, apparently formed from the MoCo *in vivo*, has also proven to be of great value in formulating a structure for the active species. The thienopterin compound urothione (Fig. 14.2) was isolated from human urine in 1940 (Koschara 1940) and structurally characterized in 1969

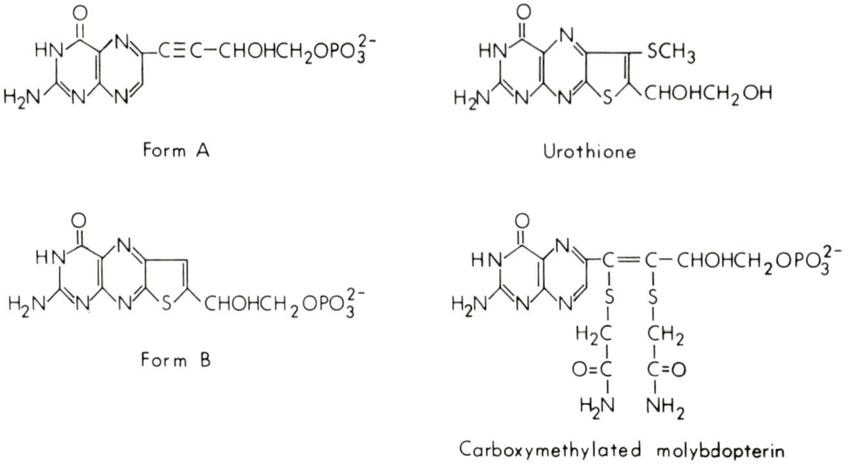

Form A

Urothione

Form B

Carboxymethylated molybdopterin

Fig. 14.2. The structures of molybdopterin derivatives.

(Goto *et al.* 1969). The obvious similarities between urothione and Form B suggested the possibility that urothione is metabolically related to MoCo. During the period when structural studies on MoCo were in progress in our laboratory, a number of patients were identified (Wadman *et al.* 1983) with symptoms of combined deficiencies of sulphite oxidase and xanthine dehydrogenase and, as a consequence, exhibiting mental retardation, neurological problems and, in most cases, ocular lens dislocation. Assays of liver biopsy material for the ability to reconstitute NR in the *Neurospora nit-1* mutant, or sulphite oxidase activity using purified demolybdoenzyme, showed that the patients lacked active MoCo (Johnson *et al.* 1980*b*). Urine samples from these patients were devoid of urothione (Johnson and Rajagopalan 1982), confirming the proposed metabolic link.

On the basis of the structural elements in the three oxidized derivatives we put forward a tentative structure for the active MoCo, as a complex of Mo with the organic constituent molybdopterin (Johnson and Rajagopalan 1982). The principal features of the proposed structure of molybdopterin were the presence of a reduced pterin ring and a 4-carbon side chain containing an enedithiol and a terminal phosphate ester. We have carried out a variety of experiments, either on cofactor freshly released from protein linkage or on stable preparations of isolated molybdopterin, to derive evidence supporting the proposed structure, and the results are summarized below.

MoCo is a complex of Mo and molybdopterin

In order to demonstrate that MoCo is a complex of the metal with molybdopterin the effects of varying the conditions used for release of MoCo from sulphite oxidase were studied (Wahl *et al.* 1984). The *nit-1* reconstitution assay showed that the reconstitutive ability of cofactor released at pH 7.0 was independent of the presence of Mo in the reconstitution mixture, whereas cofactor released at pH 8.0 was totally dependent on added Mo. The reconstituting activity of the cofactor released at pH 7.0 was abolished by sulphydryl-reactive reagents such as Hg^{2+} and arsenite, suggesting that in the active, intact cofactor the Mo is complexed to the -SH groups of molybdopterin.

Evidence for the presence of thiol groups in molybdopterin was obtained through the use of the thiol reagents dithiodipyridyl (DTDP) and iodoacetamide (IAA). Cofactor release from sulphite oxidase was inactivated in the presence of IAA at a rate dependent on time of exposure and on the concentration of IAA (Hageman and Rajagopalan 1985). In contrast, DTDP not only failed to inactivate the cofactor (presumably forming reversible mixed disulphide bonds with the thiols of the cofactor), but actually prevented inactivation by subsequent addition of IAA. The presence of 0.4 mM DTT in the reconstitution mixture (added with the *nit-1* crude extract) was sufficient to regenerate the free thiols of the cofactor from the mixed disulphide formed with DTDP. With no DTT in the *nit-1* extract, 10 μM DTDP in the reconstitution mixture completely blocked reconstitution. The protection of MoCo from IAA inactivation by DTDP indicated that the molybdopterin possesses at least one thiol group necessary for activity. Significantly, strong protection against IAA was also observed in the presence of 0.1 mM arsenite which is known to react with vicinal dithiols.

Absence of molybdopterin in the nit-1 mutant of Neurospora crassa

The oxidized Form A and Form B pterins can be readily generated from mycelia of wild type *Neurospora crassa* (Kramer *et al.* 1984). In contrast, *nit-1* cells are virtually devoid of material giving rise to these compounds. Similar mutants in *E. coli* also fail to yield these pterins (M. E. Johnson, unpublished data), and these derivatives are also absent from tissues of the MoCo-deficient patients. These findings categorically show that molybdopterin is the molecule which is lacking in these mutant organisms.

Molybdopterin is transferred to nit-1 nitrate reductase during reconstitution

Hageman and Rajagopalan (1985) isolated highly active molybdopterin from denatured sulphite oxidase and found it to be stable under anaerobic conditions for several weeks, but to be rapidly inactivated when exposed to air. The preparation was free of Mo, showing that molybdopterin can remain active even in the absence of Mo. On the basis of its phosphate content and a quantitation of nitrate reductase units reconstituted, routine preparations were greater than 50 per cent active. Using such a sample of molybdopterin it was demonstrated that after reconstitution of nitrate reductase in *nit-1* extracts and subsequent chromatography on Sephadex G-200, fractions displaying nitrate reductase activity were highly enriched in molybdopterin. The essentiality of the phosphate group of molybdopterin for biological function was shown by the finding that treatment of the isolated material with alkaline phosphatase led to total loss of reconstitutive ability. The molybdopterin preparations obtained by the procedure used in these studies were found to be free of any peptides. These data further corroborated the conclusion that molybdopterin is the sole organic constituent of MoCo and that transfer of a peptide is not required for the reconstitution of *nit-1* nitrate reductase. The molybdopterin-dependent reconstitution of the nitrate reductase in *Neurospora crassa nit-1* is diagrammatically illustrated in Fig. 14.3.

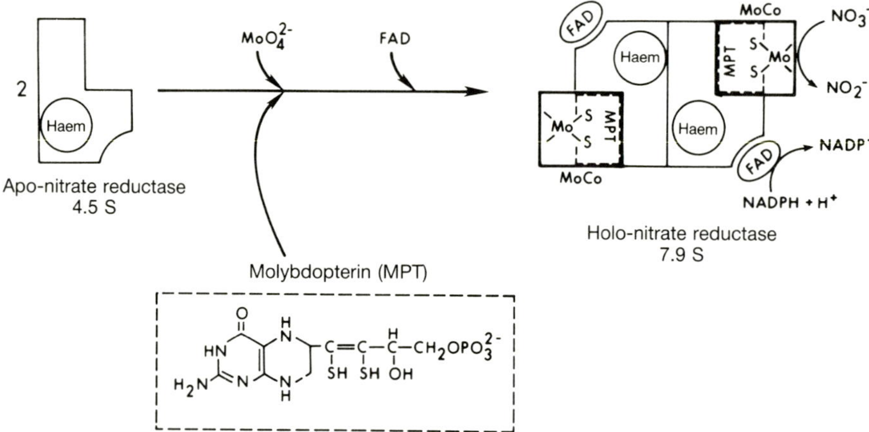

Fig. 14.3. Proposed mechanism of reconstitution of *nit-1* apo-NR by molybdopterin.

Molybdopterin is a universal cofactor

All molybdoenzymes examined appear to contain molybdopterin (Johnson *et al.* 1984). Formate dehydrogenase from *Clostridium thermoaceticum* is unique in that the active species appears to contain tungsten, rather than molybdenum (Yamamoto *et al.* 1983); nevertheless, the purified enzyme yields fluorescent species indistinguishable from Form A and Form B. The non-molybdoenzymes tested so far are devoid of molybdopterin.

The structure of molybdopterin

Since the presence of a vicinal dithiol function in molybdopterin would be expected to endow the molecule with extreme instability, it appeared that isolation of a stable, two sulphur species might depend on specific modification of the thiol groups. This expectation has been fulfilled by the successful isolation of a stable alkylated carboxamido-derivative of molybdopterin (camMPT) from sulphite oxidase and xanthine oxidase by a procedure involving treatment with iodoacetamide under mild denaturing conditions (Kramer *et al.* 1987). Structural studies on the product using mass spectroscopy, energy dispersive X-ray analysis (EDAX) and proton magnetic resonance have confirmed that molybdopterin is a 6-alkylpterin with a 4-carbon side chain which has an enedithiol at carbons 1′ and 2′, a hydroxyl at carbon 3′, and a terminal phosphate group (Fig. 14.2). The presence of two sites of alkylation per pterin was demonstrated by [1-^{14}C]iodoacetamide incorporation and by mass spectral and NMR data demonstrating two carboxamidomethyl substituents. The presence of a sulphur on C-1′ of the side chain was indicated by the absence of fluorescence in oxidized camMPT; the presence of two sulphurs in the molecule as potential sites for alkylation was demonstrated by EDAX and was also evident from the mass spectra. The mass spectral and NMR results also concurred with our prediction of the presence of the double bond between C-1′ and C-2′ of the side chain of molybdopterin, proposed initially on the basis of the structures of Form B and urothione (Johnson and Rajagopalan 1982). The monoester nature of the phosphate in MoCo, which was initially deduced from chemical studies of Form A and Form B (Johnson *et al.* 1984), has been confirmed by mass spectral analysis of oxidized camMPT (phospho), which rules out the possibility of a cyclic phosphate. The position of the phosphate ester on C-4′ rather than C-3′ of the side chain is indicated by the chemical shifts of the protons on these carbons in the phospho- relative to the dephospho-derivative.

In urothione the thiol on the carbon corresponding to C-1′ is methylated. When the original structure was proposed for the MoCo, there was no basis to either include or exclude this group in the proposed structure for molybdopterin, since the C-1′ of Form A and Form B do not contain sulphur. Although it was conjectured that methylation of the thiol occurs as a step in the cofactor degradative pathway, a – – – X was included in the speculated structure for the cofactor to indicate the possible presence of a methyl group in this position. Structural studies on camMPT led to the definitive conclusion that the methyl group of urothione is not derived from active MoCo.

In sum, the structural studies of Form A, Form B, urothione, and camMPT are consistent with the structure shown in Fig. 14.4 for MoCo.

Are there different forms of MoCo?

One proven difference among the molybdenum centers of molybdoproteins is the presence or absence of a terminal sulphide ligand on the molybdenum atom. The molybdenum-containing enzymes xanthine oxidase, aldehyde oxidase, xanthine dehydrogenase, purine hydroxylase II, and the formate dehydrogenase of *Methanobacterium formicicum* all have this terminal sulphide ligand (Bray 1975; Coughlan *et al.* 1984; Barber *et al.* 1986) which can be removed as thiocyanate

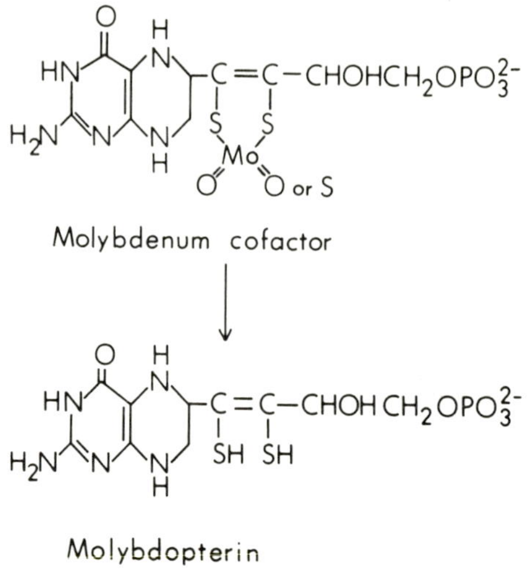

Molybdenum cofactor

Molybdopterin

Fig. 14.4. Structures of MoCo and molybdopterin. The tetrahydro form is shown, complexed to Mo by its thiol groups. Release of Mo from MoCo yields molybdopterin.

by cyanide treatment of the native oxidized forms of these enzymes (Massey and Edmondson 1970). Other molybdoproteins such as sulphite oxidase and nitrate reductase do not contain the terminal sulphur and are resistant to inactivation by cyanide in the oxidized state. Although the terminal sulphur ligand is not part of molybdopterin *per se* these differences attest to altered environments of Mo in various molybdoproteins. Thus, the possibility that other differences, perhaps involving molybdopterin itself, may exist must be considered. All molybdoenzymes so far examined have yielded fluorescent pterins under appropriate conditions, suggesting but not proving that they all contain Mo bound to molybdopterin. However, there are at least three reports which have presented evidence for the presence of variant forms of MoCo in certain enzymes.

Differences in the immediate environment of the molybdenum have been observed during EXAFS studies. Sulphite oxidase (Cramer *et al.* 1979; Cramer *et al.* 1981), xanthine dehydrogenase (Cramer *et al.* 1981), xanthine oxidase (Bordas *et al.* 1980; Tullius *et al.* 1979), and nitrate reductase (Cramer *et al.* 1984) have all given evidence of two or more thiolate ligands, as would be expected if the proposed sulphurs of the side chain of molybdopterin act as ligands for the metal (Johnson and Rajagopalan 1982). However, the EXAFS spectra of formate dehydrogenase of *Clostridium pasteurianum* showed no sulphur–molybdenum interactions (Cramer *et al.* 1985). The same workers did observe sulphur–metal interactions in another formate dehydrogenase, from *C. thermoaceticum*, which is interesting in itself since it contains tungsten in place of molybdenum but has been shown to reconstitute the apo-NR of *nit-1* in the presence of molybdate (Deaton *et al.* 1984). Whether the result for the enzyme from *C. pasteurianum* is real or is an artifact of sample preparation for EXAFS is not clear. The EXAFS data on the *C. pasteurianum* formate dehydrogenase are surprising, since the enzyme yields pterin-like fluorescence upon iodine oxidation and is able to donate molybdopterin to the apo-NR of *nit-1* (Liu and Mortenson 1984).

Another enzyme which has been reported to have molybdenum bound to a cofactor which is somewhat different from that which has been characterized from chicken liver sulphite oxidase is the formate dehydrogenase from the methanogen *Methanobacterium formicicum*. This enzyme has been shown to contain molybdenum bound to a terminal, cyanalysable sulphur (Barber *et al.* 1986), yields pterin-like fluorescence upon aerobic denaturation, and gives pterin-6-carboxylic acid upon treatment with alkaline permanganate (May *et al.* 1986). However, the latter report indicated that the MoCo from the enzyme failed to reconstitute *nit-1* nitrate reductase.

Kruger and Meyer (1986) have presented data supporting suggestions that differences exist between the MoCo from prokaryotic and eukaryotic sources. They have observed that the carbon monoxide dehydrogenase from *Pseudomonas carboxydoflava* contains eight non-covalently bound phosphate groups. Meyer and Rajagopalan (1984) found that, under defined conditions, the enzyme generates Form A and Form B chromatographically indistinguishable from those derived from sulphite oxidase. Kruger and Meyer found that when the enzyme was prepared from the bacteria grown on [^{32}P]orthophosphate approximately 50 per cent of the radioactivity from purified enzyme co-purified with FAD while the rest of the radioactivity was present in material which developed pterin-like fluorescence during aerobic incubation. This pterin-like radioactive compound was determined to have a molecular weight of 730 Da based on chromatography using a calibrated Sephadex-G15 column. Upon treatment with perchloric acid the compound was cleaved to yield two radioactive compounds, one of molecular weight 330 Da, which retained the fluorescence, and another with molecular weight 420 Da which was non-fluorescent. These authors also observed similar fluorescent compounds (molecular weight = 690–750 Da) from other prokaryotic molybdoproteins including xanthine dehydrogenase from *Clostridium barkeri* and *Veillonella alcalescens* and NR from *E. coli*, while the eukaryotic enzymes bovine milk xanthine oxidase and chicken liver sulphite oxidase only contained a 330 Da compound. Based on the presence of two FAD and two atoms of molybdenum per enzyme molecule they concluded that they were observing a new cofactor, termed bactopterin, which contained two phosphate groups, a pterin moiety, and another organic structure which is covalently bound to the pterin until removed by treatment with perchloric acid. In contrast, denatured carbon monoxide dehydrogenase from *P. carboxydoflava* was shown to reconstitute the apo-NR of *nit-1* as efficiently as bovine milk xanthine oxidase. It is not clear whether reconstitution was occurring with the entire bactopterin molecule or whether their denaturation conditions could have liberated molybdopterin in the absence of perchloric acid.

The MoCo is unique among the tightly bound enzyme prosthetic groups because of its extreme instability in the free form. This instability is a major stumbling block to the successful isolation of the active form of the cofactor. Our studies have shown that molybdopterin is likely to be a universal component of cofactors in all organisms and that the peptide-free pterin preparation has full biological activity. It must be stressed that the term 'molybdopterin' is as yet a generic one: it is possible that different molybdoenzymes will contain the pterin in one or other of the several possible oxidation states, all of which will yield a fluorescent, oxidized pterin under degradative conditions, but vary in their quantitative ability to reconstitute the *nit-1* nitrate reductase.

Virtually all molybdenum enzymes examined to date contain a MoCo which can be converted to the Form A and Form B derivatives (Johnson *et al.* 1984). This indicates that the molybdopterin structure which we have identified in the cofactor from sulphite oxidase and xanthine oxidase is the minimum common organic constituent of the universal MoCo. A 6-alkylpterin with a 4-carbon phosphorylated side chain is essential for Form A or Form B production. Moreover, the presence of the enedithiol function is directly implied by the characteristic pattern of sulphur elimination observed in conversion to Form A and Form B. However, because of the rather harsh conditions employed to produce the Form A and Form B derivatives, it is possible that certain additional subtle features unique to the cofactor could be lost in the process of forming these derivatives. The mild conditions we have used to convert molybdopterin of sulphite oxidase and xanthine oxidase to the stable camMPT species and the methodology we have developed to isolate and characterize that molecule have led to definitive structural characterization of the MoCo in these two enzymes. The techniques described should also be extremely valuable in more rigorous analyses of the structure of the MoCo in other enzymes.

The possibility also exists that a precursor of molybdopterin in a pleiotropic mutant, e.g. the *cnx H* of *A. nidulans*, could bind to apomolybdoproteins generating variably active enzyme molecules with altered physical properties. Only detailed molecular studies on the chemical characterization of the molybdopterin precursors in such mutants can provide the necessary information.

Acknowledgement

The studies from our laboratory described here were supported from the National Institutes of Health research grant GM00091.

References

Alikulov, Z. A., L'vov, N. P., Burikhanov, S. S., and Kretovich, V. L. (1980). Isolation of cofactor common to molybdenum-containing enzymes: nitrate reductase from lupine bacterioids and xanthine oxidase from milk. *Biology Bulletin of the Academy of Sciences of the USSR (New York)* **7**, 379–84.

Amy, N. K. (1981). Identification of the molybdenum cofactor in chlorate-resistant mutants of *Escherichia coli. Journal of Bacteriology* **148**, 274–82.

Bachmann, B. J. (1983). Linkage map of *Escherichia coli* K-12, edition 7. *Microbiological Reviews* **47**, 180–230.

Bagchi, S. N. and Singh, H. N. (1984). Genetic control of nitrate reduction in the cyanobacterium *Nostoc muscorum. Molecular and General Genetics* **193**, 82–4.

Barber, M. J., May, H. D., and Ferry, J. G. (1986). Inactivation of formate, dehydrogenase from *Methanobacterium formicium* by cyanide. *Biochemistry* **25**, 8150–5.

Bordas, J., Bray, R. C., Garner, C. D., Gutteridge, S., and Hasnain, S. S. (1980). X-ray absorption spectroscopy of xanthine oxidase. The molybdenum centers of the functional and the desulfo forms. *Biochemical Journal* **191**, 499–508.

Bray, R. C. (1975). Molybdenum iron–sulfur hydroxylases and related enzymes. *The Enzymes* **12**, 199–419.

Bright, S. W. J., Norbury, P. B., Franklin, J., Kirk, D. W., and Wray, J. L. (1983). A conditional-lethal *cnx*-type nitrate reductase-deficient barley mutant. *Molecular and General Genetics* **189**, 240–4.

Buchanan, R. J. and Wray, J. L. (1982). Isolation of molybdenum cofactor defective cell lines of *Nicotiana tabacum. Molecular and General Genetics* **188**, 228–34.

Claassen, V. P., Oltmann, L. F., Bus, S., v.'t Riet, J., and Stouthamer, A. H. (1981). The influence of growth conditions on the synthesis of molybdenum cofactor in *Proteus mirabilis. Archives of Microbiology* **130**, 44–9.

Coughlan, M. P., Mehra, R. K., Barber, M. J., and Siegel, L. M. (1984). Optical and electron paramagnetic resonance spectroscopic studies on purine hydroxylase II from *Aspergillus nidulans. Archives of Biochemistry and Biophysics* **229**, 596–603.

Cove, D. J. and Pateman, J. A. (1963). Indepentently segregating genetic loci concerned with nitrate reductase activity in *Aspergillus nidulans. Nature* **198**, 262–3.

Cramer, S. P., Gray, H. B., and Rajagopalan, K. V. (1979). The molybdenum site of sulfite oxidase. Structural information from X-ray absorption spectroscopy. *Journal of the American Chemical Society* **101**, 2772–4.

Cramer, S. P., Wahl, R., and Rajagopalan, K. V. (1981). Molybdenum sites of sulfite oxidase and xanthine dehydrogenase. A comparison by EXAFS. *Journal of the American Chemical Society* **103**, 7721–7.

Cramer, S. P., Solomonson, L. E., Adams, M. W. W., and Mortenson, L. E. (1984). Molybdenum sites of *Escherichia coli* and *Chlorella vulgaris* nitrate reductase. *Journal of the American Chemical Society* **106**, 1467–71.

Cramer, S. P., Liu, C.-L., Mortenson, L. E., Spence, J. T., Liu, S.-M., Yamamoto, I., and Ljungdahl, L. G. (1985). Formate dehydrogenase molybdenum and tungsten sites— observation by EXAFS of structural differences. *Journal of Inorganic Biochemistry* **32**, 119–22.

de Graaff, J., Barendsen, W., and Stouthamer, A. H. (1973). Mapping of chlorate-resistant mutants of *Citrobacter freundii* by deletion and complementation analysis. *Molecular and General Genetics* **121**, 259–69.

Deaton, J. C., Solomon, E. I., Durfor, C. N., Wetherbee, P. J., Burgess, B. K., and Jacobs, D. B. (1984). Activation of *nit-1* nitrate reductase by W-formate dehydrogenase. *Biochemical and Biophysical Research Communications* **121**, 1042–7.

Fankhauser, H., Bucher, F., and King, P. J. (1984). Isolation of biochemical mutants using haploid mesophyll protoplasts of *Hyoscyamus muticus*. IV. Biochemical characterisation of nitrate non-utilizing clones. *Plants* **160**, 415–21.

Garner, C. D. and Bristow, S. (1985). Oxomolybdenum chemistry in relation to the molybdoenzymes. In *Molybdenum enzymes* (ed. T. G. Spiro), pp. 343–410. Wiley-Interscience, New York.

Goto, M., Sakurai, A., Ohta, K., and Yamakami, H. (1969). Die structur des urothions, *Journal of Biochemistry (Tokyo)* **65**, 611–20.

Haddock, B. A. and Jones, C. W. (1977). Bacterial respiration. *Bacteriological Reviews* **41**, 47–99.

Hageman, R. V. and Rajagopalan, K. V. (1985). Characterization of molybdopterin, the organic portion of the molybdenum cofactor. In *Nitrogen fixation and CO$_2$ metabolism* (eds. P. W. Ludden and J. E. Burris), pp. 133–41. Elsevier Science Publishing Co., New York.

Hageman, R. V. and Rajagopalan, K. V. (1986). Assay and detection of the molybdenum cofactor. *Methods in Enzymology* **122**, 399–412.

Hawkes, T. R. and Bray, R. C. (1984). Quantitative transfer of the molybdenum cofactor from xanthine oxidase and from sulphite oxidase to the deficient enzymes of the *nit-1* mutant of *Neurospora crassa* to yield active nitrate reductase. *Biochemical Journal* **219**, 481–93.

Jeter, R. M., Sies, S. R., and Ingraham, J. L. (1984). Chromosomal location and function of genes affecting *Pseudomonas aeruginosa* nitrate assimilation. *Journal of Bacteriology* **157**, 673–7.

Johnson, J. L. (1980). The molybdenum cofactor common to nitrate reductase, xanthine dehydrogenase and sulphite oxidase. In *Molybdenum and molybdenum-containing enzymes* (ed. M. P. Coughlan), pp. 345–83, Pergamon Press, New York.

Johnson, J. L. and Rajagopalan, K. V. (1977). Tryptic cleavage of rat liver sulfite oxidase: isolation and characterization of molybdenum and heme domains. *Journal of Biological Chemistry* **252**, 2017–25.

Johnson, J. L. and Rajagopalan, K. V. (1982). Structural and metabolic relationship between the molybdenum cofactor and urothione. *Proceedings of the National Academy of Sciences, USA* **79**, 6856–60.

Johnson, M. E. and Rajagopalan, K. V. (1987*a*). Involvement of *chlA, E, M*, abd *N*, loci in *Escherichia coli* molybdopterin biosynthesis. Journal of Bacteriology **169**, 117–25.

Johnson, M. E. and Rajagopalan, K. V. (1987*b*). *In vitro* system for molybdopterin biosynthesis. *Journal of Bacteriology* **169**, 110–16.

Johnson, J. L., Hainline, B. E., and Rajagopalan, K. V. (1980*a*). Characterization of the molybdenum cofactor of sulfite oxidase, xanthine oxidase and nitrate reductase: identification of a pteridine as a structural component. *Journal of Biological Chemistry* **255**, 1783–6.

Johnson, J. L., Waud, W. R., Rajagopalan, K. V., Duran, M., Beemer, F. A., and Wadman, S. K. (1980*b*). Inborn errors of molybdenum metabolism: combined deficiencies of sulfite oxidase and xanthine dehydrogenase in a patient lacking the molybdenum cofactor. *Proceedings of the National Academy of Science, USA* **77**, 3715–19.

Johnson, J. L., Hainline, B. E., Rajagopalan, K. V., and Arison, B. H. (1984). The pterin component of the molybdenum cofactor: structural characterization of two fluorescent derivatives. *Journal of Biological Chemistry* **259**, 5414–22.

Ketchum, P. A. and Swarin, R. S. (1973). *In vitro* formation of assimilatory nitrate reductase: presence of the constitutive component in bacteria. *Biochemical and Biophysical Research Communications* **52**, 1450–6.

Ketchum, P. A., Cambier, H. Y., Frazier, W. A. III, Madansky, C. H., and Nason, A. (1970). *In vitro* assembly of *Neurospora* assimilatory nitrate reductase from protein subunits of a *Neurospora* mutant and the xanthine oxidizing or aldehyde oxidase systems of higher animals. *Proceedings of the National Academy of Sciences, USA* **66**, 1016–23.

Koschara, W. (1940). Urothion, ein gelber, schwefelreicher farbstoff aus menschenharn. *Hoppe-Seyler's Zeitschrift für Physiologische Chemie* **263**, 78–9.

Kramer, S., Hageman, R. V., and Rajagopalan, K. V. (1984). *In vitro* reconstitution of nitrate reductase activity of the *Neurospora crassa* mutant *nit-1*: specific incorporation of molybdopterin. *Archives of Biochemistry and Biophysics* **233**, 821–9.

Kramer, S. P., Johnson, J. L., Ribeiro, A. A., Millington, D. S., and Rajagopalan, K. V. (1987). The structure of the molybdenum cofactor. Characterization of di-(carboxami-domethyl)molybdopterin from sulfite oxidase and xanthine oxidase. *Journal of Biological Chemistry* **262**, 16357–63.

Kruger, B. and Meyer, O. (1986). The pterin (bactopterin) of carbon monoxide dehydrogenase from *Pseudomonas carboxydoflava*. *European Journal of Biochemistry* **157**, 121–8.

Liu, C.-L. and Mortenson, L. E. (1984). Formate dehydrogenase of *Clostridium pasteurianum*. *Journal of Bacteriology* **159**, 375–80.

Márton, L., Dung, T. M., Mendel, R. R., and Maliga, P. (1982). Nitrate reductase deficient cell lines from haploid protoplast cultures of *Nicotiana plumbaginifolia*. *Molecular and General Genetics* **182**, 301–4.

MacDonald, D. W. and Cove, D. J. (1974). Studies on temperature-sensitive mutants affecting the assimilatory nitrate reductase of *Aspergillus nidulans*. *European Journal of Biochemistry* **47**, 107–10.

Massey, V. (1973). Iron–sulfur flavoprotein hydroxylases. In *Iron–sulfur proteins* (ed. W. Lovenberg), pp. 301–60. Academic Press, New York.

Massey, V., and Edmondson, D. (1970). On the mechanism of inactivation of xanthine oxidase by cyanide. *Journal of Biological Chemistry* **245**, 6596–8.

May, H. D., Schauer, N. L., and Ferry, J. G. (1986). Molybdopterin cofactor from *Methanobacterium formicicum* formate dehydrognase. *Journal of Bacteriology* **166**, 500–4.

Mendel, R. R., Alikulov, Z. A., Lvov, N. P., and Muller, A. J. (1981). Presence of the molybdenum-cofactor in nitrate reductase-deficient mutant cell lines of *Nicotiana tabacum*. *Molecular and General Genetics* **181**, 395–9.

Meyer, O. and Rajagopalan, K. V. (1984). Molybdopterin in carbon monoxide oxidase from carboxydotrophic bacteria. *Journal of Bacteriology* **157**, 643–8.

Miller, J. B. and Amy, N. K. (1983). Molybdenum cofactor in chlorate-resistant and nitrate reductase-deficient insertion mutants of *Escherichia coli*. *Journal of Bacteriology* **155**, 793–801.

Munnich, A., Saudubray, J. M., Charpentier, C., Ogier, H., Coudé, F. X., Frézal, J., Yacoub, L., Harbi, A., and Snoussi, S. (1983). Multiple molybdoenzyme deficiencies due to an inborn error of molybdenum cofactor metabolism: two additional cases in a new family. *Journal of Inherited and Metabolic Diseases* **6 Suppl. 2**, 95–6.

Nason, A., Antoine, A. D., Ketchum, P. A., Frazier, W. A. III, and Lee, D. K. (1970). Formation of assimilatory nitrate reductase by *in vitro* inter-cistronic complementation in *Neurospora crassa*. *Procedings of the National Academy of Sciences, USA* **65**, 137–44.

Nason, A., Lee, K.-Y., Pan, S.-S., Ketchum, P. A., Lamberti, A., and DeVries, J. (1971). *In vitro* formation of assimilatory reduced nicotinamide adenine dinucleotide phosphate: nitrate reductase from a *Neurospora* mutant and a component of molybdenum-enzymes. *Proceedings of the National Academy of Sciences, USA* **68**, 3242–6.

Ogier, H., Saudubray, J. M., Charpentier, C., Munnich, A., Perignon, J. L., Kesseler, A., and Frezal, J. (1982). Double déficit en sulfite et xanthine oxydase, cause d'encéphalopathie due a une anomalie héréditaire du métabolisme du molybdene. *Annales de Médicine Interne* **133**, 594–6.

Pateman, J. A., Cove, D. J., Rever, B. M., and Robert, D. (1964). A common co-factor for nitrate reductase and xanthine dehydrogenase which also regulates the synthesis of nitrate reductase. *Nature* **201**, 58–60.

Pienkos, P. T., Shah, V. K., and Brill, W. J. (1977) Molybdenum cofactors from molybdoenzymes and *in vitro* reconstitution of nitrogenase and nitrate reductase. *Proceedings of the National Academy of Sciences, USA* **74**, 5468–71.

Rajagopalan, K. V. (1980). Sulphite oxidase. In *Molybdenum and molybdenum-containing enzymes* (ed. M. P. Coughlan), pp. 241–72, Pergamon Press, New York.

Richert, D. A. and Westerfeld, W. W. (1953). Isolation and identification of the xanthine oxidase factor as molybdenum. *Journal of Biological Chemistry* **203**, 915–23.

Sanderson, K. E. and Hartman, P. E. (1978). Linkage map of *Salmonella typhimurium*, edition V. *Microbiological Reviews* **42**, 471–519.

Scazzocchio, C. (1980). The genetics of the molybdenum-containing enzymes. In *Molybdenum and molybdenum-containing enymes* (ed. M. P. Coughlan), pp. 487–516, Pergamon Press, New York.

Stouthamer, A. H. (1969). A genetical and biochemical study of chlorate-resistant mutants of *Salmonella typhimurium*. *Antonie van Leeuwenhoek* **35**, 505–21.

Stouthamer, A. H. (1970). Genetics and biochemistry of reductase formation in Enterobacteriaceae. *Antonie van Leeuwenhoek* **36**, 181.

Taylor, J. L., Bedbrook, J. R., Grant, F. J., and Kleinhofs, A. (1983). Reconstitution of plant nitrate reductase by *Escherichia coli* extracts and the molecular cloning of the *chlA* gene of *Escherichia coli* K12. *Journal of Molecular and Applied Genetics* **2**, 261–71.

Tullius, T. D., Kurz, D. M., Jr., Conradson, S. D., and Hodgson, K. O. (1979). The molybdenum site of xanthine oxidase. Structural evidence from X-ray absorption spectroscopy. *Journal of the American Chemistry Society* **101**, 2776–9.

Wadman, S. K., Duran, M., Beemer, F. A., Cats, B. P., Johnson, J. L., Rajagopalan, K. V., Saudugray, J. M., Ogier, H., Charpentier, C., Berger, R., Smit, G. P. A., Wilson, J., and Krywawych, S. (1983). Absence of hepatic molybdenum cofactor: an inborn error of metabolism leading to a combined deficiency of sulphite oxidase and xanthine dehydrogenase. *Journal of Inherited and Metabolic Diseases* **6** Suppl. 1, 78–83.

Wahl, R. C., Hageman, R. V., and Rajagopalan, K. V. (1984). The relationship of Mo, molybdopterin, and the cyanolyzable sulfur in the Mo cofactor. *Archives of Biochemistry and Biphysics* **230**, 264–73.

Warner, C. K. and Finnerty, V. (1981). Molybdenum hydroxylases in *Drosophila*. II. Molybdenum cofactor in xanthine dehydrogenase, aldehyde oxidase and pyridoxal oxidase. *Molecular and General Genetics* **184**, 92–6.

Yamamoto, I., Saiki, T., Liu, S.-M., and Ljungdahl, L. G. (1983). Purification and properties of NADP-dependent formate dehydrogenase from *Clostridium thermoaceticum*, a tungsten–selenium–iron protein. *Journal of Biological Chemistry* **258**, 1826–32.

Nitrite Reduction

15. Physiology, biochemistry, and genetics of nitrite reduction by *Escherichia coli*

J. A. Cole

Physiology of nitrite reduction

The range of nitrite reductases (NiR) found either within a single bacterial species, such as *Escherichia coli*, or between different types of bacteria, reflects the various ways that nitrate or nitrite reduction can be exploited. Some of the possibilities are listed in Table 15.1. Inevitably some of the consequences of these reduction reactions are interlinked. For example, NADH-dependent nitrite reduction to ammonium by an enzyme located in the cytoplasm will provide the host with a potential nitrogen source for growth, regenerate NAD^+ from NADH, increase the pH of the environment and remove the toxic product of nitrate reduction from the cytoplasm (items a, b, d, and f in Table 15.1). Additionally, if acetate replaces ethanol or butanol as a fermentation product when nitrite is reduced, the yield of ATP per mole of substrate fermented will increase (item g) and anaerobic growth on a reduced substrate such as glycerol or mannitol might be dependent on the availability of such an electron acceptor (item i). In contrast, formate-dependent nitrite reduction by a membrane-bound electron transfer chain linked to a periplasmic NiR provides opportunities (c), (e), and (h) instead of (b), (f), or (g) (Table 15.1). We shall see that both of these examples are directly relevant to *E. coli*.

Despite statements to the contrary, anaerobic cultures of most *E. coli* strains rapidly reduce nitrite to ammonium. There is also a much slower process in which nitrite is reduced to nitrous oxide. Three pathways for nitrite reduction by K-12 strains have been characterized both biochemically and genetically, but at least four more nitrite reduction activities have been detected or postulated (Table 15.2). One of these is due to non-specific reduction of nitrite to nitrous oxide by nitrate reductase (NR) (Smith 1983) and another is probably a laboratory artifact created by combining a diaphorase activity with FAD and hexahaem cytochrome c_{552} (Liu *et al.* 1981). If this latter interpretation is incorrect, it implies that electrons can be transferred from NADH to nitrite by a branched pathway, as shown in Fig. 15.1 (see also below). Virtually nothing is

Table 15.1. Consequences of nitrite reductions to ammonium

Consequences		Notes
(a)	Generates a potential N-source	$NH_4 \rightarrow$ glutamine, glutamate
(b)	Oxidizes NAD(P)H to NAD(P)$^+$	Availability of NAD^+ might limit anaerobic growth; opposite true during aerobic growth
(c)	Oxidizes fermentation products such as formic acid	Formic acid is a moderately strong acid; can collapse membrane potential
(d)	Increases the pH of the environment	By converting an acid anion to a base. Also by oxidizing $HCOO^-$ to CO_2
(e)	Removes toxic chemical from environment	Ideally the reductase should be in the periplasm
(f)	Removes toxic product of NO_3^- reduction from cytoplasm	Cytoplasmic nitrate reductase
(g)	Generates ATP by substrate level phosphorylation	Consequence of acetate replacing ethanol as major fermentation product
(h)	Proton electrochemical gradient generated	Membrane-associated reductase required
(i)	Permits anaerobic growth on non-fermentable carbon source	Nitrite supports anaerobic growth on glycerol, lactate etc.

known about the biochemistry and genetics of nitrite assimilation in well-aerated cultures of enteric bacteria, or about nitrite reduction by ethanol and pyruvate (Pope and Cole 1984).

Given that there is such a plethora of NiR activities in *E. coli*, why do some research groups fail to detect them? One possible reason is that rates of nitrite reduction by different isolates vary by a factor of 10^3, so rapid tests used by numerical taxonomists to score superficial differences are inherently unsuitable.

The simplest test for nitrite reducing ability is to grow a healthy culture of the strain to be tested anaerobically in rich medium supplemented with glucose and nitrite. In this laboratory we routinely supplement a 1 ml overnight broth culture in a standard capped test tube with 4 ml of minimal salts containing 0.4 per cent (w/v) glucose and 2.5 mM nitrite. After 2–5 h at 37°C, no nitrite can be detected in cultures of wild-type strains (Nir$^+$ Nrf$^+$ Cys$^+$ phenotype; see Table 15.3 for an explanation of genetic symbols). If, however, the strain is defective in NADH-dependent NiR (the Nir$^-$ phenotype), the initial concentration of nitrite must be decreased to 0.25 mM for nitrite reduction by formate (Nrf$^+$ activity) to be detected within a reasonable working day. Even more care must be taken when scoring fresh isolates for their ability to reduce nitrite: some streptococci, staphylococci, and sulphate-reducing bacteria as well as enteric bacteria isolated from gastric juice require repeated subculture in media containing no more than 100 μM nitrite before reduction can be detected. Subsequently, however, these

Table 15.2. Nitrite reductase of *Escherichia coli*

Enzyme and electron donor	Product	Physiological significance or notes
Physiological significance known		
NADH-dependent nitrite reductase	NH_4^+	Most active dissimilatory NO_2^- reductase, especially during nitrate reduction. Removes NO_2^- from cytoplasm
Formate-nitrite oxidoreductase	NH_4^+	Dissimilatory reduction of extracellular nitrite, generating a proton electrochemical gradient
NADPH-dependent sulphite reductase	NH_4^+	Reduces $SO_3^=$ to $S^=$ in cysteine biosynthetic pathway
Physiological significance unknown		
FAD, diaphorase and cytochrome c_{552}-dependent reduction of NO_2^- by NADH	NH_4^+	Laboratory artefact
Reduction of NO_2^- by nitrate reductase	N_2O	None? Co-metabolism?
Pyruvate or ethanol-dependent nitrite reductase	NH_4^+	See Pope and Cole (1985)
Function known but enzyme not characterized		
Assimilatory nitrite reductase	NH_4^+	Generates NH_4^+ from NO_3^- or NO_2^- during aerobic growth

cultures readily reduce 1 mM nitrite before the stationary phase of growth is reached.

Methods designed to detect enzyme activity in cell extracts require even greater care. Although the NADPH-dependent enzyme from *E. coli* is relatively stable, its activity is so low, even in extracts of bacteria de-repressed for sulphite reductase, that it can be masked by NADPH oxidase activity (Cole and Ward 1973). No one has yet reported conditions for detecting nitrite reduction by formate with cell extracts, and the NADH-dependent enzyme is extremely labile unless it is stored in the presence of FAD, EDTA, and ascorbate (Jackson *et al.* 1981). Furthermore, NADH-NiR is only active in the presence of its reduction product, NAD^+ (Jackson *et al.* 1982).

Throughout the 1950s and 1960s there was considerable controversy concerning whether a single enzyme or two independent enzymes from *E. coli* and other enteric bacteria were required for nitrate assimilation and nitrate respiration. By 1970 the consensus view favoured a single enzyme being involved in both processes, although it was recognized that the product of the structural

Scheme 1a

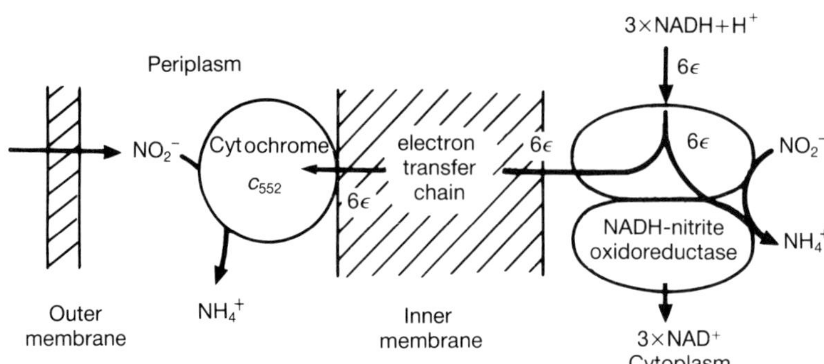

Scheme 1b

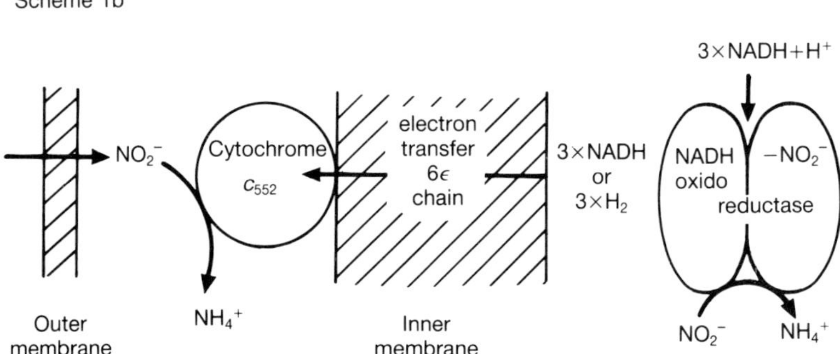

Fig. 15.1. Alternative possible electron transfer pathways from NADH or an alternative substrate to intracellular and extracellular nitrite

gene for this enzyme (the *narG* gene product) might associate with different auxillary proteins in the two processes. The recent discovery by Ingraham and his colleagues that biochemically and genetically distinct enzymes are required for nitrate assimilation and respiration by *Pseudomonas aeruginosa* is therefore particularly important (Jeter *et al.* 1984). A report that there are two homologous but independent genes for NR in *E. coli* has recently been confirmed (Bonnefoy *et al.* 1987; J. C. Wootton, personal communication) so it is possible, and indeed, likely, that there are also independent pathways for nitrite assimilation and dissimilation to ammonium in *E. coli.* Any putative assimilatory NiR remains to be identified. Two NiR from enteric bacteria have been reasonably well characterized: both are regulated by oxygen repression and nitrite induction rather than by ammonium repression, which strongly suggests that they play primarily dissimilatory roles. For the record, in recent attempts to clone the structural gene for cytochrome c_{552}, Nir⁻Nrf⁻Cys⁻ mutants were plated onto a

Table 15.3. Genes involved in nitrate or nitrite reduction by *E. coli*

Genetic symbol	Mnemonic	Significance for nitrite reduction
nirB	NADH-dependent *ni*trite *r*eductase	Structural gene for NADH-dependent nitrate reductase apoprotein
nrf	*n*itrite *r*eduction by *f*ormate	Uncharacterized genes required for formate-nitrate oxidoreductase activity
narG	*n*it*ra*te *r*eductase	Structural gene for the α subunit of the nitrate reductase apoprotein
chl	*chl*orate resistance	Genes required for the synthesis or incorporation of the molybdenum cofactor into nitrate reductase
fnr	*f*umarate and *n*itrate *r*eduction (formerly *nirA*; Cole and Ward, 1973)	Gene for a transcriptional activator protein required for the synthesis of nitrate, nitrite, fumarate and other reductases during anaerobic growth. FNR protein inactivated by O_2?
narL	*n*it*ra*te *r*eductase	Gene for a transcriptional activator protein, NARL, which activates *narG* (and other genes of the *nar* operon) in the presence of nitrate
cysG	*c*y*s*teine biosynthesis	Gene required for the synthesis of sirohaem, the prosthetic group NADPH-sulphite and NADH-nitrite reductases
fdhF	*f*ormate *d*e*h*ydrogenase	Structural gene for 80 kDa formate dehydrogenase required for formate hydrogenlyase activity
fdhN	*f*ormate *d*e*h*ydrogenase	Structural gene for one of the subunits of the membrane-bound formate dehydrogenase required for nitrate reduction (and expected to be required for nitrite reduction) by formate.

Genotypes are given by three lower-case letters in italics; corresponding phenotypes in normal script begin with a capital (upper case) letter. Protein products of the regulatory *fnr*, *crp*, and *narL* genes are referred to as the FNR, CRP, and NARL proteins, respectively.

solid medium with nitrite as a sole nitrogen source. Although no Nrf$^+$ clones were obtained anaerobically after 5 days at 37°C, healthy colonies of *E. coli* appeared when the plates were left at 18°C for a further 2 weeks. As these colonies do not appear to be revertants of the original mutants, this implies that nitrite can be assimilated by a previously uncharacterized enzyme.

Nitrite reduction to nitrous oxide is a relatively minor pathway and is of no recognized physiological significance to *E. coli*. A similar process is catalysed by most organisms that reduce nitrate, but because N_2O is an intermediate in the denitrification pathway, N_2O production is only remarkable in bacteria which normally reduce nitrite to ammonium. It is a slow process with a high K_m for nitrite. This activity correlates far better with the presence of an active NR than with any known NiR, so it appears to be a fortuitous side-reaction (Smith 1983). Nevertheless, it appears to make a significant contribution to gaseous nitrogen oxides in the atmosphere.

Gratuitous nitrite reductase activity of NADPH-dependent sulphite reductase.

The first 'nitrite reductase' to be purified and characterized was subsequently recognized to be the NADPH-dependent sulphite reductase (E.C. 1.8.1.2) required to reduce sulphite to sulphide in the cysteine biosynthetic pathway (Lazzarini and Atkinson 1961; Kemp *et al.* 1963). It is an $\alpha_8\beta_4$ dodecamer with four FMN and four FAD associated with the eight α subunits and four sirohaem moieties attached to the β subunits. Sirohaem is a novel prosthetic group so far found in only two types of enzyme, sulphite and nitrite reductases. In each case six electrons are transferred to the electron acceptor either concomitantly or in quick succession without the release of partially reduced nitrogen and sulphur compounds.

The structural genes for the sulphite reductase, *cysI* and *cysJ*, are located at 59 minutes on the *E. coli* chromosome, and their transcription is activated by the product of the *cysB* gene, which also positively regulates other genes involved in sulphate uptake or reduction to sulphide during growth in the absence of cysteine (Jones-Mortimer 1968). Mutants defective in the *cysG* gene are also unable to reduce sulphite to sulphide and therefore require cysteine (or sulphide) for growth. The *cysG* gene codes for a protein, probably a methylase, required to convert uroporphyrinogen III to factor II, the precursor of sirohaem.

The soluble, NADH-dependent nitrite reductase

Zarowny and Sanwal (1963) were the first to report that, during anaerobic growth in the presence of nitrite, some strains of *E. coli* synthesize a very active, soluble and NADH-dependent NiR (EC 1.6.6.4). The enzyme is extremely unstable *in vitro*, so many of its properties were only revealed subsequently. For example, Kemp and Atkinson (1966) reported the very low K_m for nitrite and, contrary to Zarowny and Sanwal (1963), identified ammonium as the product of nitrite reduction.

Anaerobic cultures of certain chlorate-resistant mutants are constitutive for the synthesis of this NiR, even in rich media unsupplemented with nitrite (Coleman *et al.* 1978*a*). NiR constitutes up to 4 per cent of the soluble protein in such cultures, from which it has been purified almost to homogeneity by two cycles of gel filtration and anion exchange chromatography following breakage in a French pressure cell, high-speed centrifugation and ammonium sulphate fractionation (Jackson *et al.* 1981*a*). The purified enzyme is apparently a homodimer of 88 kDa subunits, each of which contains two Fe_2S_2 iron–sulphur centres (deduced from the Fe content of the purified protein and esr spectroscopy). By comparing the amino acid sequence of the N-terminus of the purified protein with the nucleotide sequence of the *nirB* gene, *nirB* has been identified as the structural gene for the enzyme (Macdonald and Cole 1985; Jayaraman *et al.* 1987). The failure to detect an Fe_4S_4 iron–sulphur centre by ESR spectroscopy was unexpected because such signals are characteristic of some other nitrite and sulphite reductases. Nevertheless, the nucleotide sequence of the *nirB* gene also indicates the absence of an Fe_4S_4 iron–sulphur cluster. The motif cys-X-X-cys-X-X-cys-X-X-X-cys-pro-, which is characteristic of proteins with an Fe_4S_4 iron–sulphur cluster, has not been found in the amino acid sequence derived from the DNA sequence of the *nirB* gene. Both the FAD and sirohaem moieties are readily lost from the purified protein, even during gel filtration. The amber coloured enzyme can be stored in the dark at $-20°C$ in Tris buffer, ascorbate, EDTA, FAD, nitrite, and 20 per cent (w/v) glycerol without significant loss of activity and without further precautions to prevent inactivation by oxygen. All of the chemicals listed above contribute towards stability during storage. NADH is the only effective electron donor, but alternative electron acceptors to nitrite include hydroxylamine, sulphite, mammalian cytochrome *c* and a variety of dyes (Jackson *et al.* 1982).

Perhaps the most unusual property of the enzyme is its requirement for a reaction product, NAD^+, for activity (Coleman *et al.* 1978*b*). By analogy with Massey and Veeger's (1961) classical studies with lipoamide dehydrogenase, it is suggested that NAD^+ prevents the FAD moiety from being fully reduced to a catalytically inactive form. Results of a kinetic analysis for the reduction of the alternative electron acceptors have been reported (Jackson *et al.* 1981*b*, 1982).

Cytochrome c_{552} and the electrogenic formate–nitrite oxidoreductase complex

Following the discovery by Postgate (1954) of cytochrome c_3 in an obligately anaerobic sulphate-reducing bacterium, Gray *et al.* (1963) discovered a similar, low redox potential *c*-type cytochrome in anaerobic cultures of *E. coli*. By analogy with the presumed role of cytochrome c_3 in hydrogen metabolism, it was suggested that cytochrome c_{552} might also be one of the unidentified components, x_1 or x_2, of the formate hydrogenlyase complex (Peck and Gest 1957):

$$H.COO^- \longrightarrow Formate \xrightarrow{2e \longrightarrow} x_1 \longrightarrow x_2 \longrightarrow Hydrogenase \xrightarrow{2H^+} H_2$$

$$\downarrow \text{dehydrogenase (BV)}$$

$$2H^+ + CO_2$$

This became unlikely when it was realized that cytochrome c_{552} synthesis correlates far better with nitrate reductase activity than with hydrogenase activity. As the reduced cytochrome is rapidly oxidized by nitrite, both Fujita and Sato (1966b) and Cole (1968) suggested that it was more likely to be a nitrite reductase than an electron donor to hydrogenase.

Further work by Fujita and Sato (1966a,1967), subsequently confirmed by Cole (1968), established that although 98 per cent of the NADH-NiR activity was retained when bacteria were subjected to osmotic shock, the majority of the cytochrome c_{552} was released with other periplasmic proteins. This loss coincided with a loss of nitrite-dependent CO_2 evolution from glucose or an endogenous substrate, possibly formate (Fujita and Sato 1967). The purification of NADH-NiR with high specific activity but free from cytochrome c_{552} established that the cytochrome was not necessary for nitrite reduction by NADH. The results were compatible with either scheme 1a or 1b (Fig. 15.1). According to scheme 1a, NADH-NiR could either transfer electrons *via* sirohaem to nitrite generated during nitrate reduction in the cytoplasm, or *via* membrane components and cytochrome c_{552} to nitrite entering the periplasm from the environment. Scheme 1b implies that the two pathways are biochemically and genetically independent with either NADH or some other substrate acting as electron donor for the trans-membrane NiR.

Evidence that scheme 1b is correct came from the characterization of formate-dependent NiR by Chippaux and his colleagues (Abou-Jaoudé *et al.* 1979a,b), the isolation of *nirB* mutants that lack NADH-NiR but retain formate-dependent NiR activity, and from the characterization of a similar, hexahaem cytochrome *c* which is the terminal nitrite reductase of formate-dependent nitrite reduction by *Desulfovibrio* strains (Steenkamp and Peck 1980, 1981). Furthermore, Motteram *et al.* (1981) and Pope and Cole (1982) demonstrated that although no proton electrochemical gradient was established during NADH-dependent nitrite reduction, the formate-dependent activity was electrogenic. In summary, four lines of evidence indicate that cytochrome c_{552} is the terminal enzyme of a trans-membrane formate–nitrite oxidoreductase complex:

(1) The cytochrome is rapidly oxidized by nitrite and will catalyse reduced benzylviologen-dependent nitrite reduction.
(2) Synthesis of the cytochrome, both within and between strains, correlates with formate–nitrite oxidoreductase activity.
(3) Like the periplasmic nitrite reductase from *Desulfovibrio*, it is a hexahaem cytochrome *c*.
(4) Nitrite-dependent gas evolution is restored when sphaeroplasts are complemented by high concentrations of cytochrome c_{552}.

Note, however, that although each of the above correlations is consistent with the assumed role for the cytochrome, rigorous proof that it is the terminal NiR must await the cloning and sequence analysis of the structural gene and demonstration that the cloned gene is both necessary and sufficient to restore nitrite reduction by formate to Nrf⁻ mutants. Such cytochrome-deficient mutants have recently been isolated in the author's laboratory.

Components of the formate–nitrite oxidoreductase other than cytochrome c_{552} remain to be identified. Nitrite reduction is inhibited by cyanide but not by azide. It is partially sensitive to hydroxyquinoline-N-oxides, which inhibit some *b*-type cytochromes. A mutant totally deficient in both ubiquinone and menaquinone retained 70 per cent of the activity of the wild type, so quinones are unlikely to be essential components of this specialized electron transfer chain (Abou-Jaoudé *et al.* 1979*b*).

Genetic regulation of NADH-dependent nitrite reduction

Synthesis of the NADH-dependent nitrite reductase is regulated by dual control mechanisms, nitrite induction during anaerobic growth but total repression during aerobic growth (Table 15.4). Operon fusion studies have established that both of these responses are mediated at the level of transcription (Table 15.4; see also the footnote to Fig. 15.2 and Casadaban and Cohen 1979; Griffiths and Cole 1987). Although, as mentioned before, some chlorate-resistant mutants are constitutive for NiR synthesis during anaerobic growth in the absence of nitrite, such mutants remain sensitive to oxygen repression (Table 15.4). The *nirB*

Table 15.4. Regulation of *nirB*⁺ expression by oxygen and nitrite

Growth conditions	Nitrite reductase activity (nmol/min/mg protein)		β-Galactosidase of *nirB* :: *lacZ* fusion (nmol/mg dry wt/min)	
	Wild type (Chl^{S}Nar⁺)	Chl^{R}Nar⁻ mutant	Chl^{S}Nar⁺ strain	Chl^{R}Nar⁻ mutant
Aerobic	<10	<10	30	125
Anaerobic	50	700	150	610
Anaerobic + nitrite	200	700	690	630

Unlike Chl^{S}Nar⁺ strains, Chl^{R}Nar⁻ mutants are resistant to 10 mM KClO₃ during anaerobic growth and are unable to reduce nitrate due to a defect in the synthesis of the molybdenum cofactor, bactropterin. Operon fusions to *nirB* were generated using the Mud1(Ap*lac*⁺) phage (Casadaban and Cohen 1979; Griffiths and Cole 1987).

promoter requires the FNR protein, product of the *fnr* gene (Table 15.3), to be expressed. Indeed, the first NiR mutants to be described were the *nirA* mutants of Cole and Ward (1973), which were subsequently shown to be identical to the *fnr* mutants described by Lambden and Guest (1976) (Newman and Cole, 1978). The *fnr* gene is autoregulated, but only during anaerobic growth (Spiro and Guest, 1987*b*). This suggests that the FNR protein is inactive in aerated cultures, but becomes activated, possibly as a result of a conformational change, during anaerobic growth. The FNR protein shows considerable homology with the cAMP binding protein, CRP (product of the *crp* gene), which responds to conditions of starvation by forming a complex with cAMP. This complex then activates genes required for the metabolism of secondary carbon sources. Like CRP, the FNR protein contains a helix-turn-helix motif typical of many DNA binding proteins (Pabo and Sauer 1984), but there are also three major differences between these two transcriptional activator proteins:

(1) FNR lacks key amino acid residues which enable CRP to interact with its signal molecule cAMP: presumably this is why FNR and CRP respond to different physiological signals, anaerobiosis and an elevated cAMP concentration due to starvation, respectively.
(2) The amino acid terminus of FNR includes a cysteine-rich region not present in CRP which might sense the redox state of the cell.
(3) The putative DNA binding domains of FNR and CRP are different at three key amino acid residues, presumably accounting for their specificity for different promoters.

When Spiro and Guest (1987*a*) replaced these three key amino acids in the presumed DNA binding helix of FNR with those found in CRP, the modified FNR protein activated the CRP-dependent *lac* operon of *E. coli* in response to a switch from aerobic to anaerobic growth. This provided the first clear evidence that FNR is indeed a DNA binding protein. We have recently identified a possible FNR binding site in the *nirB* promoter. First, results of a deletion analysis summarized in Fig. 15.2 established that there are three regulatory domains in the *nirB* promoter. Bases between -147 and -87 with respect to the transcription start point are required for nitrite induction, and sequences between -73 and -47 are required for full promoter strength. Although only poor promoter activity is retained by the 47 bp promoter, this residual activity is still repressed by oxygen and is totally FNR-dependent (P. S. Jayaraman, unpublished results).

Binding sites for the CRP protein frequently include the motif 5′-TGTGA-3′ separated by six bases from a poorly conserved inverted motif. Jayaraman *et al.* (1988) recently noticed a similar sequence with perfect hyphenated dyad symmetry in the -47 to -35 region of the *nirB* promoter. Two lines of evidence strongly suggest that this region includes a target site at which the FNR protein activates NiR transcription during anaerobic growth. First, deletions and mutations in this region of dyad symmetry abolish or severely diminish both promoter strength and induction during anaerobic growth. Secondly, a hybrid promoter was constructed *in vitro* in which the CRP site of the galactose promoter P1 was replaced by the putative FNR binding site of the *nirB* promoter. This hybrid promoter was actively transcribed only during anaerobic growth and was dependent upon the FNR protein (Jayaraman *et al.* 1988). These results provide strong evidence that FNR is a DNA binding protein which activates nitrite

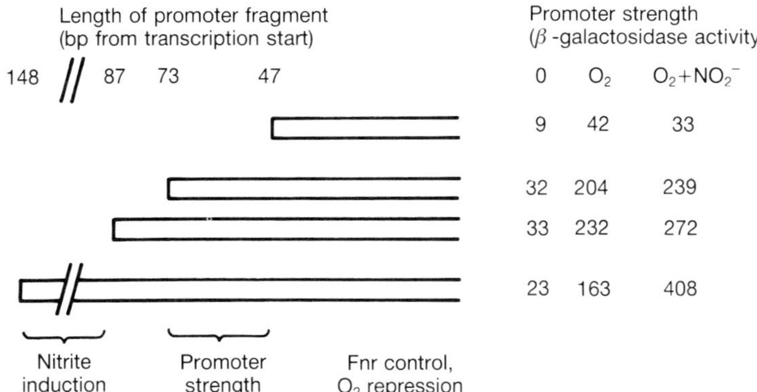

Fig. 15.2. Deletion analysis revealing three regulatory domains in the *nirB* promoter. The wild-type *nirB* promoter fused to an intact *lacZ* gene on a multicopy plasmid was shortened *in vitro* by incubation with *Bal*31 nuclease and transferred onto the chromosome to obtain strains with single copies of the *nirB*::*lacZ* operon fusions. Changes in β-galactosidase activities therefore reflect changes in the rates of transcription of the three genes in response to altered conditions of growth. Units are nmol *O*-nitrophenyl β-D-galactoside hydrolysed min/mg dry weight. Symbols are O: aerobic growth; $\emptyset_2$, anaerobic growth; $\emptyset_2 + NO_2^-$, anaerobic culture supplemented with 2.5 mM $NaNO_2^-$.

reduction as well as other aspects of anaerobic metabolism in response to oxygen starvation in *E. coli*. We do not yet know how nitrite induces *nirB* transcription, or why it is constitutive during anaerobic growth in some chlorate-resistant mutants.

Current research goals

When they first detected formate–nitrite oxidoreductase activity, Abou-Jaoudé *et al.* (1979*a,b*) assumed that *E. coli* synthesizes only one formate dehydrogenase which is involved in formate-dependent nitrate and nitrite reduction as well as formate hydrogenlyase activity. We now know that there are at least two such enzymes, one located in the cytoplasmic membrane which is involved in the electrogenic reduction of nitrate, and an independent, soluble enzyme required for formate hydrogenlyase activity. As energy is conserved during nitrite reduction by formate, we anticipated that the membrane-associated formate dehydrogenase would also be required for nitrite reduction. We were surprised, therefore, that a mutant defective in *fdhF*, the structural gene for the soluble formate dehydrogenase, is unable to use formate to reduce nitrite. Conversely, a mutant deficient in *fdhN* (the structural gene for the membrane-bound formate dehydrogenase required for nitrate reduction by formate) can reduce nitrite using formate. Further careful analysis will be required to establish whether this preliminary result is correct and to clarify the electron transfer pathway for nitrite reduction.

We have isolated a variety of mutants unable to use formate as an electron donor for nitrite reduction. Some of the mutations map in the *fnr* gene, others are located elsewhere on the chromosome. All are devoid of cytochrome c_{552} during anaerobic growth in the presence of nitrite (Cole 1988). Although many of the mutants are Lac$^+$ operon- or protein-fusion strains, expression of β-galactosidase activity is not inducible during anaerobic growth with nitrite, nor FNR-dependent. This implies either that the structural gene for cytochrome c_{552} synthesis is not regulated by oxygen repression and activated by the FNR protein, or that the mutants are defective in components other than cytochrome c_{552} itself. Attempts to identify the FNR-dependent component of the Nrf pathway are in progress.

There are three immediate but fascinating problems concerning the regulation of NADH-NiR synthesis. The first is the role of upstream activator sequences and nitrite itself in regulating transcription from the *nirB* promoter. One possibility is that nitrite can bind to and activate the NARL protein (product of the *narL* gene,

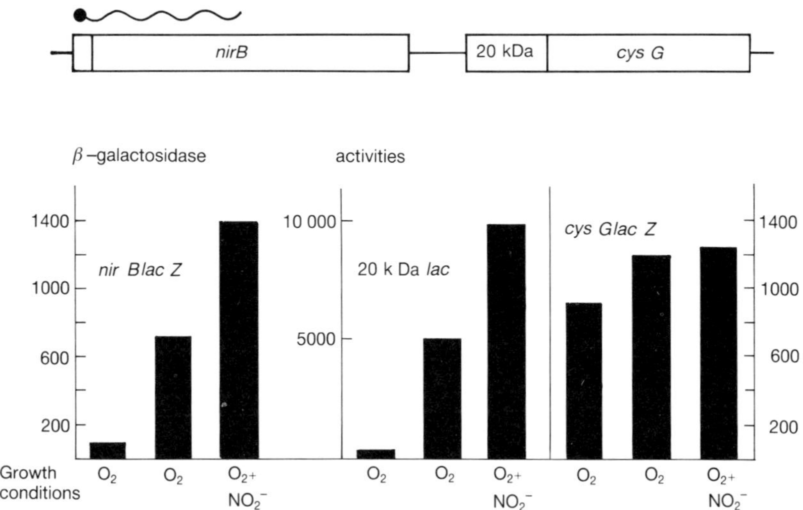

Fig. 15.3. Regulation of genes involved in nitrite reductase and sirohaem synthesis in the 74-minute region of the *E. coli* chromosome. The *lacZ* gene in a multicopy plasmid was fused *in vitro* to *nirB*, 20 kDa ORF and *cycG*. Each fusion carried a fully active *nirB* promoter. Changes in β-galactosidase activities therefore reflect changes in the rates of transcription of the three genes in response to altered conditions of growth. Units are nmol O-nitrophenyl β-D-galactoside hydrolysed min/mg dry weight. O_2: aerated culture. The black dot and wavy line indicate the transcription start point and mRNA corresponding to the *nirB* gene at 74 minute on the *E. coli* chromosome. Note that there is no gap between the 20 kDa ORF and the *cysG* gene. Other symbols and units of β-galactosidase activity are defined in the legend to Fig. 15.2.

Table 15.3) which would then activate *nirB* in a similar way to its activation of the nitrate reductase promoter in response to nitrate (Stewart 1982). Secondly, how is the synthesis of sirohaem regulated? Far more sirohaem is required during nitrite reduction by anaerobic cultures than for sulphite reduction by aerobic cultures growing with sulphate as sole sulphur source. We have recently reported that although the *cysG* gene is located very close to the *nirB* gene, it is not part of the *nirB* operon (Fig 15.3). Instead, it is regulated by its own promoter which is only slightly more active during anaerobic growth than during aerobic growth. Finally, between *nirB* and *cysG* is an open reading frame (ORF) of no known function. Although this is potentially a gene for a 20 kDa protein, there is no promoter between the carboxyl terminus of *nirB* and the *cysG* promoter which overlaps the carboxyl terminus of the 20 kDa ORF (Fig. 15.3). Nevertheless, operon and protein fusion studies have established that the 20 kDa ORF is both transcribed and translated coordinately with *nirB*, under the control of the *nirB* promoter. So what is the function of this previously unrecognized protein?

Acknowledgements

The author gratefully acknowledges the substantial contributions to research in this laboratory made by many colleagues, especially Steve Busby, Sheela Jayaraman, Tim Peakman, Lisa Page, and Lesley Griffiths. Former colleagues are co-authors of many of the cited papers.

References

Abou-Jaoudé, A., Chippaux, M., and Pascal, M.-C. (1979*a*). Formate-nitrite reduction in *Escherichia coli* K12.1. Physiological study of the system. *European Journal of Biochemistry* **95**, 309–14.

Abou-Jaoudé, A., Pascal, M.-C., and Chippaux, M. (1979*b*). Formate-nitrite reduction in *Escherichia coli* K12.2. Identification of components involved in the electron transfer. *European Journal of Biochemistry* **95**, 315–21.

Bonnefoy, V., Burini, J. F., Giordana, G., Pascal, M.-C., and Chippaux, M. (1987). Presence in the 'silent' terminus region of *Escherichia coli* K12 chromosome of cryptic gene(s) encoding a new nitrate reductase. *Molecular Microbiology* **1**, 143–50.

Casadaban, M. J. and Cohen, S. N. (1979). Lactose genes fused to exogenous promoters in one step using a Mu-lac bacteriophage: *in vivo* probe for translational control signals. *Proceedings of the National Academy of Sciences, USA* **76**, 4530–3.

Cole, J. A. (1968). Cytochrome c_{552} and nitrite reduction in *Escherichia coli*. *Biochimica et Biophysica Acta* **162**, 356–68.

Cole, J. A. (1988). Assimilatory and dissimilatory reduction of nitrate to ammonia. In *The nitrogen and sulphur cycles* (eds J. A. Cole and S. J. Ferguson), Symposium 42, pp. 281–329. The Society for General Microbiology, Cambridge University Press, U.K.

Cole, J. A. and Ward, F. B. (1973). Nitrite reductase-deficient mutants of *Escherichia coli* K12. *Journal of General Microbiology* **76**, 21–9.

Coleman, K. J., Cornish-Bowden, A., and Cole, J. A. (1978*a*). Purification and properties of nitrite reductase from *Escherichia coli* K12. *Biochemical Journal* **175**, 483–93.

Coleman, K. J., Cornish-Bowden, A., and Cole, J. A. (1978*b*). Activation of nitrite reductase from *Escherichia coli* K12 by oxidized nicotine-adenine dinucleotide. *Biochemical Journal* **175**, 495–9.

Fujita, T. and Sato, R. (1966*a*). Studies on soluble cytochromes in Enterobacteriaceae III. Localization of cytochrome c_{552} in the surface layer of cells. *Journal of Biochemistry* **60**, 568–77.

Fujita, T. and Sato, R. (1966*b*). Studies on soluble cytochromes in Enterobacteriaceae, IV. Possible involvement of cytochrome c_{552} in anaerobic nitrite metabolism. *Journal of Biochemistry* **60**, 691–700.

Fujita, T. and Sato, R. (1967). Studies on soluble cytochromes in Enterobacteriaceae, V. Nitrite-dependent gas evolution in cells containing cytochrome c_{552}. *Journal of Biochemistry* **62**, 230–8.

Gray, C. T., Wimpenny, J. W. T., Hughes, D. E., and Ranlett, M. (1963). A soluble *c*-type cytochrome from anaerobically grown *Escherichia coli* and various Enterobacteriaceae. *Biochimica et Biophysica Acta* **67**, 157–60.

Griffiths, L. and Cole, J. A. (1987). Lack of redox control of the anaerobically-induced $nirB^+$ gene of *Escherichia coli* K-12. *Archives of Microbiology* **147**, 364–9.

Jackson, R. H., Cornish-Bowden, A., and Cole, J. A. (1981*a*). Prosthetic groups of the NADH-dependent nitrite reductase from *Escherichia coli* K12. *Biochemical Journal* **193**, 861–7.

Jackson, R. H., Cole, J. A., and Cornish-Bowden, A. (1981*b*). The steady-state kinetics of the NADH-dependent nitrite reductase from *Escherichia coli* K12: nitrite and hydroxyl-amine reduction. *Biochemical Journal* **199**, 171–8.

Jackson, R. H., Cole, J. A., and Cornish-Bowden, A. (1982). The steady state kinetics of the NADH-dependent nitrite reductase from *Escherichia coli*: the reduction of single-electron acceptors. *Biochemical Journal* **203**, 505–10.

Jayaraman, P. S., Peakman, T. C., Busby, S. J. W., Quincey, R. V., and Cole, J. A. (1987). Location and sequence of the promoter of the gene for the NADH-dependent nitrite reductase of *Escherichia coli* and its regulation by oxygen, the FNR protein and nitrite. *Journal of Molecular Biology* **196**, 781–8.

Jayaraman, P. S., Gaston, K. L., Cole, J. A., and Busby, S. J. W. (1988). The *nirB* promoter of *Escherichia coli*; location of nucleotide sequences essential for regulation by oxygen, the FNR protein and nitrite. *Molecular Microbiology* **2**, 527–30.

Jeter, R. M., Sias, S. R., and Ingraham, J. L. (1984). Chromosomal location and function of genes affecting *Pseudomonas aeruginosa* nitrite assimilation. *Journal of Bacteriology* **157**, 673–7.

Jones-Mortimer, M. C. (1968). Positive control of sulphate reduction in *Escherichia coli*. Isolation, characterization and mapping of cysteineless mutants of *E. coli* K12. *Biochemical Journal* **110**, 589–95.

Kemp, J. D. and Atkinson, D. E. (1966). Nitrite reductase of *Escherichia coli* specific for reduced nicotinamide adenine dinucleotide. *Journal of Bacteriology* **92**, 628–34.

Kemp, J. D., Atkinson, D. E., Ehret, A., and Lazzarini, R. A. (1963). Evidence for the identity of the nicotinamide adenine dinucleotide phosphate-specific sulphite and nitrite reductases of *Escherichia coli*. *Journal of Biological Chemistry* **238**, 3466–71.

Lambden, P. R. and Guest, J. R. (1976). Mutants of *Escherichia coli* K12 unable to use fumarate as an electron acceptor. *Journal of General Microbiology* **97**, 145–60.

Lazzarini, R. A. and Atkinson, D. E. (1961). A TPN specific nitrite reductase from *E. coli*. *Journal of Biological Chemistry* **236**, 3330–6.

Liu, M.-C., Peck, H. D., Abou-Jaoudé, A., Chippaux, M., and Le Gall, J. (1981). A reappraisal of the role of the low potential c-type cytochrome (cytochrome c_{552}) in NADH-dependent nitrite reduction and its relationship with a co-purified NADH oxidase in *Escherichia coli* K12. *FEMS Microbiolgy Letters* **10**, 333–7.

Macdonald, H. and Cole, J. A. (1985). Molecular cloning and functional analysis of the *cysG* and *nirB* genes of *Escherichia coli* K12, two closely-linked genes required for NADH-dependent nitrite reductase activity. *Molecular and General Genetics* **200**, 328–34.

Massey, V. and Veeger, C. (1961). Studies on the reaction mechanisms of lipoyl dehydrogenase. *Biochimica et Biophysica Acta* **48**, 33–47.

Motteram, P. A. S., McCarthy, J. E. G., Ferguson, S. J., Jackson, J. B., and Cole, J. A. (1981). Energy conservation during the formate-dependent reduction of nitrite by *Escherichia coli*. *FEMS Microbiology Letters* **12**, 317–20.

Newman, B. M. and Cole, J. A. (1978). The chromosomal location and pleiotropic effects of mutations in the *nirA*[+] gene of *Escherichia coli* K12: the essential role of *nirA*[+] in nitrite reduction and in other anaerobic redox reactions. *Journal of General Microbiology* **106**, 1–12.

Pabo, C. O. and Sauer, R. T. (1984). Protein–DNA recognition. *Annual Review of Biochemistry* **53**, 293–321.

Peck, H. D. and Gest, H. (1957). Formic dehydrogenase and the hydrogenlyase enzyme complex in Coli-Aerogenes bacteria. *Journal of Bacteriology* **73** 706–21.

Pope, N. R. and Cole, J. A. (1982). Generation of a membrane potential by one of two independent pathways for nitrite reduction by *Escherichia coli*. *Journal of General Microbiology* **128**, 319–22.

Pope, N. R. and Cole, J. A. (1984). Pyruvate and ethanol as electron donors for nitrite reduction by *Escherichia coli* K12. *Journal of General Microbiology* **130**, 1279–84.

Postgate, J. R. (1954). Presence of cytochrome in an obligate anaerobe. *Biochemical Journal* **56**, xi.

Smith, M. (1983). Nitrous oxide production by *Escherichia coli* is correlated with nitrate reductase activity. *Applied and Environmental Microbiology* **45**, 1545–7.

Spiro, S. and Guest, J. R. (1987a). Activation of the *lac* operon of *Escherichia coli* by a mutant FNR protein. *Molecular Microbiology* **1**, 53–58.

Spiro, S. and Guest, J. R. (1987b). Regulation and over-production of the *fnr* gene of *Escherichia coli*. *Journal of General Microbiology* **133**, 3279–88.

Steenkemp, D. J. and Peck, H. D. (180). The association of hydrogenase and dithionite reductase of *Desulfovibrio desulfuricans*. *Biochemical and Biophysical Research Communications* **94**, 41–8.

Steenkemp, D. J. and Peck, H. D. (1981). Proton translocation associated with nitrite respiration in *Desulfovibrio desulfuricans*. *Journal of Biological Chemistry* **256**, 5450–8.

Stewart, V. (1982). Requirement of Fnr and NarL functions for nitrate reductase expression in *Escherichia coli* K12. *Journal of Bacteriology* **151**, 1320–5.

Zarowny, D. P. and Sanwal, B. D. (1963). Characterization of an NADH-specific nitrite reductase from *E. coli* K12. *Canadian Journal of Microbiology* **9**, 531–9.

16. Molecular and genetic aspects of nitrite reduction in higher plants

John L. Wray

Introduction

The nitrite reductase (NiR) enzymes catalyse the six-electron reduction of nitrite to ammonium in the third step of the nitrate assimilation pathway. In this chapter I consider mainly the developments which have occurred over the last 6 years with particular, but not exclusive, reference to the molecular and genetic aspects of the step in higher plants. It will become clear that we know little about these, although the recent molecular cloning of the NiR structural gene (Back *et al.* 1988; Chapter 18) is anticipated to be the key to rapid advances in this area. Earlier reviews on the physiology and biochemistry of nitrite reduction are available (Beevers and Hageman 1980; Hewitt and Notton 1980; Guerrero *et al.* 1981; Huffaker 1982). Rajasekhar and Oelmuller (1987) have recently reviewed the factors influencing the development of NiR.

The nitrite reductases of higher plants

The ferredoxin nitrite reductase (EC 1.7.7.1) present in leaves has been purified to apparent homogeneity from a number of plant species including spinach (Ho and Tamura 1973; Ida and Morita 1973; Ida *et al.* 1976; Vega and Kamin 1977; Ida 1977; Hirasawa and Tamura 1980; Ida and Mikami 1986), *Cucurbita pepo* (Hucklesby *et al.* 1976), barley (Serra *et al.* 1984; Ip *et al.* 1987), wheat (Small and Gray 1984), suspension culture of Paul's Scarlet rose (Gupta *et al.* 1984), and the bean, *Phaseolus angularis* (Ishiyama and Tamura 1985; Ishiyama *et al.* 1985). All enzymes examined by chemical analysis and visible and EPR spectroscopy have been shown to contain sirohaem, (an iron tetrahydroporphyrin of the isobacteriochlorin type), non-haem iron, and acid-labile sulphur which in the case of the spinach enzyme are present at the active site as a bridged sirohaem/4Fe–4S cluster (Vega and Kamin 1977; Lancaster *et al.* 1979; Chapter 17). The enzymes can use either reduced ferredoxin (physiological) or reduced methylviologen dye (non-physiological) as an electron donor. The properties of some of these highly purified higher plant nitrite reductases are shown in Table 16.1.

The enzymes are usually isolated as monomeric polypeptides of ~63 kDa with very similar amino acid composition (Table 16.2). However, Tamura and coworkers (Hirasawa and Tamura 1980; Hirasawa-Soga and Tamura 1981; Hirasawa-Soga *et al.* 1982) reported a molecular weight of ~86 kDa for the spinach enzyme (composed of components of 61 kDa and 24 kDa). Hirasawa-Soga *et al.* (1982) speculated that the larger component is a 'modified' form of the native enzyme having only methylviologen-linked activity whilst the smaller component is a 'coupling protein' which confers ferredoxin-linked activity. This is at variance, however, with the observation that the purified ~63 kDa spinach enzyme can use ferredoxin and methylviologen equally effectively as an electron donor (see for example Ida and Mikami 1986), while the spinach NiR structural gene encodes a mature protein of molecular weight 63 kDa (Chapter 18). Immunological studies on the methylviologen and ferredoxin-linked NiR activities and on the 'coupling protein' are described by Hirasawa *et al.* (1984).

Few recent studies have been carried out on the enzyme from non-green tissue. NiR activity has been assayed in cell-free extracts of the roots of a variety of plant species (Sanderson and Cocking 1964; Miflin 1974) and the enzyme has been purified from barley roots (Ida *et al.* 1974) and bean roots (Nagaoka *et al.* 1984) and shown to have a molecular weight of ~60 kDa. These enzymes are similar to the spinach leaf enzyme in a wide range of properties. However, the electrophoretically homogeneous enzyme from etiolated shoots of *P. angularis* has a molecular weight of 100 kDa and subunits of 64 kDa and 35 kDa (Ishiyama and Tamura 1985), reminiscent of some of the reports on spinach NiR discussed above.

Isoforms of NiR have been identified in a number of plants. Two forms are present in root, scutellum and etiolated shoots of maize (Hucklesby *et al.* 1972; Dalling *et al.* 1973), and in the roots of wheat (Dalling *et al.* 1972). Isoforms have also been identified in *Galium aparine*, *Chenopodium album*, pea, and tomato (Kutscherra *et al.* 1987). Two of the three isoforms found in wild oat are encoded by different genetic loci, confirming their identity as isoenzymes (Heath-Pagliuso *et al.* 1984) but the relationship between the isoforms in other systems is unclear. Isoforms are generally of the same size with similar kinetic properties, but differences in thermal stability have been noted. Differences in timing of development of the isoforms have been identified in wheat and in *Galium aparine*. Only the main isoform was dependent on nitrate and light for development (Kutscherra *et al.* 1987).

Subcellular location and electron donation

Since intact chloroplasts from a variety of species can photoreduce nitrite to ammonium (Paneque *et al.* 1963) and thence to α-amino nitrogen (Magalhaes *et*

Table 16.1. Properties of some highly purified higher plant nitrite reductases

	Spinacia oleracea	*Cucurbita pepo*	*Hordeum vulgare*	*Phaseolus angularis* green shoot	root
Molecular weight (kDa)	63	63	61–63	68	62
Subunits	1	nd	1	1	1
$S_{20,w}$(s)	4.38	4.9	nd	nd	nd
Stokes radius (nm)	3.3	nd	nd	nd	nd
Isoelectric point (pH)	nd	nd	4.7	nd	nd
Amino acid residues	562	565	575	nd	nd
Specific activity (μmol/min/mg)	207	85	102	91	73
Turnover number (s^{-1})	nd	89	nd	nd	nd
Optimum pH	7.5	7.5	6.0–6.5	nd	7.5
K_m NO$_2^-$ (μM)	360	1	250	nd	294
K_m ferredoxin (μM)	6	nd	0.04	nd	nd
K_m methyl viologen (μM)	120	nd	120	nd	nd
Prosthetic groups (mol/mol):					
(4Fe–4S)	1	nd*	nd*	nd	nd
Sirohaem	1	nd*	nd*	nd*	nd*
E_m^1 (mV):					
(4Fe–4S)	−550 (pH = 9.0); n = 1?	−570 (pH = 8.1); n = 1?	−517 (pH = 8.0); n = 1	nd	nd
Sirohaem	−50 (pH = 7.8); n = 1	−120 (pH = 8.5); n = 1	+80 (pH = 8.0); n = 1	nd	nd
Spectral properties: absorption maxima (nm)	280; 389; 574; 690	280; 384; 572; 635; 697	280; 392; 574; 693	275; 393; 535; 571	280; 397; 535; 573
A_{280}/A_{soret}	1.54	2.0	2.24	3.9	3.3
ε(soret)(mM^{-1} . cm^{-1})	60.9	37.9	33.5	nd	nd
References	Back *et al.* (1988) Ida and Mikami (1986) Lancaster *et al.* (1979) Stoller *et al.* (1977) Vega and Kamin (1977)	Cammack *et al.* (1978) Hucklesby *et al.* (1976) Vega *et al.* (1980)	Ip *et al.* (1987) Serra *et al.* (1982)	Ishiyama *et al.* (1985)	Nagaoka *et al.* (1984)

nd = not determined; nd* = present, but no quantitative data available.

Table 16.2. Amino acid composition of some higher plant nitrite reductases

Amino acid	*Cucurbita pepo* (residues/ 62 kDa)	*Spinacea oleracea* (residues/ 61 kDa)	(residues/ 63 kDa)	*H. vulgare* cv. Golden Promise (residues/62 kDa)
Aspartic acid	63	64	57	51
Threonine	27	19	20	22
Serine	32	28	30	29
Glutamic acid	64	70	67	69
Proline	31	34	31	31
Glycine	48	47	46	65
Alanine	39	39	34	42
Valine	53	46	48	46
Half-cystine	10	8	12	10
Methionine	10	14	16	7
Isoleucine	20	27	30	26
Leucine	56	49	48	62
Tyrosine	21	12	8	14
Phenylalanine	19	20	18	20
Lysine	31	38	36	36
Histidine	11	12	11	10
Tryptophan	–	5	7	6
Arginine	30	32	39	29
Total	565	564	558	575
References	Hucklesby *et al.* (1976)	Kamin and Vega (1977)	Ida and Mikami (1986)	Ip *et al.* (1987)

al. 1974; Anderson and Done 1978) and the enzyme's physiological electron donor in leaves is reduced ferredoxin (Joy and Hageman 1966) it is usually concluded that the chloroplast is the sole (functional) site of NiR within the leaf cell. This is supported by cell fractionation studies (for example Dalling *et al.* 1972; Wallsgrove *et al.* 1979). The NiR of non-green tissue, such as roots, is localized in plastids (Dalling *et al.* 1972; Miflin 1974; Oaks and Hirel 1985). The root enzyme, like the leaf enzyme, can use the non-physiological electron donor, reduced methylviologen, but the nature of the physiological electron donor has been obscure until recently. Suzuki *et al.* (1985) have identified a non-haem iron-containing protein (root electron carrier) in young maize roots which has antigenic similarities to spinach leaf ferredoxin. A pyridine nucleotide reductase which shares identical immunological determinants with the spinach ferredoxin-NADP$^+$ reductase was also identified in young maize roots. In the presence of this root pyridine nucleotide reductase and root electron carrier either NADPH or NADH served as the primary electron donor for glutamate synthesis (*via* glutamate synthase) but were not able to support nitrite reduction. Either a component of the electron transport chain is missing or, as argued by Hucklesby

(1986), there may be a disparity in potential between NAD(P)H ($E_0 = -0.321$ mv) and the root electron carrier (E_0 unknown). A protein isolated from tobacco suspension cells, and perhaps the equivalent of the root electron carrier (Ninomiya and Sato 1984), is able to mediate the reduction of nitrite with NADPH as electron donor using spinach leaf NADPH-ferredoxin reductase. The reduced protein has an epr signal at $g = 1.93$, suggesting that it is a ferredoxin-like protein, but it is larger than ferredoxin (molecular weight 19.5 kDa) and is unable to carry out the photoreduction of $NADP^+$.

Immunological studies on nitrite reductase

Polyclonal antibodies to NiR have been produced for pea (Gupta *et al.* 1984), wheat (Small and Gray 1984), spinach (Hirasawa *et al.* 1984; Ida, 1987), *Phaseolus angularis* (Ishiyama *et al.* 1985), and barley (Ip *et al.* 1987). Ishiyama *et al.* (1985) concluded that the NiR from green shoots, etiolated shoots, and roots of *P. angularis* have antigenic determinants in common, and similar conclusions with regard to the spinach leaf and root enzymes had been reached a little earlier by Hirasawa *et al.* (1984). Ida (1987) demonstrated a close antigenic relationship between spinach and other dicotyledonous enzymes whereas the monocotyledonous NiRs were more distantly related. Similarly, barley NiR is more closely related to other monocotyledenous NiR proteins (D. W. Kirk and J. L. Wray, unpublished). Use of antibodies in the study of development, *in vitro* translation products and molecular cloning of the NiR structural gene is discussed below.

Synthesis of nitrite reductase and its regulation

Early attempts to identify the genome from which NiR mRNA was derived depended on the use of inhibitors of protein synthesis on cytoplasmic or chloroplastic ribosomes and were inconclusive (Schrader *et al.* 1967; Sawhney and Naik 1973; Sluiters-Scholten 1973). The studies of Heath-Pagliuso *et al.* (1984), however, suggest that the enzyme is encoded by the nuclear DNA.

NiR is not unique in this respect. Numerous chloroplast proteins are encoded by the nuclear genome (Whitfield and Bottomley 1983) and are synthesized in the cytoplasm on 80 S ribosomes (Ellis 1981). Such proteins are usually synthesized as larger molecular weight precursors containing an amino terminal sequence, the transit peptide (Chua and Schmidt 1979), which is proteolytically cleaved to generate the mature functional protein during or after transport into the chloroplast (Grossman *et al.* 1982; Robinson and Ellis 1984). Evidence from the *in vitro* translation of polyA$^+$ RNA shows that NiR shares these properties of nuclear-encoded chloroplast proteins. Thus in wheat (Small and Gray 1984), pea (Gupta and Beevers 1985, 1987), and rice (Ogawa and Ida 1987) immunoprecipitation of *in vitro* translation products with specific NiR antisera reveals the synthesis of a peptide with a higher molecular weight than the native enzyme. Like the precursor of the small subunit of ribulose-1,5-bisphosphate carboxylase (pS) (Robinson and Ellis 1984) this *in vitro* synthesized peptide can be processed in a two-step reaction to a peptide of the same size as that of the native enzyme by a protein extract from chloroplasts (Gupta and Beevers 1987).

This evidence for a precursor protein has recently been confirmed by the molecular cloning of spinach NiR cDNA species using oligonucleotide probes based on partial amino acid sequence data and through immunoscreening of a cDNA library in the expression vector, λgt11. These studies show that the precursor protein for NiR is 594 amino acids long (molecular weight 66,394) and has a 32 amino acid extension at the N-terminal end of the mature protein (molecular weight 62,883) which probably serves as the transit peptide (Back *et al.* 1988). Comparison of the transit peptide sequence with that of some others showed few conserved residues and little evidence for the three major blocks of amino acid homology shared by the transit peptide of the light-harvesting chlorophyll *a/b* binding protein precursor and pS (Karlin-Neumann and Tobin 1986).

Transit peptides are believed to contain the information for chloroplast-specific import, since that of pS can direct the import of neomycin phosphotransferase II (Van den Broeck *et al.* 1985; Schreier *et al.* 1985; Kuntz *et al.* 1986) and the ferredoxin and plastocyanin transit peptides can direct the import of the yeast mitochondrial superoxide dismutase (Smeekens *et al.* 1987). However the mechanisms involved in targeting to sub-chloroplast locations (inner and outer envelope, envelope space, stroma, thylakoid membrane, and thylakoid lumen) are poorly understood. Recent experiments in which the transit peptides for the ferredoxin (stromal-located) and plastocyanin (thylakoid-located) precursors were exchanged to generate chimaeric plastocyanin transit peptide-mature ferredoxin and ferredoxin transit peptide-mature plastocyanin constructs (Smeekens *et al.* 1986) and deletion analysis of the pS transit peptide (Reiss *et al.* 1987) suggest that the transit peptide is also a major determinant in the localization of proteins within the chloroplast. Approaches of this type will be useful in the study of the role of the NiR transit peptide.

Recently, Pain *et al.* (1988) have used an anti-idiotypic antibody approach to identify a major 30 kDa integral membrane protein of the chloroplast envelope as a receptor for the import of pS. They raised antibodies against a chemically synthesized peptide representing a 30 residue long carboxyl terminal portion of the transit peptide of pea pS (Coruzzi *et al.* 1983). A second set of antibodies raised against the transit peptide antibodies, and presumably containing anti-idiotypic antibodies directed against the transit peptide-binding site of the putative import receptor, was used to demonstrate that this receptor is located in contact sites between the outer and inner membrane of the chloroplast envelope (Pain *et al.* 1988). It is unlikely that each protein has its own unique import receptor but there is some circumstantial evidence to indicate that more than one class exists (discussed in Schmidt and Mishkind 1986). Now that the sequence of the transit peptide of NiR is available (Chapter 18) it should be possible to identify its import receptor and determine whether or not it is the same as for pS. Of significance here perhaps is the fact that the primary sequences of the transit peptides of these two proteins bear little similarity.

Nitrate and light both influence the development of NiR activity. Whether nitrate itself, or nitrite produced as a result of nitrate reduction, is the true 'inducer' of NiR has been considered by several workers, either by studying directly the ability of nitrite to 'induce' NiR activity or by the use of the molybdate analogue, tungstate (Wray and Filner, 1970) to block reduction of applied nitrate to nitrite. These studies show that in *Lemna minor* (Stewart 1968; Joy 1969), in non-photosynthetic tobacco XD cell cultures (Kelker and Filner 1971), and in maize (Rao *et al.* 1981) nitrite is able to act as an 'inducer' of NiR activity. Whether it acts in this way during normal growth is unclear but there is some evidence from *Lemna* for sequential 'induction' of NR and NiR activities upon nitrate addition to ammonium-grown plants (Stewart 1968).

In contrast to nitrate reductase, where ample evidence demonstrates that end-products (ammonium/amino acids) can either promote [e.g. in maize (Schrader and Hageman 1967; MacKown *et al.* 1982)] or inhibit [e.g. in barley (Smith and Thompson 1971) and in wheat (Vijayaraghavan *et al.* 1979)] nitrate 'induction' there is little information available on end-product control of NiR levels in higher plants. In *Lemna* (Joy 1969) and tobacco XD cell cultures (Kelker and Filner 1971), not the best examples of green, differentiated and photosynthetically functional higher plants, ammonium and casein hydrolysate, respectively, inhibit the development of NiR (and NR) activity in response to nitrate. Recently Rajasekhar and Mohr (1986b) demonstrated that in dark-grown mustard cotyledons ammonium, given together with nitrate, strongly inhibits the nitrate-'induced' development of NR and NiR whereas, in continuous far-red light, ammonium given together with nitrate strongly stimulates the nitrate-induced development of NR but has no effect on NiR. Further, in the absence of nitrate appearance of NR is 'induced' by ammonium in the dark as well as in continuous far-red light, whereas NiR levels are not affected. These results suggest that NR and NiR levels are regulated differently, at least in mustard. However, the molecular mechanisms underlying these interactions are unknown and whether nitrate uptake was also affected was not examined.

The interaction between nitrate and light in the regulation of NiR levels has been examined by several workers. Red–far red light pulse experiments in maize (Rao *et al.* 1980, 1981; Sharma and Sopory 1984) and mustard (Rajasekhar and Mohr) indicate the involvement of phytochrome. The operation of a blue light photoreceptor has been suggested for *Sorghum bicolor* (Rajasekhar and Sopory 1985). In the absence of both light and nitrate NiR activity is low in mustard (Rajasekhar and Mohr 1986*a*) and in barley (E. Duncanson and J. L. Wray, unpublished) while high levels of NiR activity are found in plants treated with both light and nitrate (Gupta and Beevers 1983; Rajasekhar and Mohr, 1986*a*; E. Duncanson and J. L. Wray, unpublished). In maize (Sharma and Sopory 1984) and mustard (Rajasekhar and Mohr 1986*a*) nitrate addition increases NiR activity in dark-grown plants and the addition of light either at the same time or subsequent to nitrate addition leads to further increases in NiR activity. Light alone has no effect on the development of NiR activity (Schuster *et al.* 1987). Taken together these data suggest that the light (phytochrome) effect cannot express itself in the absence of nitrate and that nitrate (or perhaps nitrite?) is the true 'inducer'.

Of some interest is the observation that in maize leaves (Sharma and Sopory 1984) and in mustard cotyledons (Schuster *et al.* 1987) a red or far-red light treatment given some hours before nitrate addition can strongly enhance subsequent nitrate 'induction' of NiR activity. That is, the light signal can be stored. Signal storage is proposed to be a means of enabling the plant to maintain the appropriate levels of NiR (and indeed also of NR) during the dark period of the natural light/dark cycle. Schuster *et al.* (1987) argue that the formation and action of the stored signal are a bypass to the process of direct signal transduction. Since photo-oxidative damage of the chloroplasts abolishes the action of nitrate and light in NiR and NR development in mustard cotyledons, Rajasekhar and Mohr (1986*a*) have postulated that a 'plastidic signal' is a prerequisite for the nitrate 'induced' and phytochrome-modulated appearance of NiR and NR. However, such damage also affects accumulation of NiR directly (Oelmuller *et al.* 1988). The three factors are postulated to act in a hierarchy: without the plastidic signal nitrate cannot perform enzyme 'induction'; without nitrate the light effect cannot express itself (Rajasekhar and Mohr, 1986*b*).

Increases in NiR activity are correlated with increases in NiR cross-reacting material in wheat (Small and Gray 1984), pea (Gupta and Beevers 1984), rice (Ogawa and Ida 1987) and barley (E. Duncanson and J. L. Wray, unpublished). Poly(A)$^+$ RNA from nitrate deficient, dark-grown wheat (Small and Gray 1984) and pea (Gupta and Beevers 1985) does not support the synthesis of NiR whilst that isolated from nitrate-treated, light-grown plants does. These results suggest that nitrate and light modulate the synthesis of NiR by regulating the availability of NiR mRNA. Evidence that nitrate at least acts in this manner is provided by the use of the cloned spinach NiR cDNA gene in Northern blot analysis. Application of nitrate to ammonium/light-grown plants leads to a substantial increase in NiR-specific mRNA (Back *et al.* 1988; Chapter 18). While this increase in NiR-mRNA in response to nitrate may be due to an increase in the rate of transcription of the NiR structural gene other possibilities, such as altered RNA processing or decreased mRNA degradation, exist and will need to be eliminated.

How phytochrome interacts with the NiR gene system is unknown. It is possible that the phytochrome effect is indirect through effects on, say, the accumulation of the inducer nitrate, as has been suggested to occur in pea terminal buds (Jones and Sheard 1975). However the fact that red light is able to further increase NiR activity when 'induction' by nitrate is saturated (Sharma and Sopory 1984) perhaps argues against this point. Other phytochrome-regulated genes carry a *cis*-acting light responsive element in the 5'-upstream region of the gene (Kuhlemeier *et al.* 1987) and a pea nuclear protein which interacts with this sequence and presumably mediates phytochrome regulation has been identified (Green *et al.* 1987). Some thoughts on the nature of the phytochrome–signal transduction chain are discussed by Sharma (1985). Clearly the availability of the cloned NiR gene (Back *et al.* 1988; Chapter 18), gene transfer systems and reporter genes (for example Jefferson 1987) now allows the search for *cis*-acting elements responsive to nitrate, light and end products and for protein factors which mediate their effects.

Whether higher plants, like filamentous fungi (Chapters 6, 19, 20 and 22), possess a system involving the product of a *nir*A-type gene and functional NR molecules which regulates expression of the NR and NiR structural genes is presently unclear and, indeed, largely unexplored, as is the existence of an *areA*-type gene mediating ammonium repression. In *N. tabacum* two *cnxA* alleles (Cnx68 and Cnx101) produce constitutive levels of NR apoprotein and also show an increased constitutive level of NiR, as do the *nia* allele, Nia102, and *cnxB* alleles (Mendel and Müller 1979; Mendel *et al.* 1984; Schiemann and Müller 1985; Chapter 11). In barley some mutations in the NADH-NR apoprotein *nar-1* locus affect the regulation of NR mRNA accumulation, suggesting perhaps that some form of feedback regulation present in the wild type is absent in these mutants (Chapter 13). There is, however, little other evidence in the literature (Wray 1986) for an autoregulatory role for NR in higher plants and there is of course no *a priori* reason to suppose that the regulatory networks operating in higher plants are the same as those in lower eukaryotes. Indeed, the involvement of light in the regulation of the higher plant enzymes and the different modes of ammonium control indicated above suggests that they are different, at least in part.

Genetics of nitrite reduction

Little information is available on the genetics of nitrite reduction. Inheritance studies in oats suggests that the enzyme is nuclear-encoded (Heath-Pagliuso *et al.* 1984) but mutant analysis of the type carried out on the nitrate reduction step has not been explored in any detail so far. This is undoubtedly due to the fact that a positive selection scheme of the type utilized so successfully in the isolation of nitrate reduction mutants (Chapter 11) is not available. Consequently attempts to isolate mutants defective in nitrite reduction will depend on alternative procedures in which the variant phenotypes of individual cells or plants is examined. Although such procedures have been utilized successfully in the isolation of nitrate reduction mutants at the whole plant level (Warner *et al.* 1977) they are more tedious and time-consuming to operate than selection for resistance to chlorate.

The first plant variants defective in nitrite reduction were described in *Haplopappus gracilis* by Gilissen *et al.* (1985). They selected a callus (Ao) which was able to grow in the presence of normally toxic levels of asparagine. After X-irradiation this line was plated on medium containing alanine as nitrogen source and calli growing at wild type rates were selected. Two lines failed to grow on medium containing nitrate as a nitrogen source and excreted nitrite. While *in vitro* NR activity was similar to the wild type, both *in vivo* NR and *in vitro* NiR activities were only 20–40 per cent of the wild type level. Unfortunately attempts at plant regeneration were unsuccessful and the molecular basis of the defects is unknown. Clearly whole plant mutants defective in nitrite reduction are of particular significance since they allow not only the characterization of the number and type of genes which determine this step in nitrate assimilation, but also the consequences of such mutations in the intact differentiated organism. A further consideration is that some types of mutation may only be expressed, and thus can only be selected for, at the whole-plant level.

In order to isolate whole-plant barley mutants defective in nitrite reduction we have devised a selection scheme based on the rationale that such mutants should be identifiable by virtue of the fact that they would be expected to accumulate nitrite after nitrate treatment (Duncanson *et al.* 1987). A possible constraint that was anticipated was that nitrite is reportedly toxic to higher plants. A wide variety of mutations might be expected, *a priori*, to block nitrite reduction, viz. within the NiR structural gene affecting either catalytic function or the transit peptide; in processing of the precursor protein; in recognition of the chloroplast/plastid envelope; in import into, or targeting within, the chloroplast/plastid; in electron donation and in some aspect of regulation. Clearly, some of these mutations might be expected to be lethal.

In practice, bulk M_2 seed of barley, mutagenized with azide in the M_1, was sown in vermiculite at a density sufficient to allow the development of approximately 200 seedlings/tray. The seeds were treated with half-Hoagland nutrient solution lacking nitrate and germinated in the dark for 5 days. After transfer to the light for 2 days to allow the seedlings to green fully they were treated for 18 h with half-Hoagland nutrient solution containing 15 mM potassium nitrate. In order to identify individual seedlings which accumulated nitrite the tip of each seedling (~ 10 mg tissue) (a pool of ~ 50 tips at a time) was placed in a mortar and ground with 1 ml water, 1 ml 1 per cent sulphanilamide in 3M HCl, and 1 ml 0.02 per cent N-ethylenediamine dihydrochloride (Snell and Snell 1949). After centrifugation of the homogenate the supernatant was examined. A pink/red colour indicates the presence of nitrite. Those pools that give a positive nitrite reaction would be expected to contain at least one nitrite accumulator amongst the 50 seedlings, which would then need to be rechecked individually. Wild type plants do not accumulate nitrite under these conditions.

Over 250 000 M_2 seedlings were screened using this procedure (Table 16.3). As anticipated most pools of 50 tips were nitrite-negative but some 'pools' did produce a positive nitrite reaction. In all cases when we rechecked the 50 individual seedlings we were able to identify one or more individuals which accumulated nitrite. There were no problems with false-positives; however the level of nitrite accumulated by selected individual seedlings differed. Some had low levels of accumulated nitrite (~ 0.3 μmol/g fresh weight of tissue) while others had very high levels (~ 200 μmol/g fresh weight of tissue). The overall frequency of detection of nitrite accumulators was around $1:17\,000$. Screening was found to be a labour-intensive, time-consuming task.

Table 16.3. Number and type of bulk M_2 barley seed screened for nitrite accumulation

Cultivar	Number screened	Number of accumulators identified	Frequency
Golden Promise	118,547	5	1:24,000
Apex	8,421	0	–
Mink	1,234	0	–
Klaxon	41,767	3	1:14,000
Patty	83,693	7	1:12,000

The screening procedure is indicated in the text.

Selected plants require to be maintained on a reduced N source since they would not be expected to be able to assimilate their preferred N source, nitrate. We have maintained them on a hydroponic nutrient system with 1 mM glutamine as sole N source as described previously for the maintenance of mutants defective in nitrate reduction (Bright *et al.* 1983). Nutrient medium is changed every other day. However, not all plants reached maturity. Some of the selections died within 5 weeks of transfer to the hydroponic system, despite the fact that they were able to produce up to 6 leaves. The cause of death is unknown, but is correlated with accumulation of high levels of nitrite during the screen, and with higher levels of microbial contamination of their root systems during hydroponic growth. Initial studies indicate that seedlings derived from selfed-seed of the surviving selections (which accumulated low levels of nitrite) possess only 50 per cent of the wild type level of NiR activity and cross-reacting material. They do not, however, accumulate nitrite under the original isolation (screen) conditions (E. Duncanson and J. L. Wray, unpublished). We are currently examining them further.

In order to circumvent the possible problem of nitrite toxicity we are also screening M_2 seed derived from individual (numbered) ears of M_1 plants. In practice 4–5 seed from each individual M_1 ear are screened essentially as described above. However, the seed is not broadcast into the trays, but placed in numbered squares in a grid laid on top of the vermiculite. In this way the ear from which homozygous, nitrite-accumulating seedlings are derived can be identified. There is then a better than 90 per cent probability that at least one other seed on that ear carries the mutation in the heterozygous state (A. Kleinhofs, personal

Table 16.4. Number and type of M_1 ears screened for nitrite accumulation

Cultivar	Number of M_1 ears screened	Number of M_2 seed screened	Number of M_1 ears carrying an accumulator	Number of M_2 seedling accumulators	Frequency
Tweed	2960	14,398	0	0	—
Doublet	1176	5656	1	1	1:5600
Digger	1448	5504	2	2	1:2750
Natasha	1336	6329	0	0	—
Klaxon	336	1543	1	1	1:1500
Vista	4704	22,293	0	0	—
Golden Promise	6188	28,788	2	4	1:7000
Corniche	2991	10,867	0	0	—

The screening procedure is indicated in the text.

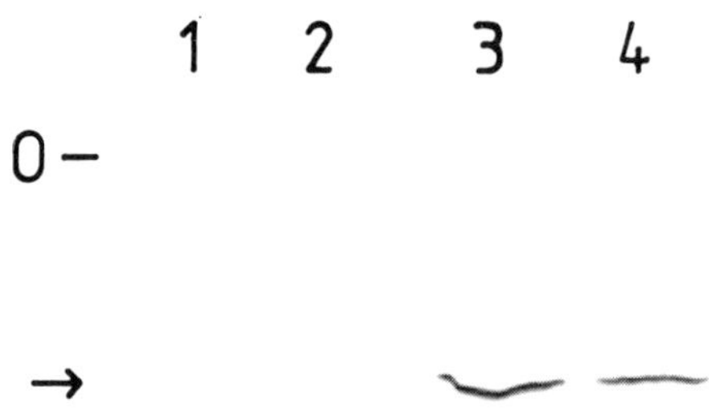

Fig. 16.1. Western blot analysis of NiR-CRM present in leaf tissue of M_2 seedlings derived from M_1 ear, GP/4169, of the barley cultivar, Golden Promise, which either accumulate nitrite (lanes 1 and 2) or do not accumulate nitrite (lanes 3 and 4) after nitrate treatment. Plants were grown as described in the text. Leaf tissue (~ 7 mg) was extracted in 50 μl buffer (50 mM tris-HCl, pH 7.5 containing 10 mM EDTA, 10 mM β-mercaptoethanol and 10 per cent glycerol). After centrifugation at $11,600 \times g$ for 10 min, 40 μl of supernatant was subjected to SDS-PAGE according to Laemmli (1970). After electroblotting, the nitrocellulose filter was developed using specific barley NiR polyclonal antiserum (Ip *et al.*, 1987) as first antibody and goat anti-rabbit IgG–alkaline phosphatase conjugate as second antibody. Substrate was nitroblue tetrazolium/5-bromo-4-chloro-3-indolyl phosphate. $0 = $ electrophoretic origin. The arrow indicates the position, after electrophoresis, of purified barley NiR (63 kDa).

communication). In this way the mutation can be maintained in the heterozygous state. Subsequent analysis can be performed on the homozygous-recessive nitrite accumulating segregants derived from self-fertilization of these heterozygous individuals.

We have screened over 21 000 M_1 ears (over 95 000 M_2 seed) (Table 16.4) so far and have identified six ears which carry a nitrite-accumulator. The results of Western blot analysis of the seedling tissue derived from the seeds screened from one such ear, GP/4169 (generated from the cultivar Golden Promise), are shown in Fig. 16.1. Of the four seeds sown two gave rise to seedlings which accumulated nitrite and which had barely detectable levels of NiR cross-reacting material. The genetic and biochemical basis for the loss of NiR protein in these seedlings is currently under study.

The investigations described above demonstrate for the first time that it is possible to identify whole-plants defective in nitrite reduction by screening M_2 populations for the accumulation of nitrite after nitrate treatment. Further development of these studies, together with others on the molecular cloning of the barley NiR structural gene, will complement other work on the biochemical and molecular genetics of nitrate uptake (Chapter 2) and nitrate reduction (Chapter 13) in barley and together will allow a molecular and genetic characterization of its nitrate assimilation pathway.

References

Anderson, J. W. and Done, J. (1978). Light-dependent assimilation of nitrite by isolated pea chloroplasts. *Plant Physiology* **61**, 692–7.

Back, E., Burkhardt, W., Moyer, M., Privalle, L., and Rothstein, S. (1988). Isolation of cDNA clones coding for spinach nitrite reductase: complete sequence and nitrate induction. *Molecular and General Genetics* **212**, 20–6.

Beevers, L. and Hageman, R. H. (1980). Nitrate and nitrite reduction. In *The biochemistry of plants*, Vol. 5; Amino acids and derivatives (ed. B. J. Miflin), pp. 115–68. Academic Press, New York.

Bright, S. W. J., Norbury, P. B., Franklin, J., Kirk, D. W., and Wray, J. L. (1983). A conditional-lethal *cnx*-type nitrate reductase-deficient barley mutant. *Molecular and General Genetics* **189**, 240–4.

Cammack, R., Hucklesby, D. P., and Hewitt, E. J. (1978). Electron-paramagnetic-resonance studies of the mechanism of leaf nitrite reductase. *Biochemical Journal* **171**, 519–26.

Chua, N.-H. and Schmidt, G. W. (1979). Transport of proteins into mitochondria and chloroplasts. *Journal of Cell Biology* **81**, 461–83.

Coruzzi, G., Broglie, P., Cashmore, A., and Chua, N.-H. (1983). Nucleotide sequence of two pea cDNA clones encoding the small subunit of ribulose 1,5-bisphosphate carboxylase and the major chlorophyll *a/b*-binding thylakoid polypeptide. *Journal of Biological Chemistry* **258**, 1399–402.

Dalling, M. J., Tolbert, N. E. and Hageman, R. H. (1972). Intracellular location of nitrate reductase and nitrite reductase. *Biochimica et Biophysica Acta* **283**, 505–12.

Dalling, M. J., Hucklesby, D. P., and Hageman, R. H. (1973). A comparison of nitrite reductase enzymes from green leaves, scutella and roots of corn (*Zea mays* L.). *Plant Physiology* **51**, 481–4.

Duncanson, E., Wilshin, A., Holyoake, A., Kirk, D. W. and Wray, J. L. (1987). Genetic analysis of nitrite reduction in barley. In *Abstracts of AFRC Meeting on Plant and Soil Nitrogen Metabolism*, P22. AFRC, London.

Ellis, R. J. (1981). Chloroplast proteins: synthesis, transport and assembly. *Annual Review of Plant Physiology* **32**, 111–37.

Gilissen, L. J. W., Barneix, A. J., van Staveren, M., and Breteler, H. (1985). Variant cell lines of *Haplopappus gracilis* with disturbed activities of nitrate reductase and nitrite reductase. *Plant Physiology* **78**, 658–60.

Green, P. J., Kay, S. A., and Chua, N.-H. (1987). Sequence-specific interactions of a pea nuclear factor with light-responsive elements upstream of the *rbc-3A* gene. *EMBO Journal* **16**, 2543–9.

Grossman, A. R., Bartlett, S. G., Schmidt, G. W., Mullet, J. E., and Chua, N.-H. (1982). Optimal conditions for post-translational uptake of proteins by isolated chloroplasts. *In vitro* synthesis and transport of plastocyanin, ferredoxin-NADP$^+$ oxidoreductase and fructose-1,6-bisphosphatase. *Journal of Biological Chemistry* **257**, 1558–63.

Guerrero, M. G., Vega, J. M., and Losada, M. (1981). The assimilatory nitrate-reducing system and its regulation. *Annual Review of Plant Physiology* **32**, 169–204.

Gupta, S. C. and Beevers, L. (1983). Environmental influences on nitrite reductase activity in *Pisum sativum* L. seedlings. *Journal of Experimental Botany* **34**, 1455–62.

Gupta, S. C. and Beevers, L. (1985). Regulation of synthesis of nitrite reductase in pea leaves: *in-vivo* and *in-vitro* studies. *Planta* **166**, 89–95.

Gupta, S. C. and Beevers, L. (1987). Regulation of nitrite reductase. Cell-free translation and processing. *Plant Physiology* **83**, 750–4.

Gupta, S. C., Fletcher, J., and Beevers, L. (1984). Purification and characterisation of nitrite reductase from cell-suspension cultures of Pauls Scarlet Rose and its cross-reactivity to antiserum prepared against pea nitrite reductase. *Zeitschrift für Pflanzenphysiologie* **114**, 321–9.

Heath-Pagliuso, S., Huffaker, R. C., and Allard, R. W. (1984). Inheritance of nitrite reductase and regulation of nitrate reductase, nitrite reductase and glutamine synthetase isozymes. *Plant Physiology* **76**, 353–8.

Hewitt, E. J. and Notton, B. A. (1980). Nitrate reductase systems in eukaryotic and prokaryotic organisms. In *Molybdenum and molybdenum-containing enzymes* (ed. M. Coughlan), pp. 273–325. Pergamon Press, Oxford.

Hirasawa, M. and Tamura, G. (1980). Ferredoxin-dependent nitrite reductase from spinach leaves. *Agricultural and Biological Chemistry* **44**, 749–58.

Hirasawa, M., Fukushima, K., Tamura, G., and Knaff, D. B. (1984). Immunochemical characterisation of nitrite reductases from spinach leaves, spinach roots and other plants. *Biochimica et Biophysica Acta* **291**, 145–54.

Hirasawa-Soga, M., Horie, S., and Tamura, G. (1982). Further characterization of ferredoxin nitrite reductase and the relation between the enzyme and methyl viologen-dependent nitrite reductase. *Agricultural and Biological Chemistry* **46**, 1319–28.

Hirasawa-Soga, M. and Tamura, G. (1981). Some properties of ferredoxin-nitrite reductase from *Spinacia oleracea*. *Agricutural and Biological Chemistry* **45**, 1615–20.

Ho, C.-H. and Tamura, G. (1983). Purification and properties of nitrite reductase from spinach leaves. *Agricultural and Biological Chemistry* **37**, 37–44.

Hucklesby, D. P. (1986). Nitrite reduction in leaf and root. In *Inorganic nitrogen metabolism* (eds. W. R. Ullrich, P. J. Aparacio, P. J. Syrett and F. Castillo), pp. 123–5. Springer-Verlag, Berlin.

Hucklesby, D. P., Dalling, M. J. and Hageman, R. H. (1972). Some properties of two forms of nitrite reductase from corn (*Zea mays* L.) scutellum. *Planta* **104**, 220–3.

Hucklesby, D. P., James, P. M., Banwell, M. J. and Hewitt, E. J. (1976). Properties of nitrite reductase from *Cucubita pepo*. *Phytochemistry* **15**, 599–603.

Huffaker, R. C. (1982). Biochemistry and Physiology of leaf proteins. In *Encyclopaedia of plant physiology*, New series, Vol. 14A (ed. D. Boulter and B. Parthier), pp. 370–400. Springer-Verlag, Berlin.

Ida, S. (1977). Purification to homogeneity of spinach nitrite reductase by ferredoxin–Sepharose affinity chromatography. *Journal of Biochemistry* **82**, 915–18.

Ida, S. (1987). Immunological comparisons of ferredoxin-nitrite reductases from higher plants. *Plant Science* **49**, 111–16.

Ida, S. and Mikami, B. (1986). Spinach ferredoxin–nitrite reductase: a purification procedure and characterization of chemical properties. *Biochimica et Biophysica Acta* **871**, 167–76.

Ida, S. and Morita, Y. (1973). Purification and general properties of spinach leaf nitrite reductase. *Plant and Cell Physiology* **14**, 661–71.

Ida, S., Mori, E., and Morita, Y. (1974). Purification, stabilisation and characterisation of nitrite reductase from barley roots. *Planta* **121**, 213–24.

Ida, S., Kobayakawa, K., and Morita, Y. (1976). Ferredoxin–Sepharose affinity chromatography for the purification of assimilatory nitrite reductase. *FEBS Letters* **65**, 305–8.

Ip, S. M., Kerr, J., and Wray, J. L. (1987). Purification, characterisation and immunology of barley-leaf nitrite reductase. In *Abstracts of the Second International Symposium on Nitrate Assimilation—Molecular and Genetic Aspects*, B21, St. Andrews, Scotland.

Ishiyama, Y. and Tamura, G. (1985). Isolation and partial characterization of homogeneous nitrite reductase from etiolated bean shoots (*Phaseolus angularis* W. P. Wright). *Plant Science Letters* **37**, 251–6.

Ishiyama, Y., Shinoda, I., Fukushima, K., and Tamura, G. (1985). Some properties of ferredoxin–nitrite reductase from green shoots of bean and an immunological comparison with nitrite reductase from roots and etiolated shoots. *Plant Science* **39**, 189–95.

Jefferson, R. A. (1987). Assaying chimeric genes in plants. The GUS gene fusion system. *Plant Molecular Biology Reporter* **5**, 387–405.

Jones, R. W. and Sheard, R. W. (1975). Phytochrome, nitrate movement, and induction of nitrate reductase in etiolated pea terminal buds. *Plant Physiology* **55**, 954–9.

Joy, K. W. (1969). Nitrogen metabolism of *Lemna minor* II. Enzymes of nitrate assimilation and some aspects of their regulation. *Plant Physiology* **44**, 849–53.

Joy, K. W. and Hageman, R. H. (1966). The purification and properties of nitrite reductase from higher plants and its dependence on ferredoxin. *Biochemical Journal* **100**, 263–73.

Karlin-Neumann, G. A. and Tobin, E. M. (1986). Transit peptides of nuclear-encoded chloroplast proteins share a common amino acid framework. *EMBO Journal* **5**, 9–13.

Kelker, H. C. and Filner, P. (1971). Regulation of nitrite reductase and its relationship to the regulation of nitrate reductase in cultured tobacco cells. *Biochimica et Biophysica Acta* **252**, 69–82.

Kuhlemeier, C., Green, P. J., and Chua, N.-H. (1987). Regulation of gene expression in higher plants. *Annual Review of Plant Physiology* **38**, 221–57.

Kuntz, M., Simons, A., Schell, J., and Schreier, P. H. (1986). Targeting of protein to chloroplasts in transgenic tobacco by fusion to mutated transit peptide. *Molecular and General Genetics* **205**, 454–60.

Kutscherra, M., Jost, W., and Schlee, D. (1987). Isoenzymes of nitrite reductase in higher plants-occurrence, purification, properties and alteration during ontogenesis. *Journal of Plant Physiology* **129**, 383–93.

Lancaster, J. R., Vega, J. M., Kamin, H., Orme-Johnson, N. R., Orme-Johnson, W. H., Krueger, R. J., and Siegel, L. M. (1979). Identification of the iron–sulphur centre of spinach ferredoxin–nitrite reductase as a tetranuclear centre and preliminary EPR studies of mechanism. *Journal of Biological Chemistry* **254**, 1268–72.

Laemmli, U. K. (1970). Cleavage of structural proteins during the assembly of the head of bacteriophage T4. *Nature* **277**, 680–8.

Mackown, C. T., Volk, R. J., and Jackson, W. A. (1982). Nitrate assimilation by decapitated corn root systems: effects of ammonium during induction. *Plant Science Letters* **24**, 295–302.

Magalhaes, A. C., Neyra, C. A. and Hageman, R. H. (1974). Nitrate assimilation and amino nitrogen synthesis in isolated spinach chloroplasts. *Plant Physiology* **53**, 411–15.

Mendel, R. R. and Müller, A. J. (1979). Nitrate reductase-deficient mutant cell lines of *Nicotiana tabacum*. Further biochemical characterisation. *Molecular and General Genetics* **177**, 145–53.

Mendel, R. R., Buchanan, R. J., and Wray, J. L. (1984). Characterization of a new type of molybdenum cofactor-mutant in cell cultures of *Nicotiana tabacum*. *Molecular and General Genetics* **195**, 186–9.

Miflin, B. J. (1974). The location of nitrite reductase and other enzymes related to amino acid biosynthesis in the plastids of root and leaves. *Plant Physiology* **54**, 550–5.

Nagaoka, S., Hirasawa, M., Fukushima, K., and Tamura, G. (1984). Methyl viologen-linked nitrite reductase from bean roots. *Agricultural and Biological Chemistry* **48**, 1179–88.

Ninomiya, Y. and Sato, S. (1984). A ferredoxin-like electron carrier from non-green cultured tobacco cells. *Plant and Cell Physiology* **25**, 453–8.

Oaks, A. and Hirel, B. (1985). Nitrogen metabolism in roots. *Annual Review of Plant Physiology* **36**, 345–65.

Oelmuller, R., Schuster, C. and Mohr, H. (1988). Physiological characterisation of a plastidic signal required for nitrate-induced appearance of nitrate and nitrite reductase. *Planta* **174**, 75–83.

Ogawa, M. and Ida, S. (1987). Biosynthesis of ferredoxin nitrite reductase in rice seedlings. *Plant and Cell Physiology* **28**, 1501–8.

Pain, D., Kanawar, Y. S., and Blobel, G. (1988). Identification of a receptor for protein import into chloroplasts and its localisation to envelope contact zones. *Nature* **331**, 232–7.

Paneque, A., del Campo, F. F. and Losada, M. (1963). Nitrite reduction by isolated chloroplasts in light. *Nature* **198**, 90–1.

Rajasekhar, V. K. and Sopory, S. K. (1985). The blue light effect and its interaction with phytochrome in the control of nitrite reductase activity in *Sorghum bicolor* Wild. *New Phytologist* **101**, 251–8.

Rajasekhar, V. K. and Mohr, H. (1986a). Appearance of nitrite reductase in cotyledons of the mustard (*Sinapis alba* L.) seedling as affected by nitrate, phytochrome and photooxidative damage of plastids. *Planta* **168**, 369–76.

Rajasekhar, V. K. and Mohr, H. (1986b). Effect of ammonium and nitrate on growth and appearance of nitrate reductase and nitrite reductase in dark- and light-grown mustard seedlings. *Planta* **169**, 594–9.

Rajasekher, V. K. and Oelmuller, R. (1987). Regulation of induction of nitrate reductase and nitrite reductase in higher plants. *Physiologia Plantarum* **71**, 517–21.

Rao, L. V. M., Rajasekhar, V. K., Sopory, S. K. and Guha-Mukherjee, S. (1981). Phytochrome regulation of nitrite reductase—a chloroplast enzyme—in etiolated maize leaves. *Plant and Cell Physiology* **22**, 577–82.

Reiss, B., Wasmann, C. C., and Bohnert, H. J. (1987). Regions in the transit peptide of SSU essential for transport into chloroplasts. *Molecular and General Genetics* **209**, 116–21.

Robinson, C. and Ellis, R. J. (1984). Transport of proteins into chloroplasts. The precursor of small subunit of ribulose bisphosphate carboxylase is processed to the mature size in two steps. *European Journal of Biochemistry* **142**, 343–6.

Sanderson, G. W. and Cocking, E. C. (1964). Enzymic assimilation of nitrate in tomato plants II. Reduction of nitrite to ammonia. *Plant Physiology* **39**, 423–31.

Sawhney, S. L. and Naik, M. S. (1973). Effect of CM and CH on the synthesis of nitrate reductase and nitrite reductase in rice leaves. *Biochemical and Biophysical Research Communications* **51**, 67–73.

Schiemann, J. and Müller, A. J. (1985). Detection of nitrate reductase cross-reacting material in wild-type and mutant cells of *Nicotiana tabacum*. *Biochemie und Physiologie der Pflanzen* **180**, 63–74.

Schmidt, G. W. and Mishkind, M. L. (1986). The transport of proteins into chloroplasts. *Annual Review of Plant Physiology* **55**, 879–912.

Schrader, L. E. and Hageman, R. H. (1967). Regulation of nitrate reductase activity in corn (*Zea mays* L.) seedlings by endogenous metabolites. *Plant Physiology* **42**, 1750–6.

Schrader, L. E., Beever, L. and Hageman, R. H. (1967). Differential effects of CM on the induction of nitrate and nitrite reductase in green leaf tissue. *Biochemical and Biophysical Research Communications* **26**, 14–17.

Schreier, P. H., Seftor, E. A., Schell, J., and Bohnert, H. J. (1985). The use of nuclear-encoded sequences to direct the light-regulated synthesis and transport of a foreign protein into plant chloroplasts. *EMBO Journal* **4**, 25–32.

Schuster, C., Oelmuller, R., and Mohr, H. (1987). Signal storage in phytochrome action on nitrate-mediated induction of nitrate and nitrite reductases in mustard seedling cotyledons. *Planta* **171**, 136–43.

Serra, J. L., Ibarlucea, J. M., Arizmendi, J. M., and Llama, M. J. (1982). Purification and properties of the assimilatory nitrite reductase from barley (*Hordeum vulgare*) leaves. *Biochemical Journal* **201**, 167–70.

Sharma, R. (1985). Phytochrome regulation of enzyme activity in higher plants. *Photochemistry and Photobiology* **41**, 747–55.

Sharma, A. K. and Sopory, S. K. (1984). Independent effects of phytochrome and nitrate on nitrate reductase and nitrite reductase activities in maize. *Photochemistry and Photobiology* **39**, 491–3.

Sluiters-Scholten, C. M. T. (1973). Effect of CM and CH on the induction of nitrate reductase and nitrite reductase in bean leaves. *Planta* **113**, 229–40.

Small, I. S. and Gray, J. C. (1984). Synthesis of wheat leaf nitrite reductase *de novo* following induction with nitrate and light. *European Journal of Biochemistry* **145**, 291–7.

Smeekens, S., Bauerie, C., Hageman, J., Keegstra, K., and Weisbeek, P. (1986). The role of the transit peptide in the routing of precursors toward different chloroplast compartments. *Cell* **46**, 365–75.

Smeekens, S., van Steeg, H., Bauerle, C., Bettenbroek, H., Keegstra, K., and Weisbeek, P. (1987). Import into chloroplasts of a yeast mitochondrial protein directed by ferredoxin and plastocyanin transit peptides. *Plant Molecular Biology* **9**, 377–88.

Smith, F. W. and Thompson, J. F. (1971). Regulation of nitrate reductase in excised barley roots. *Plant Physiology* **48**, 219–23.

Snell, F. D. and Snell, C. T. (1949). Nitrites. In *Colorimetric methods of analysis* (eds. F. D. Snell and C. T. Snell), pp. 802–7. Van Nostrand, New York.

Stewart, G. R. (1968). The effect of cycloheximide on the induction of nitrate and nitrite reductase in *Lemna minor* L. *Phytochemistry* **7**, 1139–42.

Stoller, M. L., Malkin, R., and Knaff, D. M. (1977). Oxidation–reduction properties of photosynthetic nitrite reductase. *FEBS Letters* **81**, 271–4.

Suzuki, A., Oaks, A., Jacquot, J.-P., Vidal, J., and Gadal, P. (1985). An electron transport system in maize roots for reactions of glutamate synthase and nitrite reductase. *Plant Physiology* **78**, 374–8.

Van den Broeck, G., Timko, M. P., Kausch, A. P., Cashmore, A. P., Van Montagu, M., and Herrera-Estrella, L. (1985). Targeting of foreign proteins to chloroplasts by fusion of the transit peptide from the small subunit of ribulose-1,5-bisphosphate carboxylase. *Nature* **313**, 358–63.

Vega, J. M. and Kamin, H. (1977). Spinach nitrite reductase: Purification and properties of a siroheme-containing iron–sulphur enzyme. *Journal of Biological Chemistry* **252**, 896–909.

Vega, J. M., Cardenas, J., and Losada, M. (1980). Ferredoxin–nitrite reductase. *Methods in Enzymology* **69**, 255–70.

Vijayaraghavan, S. J., Sopory, S. K., and Guha-Mukherjee, S. (1979). Ammonium stimulation of nitrate reductase induction in excised leaves of wheat (*Triticum aestivum*). *Zeitschrift für Pflanzenphysiologie* **93**, 395–402.

Wallsgrove, R. M., Lea, P. J., and Miflin, B. J. (1979). Distribution of the enzymes of nitrogen assimilation within the pea leaf cell. *Plant Physiology* **63**, 232–6.

Warner, R. L., Lin, C. J., and Kleinhofs, A. (1977). Nitrate reductase-deficient mutants in barley. *Nature* **269**, 406–7.

Whitfield, P. R. and Bottomley, W. (1983). Organization and structure of chloroplast genes. *Annual Review of Plant Physiology* **34**, 279–310.

Wray, J. L. (1986). The molecular genetics of higher plant nitrate assimilation. In *A genetic approach to plant biochemistry* (eds. A. Blonstein and P. J. King), pp. 101–57. Springer-Verlag, Vienna, Austria.

Wray, J. L. and Filner, P. (1970). Structural and functional relationships of enzyme activities induced by nitrate in barley. *Biochemical Journal* **119**, 715–25.

17. Structure and function of spinach ferredoxin-nitrite reductase

Lewis M. Siegel and James O. Wilkerson

Introduction

Reduction of nitrate to ammonia occurs in two enzymatic steps. The first step, the 2-electron reduction of nitrate to nitrite, is catalysed by the molybdoenzyme nitrate reductase. The second step, the 6-electron reduction of nitrite to ammonia (reaction 1)

$$NO_2^- + 6e^- + 7H^+ \rightarrow NH_3 + 2H_2O \tag{1}$$

is catalysed by a haem-containing enzyme, usually termed 'assimilatory' nitrite reductase (NiR) (Guerrero *et al.* 1981).

Unfortunately, the term 'assimilatory', as applied to NiR, has come to mean 'ammonia producing' rather than describing a true physiological function. Thus, while the NH_3-producing NiR of higher plants, algae, and fungi do appear to be involved in producing NH_3 from NO_3^- only in quantities sufficient for the biosynthetic needs of the organism (i.e. nitrate assimilation), the NH_3-producing NiR of *Escherichia coli* and a number of anaerobic bacteria actually function in nitrate respiration, NO_2^- serving as an electron sink for excess reducing equivalents produced by fermentative metabolism as NADH or formate (Cole *et al.* 1974; Abou-Jaoudé *et al.* 1979). The latter process results in NH_3 production far in excess of the biosynthetic needs of the organism. Production of the second group of enzymes is regulated by absence of oxygen and presence of NO_3^-/NO_2^- (Cole *et al.* 1974; Griffiths and Cole 1987; Chapter 15), whereas the level of the first group of enzymes is sensitive to the availability of reduced nitrogen compounds as well as to that of NO_3^-/NO_2^- (Guerrero *et al.* 1981). The situation is further complicated by the existence in nature of an entirely distinct group of 'dissimilatory' NiR (generally enzymes containing Cu or *d*-type haems) which also serve in anaerobic nitrate respiration in bacteria. These dissimilatory' NiR catalyse reduction of NO_2^- to the gaseous products NO and N_2O, rather than to NH_3.

Sirohaem (4Fe–4S) and hexahaem nitrite reductases

There appear to be two chemically distinct groups of NH_3-producing NiR in nature. The first type, found in higher plants (Hucklesby *et al.* 1976; Vega and Kamin 1977; Cammack *et al.* 1978; Hirasawa and Tamura 1980; Ida and Mikami 1986; Hirasawa *et al.* 1987), algae (Zumft 1972; Ho *et al.* 1976; Romero *et al.* 1987), fungi (Vega *et al.* 1975; Prodouz and Garrett 1981), and in *E. coli* (Jackson *et al.* 1981), contains the novel haem prosthetic group termed 'sirohaem' (Fig. 17.1) (Murphy *et al.* 1974) as well as iron–sulphur (Fe/S) centres, generally of the 4Fe–4S type (Lancaster *et al.* 1979), although the *E. coli* NADH-NiR may contain a 2Fe–2S centre (Cammack *et al.* 1982). The plant and algal enzymes contain no other prosthetic groups, and utilize reduced ferredoxin (Fd_r) as their physiological electron donor. The fungal and *E. coli* enzymes utilize NADPH and NADH, respectively, as physiological electron donors, and contain FAD in addition to sirohaem and Fe/S centres.

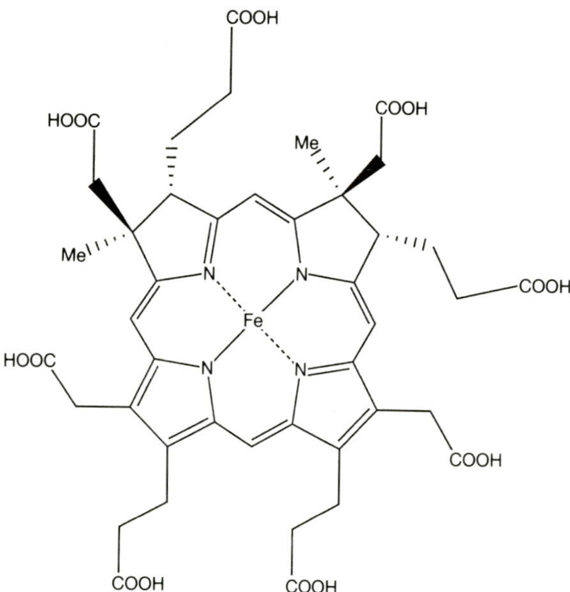

Fig. 17.1. Structure of sirohaem (after Scott, *et al.* 1978).

The second group of NH_3-producing NiR, found in *E. coli* and a number of other facultative or obligate anaerobic bacteria, contains six *c*-type haem groups on a single polypeptide chain (Liu and Peck 1981; Liu *et al.* 1983, 1988; Blackmore *et al.* 1986; Kajie and Anraku 1986). The hexahaem NiR may be membrane-associated and probably derive their electrons from NADH or formate via the bacterial respiratory chain (Abou-Jaoudé *et al.* 1979). All of the hexahaem NiR function in nitrate respiration. The sirohaem-Fe/S NiR generally function in nitrate assimilation, although the *E. coli* NADH-NiR is clearly involved in anaerobic nitrate respiration (Cole *et al.* 1974). It is of interest that both types of NH_3-producing NiR are found in anaerobically-grown *E. coli* (Cole 1982).

General catalytic properties of nitrite reductases

Whatever their physiological electron donors, all NH_3-producing NiR can be studied with reduced viologens as the source of the six electrons in reaction 1; reduced methylviologen (MV^+) and benzylviologen (BV^+), with redox potentials similar to Fd_r and NAD(P)H, respectively, have been widely utilized. The low-potential reductant dithionite ($S_2O_4^{2-}$) can also serve as an electron donor, although with much lower catalytic efficiency than the reduced viologens.

All NH_3-producing NiR can reduce NH_2OH as well as nitrite to NH_3. This phenomenon led to early speculation that NH_2OH, in solution, was a true intermediate in reaction 1. However, in all NH_3-producing NiR, the K_m for NH_2OH (generally $\sim 10^{-2}$ M) are about 2–3 orders of magnitude higher than the K_m for nitrite (generally $\sim 10^{-5}$ M when assayed in the absence of dithionite, *vide infra*), while the V_{max} values with NH_2OH are generally of the same magnitude as those with NO_2^-. Thus it is kinetically impossible for 'free' NH_2OH to serve as an intermediate, although it is reasonable to suspect that a tightly bound form of N at the oxidation level of NH_2OH is an intermediate in the 6-electron reduction of NO_2^- to NH_3.

Spinach NiR has been shown to catalyse the 6-electron reduction of sulphite to sulphide (reaction 2) in addition to NO_2^- and NH_2OH reduction (Krueger and Siegel 1982).

$$SO_3^{2-} + 6e^- + 7H^+ \rightarrow HS^- + 3H_2O \tag{2}$$

The K_m for SO_3^{2-} ($\sim 10^{-3}$ M) with NiR is much higher than that for NO_2^-, while the V_{max} with SO_3^{2-} is much less. The ability of most NH_3-producing NiR to reduce SO_3^{2-} has not been carefully examined, but the fact that in a number of cases apparent K_m values for NO_2^- determined in assay mixtures containing $S_2O_4^{2-}$ and MV^+ are much greater than K_m values obtained when MV^+ reduced by other means is employed (compare values obtained by Krueger and Siegel 1982, with those of Ida and Mikami 1986), suggests that there is a widespread affinity of NiR for compounds related to $S_2O_4^{2-}$ at the NO_2^- binding site. The principal such compound generated in solutions which are oxidizing $S_2O_4^{2-}$ is SO_3^{2-}.

Spinach NiR, with its ability to catalyse reduction of both SO_3^{2-} and NO_2^-, as well as NH_2OH, exhibits the same range of specificity as a second group of sirohaem-(4Fe–4S) enzymes, the assimilatory sulphite reductases (SiR). In contrast to spinach NiR, the SiR from spinach (Krueger and Siegel 1982) and *E. coli* (Siegel *et al.* 1974, 1982) exhibit much higher affinity for SO_3^{2-} than for NO_2^-, and the SiR can attain much higher V_{max} values for SO_3^{2-} reduction (but far lower V_{max} for NO_2^- reduction) than can spinach NiR. However, both types of enzyme, when saturated with substrate, yield higher turnover numbers with NO_2^- than with SO_3^{2-}. These findings suggested that study of the active centre of SiR could yield information which might be pertinent to an understanding of the mechanistic basis for catalysis in the catalytically similar sirohaem-(4Fe–4S) NiR enzymes.

Chemical structure of ferredoxin-nitrite reductases

Ferredoxin-NiR, as isolated in most laboratories from spinach and other higher plants, is a single polypeptide chain of molecular weight 63 kDa (reported range 60–63 kDa) (Vega and Kamin 1977; Ida and Mikami 1986). The gene for spinach NiR has been cloned and sequenced (Back *et al.* 1988; Chapter 18) and shows a 66 kDa open reading frame, which seems to consist of a 3 kDa 'signal' sequence followed by the 63 kDa region which corresponds to the catalytically active NiR itself. The signal sequence is presumably involved in targeting NiR to the chloroplast and is cleaved once the enzyme has achieved this location. The 63 kDa NiR is catalytically active with both Fd_r and MV^+ as electron donors for NO_2^- reduction. The ratio of V_{max} values with Fd_r *vs* MV^+ has been reported variously as 1.0 (Fry *et al.* 1982; Ida and Mikami 1986; we agree with this value) and 3.4 (Krueger and Siegel 1982). The 63 kDa NiR contains one mol of sirohaem and one mol of 4Fe–4S cluster per 63 kDa of enzyme (Ida and Mikami 1986). (Some workers who have isolated the 63 kDa form of NiR have found that it contains an incomplete complement of prosthetic groups; thus Lancaster *et al.* (1979) reported that their preparation of spinach NiR contained 0.5–0.6 sirohaem and 4Fe–4S (both measured chemically and by EPR) per 61 kDa of protein, as well as a similar number of NO_2^- and cyanide binding sites.)

Hirasawa *et al.* (1984, 1987; Hirasawa and Tamura 1980; Hirasawa-Soga *et al.*, 1982) have isolated an 86 kDa form of NiR from spinach leaves. This form of NiR is composed of a dissociable 24 kDa subunit and the 63 kDa NiR polypeptide isolated in other laboratories. The 86 kDa form of NiR is reported to have a greater specificity for Fd_r (as opposed to MV^+) than the 63 kDa form of the enzyme (Hirasawa-Soga *et al.* 1982). An 86 kDa NiR composed of 25 and 63 kDa subunits has also been isolated from *Chlamydomonas reinhardtii* by Romero *et al.* (1987). This enzyme, with a Fd_r/MV^+ nitrite reduction activity ratio of ~ 1.0, loses its 25 kDa subunit upon storage at 4°C for one week. After rechromatography on a gel filtration column, the resulting 63 kDa NiR exhibits a Fd_r/MV^+ nitrite reduction activity ratio of 0.58. Analytical data provided by Hirasawa *et al.* (1987) for the 86 kDa and 63 kDa forms of spinach NiR suggest that the former may contain more than one sirohaem. The analytical results obtained by chemical, EPR, and ligand binding measurements by Hirasawa *et al.* (1987) are widely divergent, however, and this conclusion must be considered problematic at this time, particularly in view of the fact that Romero *et al.* (1987) have reported no more than one sirohaem per mol of the 86 kDa *Chlamydomonas* NiR. Although the role of the 24 kDa subunit in enhancement of specificity for Fd_r in NiR remains to be determined, it must be remembered that the 63 kDa form of NiR, which is the object of the detailed spectroscopic studies to be reported below, is highly active catalytically in catalysing the 6-electron reduction of NO_2^- to NH_3 (with *both* Fd_r and MV^+ as electron donors) and therefore must contain the active site for binding and reduction of NO_2^-.

The sirohaem (4Fe–4S) active centre of *E. coli* sulphite reductase

The haemoprotein subunit of *E. coli* sulphite reductase (SiR-HP) contains one sirohaem and one 4Fe–4S cluster per ~ 60 kDa polypeptide chain (Siegel *et al.* 1982). This subunit, which exists as a monomer in solution, catalyses the 6-electron reductions of both NO_2^- and SO_3^{2-} when provided with MV^+ as electron donor. Thus, SiR-HP shares important properties with ferredoxin-NiR. (In intact *E. coli*, the SiR-HP subunit exists as part of a complex with a flavoprotein, termed SiR-FP, which accepts electrons from the physiological electron donor, NADPH, and transfers them to the site of SO_3^{2-}/NO_2^- reduction on SiR-HP (Siegel *et al.* 1973, 1974; Siegel and Davis 1974)). Spinach leaves contain a ferredoxin-linked sulphite reductase which is similar in many of its properties to the *E. coli* SiR-HP subunit, but exists in solution primarily as a dimer of 69 kDa polypeptide chains, each of which contains one sirohaem and one 4Fe–4S centre (Krueger and Siegel 1982).

The active centre of SiR-HP has been extensively characterized spectroscopically; an amino acid sequence has been obtained by isolation and sequencing of the *E. coli* gene coding for SiR-HP (Ostrowski *et al.* 1987); and a preliminary X-ray crystal structure of SiR-HP in the oxidized state (SiR-HP^0) has been obtained at 3.0 Å resolution (McRee *et al.* 1986). As will be shown below, the spectroscopic signatures of the active centres of SiR-HP and spinach NiR share so many properties that it is likely that the structural arrangement of the prosthetic groups at the active centre is similar, if not identical, in the two enzymes. A schematic diagram of the active centre of SiR-HP^0 is shown in Fig. 17.2, while a space-filling model for the possible arrangement of atoms in this centre is shown in Fig. 17.3.

The X-ray crystal structure for SiR-HP^0, although obtained only at 3.0 Å resolution, shows that the sirohaem Fe is located 4.4 Å from the nearest cluster Fe. These two Fe atoms appear to be bridged by a common ligand, the electron density of which is consistent with a cysteine S (or possibly a serine O) deriving from the polypeptide chain. (A bridging N atom is eliminated by ENDOR data (Cline *et al.* 1985), which shows evidence for only the sirohaem pyrrole and pyrroline nitrogens as N ligands to the sirohaem Fe.) The 4Fe–4S cluster cubane occupies approximately 50% of one face of the sirohaem macrocycle, and the two prosthetic groups are situated so close together that one of the cubane S atoms (adjacent to the cluster Fe bridged to the sirohaem Fe) is in van der Waal's contact with the edge of the sirohaem ring. There are thus two possible pathways for rapid

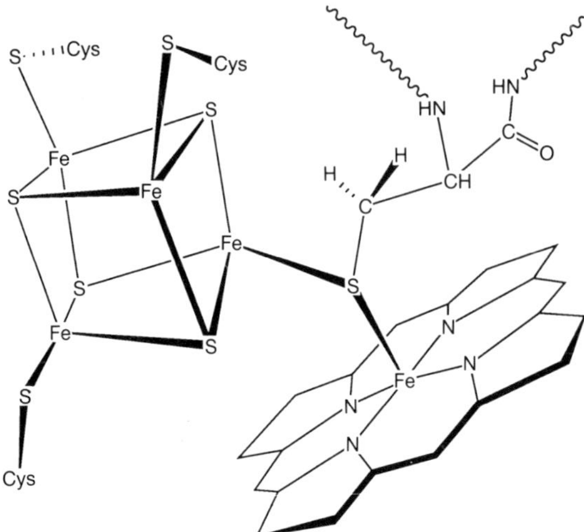

Fig. 17.2. Schematic model of the active centre of oxidized *E. coli* sulphite reductase haemoprotein based on the X-ray crystallographic data of McRee *et al.* (1986) and the electron-nuclear double resonance data of Cline *et al.* (1985). The atomic distances are not drawn to scale. It is assumed that the bridging ligand is a cysteine S rather than a serine O.

inner sphere electron transfer between the 4Fe–4S cluster and the sirohaem in SiR-HP, one through the haem Fe (*via* the bridging cysteine S ligand), and one through the ring π-electron system (*via* a cubane S). In oxidized SiR-HP, the sixth coordination position of the sirohaem appears to be unoccupied and is relatively exposed to solvent. It is likely, then, that reducible substrates bind to the sirohaem on the side of the ring distal to the 4Fe–4S cluster.

Resonance Raman spectra of SiR-HP^0 (J. F. Madden, S. Han, L. M. Siegel, and T. G. Spiro, submitted) show evidence for two S-isotope-sensitive Fe-S vibrations, at 356 and 395 cm^{-1}, excited by laser irradiation in the sirohaem Soret band region. These bands do not appear to be due to the 4Fe–4S cluster itself and may be accounted for by the symmetric and assymetric stretches of the Fe_{haem}–S–$Fe_{cluster}$ bridge. The cluster vibrations, which can be excited independently of the sirohaem vibrations by irradiation of SiR-HP^0 at 457 nm (near a minimum in the sirohaem absorption spectrum), are atypical for 4Fe4S clusters. The SiR-HP^0 cluster vibrations are in fact similar to those reported for the 3Fe–4S centre of aconitase, a result which indicates that one of the cluster Fe atoms in SiR-HP^0 is in a distinctly different environment from the other three cluster Fe atoms. When complexes of SiR-HP with haem ligands such as NO, CO, or CN$^-$ were studied by resonance Raman spectroscopy, it was found that the cluster vibrations were similar to those in SiR-HP^0, a result which again suggests that these ligands do not bridge the sirohaem Fe and the cluster (displacing the endogenous bridging ligand), but rather must bind to the side of the sirohaem *distal* to the cluster.

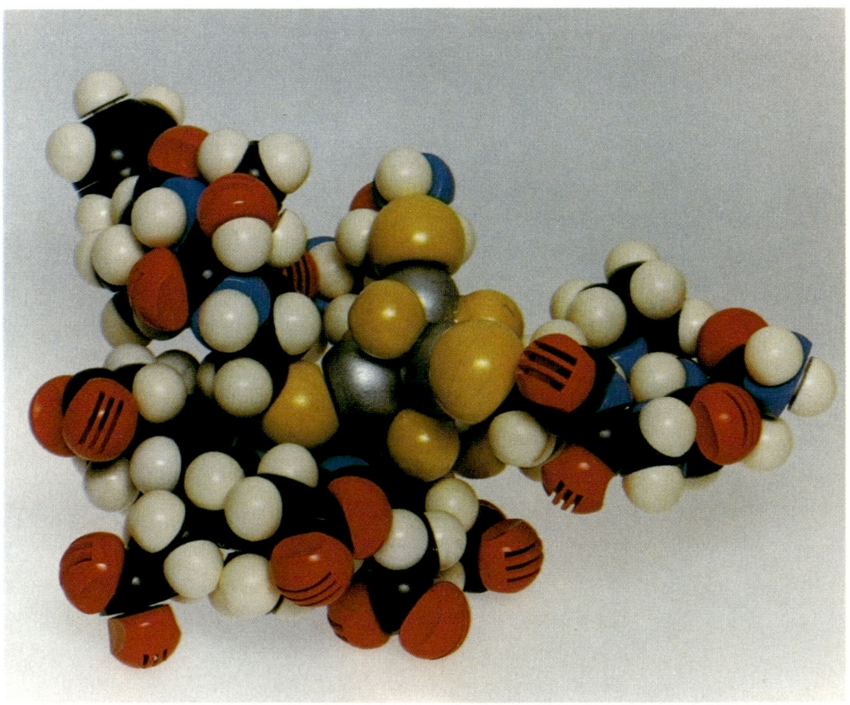

Fig. 17.3. Space-filling model of the proposed active-site structure of oxidized *E. coli* sulphite reductase haemoprotein. Sirohaem lies at the bottom of the figure, with the plane of the ring oriented approximately perpendicular to the page. The ring is assumed to be planar, and several peripheral carboxylate substituents are visible. The bridging cysteinyl sulphur atom is at left centre of the figure (yellow), and the 4Fe–4S cluster is above and slightly to the right of centre of the sirohaem ring. Portions of the polypeptide chain containing the proposed cluster cysteinyl ligands are shown to the right and left of the cluster. Their orientation is not yet fixed in the crystallographic structure. Note the close contact between one of the cluster S atoms and the sirohaem macrocycle.

Mossbauer spectroscopy of SiR-HP^0 shows that the spectrum of the formally diamagnetic (S = 0) (4Fe–4S)$^{+2}$ cluster is magnetically split by the paramagnetic (S = 5/2) high spin Fe^{III} of the sirohaem (Christner *et al.* 1981). This result requires electronic overlap between the two prosthetic groups, as implied in Fig. 17.2 and 17.3, since through-space (dipolar) splitting could occur only if *both* the cluster and the haem were paramagnetic (S $\neq$ 0). It is of interest that magnetic splitting (or broadening) of the S = 0 4Fe–4S cluster Mossbauer spectrum by a paramagnetic haem persists in a variety of complexes of SiR-HP with haem ligands, such as SiR-HP^0-CN$^-$, SiR-HP^0-S^{2-}, and SiR–HP^{1-}–NO, where in all cases the sirohaem contains low spin (S = $\frac{1}{2}$) Fe^{III} (Christner *et al.* 1983*a,b*, 1984). Thus ligation of the haem does not break its electronic overlap with the cluster. Mossbauer (Christner *et al.* 1983*a*, 1984), ENDOR (Cline *et al.* 1986) and EPR studies (Janick and Siegel 1982, 1983) also indicate that the exchange coupling between the prosthetic groups is maintained in SiR-HP reduced by one and two electrons, respectively. These results indicate that the close interaction between the sirohaem and 4Fe4S cluster shown in Fig. 17.2 and 17.3 for SiR-HP^0 is maintained in a wide variety of reduction and ligation states of the enzyme; it is therefore probably also of significance during the catalytic cycle of the enzyme.

The sirohaem–(4Fe–4S) active centre of spinach nitrite reductase

Although far fewer spectroscopic studies have been performed with NiR than with SiR-HP, and no crystallographic data are available for NiR, several lines of evidence indicate that the type of active centre pictured in Fig. 17.2 and 17.3 for SiR-HP^0 is also present in spinach NiR^0 (Siegel *et al.* 1987).

Mossbauer spectroscopy

NiR^0 exhibits Mossbauer spectra at both 4°K and 190°K almost identical to those of SiR-HP^0, indicating that the sirohaem contains high spin Fe^{III} and the cluster is in the + 2 oxidation state (two Fe^{II} and two Fe^{III}) in both enzymes, and that there is electronic overlap between cluster and sirohaem in both cases. EPR spectra of both enzymes (Lancaster *et al.* 1979; Siegel *et al.* 1982) show the presence of high spin Fe^{III} haem in a relatively rhombic environment.

ENDOR spectroscopy

This type of spectroscopy can identify atoms with non-zero nuclear spin which interact with the paramagnetic Fe atoms responsible for an EPR signal. When an epr spectrum is highly anisotropic, as is the case with the Fe(III) sirohaem signal of SiR-HP^0 and NiR^0, information on axial ligands can be obtained in favourable situations. Both enzymes show nearly identical ^{1}H hyperfine couplings to the sirohaem Fe^{III} in the g_z direction of the EPR spectrum (i.e. perpendicular to the plane of the haem). The strongest of these couplings are consistent with those expected for β-CH_2 and α-CH protons of a bridging cysteine (or serine) ligand bound to the sirohaem Fe (Cline *et al.* 1985; McRee *et al.* 1986; Siegel *et al.* 1987). Neither enzyme exhibits ^{1}H ENDOR couplings which would be expected for an H_2O ligand to the haem Fe, nor ^{14}N couplings expected for non-pyrrole or pyroline N-ligands. These results suggest that a 5-coordinate high spin Fe^{III} sirohaem bridged through a cysteine S atom to the 4Fe–4S cluster, as shown in Fig. 17.2 and 17.3, is characteristic of NiR^0 as well as SiR-HP^0.

Novel EPR spectra in fully reduced nitrite and sulphite reductases

Evidence for the persistence of exchange coupling between sirohaem and the 4Fe–4S cluster in both enzymes when both prosthetic groups have been fully reduced is derived from the novel EPR spectra exhibited by both SiR-HP^{2-} (Janick and Siegel 1982, 1983) and NiR^{2-} (Wilkerson *et al.* 1983). These spectra are due to magnetic perturbation of the $S = \frac{1}{2}$ $(4Fe$–$4S)^{+1}$ cluster, which normally would be expected to give a '$g = 1.94$' type of EPR signal, by $S = 1$ or 2 ferrohaem. Thus, the predominant EPR species observed in NiR^{2-} exhibits $g_{x,y,z} \simeq 5$, 3, and > 2.0, and is plausibly explained by exchange interaction between high spin ($S = 2$) Fe^{II} sirohaem and the reduced ($S = \frac{1}{2}$) 4Fe–4S cluster. (This '$g_1 = 5$' type of EPR species is also seen in spectra of SiR-HP^{2-} as a minority species. Addition of small amounts of weak field haem ligands, such as Cl^-, or of small concentrations of certain chaotropes, such as guanidinium salts, to SiR-HP^{2-}, converts most of the enzyme into a state which yields the '$g_1 = 5$' type of epr species). The predominant EPR species observed in SiR-HP^{2-} (in phosphate buffer) exhibits three g values in the range > 2.0 to 2.7. A similar type of EPR signal is seen as a minority species in spectra of NiR^{2-} in phosphate buffer. Christner *et al.* (1974) have postulated, from the position of the Mossbauer parameters of Fe^{II} sirohaem in reduced SiR-HP, that the Fe^{II} sirohaem in this type of EPR species may be in the unusual $S = 1$ state. This type of spin state has not previously been observed in haemoproteins; it is normally seen in model haem compounds only when Fe^{II} haem has no axial ligands. If this were the case in SiR-HP^{2-} and NiR^{2-} as well, it would suggest that the Fe_{haem}–S–$Fe_{cluster}$ bridge may be broken in the enzyme species which give rise to this type of EPR signal. However, the persistence of exchange coupling, needed to explain perturbation of the $S = \frac{1}{2}$ reduced cluster EPR signal by the $S = 1$ ferrohaem, suggests that the interaction between the cluster cubane S and the sirohaem ring persists in this form of reduced enzyme. That these unusual EPR spectra are in fact due to magnetic perturbation of the reduced cluster spectrum by a paramagnetic haem is demonstrated by the fact that complexation of the Fe^{II} sirohaem with strong field ligands, such as CN^- or CO, which converts the haem to the low spin $S = 0$ state, causes both SiR-HP^{2-} (Siegel *et al.* 1982; Janick and Siegel 1983) and NiR^{2-} (Aparicio *et al.* 1975; Wilkerson *et al.* 1983) to yield EPR spectra of the classical '$g = 1.94$' type.

Possible ligands to the 4Fe–4S centres of nitrite and sulphite reductases

If one assumes that the structure of the sirohaem/4Fe–4S cluster active centre is like that shown in Fig. 17.2 in both SiR-HP and NiR, as suggested by the spectroscopic and crystallographic data already cited, a comparison of the amino acid sequences of the two enzymes, obtained from the cloned gene sequences (and confirmed with partial protein sequence data) (Ostrowski *et al.* 1987; Back *et al.* 1988) becomes instructive. Although there is only a moderate degree of homology in the overall sequences of *E. coli* SiR-HP and spinach NiR, a comparison of those sequences in the two enzymes which contain cysteine residues has yielded information as to the probable ligands of the 4Fe–4S clusters. Each cluster can be expected to contain four cysteine ligands, one for each of the cubane Fe atoms. Although there are 12 cysteine residues in NiR and six in SiR-HP, only four of these cysteines are found in sequences which show any degree of homology in the two enzymes. These cysteine-containing sequences can be arranged into two groups, each containing two of the four cysteine residues:

Sequence A:
    NiR: met-his-tyr-thr-gly-*cys*-pro-asn-ser-*cys*-gly
    SiR-HP: met-arg-val-thr-gly-*cys*-pro-asn-gly-*cys*-gly

Sequence B:
    NiR: ala-*cys*-thr-gly-ser-trp-phe-*cys*-gly-trp-ala-ile-ile-glu
    SiR-HP: ala-*cys*-val-ser-phe-pro-thr-*cys*-pro-leu-ala-met-ala-glu

The A sequences are located toward the C-termini in both enzymes and show near identity in the octapeptide containing the two cysteine residues. This type of sequence predicts a very sharp turn in the polypeptide chain between the two cysteines, and such a 'hairpin' turn has recently been detected in the crystallographic structure of SiR-HP⁰ between the cysteine residues ligated to the 4Fe–4S cluster irons most distal to the sirohaem macrocycle. Thus, two of the four potential cysteine ligands of the cluster have been identified (although the direction of the polypeptide chain in this region is uncertain).

The B sequences are located 32–36 residues toward the N-terminus from sequence A in both SiR-HP and NiR. Both B sequences contain cysteines separated by five amino acid residues, and contain identical residues immediately preceding the first cysteine and three and six residues, respectively, from the second cysteine. The second type of identity at three residue intervals may suggest a common set of helical structure contacts in the two enzymes. However, it is striking that there is in fact little homology between the two enzymes in the five amino acid residues which connect the two cysteines of B sequences. If one considers that one of the two cysteine residues in sequence B is the likely bridging ligand between the sirohaem and 4Fe4S cluster, this lack of homology is indeed puzzling. This result may be explained if one recognizes that the sirohaem macrocycle and the 4Fe–4S cluster are in fact very closely packed (see Fig. 17.3) in the crystal structure of SiR-HP^0. It may be that the purpose of the five residue sequence joining the cysteines ligated to the cluster Fe atoms nearest the sirohaem macrocycle may in fact be to form a loop of amino acids which simply must 'get out of the way' of the closely apposed prosthetic groups. If the substrate binding site is on the opposite side of the sirohaem macrocycle from the 4Fe4S cluster, this loop of amino acids might be expected to play little or no role in the catalytic process other than to permit the appropriate degree of steric interaction between the prosthetic groups. A possible model of the active centre of SiR-HP^0 (and by inference NiR^0), using the amino acid sequence data, together with the previously cited crystallographic and spectroscopic information, is shown in Fig. 17.3.

Why sirohaem?

Sirohaem (Fig. 17.1) has been isolated from a variety of SiR (Murphy and Siegel 1973; Murphy *et al.* 1973*a,b*; Siegel *et al.* 1973) as well as from several NiR. Its porphyrin macrocycle, termed sirohydrochlorin, is unusual in that it contains two adjacent pyrrole rings which are partly saturated (Murphy *et al.* 1973*b*; Deeg *et al.* 1977; Battersby *et al.* 1978; Scott *et al.* 1978). This type of macrocycle is formally termed an isobacteriochlorin (iBC). Since SiR and NiR are the only known enzymes which contain sirohaem, it is reasonable to ask why nature has selected this novel type of compound for this purpose. Studies with model iBC have shown that metal-centred redox properties are not very different in metallo-porphyrins, chlorins, and iBC (Stolzenberg *et al.* 1981; Chang 1982). The utility of sirohaem for catalysis of the multielectron reductions of NO_2^- and SO_3^{2-} must lie, therefore, in other properties of the iBC macrocycle.

One possibility is that the necessity for sirohaem is due to steric considerations which arise from the catalytic importance of the closely apposed haem–4Fe–4S structure in these two enzymes. First, it is truly difficult to position a 4Fe–4S cluster Fe 4.4 Å away from a haem Fe, particularly in the presence of a covalent bridging ligand, without having extreme crowding between one or more cubane atoms and the haem macrocycle. As it is, a very large Fe_{haem}–S–$Fe_{cluster}$ bond angle of $\sim 140°$ is required to permit a cysteine S to serve as the bridging ligand, even with a rather long Fe_{haem}–S bond distance of ~ 2.5 Å (McRee *et al.* 1986). As seen in Fig. 17.3, there is no way to avoid a direct (within the van der Waals' radii of the atoms) contact between one of the cubane S atoms and the sirohaem macrocycle itself when Fe atoms of the two prosthetic groups are constrained to be so close to one another. (This contact, as stated previously, may be useful to provide a second pathway of electron transfer between the enzyme prosthetic groups). 'Normal' haems, with fully oxidized porphyrin macrocycles, are nearly planar, whereas iBC like sirohaem can easily form ruffled macrocycles (Kratky *et al.* 1981; Barkagia *et al.* 1982; Suh *et al.* 1984). Such a bending of the sirohaem macrocycle about the central Fe atom may be essential to permit the close approach of the sirohaem Fe to the cluster seen in the SiR-HP^0 crystal structure. Such a close approach could obviously facilitate rapid electron transfer between the centres. Such rapid electron transfer can be expected to be essential to efficient catalysis of multielectron transfer reactions, such as SO_3^{2-} and NO_2^- reduction, without release of labile intermediates from the enzyme active centre into solution.

A second property of iBC that may be relevant to the multielectron reduction process is the ease with which they can be oxidized to form π-cation radicals (Richardson *et al.* 1979; Stolzenberg *et al.* 1980; Chang *et al.* 1981; Fujita and Fajer 1983; Fajer *et al.* 1985). Thus, several groups have shown that Fe^{II} iBC ligated with CO can be oxidized to the Fe^{II}–CO–iBc π-cation radical more easily than the Fe^{II} can be oxidized to Fe^{III} in the complex. Young and Siegel (1988) have recently detected near stoichiometric formation of Fe^{III} sirohaem π-cation radicals upon treatment of either free SiR-HP^0 or its sulphite complex with the oxidant porphyrexide. We have also detected oxidation of the nitrite complex of spinach NiR^0 to a species which yields an optical spectrum characteristic of a Fe^{III} sirohaem π-cation radical upon treatment with excess ferricyanide. In all cases, these π-cation radical species of enzyme-bound sirohaem can be reduced, indicating that these oxidations are not irreversible. (In the case of the SiR-HP Fe^{III} sirohaem π-cation radical species, dihydroascorbate reduces these to the Fe^{III} sirohaem state characteristic of the native enzyme and its sulphite complex, respectively. In the case of the NiR-NO_2^- complex π-cation radical species, dithionite reduces this to the NiR Fe^{II} sirohaem-NO complex). Although we have no evidence yet as to whether such 'superoxidized' states of sirohaem are involved in the catalytic cycles of SiR and NiR, it is tempting to consider that donation of an 'extra' electron to partially reduced substrates during the course of a 6-electron reduction process could facilitate the rate of the overall reduction reaction (*vide infra*). In this regard, the close apposition to the haem of a 4Fe4S cluster cubane S atom with its unshared π-electron pair directed toward the sirohaem macrocycle π-electron system could be expected to contribute extra stability to a sirohaem π-cation radical in SiR and NiR.

Redox properties of the sirohaem and 4Fe–4S prosthetic groups

The overall reduction of NO_2^- to NH_3 is a thermodynamically favourable process, with $E'_0 = +0.32$ V (*vs* SHE) for the average electron transfer. Since electrons are derived from physiological donors (e.g. Fd$_r$, $E'_0 = -0.45$ V; NADPH, $E'_0 = -0.32$ V) of much greater reducing power than this, the process should be limited only by kinetic factors or specific thermodynamically difficult electron transfer steps involving N compounds intermediate in oxidation state between NO_2^- and NH_3. The redox potentials of sirohaem (Fe^{III}/Fe^{II}) and the 4Fe–4S cluster

$(+2/+1)$ estimated for unligated SiR-HP $(E'_{0,haem} = -0.34$ V; $E'_{0,cluster} = -0.41$ V) (Janick and Siegel 1982) indicate that these groups are able to donate electrons at a thermodynamic pressure roughly equivalent to that of the types of electron donors normally used to assay nitrite or sulphite reduction. There is evidence that the SiR-HP sirohaem and cluster potentials may be significantly altered when the sirohaem is ligated with e.g. CN^-, CO, or S^{2-} (Siegel *et al.* 1982; Christner *et al.* 1984). The redox potentials reported in the literature for spinach (Stoller *et al.* 1977) and *Cucurbita* (Cammack *et al.* 1978) NiR suggest that the sirohaem potential is much more positive in this enzyme (reported $E'_{0,haem} = -0.05$ to -0.12 V) than in SiR-HP, while the potential for the 4Fe–4S cluster may be much more negative $(E'_{0,cluster} \simeq -0.55$ V). The potentials given for the cluster may not be relevant to the unligated enzyme, however, since dithionite was used as reductant (which can lead to generation of the sirohaem ligands SO_3^{2-} and S^{2-}) and only a '$g + 1.94$' type of EPR signal, not characteristic of the unligated enzyme, was measured.

We have recently re-examined the redox potentials of the prosthetic groups of spinach NiR in phosphate buffer at pH 7.7, and have found that the sirohaem $E'_0 = -0.39 \pm 0.03$ V (measured by loss of the high spin Fe^{III} sirohaem EPR signal) and the cluster $E'_0 = -0.45 \pm 0.03$ V (the '$g_1 = 5$' and $g_1 = 2.4$ EPR species behaved identically in the potentiometric titration). Thus, in our hands, the redox potentials of the prosthetic groups in SiR-HP and NiR are very similar. (We at present have no explanation for the much more positive sirohaem potentials obtained by previous workers.)

The Fe(II)sirohaem-NO complex as an intermediate in nitrite reduction

Potentiometric titrations conducted with the NiR^0-NO_2^- complex show that this complex is reduced to a species yielding the Fe^{II}sirohaem-NO EPR signal at an $E'_0 = +0.10$ V, and that this species undergoes apparent multielectron reduction to yield the unligated Fe^{III}sirohaem NiR at a very sharp redox potential of -0.26 V. Thus, the thermodynamically difficult step or steps in the reduction of NO_2^- to NH_3 as catalysed by NiR would appear to occur during the reduction of bound NO to NH_3. This result is consistent with the finding that the predominant enzymatic species present during turnover when either NiR (Cammack *et al.* 1978, 1982; Lancaster *et al.* 1979; Fry *et al.* 1980; Hirasawa-Soga *et al.* 1983) or SiR-HP (Janick and Siegel 1983) are mixed with NO_2^- and reductants is in fact the Fe^{II}sirohaem-No complex in which the 4Fe–4S cluster remains oxidized $(+2$ state). This species is the enzyme form which persists in solution whenever turnover occurs in the presence of reductants and excess NO_2^-. These results suggest that addition of the next electron to what is in effect a two-electron reduced enzyme–NO_2^- complex (one electron is on the haem, the other is on a dehydrated nitrite) is likely to be the rate limiting step in NO_2^- reduction to NH_3.

Mechanism of the non-enzymatic reduction of nitrite to ammonia

Insights into possible 'bottlenecks' in the process of NO_2^- reduction to NH_3 have come from consideration of studies of the non-enzymatic reduction process as catalysed by a variety of metals, including model haem compounds. (It must be recognized that this process need not occur without release of intermediates, and that complex reactions involving two or more bound or free nitrogen species may occur, while such reactions are unlikely in the enzyme-catalysed reaction). Murphy *et al.* (1982) and Bottomley and Mukaida (1982) found that Fe, Os, and Ru complexes containing bound NO^+ (dehydrated NO_2^-) and NO ligands could be stoichiometrically reduced to bound NH_3 *via* a sequence of intermediates. A number of simple complexes are capable of electrocatalytically reducing NO to NH_3, but with relatively low efficiency; NH_2OH is the main product (Nakashima *et al.* 1987). Of more potential relevance to the enzyme-catalysed reaction is the fact that Barley *et al.* (1986, 1987) have shown that NO_2^- itself can be electrochemically reduced to NH_3 in the presence of water soluble haems such as Fe-*meso*-tetra-kis(p-sulfonatophenyl)porphyrin. The haem reacts catalytically in this process, undergoing multiple turnovers. NH_2OH appears to be an intermediate in the reaction, and N_2O is a minor by-product. Although the electrochemical profile is highly pH-dependent, as might be expected for a reaction (reaction 1) which involves multiple H^+ as well as e^- additions, the reaction appears to proceed in the following fashion in aqueous media:

1. At pH 4–7, Fe^{II}haem reacts with NO_2^- to form a complex which has been designated $Fe^{II}(P)$-NO^+ (where P = porphyrin). This process, which involves the dehydration of NO_2^-, is presumably an acid–base catalysed reaction.
2. This complex is reduced by two sequential one electron steps. The first, which leads to $Fe^{II}(P)$-NO, is pH-independent, with an $E_{1/2} = +0.6$ V (*vs* SHE).
3. The second reduction step, which leads formally to $Fe^{II}(P)$-NO^-, is pH-independent above pH 2.6, and exhibits an $E_{1/2} = -0.4$ V (*vs* SHE) at pH 4.5.
4. Further reduction of $Fe^{II}(P)$-NO^- is kinetically slow at the electrode at pH ≥ 4, and clearly involves at least one step with an effective $E_{1/2}$ more negative than -0.4 V (*vs* SHE). However, the extent and rate of the reduction increases sharply as the pH is lowered from 4 to 2.9. Under these conditions, multielectron reduction of NO_2^- occurs at an $E_{1/2} = -0.4$ V (*vs* SHE), i.e. reduction of $Fe^{II}(P)$-NO to $Fe^{II}(P)$-NO^- becomes the limiting process when a substantial supply of protons is available.

Studies of the redox chemistry of NO ligated to various Fe-porphyrins in aprotic non-aqueous media (Olson *et al.* 1982; Fujita and Fajer 1983; Lancon and Kadish 1983) have shown, as a complement to the above results, that the reversible one-electron oxidation of $Fe^{II}(P)$-NO [where P = octaethylporphyrin (OEP) or tetraphenylporphyrin (TPP)] occurs at very positive potentials

$[E_{1/2} = +0.7–0.9$ V (*vs* SHE)], and the reversible one-electron reduction of $Fe^{II}(P)$-NO occurs only at much more negative potentials $[E_{1/2} = -0.7$ to -0.9 V (*vs* SHE)]. However, addition of a second electron to $Fe^{II}(P)$-NO in aprotic solvents is even more difficult, with an $E_{1/2} \simeq -1.6$ V (*vs* SHE).

It would appear that it is the latter reaction step, involving what amounts to addition of a third electron to NO_2^-, which must be modulated both thermodynamically and kinetically by an enzyme active site designed to efficiently catalyse reaction 1. One would expect such an active site to provide a high effective concentration of protons, *via* appropriate H-bonding amino acid side chains, at the locus of NO^- binding. In addition, factors which could enhance the basicity of $Fe^{II}(P)$-NO^-, such as the presence of an electron donating distal haem ligand, would be expected to enhance both protonation and electron acceptance of the bound NO^-.

It may also be possible, if these conditions have been met, for an enzyme active site to further facilitate reduction of bound NO^- by providing a source of additional electrons capable of rapidly converting the bound species to NH_2OH. Under these circumstances, reduction of $Fe^{II}(P)NO$, rather than of $Fe^{II}(P)$-NO^-, could become rate-limiting in the enzyme-catalysed process, as it is in the non-enzymatic haem-catalysed electrochemical process at low pH in aqueous media. In this regard, it is interesting that Feng and Ryan (1987) have recently prepared a complex between $Fe^{II}(TPP)$ and NH_2OH. This complex is stable only at low temperature, but cyclic voltametry of $Fe^{II}(TPP)(NH_2OH)_2$, at $-35°C$ in aprotic solvents, yields a 3-electron oxidative wave, with the relatively favourable potential $E_{1/2} = +0.4$ V (*vs* SHE), to form $Fe^{II}(TPP)$-NO and NH_2OH. This result indicates that while addition of one electron to $Fe^{II}(P)$-NO may be thermodynamically difficult, addition of three electrons may be far easier.

Possible mechanism for enzymatic catalysis of nitrite reduction to ammonia

Studies cited above indicate that the Fe^{II}sirohaem-NO complex, with the 4Fe–4S cluster in the oxidized ($+2$) state, is in fact the principal turnover species observed during catalysis of NO_2^- reduction to NH_3 by sirohaem-(4Fe–4S) enzymes. It is clear that the redox potential of the 4Fe-4S centres in these enzymes is sufficiently negative to permit electron donation to a Fe^{II}haem-bound NO ligand, if this potential is in the range found with the aqueous haem-NO complexes.

Recent resonance Raman studies with SiR-HP[1−]-NO (S. Han, J. F. Madden, L. M. Siegel, and T. G. Spiro, submitted for publication) indicate that the bound NO ligand is hydrogen-bonded to an amino acid side chain at the active centre of the enzyme. The NO ligand appears to be bent about the Fe–N–O bond, a result which suggests that it has considerable enhancement of electron density at the oxygen atom (see Fujita and Fajer 1983). EPR spectra of the SiR-HP[1−]-NO are predominantly rhombic (Janick *et al.* 1983), a result which suggests that the sirohaem Fe may be 6-coordinate in this species. The fact that the strength of magnetic exchange coupling between the 4Fe–4S cluster and the sirohaem Fe is even greater in SiR-HP^{1-}-NO than in SiR-HP^0, is consistent with the idea that the Fe$_{haem}$–S–Fe$_{cluster}$ bridge is intact in the species containing bound NO. Thus, the presence of the π-electron-rich bridging S atom may well be responsible for increasing the basicity of the bound NO ligand; this basicity would be expected to be further enhanced if the cluster is reduced.

It appears, then, that the sirohaem-(4Fe–4S) active centre of NiR and SiR-HP contains at least some of the features expected to promote electron addition to bound NO and possible NO$^-$/NOH intermediates. However, the finding that enzyme-bound sirohaem, like other iBC, can readily undergo oxidation to the π-cation radical state, suggests the possibility that the sirohaem-(4Fe–4S) type of active centre may also be designed to effectively bypass the thermodynamically difficult one- and two-electron reductions of bound NO (i.e. sharply limit the lifetimes of such intermediates as bound -NOH and -NHOH) by making it possible for the fully reduced enzyme to add three electrons nearly simultaneously to the bound NO, converting it to bound NH$_2$OH (or a dehydrated form of NH$_2$OH). In this scenario, addition of one electron to the stable turnover species Fe^{II}(iBC)(NO)(4Fe–4S)$^{2+}$, which converts it transiently to Fe^{II}(iBC)(NO)(4Fe–4S)$^{1+}$, permits the latter species to undergo internal oxidation–reduction, in the presence of an efficient proton donor, to form the species Fe^{III}(iBC)$^+$(NH$_2$OH)(4Fe–4S)$^{2+}$. (Dehydrated forms of the latter species, containing the -NH ligand, are also possible.)

This consideration, taken together with the finding in our laboratory that the Fe^{III}sirohaem form of NiR and SiR-HP reacts too slowly with NO$_2^-$ to be active in the catalysis of NO$_2^-$ reduction, suggests the following type of reaction mechanism as a working hypothesis for NO$_2^-$ reduction as catalysed by Fd$_r$-NiR and other sirohaem NiR (C = 4Fe–4S cluster; iBC = isobacteriochlorin macrocycle of sirohaem):

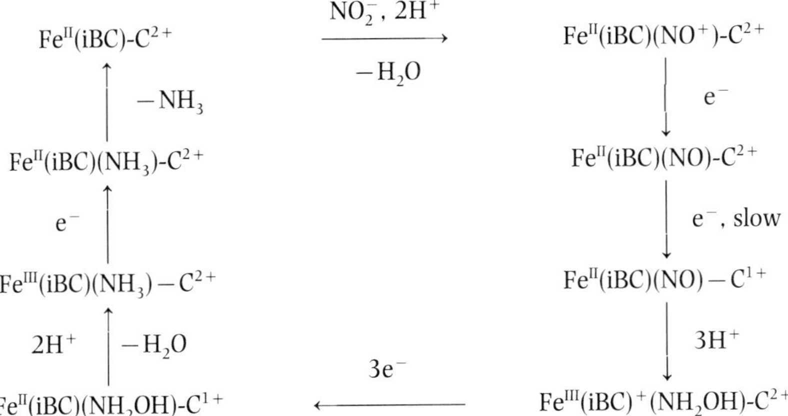

This mechanism is, of course, speculative and requires experimental verification.

While we have focused on the special role of the sirohaem-(4Fe–4S) active centre in catalysis of NO_2^- reduction, since this is the type of centre found in ferredoxin-NiR, it is important to keep in mind the fact that efficient enzymatic reduction of NO_2^- to NH_3 does not require sirohaem or 4Fe–4S prosthetic groups. An entirely different type of NH_3-producing NiR has a complicated active centre which contains six c-type haem groups on a single polypeptide chain of ~ 70 kDa (Liu and Peck 1981; Liu *et al.* 1983, 1988; Blackmore *et al.* 1986; Kajie and Anraku 1986). There is evidence (Liu *et al.* 1980, 1987; Sadana *et al.* 1986; Blackmore *et al.* 1987) that one of the haems, which is high spin in the enzyme as isolated and thus probably contains the site of NO_2^- (and CO) binding, is magnetically coupled (through an unknown mechanicm) to a low spin haem. Two other haems also appear to be magnetically coupled. There is a sharp distinction in the rate of reduction of the ligand-binding haem pair from that of the other haems, so that it is possible that the hexahaem NiR must be fully reduced before it can bind ligands (Blackmore and Brittain 1986). The Fe^{II}haem-NO complex in this enzyme also seems to be the major species present upon reaction of enzyme with reductant and excess NO_2^- (Liu *et al.* 1980, 1987; Sadana *et al.* 1986). Thus, the problem of 'storage' of multiple electrons needed to achieve rapid electron transfer to enzyme-bound NO may be overcome by the presence of a sufficient number of closely apposed reduced prosthetic groups in this type of enzyme.

Acknowledgement

This work has been supported by research grants (to L.M.S.) from the National Institutes of Health (GM-32210) and from the Veterans Administration (Project 215406554).

References

Abou-Jaoudé, A., Pascal, M. C., and Chippaux, M. 1979. Formate-nitrite reduction in *Escherichia coli* K12. 2. Identification of the components involved in the electron transfer. *European Journal of Biochemistry* **95**, 315–21.

Aparicio, P. J., Knaff, D. B., and Malkin, R. (1975). The role of an iron–sulfur center and siroheme in spinach nitrite reductase. *Archives of Biochemistry and Biophysics* **169**, 102–7.

Back, E., Burkhart, W., Moyer, M., Privalle, L., and Rothstein, S. (1988). Isolation of cDNA clones coding for spinach nitrite reductase: complete sequence and nitrite induction. *Molecular and General Genetics* **212**, 20–6.

Barkigia, K. M., Fajer, J., Chang, C. K., and Williams, G. J. B. (1982). Crystal and molecular structure of the isobacteriochlorin 3,7-dimethyl-3′,7′-dihydro-2,2′,8,8′,12,13,17,18-

octaethylporphyrin: a model for siroheme and sirohydrochlorin. *Journal of the American Chemical Society* **104**, 315–17.

Barley, M. H., Takeuchi, K. J., and Meyer, T. J. (1986). Electrocatalytic reduction of nitrite to ammonia based on a water-soluble iron porphyrin. *Journal of the American Chemical Society* **108**, 5876–85.

Barley, M. H., Rhodes, M. R., and Meyer, T. J. (1987). Electrocatalytic reduction of nitrite to nitrous oxide and ammonia based on the N-methylated cation iron-porphyrin complex Fe(III)(H$_2$O)(TMPYP)$^{+5}$. *Inorganic Chemistry* **26**, 1746–50.

Battersby, A. R., McDonald, E., Thompson, M., and Kykhovsky, V. Ya. (1978). Biosynthesis of vitamin B$_{12}$: Proof of A–B structure for sirohydrochlorin by its specific incorporation into cobyrinic acid. *Journal of the Chemical Society Communications* **1978**, 150–1.

Blackmore, R. and Brittain, T. (1986). Kinetic studies on the nitrite reductase of *Wolinella succinogenes*. *Biochemical Journal* **233**, 553–7.

Blackmore, R., Robertson, A. M., and Brittain, T. (1986). The purification and some equilibrium properties of the nitrite reductase of the bacterium *Wolinella succinogenes*. *Biochemical Journal* **233**, 547–52.

Blackmore, R. S., Brittain, T., Gadsby, P. M., Greenwood, C., and Thomson, A. J. (1987). Electron paramagnetic resonance and magnetic circular dichroism studies of as hexaheme nitrite reductase from *Wolinella succinogenes*. *FEBS Letters* **219**, 244–8.

Bottomley, F. and Mukaida, M. (1982). Electrophilic behavior of nitrosyls: preparation and reactions of 6-coordinate ruthenium tetra(pyridine) nitrosyl complexes. *Journal of the Chemical Society, Dalton Transactions* **1982**, 1933–7.

Cammack, R., Hucklesby, D. P., and Hewitt, E. J. (1978). Electron paramagnetic resonance studies of the mechanism of leaf nitrite reductase: signals from the iron–sulfur centre and haem under turnover conditions. *Biochemical Journal* **171**, 519–26.

Cammack, R., Jackson, R. H., Cornish-Bowden, A., and Cole, J. A. (1982). Electron spin resonance studies of the NADH-dependent nitrite reductase from *E. coli* K12. *Biochemical Journal* **207**, 333–9.

Chang, C. K. (1982). Hemes of hydroporphyrins, In *The biological chemistry of iron* (ed. H. B. Dunford), pp. 313–34. D. Riedel, Philadelphia.

Chang, C. K., Hanson, L. K., Richardson, P. F., Young, R., and Fajer, J. (1981). π-cation radicals of ferrous and free base isobacteriochlorins: models for siroheme and sirohydrochlorin. *Proceedings of the National Academy of Science, USA* **78**, 2652–6.

Christner, J. A., Munck, E., Janick, P. A., and Siegel, L. M. (1981). Mossbauer spectroscopic studies of *Escherichia coli* sulfite reductase: evidence for coupling between the siroheme and Fe$_4$S$_4$ cluster prosthetic groups. *Journal of Biological Chemistry* **256**, 2098–101.

Christner, J. A., Munck, E., Janick, P. A., and Siegel, L. M. (1983*a*). Mossbauer evidence for exchange-coupled siroheme and 4Fe–4S prosthetic groups in *Escherichia coli* sulfite reductase: studies of the reduced states and a nitrite turnover complex. *Journal of Biological Chemistry* **258**, 11147–56.

Christner, J. A., Janick, P. A., Siegel, L. M., and Munck, E. (1983*b*). Mossbauer studies of *Escherichia coli* sulfite reductase complexes with carbon monoxide and cyanide: exchange coupling and intrinsic properties of the 4Fe–4S cluster. *Journal of Biological Chemistry* **258**, 11157–64.

Christner, J. A., Munck, E., Kent, T. A., Janick, P. A., Salerno, J. C., and Siegel, L. M. (1984). Exchange coupling between siroheme and 4Fe–4S cluster in *E. coli* sulfite reductase: Mossbauer studies and coupling models for a 2-electron reduced enzyme state and complexes with cyanide. *Journal of the American Chemical Society* **106**, 6786–94.

Cline, J. F., Janick, P. A., Siegel, L. M., and Hoffman, B. M. (1985). Electron-nuclear double resonance studies of oxidized *Escherichia coli* sulfite reductase: ^{1}H, ^{14}N, and ^{57}Fe measurements. *Biochemistry* **24**, 7942–7.

Cline, J. F., Janick, P. A., Siegel, L. M., and Hoffman, B. M. (1986). ^{57}Fe and ^{1}H ENDOR of three doubly reduced states of *Escherichia coli* sulfite reductase. *Biochemistry* **25**, 4647–54.

Cole, J. A. (1982). Independent pathways for the anaerobic reduction of nitrite to ammonia by *Escherichia coli*. *Biochemical Society Transactions* **10**, 476–8.

Cole, J. A., Coleman, K. J., Compton, B. E., Kavanagh, B. M., and Keevil, C. W. (1974). Nitrite and ammonia assimilation by anaerobic continuous cultures of *Escherichia coli*. *Journal of General Microbiology* **85**, 11–22.

Deeg, R., Kriemler, H. P., Bergmann, K. H., and Muller, G. (1977). Cobyrinic acid biosynthesis: novel methylated hydroporphyrins and their role in cobyrinic acid formation, *Hoppe-Seyler's Zeitschrift für Physiologische Chemie* **358**, 339–42.

Feng, D. W. and Ryan, M. D. (1987). Electrochemistry of nitrite reductase model compounds. 3. Formation and characterization of a bis(hydroxylamine)tetraphenyl (porphyrinato)iron(II) complex. *Inorganic Chemistry* **26**, 2480–3.

Fry, I. V., Cammack, R., Hucklesby, D. P., and Hewitt, E J. (1980). Stability of the nitrosyl-siroheme complex of plant nitrite reductase, investigated by EPR spectroscopy, *FEBS Letters* **111**, 377–80.

Fry, I. V., Cammack, R., Hucklesby, D. P., and Hewitt, E. J. (1982). Kinetics of leaf nitrite reductase with methyl viologen and ferrodoxin under controlled redox conditions. *Biochemical Journal* **205**, 235–8.

Fujita, E. and Fajer, J. (1983). Models for nitrite reductases: redox chemistry of iron nitrosyl porphyrins, chlorins, and isobacteriochlorins and π cation radicals of cobalt nitrosyl isobacteriochlorins. *Journal of the American Chemical Society* **105**, 6743–5.

Fujita, E., Chang, C. K., and Fajer, J. (1985). Cobalt (II) nitrosyl cation radicals of porphyrins, chlorins, and isobacteriochlorins: models for nitrite and sulfite reductases and implications for A_{lu} heme radicals. *Journal of the American Chemical Society* **107**, 7665–9.

Griffiths, L. and Cole, J. A. (1987). Lack of redox control of the anaerobically-induced *nir* B$^+$ gene of *Escherichia coli* K12. *Archives fur Microbiologie* **147**, 364–9.

Guerrero, M. G., Vega, J. M., and Losada, M. (1981). The assimilatory nitrate reducing system and its regulation. *Annual Review of Plant Physiology* **32**, 169–204.

Hirasawa, M. and Tamura, G. (1980). Ferredoxin-dependent nitrite reductase from spinach leaves. *Agricultural and Biological Chemistry* **44**, 749–58.

Hirasawa, M., Fukushima, K., Tamura, G., and Knaff, D. B. (1984). Immunochemical characterization of nitrite reductases from spinach leaves, spinach roots, and other higher plants. *Biochemica et Biophysica Acta* **791**, 145–54.

Hirasawa, M., Shaw, R. W., Palmer, G., and Knaff, D. B. (1987). Prosthetic group content and ligand binding properties of spinach nitrite reductase. *Journal of Biological Chemistry* **262**, 12428–33.

Hirasawa-Soga, M., Horie, S., and Tamura, G. (1982). Further characterization of ferredoxin-nitrite reductase and the relationship between the enzyme and methyl viologen-dependent nitrite reductase. *Agricultural and Biological Chemistry* **46**, 1319–29.

Hirasawa-Soga, M., Tamura, G., and Horie, S. (1983). Spectrophotometric and electron spin resonance studies on the substrate interactions of ferredoxin-linked nitrite reductase from spinach. *Journal of Biochemistry* **94**, 1833–40.

Ho, C. H., Ikawa, T., and Nikizawa, K. (1976). Purification and properties of a nitrite reductase from *Porphyra yezoensis*, *Plant and Cell Physiology* **17**, 417–30.

Hucklesby, D. P., James, D. M., Banwell, M. J., and Hewitt, E. J. (1976). Properties of nitrite reductase from *Curburbita pepo*. *Phytochemistry* **15**, 599–603.

Ida, S. and Mikami, B. (1986). Spinach ferredoxin-nitrite reductase: a purification procedure and characterization of chemical properties. *Biochimica et Biophysica Acta* **871**, 167–76.

Jackson, R. H., Cornish-Bowden, A., and Cole, J. A. (1981). Prosthetic groups of the NADH-dependent nitrite reductase from *Escherichia coli* K12, *Biochemical Journal* **193**, 861–7.

Janick, P. A. and Siegel, L. M. (1982). Electron paramagnetic resonance and optical spectroscopic evidence for interaction between siroheme and Fe_4S_4 prosthetic groups in *Escheria coli* sulfite reductase hemoprotein subunit. *Biochemistry* **21**, 3538–47.

Janick, P. A. and Siegel, L. M. (1983). Electron paramagnetic resonance and optical spectroscopic evidence for interaction between siroheme and Fe_4S_4 groups in complexes of *Escherichia coli* sulfite reductase hemoprotein with added ligands. *Biochemistry* **22**, 504–14.

Janick, P. A., Rueger, D. C., Krueger, R. J., Barber, M. J., and Siegel, L. M. (1983). Characterization of complexes between *Escherichia coli* sulfite reductase hemoprotein and its substrates sulfite and nitrite. *Biochemistry* **22**, 396–408.

Kajie, S. and Anraku, Y. (1986). Purification of a hexaheme cytochrome c_{552} from *Escherichia coli* K12 and its properties as a nitrite reductase. *European Journal of Biochemistry* **154**, 457–63.

Kratky, C., Angst, C., and Johansen, J. E. (1981). Conformational coupling between ring A and ring B in isobacteriochlorins. *Angewandte Chemie, International Edition in English* **20**, 211–12.

Krueger, R. J. and Siegel, L. M. (1982). Spinach siroheme enzymes: isolation and characterization of ferredoxin-sulfite reductase and comparison of properties with ferredoxin-nitrite reductase. *Biochemistry* **21**, 2892–904.

Lancaster, J. R., Vega, J. M., Kamin, H., Orme-Johnson, N. R., Orme-Johnson, W. H., Krueger, R. J., and Siegel, L. M. (1979). Identification of the iron–sulfur center of spinach ferredoxin-nitrite reductase as a tetranuclear center, and preliminary EPR studies of mechanism. *Journal of Biological Chemistry* **254**, 1268–72.

Lancon, D. and Kadish, K. M. (1983). Electrochemical and spectral characterization of iron mono- and dinitrosyl porphyrins. *Journal of the American Chemical Society* **105**, 5610–17.

Liu, M. C. and Peck, H. D., Jr. (1981). The isolation of a hexaheme cytochrome from *Desulfovibrio desulfuricans* and its identification as a new type of nitrite reductase. *Journal of Biological Chemistry* **256**, 13159–64.

Liu, M. C., DerVartanian, D. V., and Peck, H. D., Jr. (1980). On the nature of the oxidation–reduction properties of nitrite reductase from *Desulfovibrio Desulfuricans*. *Biochemical and Biophysical Research Communications* **96**, 278–85.

Liu, M. C., Liu, M. Y., Payne, W. J., Peck, H. D., Jr., and LeGall, J. (1983). *Wolinella succinogenes* nitrite reductase: purification and properties. *FEMS Microbiology Letters* **19**, 201–6.

Liu, M. C., Liu, M. Y., Payne, W. J., Peck, H. D., Jr., LeGall, J., and DerVartanian, D. V. (1987). Comparative EPR studies onthe nitrite reductases from *Escherichia coli* and *Wolinella succinogenes*. *FEBS Letters* **218**, 227–30.

Liu, M. C., Bakel, B. W., Liu, M. Y., and Dao, T. N. (1988). *Vibro fischeri* nitrite reductase and its characterization as a hexaheme *c*-type cytochrome. *Archives of Biochemistry and Biophysics* **262**, 259–65.

McRee, D. E., Richardson, D. C., Richardson, J. S., and Siegel, L. M. (1986). The heme and Fe_4S_4 cluster in the crystallographic structure of *Escherichia coli* sulfite reductase. *Journal of Biological Chemistry* **261**, 10277–81.

Murphy, M. J. and Siegel, L. M. (1973). Siroheme and sirohydrochlorin: the basis for a new type of porphyrin-related prosthetic group common to both assimilatory and dissimilatory sulfite reductases. *Journal of Biological Chemistry* **248**, 6911–9.

Murphy, M. J., Siegel, L. M., Kamin, H., DerVartanian, D. V., Lee, J. P., LeGall, J., and Peck, H. D., Jr. (1973*a*). An iron tetrahydroporphyrin group common to both assimilatory and dissimilatory sulfite reductases. *Biochemical and Biophysical Research Communications* **54**, 82–8.

Murphy, M. J., Siegel, L. M., Kamin, H., and Rosenthal, D. (1973*b*). NADPH-sulfite reductase of enterobacteria. II. Identification of a new class of heme prosthetic group: an iron tetrahydroporphyrin (isobacteriochlorin type) with eight carboxylic acid groups. *Journal of Biological Chemistry* **248**, 2801–14.

Murphy, M. J., Siegel, L. M., Tove, S. R., and Kamin, H. (1974). Siroheme: a new prosthetic group participating in six-electron reduction reactions catalyzed by sulfite and nitrate reductases. *Proceedings of the National Academy of Sciences, USA* **71**, 612–16.

Murphy, W. R., Jr., Takeuchi, K. J., and Meyer, T. J. (1982). Interconversion of nitrite and ammonia: progress toward a model for nitrite reductase. *Journal of the American Chemical Society* **104**, 5817–18.

Nakashima, N., Matsushi, K., and Yasukoucha, K. (1987). Electrocatalytic reduction of nitrate and nitrite to hydroxylamine and ammonia using metal cyclams. *Journal of Electrochemistry* **224**, 199–209.

Olson, L. W., Schaeper, D., Lancon, D., and Kadish, K. M. (1982). Characterization of several iron nitrosyl porphyrins. *Journal of the American Chemical Society* **104**, 2042–4.

Ostrowski, J., Back, E. W., Madden, J., Siegel, L. M., and Kredich, N. M. (1987). Comparison of deduced amino acid sequences of α and β subunits of *S. typhimurium* NADPH-sulfite reductase with those of spinach nitrite reductase and cytochrome P450 oxidoreductase. *Federation Proceedings* **46**, 2231 (abstr 1780).

Prodouz, K. N. and Garrett, R. H. (1981). *Neurospora crassa* NAD(P)H-nitrite reductase: studies on its composition and structure. *Journal of Biological Chemistry* **256**, 9711–17.

Richardson, P. F., Chang, C. K., Spaulding, L. D., and Fajer, J. (1979). Radicals of isobacteriochlorins: models of siroheme and sirohydrochlorin. *Journal of the American Chemical Society* **101**, 7736–8.

Romero, L. C., Galvan, F., and Vega, J. M. (1987). Purification and properties of the siroheme-containing ferredoxin-nitrite reductase from *Chlamydomonas reinhardtii*. *Biochimica et Biophysica Acta* **914**, 55–63.

Sadana, J. C., Khan, B. M., Fry, I. V., and Cammack, R. (1986). Electron paramagnetic resonance studies of heme *c* and its nitrosyl derivative in *Vibrio (Achromobacter) fischeri* nitrite reductase. *Biochemical and Cellular Biology* **64**, 394–9.

Scott, A. I., Irwin, A. J., Siegel, L. M., and Shoolery, J. N. (1978). Sirohydrochlorin: prosthetic group of sulfite and nitrite reductases and its role in the biosynthesis of vitamin B_{12}. *Journal of the American Chemical Society* **100**, 7987–94.

Siegel, L. M. and Davis, P. S. (1974). NADPH-sulfite reductase of enterobacteria. IV. The *Escherichia coli* hemoflavoprotein: subunit structure and dissociation into hemoprotein and flavoprotein components. *Journal of Biological Chemistry* **249**, 1587–98.

Siegel, L. M., Murphy, M. J., and Kamin, H. (1973). NADPH-sulfite reductase of enterobacteria. I. The *Eschericia coli* hemoflavoprotein: molecular parameters and prosthetic groups. *Journal of Biological Chemistry* **248**, 251–64.

Siegel, L. M., Davis, P. S., and Kamin, H. (1974). NADPH-sulfite reductase of enterobacteria. III. The *Escherichia coli* hemoflavoprotein: catalytic parameters and the sequence of electron flow. *Journal of Biological Chemistry* **249**, 1572–86.

Siegel, L. M., Rueger, D. C., Barber, M. J., Krueger, R. J., Orme-Johnson, N. R., and Orme-

Johnson, W. H. (1982). *Escherichia coli* sulfite reductase hemoprotein subunit: prosthetic groups, catalytic parameters, and ligand complexes. *Journal of Biological Chemistry* **257**, 6343–50.

Siegel, L. M., Wilkerson, J. O., and Janick, P. A. (1987). Structural studies on the siroheme (4Fe–4S) cluster active centers of spinach ferredoxin-nitrite reductase and *Escherichia coli* sulfite reductase. In *Inorganic nitrogen metabolism* (eds W. R. Ullrich, P. Aparicio, P. J. Syrett, and F. Castillo), pp. 118-22. Springer-Verlag, Berlin.

Stoller, M. L., Malkin, R., and Knaff, D. B. (1977). Oxidation-reduction properties of photosynthetic nitrite reductase. *FEBS Letters* **81**, 271–4.

Stolzenberg, A. M., Spreer, L O., and Holm, R. J. (1980). Octaethylisobacteriochlorin, a model of the sirohydrochlorin ring system of siroheme enzymes: preparation and spectroscopic and electrochemical properties of the free base and its Zn(II) complex. *Journal of the American Chemical Society* **102**, 364–9.

Stolzenberg, A. M., Strauss, S. H., and Holm, R. J. (1981). Iron (II, III)-chlorin and isobacteriochlorin complexes, models of the heme prosthetic groups in nitrite and sulfite reductases: means of formation and spectroscopic and redox properties. *Journal of the American Chemical Society* **103**, 4763–78.

Suh, M. O., Swepston, P. N., and Ibers, J. A. (1984). Direct preparation of a siroheme model compound: (5,10,15,20-tetramethylisobacteriochlorinato) nickel (II). *Journal of the American Chemical Society* **106**, 5114–71.

Vega, J. M. and Kamin, H. (1977). Spinach nitrite reductase: purification and properties of a siroheme-containing iron-sulfur enzyme. *Journal of Biological Chemistry* **252**, 896–909.

Vega, J. M., Garrett, R. H., and Siegel, L. M. (1975). Siroheme: a prosthetic group of the *Neurospora crassa* assimilatory nitrite reductase. *Journal of Biological Chemistry* **250**, 7980–9.

Wilkerson, J. O., Janick, P. A., and Siegel, L. M. (1983). Electron paramagnetic resonance and optical spectroscopic evidence for interaction between siroheme and Fe_4S_4 prosthetic groups in spinach ferredoxin-nitrite reductase. *Biochemistry* **22**, 5048–54.

Young, L. J. and Siegel, L. M. (1988). Superoxidized states of *Escherichia coli* sulfite reductase heme protein subunit. *Biochemistry* **27**, 5984–90.

Zumft, W. G. 1972. Ferredoxin-nitrite oxidoreductase from *Chlorella*: purification and properties. *Biochimica et Biophysica Acta* **276**, 363–75.

18. Molecular cloning of spinach nitrite reductase

E. Back, W. Burkhart, M. Moyer, L. Privalle, and S. Rothstein

Introduction

Nitrate, the predominant form of nitrogen taken up by most plants, enters through the roots and is taken into the cytoplasm by a transport system that is nitrate-inducible, requires continual protein synthesis for maximal function, and has a fairly short-term turnover (Clarkson 1986). In most plants, the bulk of the nitrate is then transported through the xylem to the leaves. Prior to its incorporation into amino acids, nitrate is reduced to nitrite by nitrate reductase (NR; E.C.1.6.6.1) and nitrite is then reduced to ammonium by nitrite reductase (NiR; E.C.1.7.7.1)). In leaves and roots, NR is found in the cytoplasm, whereas NiR is localized in the chloroplasts of leaves and in the plastids of roots (see Oaks and Hirel 1985). Since NiR is a nuclear-encoded protein, it must be transported from the cytoplasm to the chloroplasts. The mechanism by which nitrite, the cytotoxic end product of the first reduction step, is effectively transported from the cytoplasm into the chloroplast is unknown.

In most plants studied, the reduction of nitrate takes place primarily in green tissues, but some reduction in non-green tissues has been shown to occur in many plants (for general review see Haynes 1986). In those plants studied, the NR and NiR polypeptides isolated from non-green tissues seem to be different from their counterparts in green tissues (Hucklesby *et al.* 1972; Dalling *et al.* 1973; Ishiyama and Tamura 1985). However, in spinach there is some immunological evidence that leaf and root NiR are actually the same protein (Hirasawa *et al.* 1984). Reduced ferredoxin is the physiological electron donor in leaves, providing the six electrons necessary for the reduction of nitrite to ammonium (Vega *et al.* 1980). A ferredoxin-like protein has been identified as electron donor in non-green cultured tobacco cells (Ninomiya and Sato 1984) and in roots (Suzuki *et al.* 1985; Oji *et al.* 1985). Ishiyama *et al.* (1985) found an electron donor in aetiolated bean stems with an absorption spectrum different from that of ferredoxin and the ferredoxin-like protein found in cultured tobacco cells.

NiR has been purified to homogeneity from green tissues of several species (reviewed in Guerrero *et al.* 1981; Serra *et al.* 1982; Small and Gray 1984;

Ishiyama *et al.* 1985; Ida and Mikami 1986) and from aetiolated bean shoots (Ishiyama and Tamura 1985). The enzyme from green tissues appears to be a monomer with a molecular weight of 60–70 kDa, depending on the species. It contains one sirohaem and one tetranuclear iron–sulphur centre as prosthetic groups (Lancaster *et al.* 1979; Vega *et al.* 1980). Immunological comparison of NiR from different plant sources shows cross-reactivity correlating roughly with the phylogenetic distances among higher plants (Ishiyama *et al.* 1985; Ida 1987).

The expression of NiR, as well as NR, has been shown to be induced by nitrate (Beevers and Hageman 1969; Hewitt *et al.* 1976; Rajasekhar and Mohr 1986) and by nitrite (Ferguson 1969; Kelker and Filner 1971; Stewart 1972). Nitrate and light are needed for maximal synthesis of the NiR enzyme and the regulation is a *de novo* synthesis rather than an activation–inactivation mechanism (Gupta and Beevers 1984; Rajasekhar and Mohr 1986; Gupta and Beevers 1987). Nitrate given in complete darkness induces some NiR activity while light alone has no influence on NiR enzyme activity (Rajasekhar and Mohr 1986). In mustard (*Sinapis alba* L.) cotyledons, nitrate induction of NiR activity is first detectable 36 h after sowing (Rajasekhar and Mohr 1986). This nitrate induction is modulated *via* the far-red-absorbing form of phytochrome, but only occurs 42 h after sowing. Thus, nitrate, as an inducer proper, and light, as a modulator, act independently of each other in controlling the appearance of NiR activity.

The NiR mRNA is also induced by nitrate and light over several days, whereas nitrate or light alone induces the production of very little and none, respectively, *in vitro* translatable NiR mRNA (Gupta and Beevers 1987; this chapter).

Isolation of the nitrite reductase gene

The molecular analysis of the NiR gene is of great importance for studying its regulation; furthermore, the primary amino acid sequence of the NIR protein can be deduced from the DNA sequence. This information is a necessary prerequisite for the understanding of the catalytic site(s) of NiR and the interaction between cofactors and polypeptide. It also allows the study of the mechanism for the transport of the NiR protein from the cytoplasm, where it is synthesized, to the chloroplast, where it is localized.

We employed two methods to isolate the NiR gene from spinach. A cDNA library was constructed using mRNA isolated from nitrate-treated spinach and probed either with oligonucleotides based on partial amino acid sequences of NiR or with the antibodies raised against purified NiR.

NiR was isolated from spinach leaves following the procedure of Vega and Kamin (1977). This protein preparation when subjected to SDS polyacrylamide gel electrophoresis gave a single polypeptide band of 61 kD and was at least 90% pure. Antibodies raised in rabbits against the non-denatured NiR protein were incubated with total spinach protein immobilized on a nitrocellulase filter (Western blot) and reacted with two polypeptides. One of these was the correct molecular weight for NiR while the other had an apparent molecular weight of 70 kD. Several peptide fragments of the purified NiR protein were partially sequenced (Back *et al.* 1988), and based on their amino acid sequences, the following mixed oligonucleotide probes were synthesized:

'oligoNiR1' (20-mer; 64-fold redundancy)

Lys	His	Asn	Lys	Asp	Asp	Ile
5' - AAA	CAC	AAC	AAA	GAC	GAC	AT - 3'
G	T	T	G	T	T	

'oligoNiR2' (18-mer; 64-fold redundancy)

Met	Glu	Glu	Val	Asp	Lys
3' - TAC	CTT	CTT	CAA	CTA	TTT - 5'
	C	C	C	G	C
			G		
			T		

'oligoNiR3' (14-mer; 32-fold redundancy)

Met	Asp	Asn	Val	Arg
5' - ATG	GAC	AAC	GTA	AG - 3'
	T	T	C	C
			G	
			T	

A cDNA library was constructed in the expression vector λgt11 from poly(A)$^+$ RNA from leaves of nitrate-induced spinach (Back *et al.* 1988) and screened for the presence of NiR cDNAs with the NiR antiserum (Young and Davis 1983) and the mixed oligonucleotide probes (Wood *et al.* 1985). Using this antiserum two NiR cDNA clones (CIB400 and CIB402) were isolated from a total of 120,000 recombinant phages. In a parallel experiment using the mixed oligonucleotide probes one NiR cDNA clone (CIB401) was isolated from a total of 200 000 recombinant phages. In addition to the two NiR cDNAs isolated with the NiR-antiserum, seven cross-hybridizing cDNA clones were isolated that did not hybridize with the NiR oligonucleotide probes. These clones seem to represent a gene or gene family that may code for the 70 kDa protein of unknown function which reacts with our NiR antiserum.

The three isolated NiR cDNA clones (pCIB400, pCIB401, and pCIB402) had an identical restriction enzyme pattern and varied only slightly in length. The map of one of the cDNA clones, pCIB400, is shown at the top of Fig. 18.1.

Using the NiR cDNA as a probe, the gene coding for NiR was isolated from a genomic library of spinach DNA (Maniatis *et al.* 1982). Restriction maps were prepared for six independently isolated genomic clones, all of which proved to be overlapping clones of the same genomic DNA region. This finding is consistent with a single NiR gene being present per haploid genome and has been confirmed

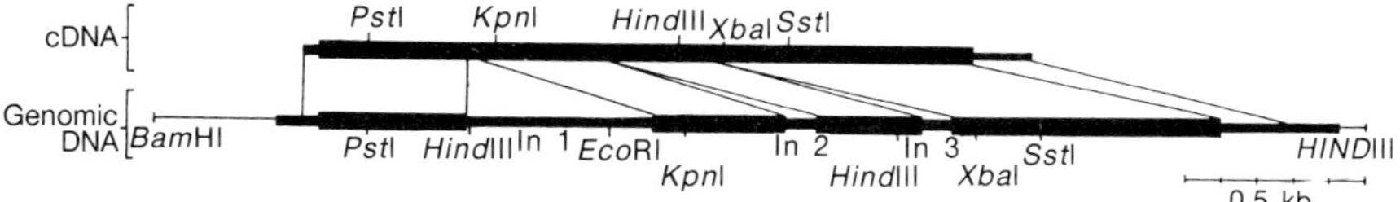

Fig. 18.1. Restriction map showing the relationship between the NiR cDNA clone (pCIB400) and the NiR gene. The wide bar represents the translated region, the narrow bar the transcribed but untranslated regions. The dideoxy chain termination method of Sanger *et al.* (1977) was used to determine the DNA sequence. The coding region in the NiR gene is interrupted by three introns (In1, In2, and In3) of 531 bp, 93 bp and 93 bp, respectively.

by Southern blots (data not shown). The physical map of the genomic NiR clone is shown at the bottom of Fig. 18.1.

The entire sequence of the NiR cDNA was determined and is shown in Fig. 18.2. The DNA sequence of the genomic clone is currently being established and is near completion. The cDNA is 2015 bp long and has an open reading frame of 594 amino acids starting with a potential translation initiation codon. This ATG is also surrounded by a plant translation initiation consensus sequence (Lütcke *et al.* 1987). The protein sequence derived from the DNA sequence confirmed the presence of all three amino acid sequences determined by direct protein sequencing (underlined sequences in Fig. 18.2), with the exception of one amino acid. At position 221 the sequenced protein shows a valine rather than an isoleucine. This difference is probably due to a point mutation in the DNA, reflecting allelic variation in the protein between different spinach cultivars. In the NiR gene the coding region consists of four exons interrupted by three introns. No poly(A) tail is found in any of the three NiR cDNAs. Moreover, sequencing the 3'-end of the three NiR cDNA inserts demonstrated that each one ends at a different location (see Fig. 18.2). We have not determined where the NiR mRNA ends *in vivo*. Based on preliminary data for the transcription start site and a NiR mRNA length of 2.3 kb (see Fig. 18.4) one would expect the polyadenylation signal sequence to be just downstream from the end of our sequenced NiR cDNA. However, in the downstream genomic DNA region no animal consensus polyadenylation signal sequence (AATAAA) is present. Several other plant genes have been reported to deviate from the AATAAA consensus sequence (reviewed in Heidecker and Messing 1986).

In summary, we have cloned an almost full length copy of the NiR mRNA from spinach leaves and isolated a copy of the entire NiR gene including its putative regulatory sequences.

Amino acid sequence of nitrite reductase

The cloned NiR cDNA codes for a precursor protein that is 594 amino acids long, with a deduced molecular weight of 66 394. To determine the length of the

288

```
-60                                                                                      -1                                                                                        60
CATCATCTTCATCTTCATCTTCATCATTCATAGTTGCAAGAAACAGAGCAACCAAAAAAAATGGCATCACTTCCAGTCAACAAGATCATACCATCATCAACGACATTACTGTCATCGTCG
                                                                 MetAlaSerLeuProValAsnLysIleIleProSerSerThrThrLeuLeuSerSerSer
                                                                 1                                                                                                               20

 61                                                                                                                                                                             180
AACAACAACAGAAGAAGAAATAACTCATCAATTCGATGCCAGAAGGCGGTTTCACCCGCGGCAGAAACGGCTGCAGTGTCGCCGTCTGTGGACGCGGCGAGGCTGGAGCCGAGAGTGGAG
AsnAsnAsnArgArgArgAsnAsnSerSerIleArgCysGlnLysAlaValSerProAlaAlaGluThrAlaAlaValSerProSerValAspAlaAlaArgLeuGluProArgValGlu
21                                                                                                                                                                             60

181                                                                                                                                                                            300
GAGAGAGATGGGTTTTGGGTATTGAAGGAGGAATTTAGGAGTGGGATTAACCCAGCTGAGAAAGTTAAGATTGAGAAAGACCCAATGAAGTTGTTTATTGAGGATGGGATTAGTGATCTT
GluArgAspGlyPheTrpValLeuLysGluGluPheArgSerGlyIleAsnProAlaGluLysValLysIleGluLysAspProMetLysLeuPheIleGluAspGlyIleSerAspLeu
61                                                                                                                                                                             100

301                                                                                                                                                                            420
GCTACTTTGTCAATGGAGGAAGTTGATAAATCTAAGCATAATAAGGATGATATTGATGTTAGACTCAAGTGGCTTGGACTTTTCCATCGCCGTAAACATCACTATGGGAGATTCATGATG
AlaThrLeuSerMetGluGluValAspLysSerLysHisAsnLysAspAspIleAspValArgLeuLysTrpLeuGlyLeuPheHisArgArgLysHisHisTyrGlyArgPheMetMet
101                                                                                                                                                                            140

421                                                                                                                                                                            540
AGGTTGAAGCTGCCGAATGGGGTAACAACGAGTGAGCAGACACGGTACCTAGCAAGCGTGATCAAGAAGTACGGAAAAGATGGATGTGCGGATGTAACAACAAGGCAAAACTGGCAAATT
ArgLeuLysLeuProAsnGlyValThrThrSerGluGlnThrArgTyrLeuAlaSerValIleLysLysTyrGlyLysAspGlyCysAlaAspValThrThrArgGlnAsnTrpGlnIle
141                                                                                                                                                                            180

541                                                                                                                                                                            660
AGAGGAGTTGTTCTGCCTGATGTGCCAGAGATCATCAAAGGGCTGGAATCCGTTGGTCTTACCAGCTTACAGAGTGGGATGGACAATGTAAGGAACCCTGTAGGTAACCCTCTTGCAGGG
ArgGlyValValLeuProAspValProGluIleIleLysGlyLeuGluSerValGlyLeuThrSerLeuGlnSerGlyMetAspAsnValArgAsnProValGlyAsnProLeuAlaGly
181                                                                                                                                                                            220

661                                                                                                                                                                            780
ATTGACCCTCATGAAATTGTTGACACCCGACCTTTTACCAACCTAATTTCCCAATTTGTCACTGCCAATTCGCGTGGAAACCTTTCTATTACCAATCTGCCAAGGAAGTGGAATCCATGT
IleAspProHisGluIleValAspThrArgProPheThrAsnLeuIleSerGlnPheValThrAlaAsnSerArgGlyAsnLeuSerIleThrAsnLeuProArgLysTrpAsnProCys
221                                                                                                                                                                            260

781                                                                                                                                                                            900
GTTATTGGGTCCCATGATCTTTATGAGCATCCACACATCAATGACCTTGCTTACATGCCTGCTACAAAGAATGGGAAATTCGGGTTTAATTTGTTGGTTGGAGGATTCTTTAGCATCAAA
ValIleGlySerHisAspLeuTyrGluHisProHisIleAsnAspLeuAlaTyrMetProAlaThrLysAsnGlyLysPheGlyPheAsnLeuLeuValGlyGlyPhePheSerIleLys
261                                                                                                                                                                            300

901                                                                                                                                                                           1020
AGATGTGAAGAGGCAATCCCACTAGACGCTTGGGTCTCAGCAGAAGATGTGGTTCCTGTATGCAAAGCTATGCTTGAAGCTTTCAGGGACCTTGGCTTTAGAGGAAACAGGCAGAAGTGC
ArgCysGluGluAlaIleProLeuAspAlaTrpValSerAlaGluAspValValProValCysLysAlaMetLeuGluAlaPheArgAspLeuGlyPheArgGlyAsnArgGlnLysCys
301                                                                                                                                                                           340
```

1021
AGAATGATGTGGCTTATTGATGAGCTTGGTATGGAAGCATTCAGGGGAGAGGTTGAGAAGAGAATGCCTGAGCAAGTTCTAGAAAGAGCATCCTCAGAAGAGCTGGTTCAGAAGGACTGG
ArgMetMetTrpLeuIleAspGluLeuGlyMetGluAlaPheArgGlyGluValGluLysArgMetProGluGlnValLeuGluArgAlaSerSerGluGluLeuValGlnLysAspTrp
341   380

1141
GAGAGAAGAGAATACTTAGGAGTTCACCCTCAGAAACAACAAGGACTTAGCTTTGTGGGTCTCCACATTCCTGTGGGCCGTCTGCAAGCTGATGAGATGGAAGAGTTAGCCCGTATAGCT
GluArgArgGluTyrLeuGlyValHisProGlnLysGlnGlnGlyLeuSerPheValGlyLeuHisIleProValGlyArgLeuGlnAlaAspGluMetGluGluLeuAlaArgIleAla
381   420

1261
GATGTGTATGGATCAGGGGAGCTCCGTCTGACAGTAGAGCAGAACATAATCATCCCAAATGTTGAAAACTCAAAGATAGATTCACTACTAAACGAGCCTCTGTTAAAAGAGCGTTACTCC
AspValTyrGlySerGlyGluLeuArgLeuThrValGluGlnAsnIleIleIleProAsnValGluAsnSerLysIleAspSerLeuLeuAsnGluProLeuLeuLysGluArgTyrSer
421   460

1381
CCTGAACCACCCATCTTGATGAAGGGGCTTGTGGCCTGTACGGGGAGCCAATTTTGTGGACAAGCCATTATCGAGACCAAGGCTAGGGCACTCAAGGTGACAGAAGAGGTACAACGACTA
ProGluProProIleLeuMetLysGlyLeuValAlaCysThrGlySerGlnPheCysGlyGlnAlaIleIleGluThrLysAlaArgAlaLeuLysValThrGluGluValGlnArgLeu
461   500

1501
GTGTCTGTAACACGGCCTGTTAGGATGCATTGGACCGGGTGTCCTAATAGTTGTGGTCAAGTACAAGTGGCTGATATTGGGTTCATGGGTTGCATGACTAGGGATGAGAACGGTAAGCCT
ValSerValThrArgProValArgMetHisTrpThrGlyCysProAsnSerCysGlyGlnValGlnValAlaAspIleGlyPheMetGlyCysMetThrArgAspGluAsnGlyLysPro
501   540

1621
TGTGAAGGAGCTGATGTGTTTGTAGGAGGACGTATAGGAAGTGACTCGCATCTAGGAGACATTTACAAGAAGGCAGTCCCATGTAAAGATTTGGTGCCTGTTGTTGCTGAGATATTGATC
CysGluGlyAlaAspValPheValGlyGlyArgIleGlySerAspSerHisLeuGlyAspIleTyrLysLysAlaValProCysLysAspLeuValProValValAlaGluIleLeuIle
541   580

1741
AACCAATTCGGTGCTGTTCCTAGGGAGAGGGAAGAGGCAGAGTAGTAGCTAGACTGTTTTGGGTGCCTGTTCTTGTTAACTGTTATCGGTATTCGGTAATTACTTGTAATATTTGCATTT
AsnGlnPheGlyAlaValProArgGluArgGluGluAlaGlu
581   CIB402   CIB400

1861
TTTTTCAAGCATATAATTAAATTGCATAAAGATCCCTTGTATGTCTGCATAACAAGATACTCAGTTATGTAATGTCAATAGCAGGTTTACTTTGTTTATTCAATAGGCACTGTGAAAGGG
                       CIB401
1981
AAAGTTCATTATTCATTTCTCA

Fig. 18.2. DNA sequence and deduced amino acid sequence of the NiR cDNA clone pCIB400. Amino acids confirmed by protein sequencing are underlined. At position 221 the sequenced protein shows a valine rather than an isoleucine (see text). The mature NiR protein starts with the cysteine at position 33 (*arrow*). The nucleotide residues from position 1957 to 2002 are from the cDNA clone pCIB401. The 3′-ends of the three cDNA sequenced are at position 1887 (pCIB402), 1956 (pCIB400), and 2002 (pCIB401).

mature protein, the N-terminus of the isolated NiR protein was sequenced. According to the sequencing data about 25 per cent of the protein started with the cysteine at position 33, 50 per cent with the alanine at position 36 and 25 per cent with the valine at position 37. The most likely interpretation of these data is that the mature NiR protein starts with a cysteine at position 33 (arrow in Fig. 18.2). The two other start sites are probably artifacts resulting from protease activity that was present during protein isolation with several amino acids being digested away at the N-terminal end. Starting with the cysteine at position 33 the mature NiR protein is 562 amino acids long and has a deduced molecular weight of 62 883. This value corresponds well with the apparent molecular weight of 61 kDa observed by us (data not shown) and others (Ho and Tamura 1973; Vega and Kamin 1977). The amino acid composition of the mature NiR protein predicted from the DNA sequence is very similar to that determined for the protein by Vega and Kamin (1977).

Given the primary amino acid sequence for NiR one can hypothesize which amino acids are involved in the active centre of the enzyme. This subject is discussed in detail in Chapter 17 and, therefore, is mentioned only briefly here. NiR has as cofactors a 4Fe–4S centre (Aparicio *et al.* 1975; Lancaster *et al.* 1979) and a sirohaem (Murphy *et al.* 1974) which are both part of the active centre (Cammack *et al.* 1978). Four cysteine sulphurs are expected to bind the tetranuclear iron cluster, with four bridging sulphur ligands, to the protein. One of these cysteine sulphurs is also probably bound to the sirohaem iron (Siegel *et al.* 1987). The four cysteines in positions 473, 479, 514, and 518 are probably involved in the binding of the cofactors; this is supported by a comparison of the sequence of NiR with that of the β-subunit of NADPH-sulphite reductase from *Salmonella typhimurium* (Ostrowski *et al.* 1987) and from X-ray crystallography (McRee *et al.* 1986) of the β-subunit of NADPH-sulphite reductase from *E. coli* revealing conservation of these cysteines (Ostrowski *et al.* in preparation).

The transit peptide sequence of nitrite reductase

The NiR precursor protein is 32 amino acids longer than the mature protein (Fig. 18.2). These additional amino acids precede the mature protein and are likely to act as a transit peptide sequence in directing the nuclear-encoded NiR protein into the chloroplast. Nuclear-encoded chloroplast proteins in general seem to be synthesized as larger molecular weight precursors containing an amino-terminal sequence which is cleaved from the mature protein during or after transport into the chloroplast (reviewed by Chua and Schmidt 1979; Schmidt *et al.* 1981; Grossman *et al.* 1982; Smeekens *et al.* 1985, 1986). In wheat, the 60.5 kDa molecular weight NiR is synthesized as a 64 kDa precursor (Small and Gray 1984). *In vitro* translated NiR from pea has a molecular weight of 67 kDa while the native NiR is 62 kDa (Gupta and Beevers 1987). A proteinaceous extract prepared from pea chloroplasts (Robinson and Ellis 1984*a*) is able to process the 67 kDa form to the *in vivo* NiR length through a 65 kDa intermediate (Gupta and Beevers 1987). Processing through an intermediate has been observed earlier for several nuclear-encoded chloroplastic precursor proteins (Mishkind *et al.* 1985; Robinson and Ellis 1984*b*).

Transit peptide sequences have been shown to be necessary and sufficient for the transport of proteins into chloroplasts (Van den Broeck *et al.* 1985; Schreier *et al.* 1985; Broglie *et al.* 1984). Furthermore, a transit peptide coded for by a gene from one plant can be recognized by chloroplasts from distantly related species (Chua and Schmidt 1978; Coruzzi *et al.* 1983; Mishkind *et al.* 1985; Van den Broeck *et al.* 1985; Schreier *et al.* 1985; Broglie *et al.* 1984). Nevertheless, little amino acid homology is seen among transit peptides from different species. A common amino acid framework among nuclear-encoded chloroplast proteins has been hypothesized (Karlin-Neumann and Tobin 1986). However, a comparison of this amino acid framework with the transit peptides from NiR and from plastocyanin (Smeekens *et al.* 1985, 1986) shows very few conserved residues. While all transit peptides known so far are positively charged, this alone cannot determine if a protein will be transported into chloroplasts, since mitochondrial transit peptides are also positively charged.

Induction of NiR mRNA by nitrate

Nitrate induction of NiR expression is well documented for a number of plant species (Beevers and Hageman 1969; Hewitt *et al.* 1976; Gupta and Beevers 1984; Rajasekhar and Mohr 1986; Gupta and Beevers 1987). However, the mechanism of induction is still unknown. To investigate the induction of NiR mRNA by nitrate the isolated NiR cDNA was used as a probe. Our test plants were spinach seedlings grown on ammonium as the sole source of nitrogen (Back *et al.* 1988) for approximately 5 weeks (2–4 leaf stage), after which NiR mRNA induction was initiated by watering the spinach seedlings with 25 mM $Ca(NO_3)_2$; control seedlings were treated with water. The plants were kept under continuous light for 100 h and poly(A)$^+$ RNA was then isolated from the leaves. The RNA was used for a Northern blot analysis with ^{32}P-labelled NiR cDNA as a probe. The autoradiograph of the Northern blot hybridization is shown in Fig. 18.3. A single NiR mRNA of about 2.3 kb in length is present in the poly(A)$^+$ RNA of spinach leaves. The NiR mRNA band was increased considerably in samples that were treated with nitrate for 26 h or 100 h (lanes b and d) compared to controls (lanes a and c). A significant amount of NiR mRNA is, however, present in uninduced RNA samples, possibly reflecting a constitutive level of NiR mRNA in leaf cells. Alternatively, we cannot rule out the possibility that, under the experimental conditions used, trace amounts of nitrate were available to the plants, inducing low levels of NiR mRNA. When the same filter was rehybridized with a heterologous actin cDNA probe from soybean all lanes showed roughly the same hybridization signal (Fig. 18.3B). These results show that nitrate induction leads

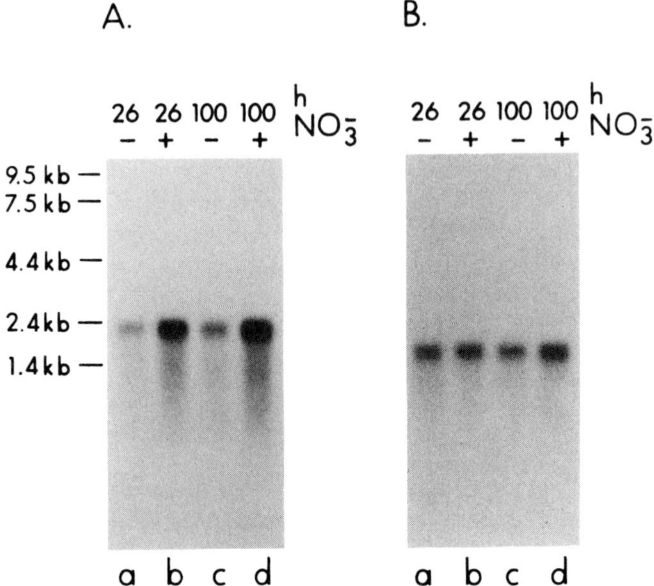

Fig. 18.3. Induction of NiR mRNA by nitrate. Poly(A)$^+$ RNA (10 μg) from plants treated with nitrate (+) or with H$_2$O (−) were loaded on a 0.8% agarose gel, electrophoresed and transferred to nitrocellulose. (A) The nitrocellulose filter was hybridized to ^{32}P-labelled insert DNA from the NiR cDNA pCIB400 (3×10^6 cpm/ml; 2×10^6 cpm/μg) and washed under high stringency conditions in $0.1 \times$ SSC/0.1 per cent SDS at 60°C ($1 \times$ SSC $= 0.15$ M NaCl, 0.015 M Na citrate). (B) The same filter was rehybridized to ^{32}P-labelled insert DNA of the actin gene pSAc3 (Shah *et al.* 1982) from soybean (3×10^5 cpm/ml; 10^8 cpm/μg). The filter was washed under low stringency conditions in $2 \times$ SSC/0.1 per cent SDS at 50°C. Molecular size markers were a RNA ladder from Bethesda Research Laboratories, Life Technologies, Inc.

to a marked increase in the steady-state level of NiR mRNA after 100 h. The NiR mRNA level may continue to increase over a more extended time period. Whether this is due to changes in transcription rate, altered RNA processing, or decreased mRNA degradation remains to be determined.

Recently, it has been shown in barley (Cheng *et al.* 1986 Chapter 13) and squash (Crawford *et al.* 1986, Chapter 21) that the steady-state level of NR mRNA is also increased in response to nitrate induction.

The isolated NiR cDNA clone can be used in the analysis of several important molecular processes. The regulatory elements for NiR can be studied in an attempt to elucidate the mechanisms of nitrate-induced expression of the gene. It should also be possible to see whether the same NiR gene is expressed in the leaves and roots of spinach, as well as in other plant species. Furthermore, the functional dissection of the NiR transit peptide sequence might permit better understanding of the mechanism of transport of proteins into chloroplasts. Finally, nitrite reduction has been the object of considerable biochemical interest and it should now be possible through the use of *in vitro* mutagenesis and expression of this gene in bacteria to delineate the amino acids important for enzymatic function.

Acknowledgements

We thank Shericca Williams and Steve Lombardi for synthesizing oligonucleotides, Bill Dunne for plasmid DNA isolation, and Richard Meagher for providing the actin cDNA clone pSAc3 from soybean.

References

Aparicio, P. J., Knaff, D. B., and Malkin, R. (1975). The role of an iron–sulfur center and siroheme in spinach nitrite reductase. *Archives of Biochemistry and Biophysics* **169**, 102–7.

Back, E., Burkhart, W., Moyer, M., Privalle, L., and Rothstein, S. (1988). Isolation of cDNA clones coding for spinach nitrite reductase: complete sequence and nitrate induction. *Molecular and General Genetics* **212**, 20–6.

Beevers, L. and Hageman, R. H. (1969). Nitrate reduction in higher plants. *Annual Review of Plant Physiology* **20**, 495–522.

Broglie, R., Coruzzi, G., Fraley, R. T., Rogers, S. G., Horsch, R. B., Niedermeyer, J. G., Fink, C. L., Flick, J. S., and Chua, N. H. (1984). Light regulated expression of a pea ribulose-1,5-biphosphate corboxylase small subunit gene in transformed plant cells. *Science* **224**, 838–43.

Cammack, R., Hucklesby, D. P., and Hewitt, E. J. (1978). Electron-paramagnetic-resonance studies of the mechanism of leaf nitrite reductase: signals from the iron-sulfer center and haem under turnover conditions. *Biochemical Journal* **171**, 519–26.

Cheng, C. L., Dewdney, J., Kleinhofs, A., and Goodman, H. M. (1986). Cloning and nitrate induction of nitrate reductase mRNA. *Proceedings of the National Academy of Science, USA* **83**, 6825–8.

Chua, N. H. and Schmidt, G. W. (1978). Post-translational transport into intact chloroplasts of a precursor tro the small subunit of ribulose-1,5-bisphosphate carboxylase. *Proceedings of the National Academy of Science, USA* **75**, 6110–14.

Chua, N. H. and Schmidt, G. W. (1979). Transport of proteins into mitochondria and chloroplasts. *Journal of Cell Biology* **81**, 461–83.

Clarkson, D. T. (1986). Regulation of the absorption and release of nitrate by plant cells: a review of current ideas and methodology. In *Fundamental, ecological and agricultural aspects of nitrogen metabolism in higher plants* (eds H. Lambers, J. Neeteson, and I. Stulen), pp. 3–28. Nijhoff, M. and Junk, W., Den Haag, Netherlands.

Coruzzi, G., Broglie, R., Cashmore, A., and Chua, N. H. (1983). Nucleotide sequences of two pea cDNA clones encoding the small subunit of ribulose bisphosphate carboxylase and the major chlorophyll a/b-binding thylakoid polypeptide. *Journal of Biological Chemistry* **258**, 1399–402.

Crawford, N. M., Campbell, W. H., and Davis, R. W. (1986). Nitrate reductase from squash: cDNA cloning and nitrate regulation. *Proceedings of the National Academy of Science, USA* **83**, 8073–6.

Dalling, M. J., Hucklesby, D. P., and Hageman, R. H. (1973). A comparison of nitrite reductase from green leaves, scutella and roots of corn. *Plant Physiology* **51**, 481–4.

Ferguson, A. R. (1969). The nitrogen metabolism of *Spirodela oligorrhiza*. II. Control of the enzymes of nitrogen assimilation. *Planta* **88**, 353–63.

Grossman, A. R., Bartlett, S. G., Schmidt, G. W., Mullett, J. E., and Chua, N. H. (1982). Optimal conditions for post-translational uptake of proteins by isolated chloroplasts. *In vitro* synthesis and transport of plastocyanin, ferredoxin-NADP$^+$ oxidoreductase and fructose-1,6-bisphosphatase. *Journal of Biological Chemistry* **257**, 1558–63.

Guerrero, M. G., Vega, J. M., and Losada, M. (1981). The assimilatory nitrate-reducing system and its regulation. *Annual Review of Plant Physiology* **32**, 169–204.

Gupta, S. C. and Beevers, L. (1984). Synthesis and degradation of nitrite reductase in pea leaves. *Plant Physiology* **75**, 251–2.

Gupta, S. C. and Beevers, L. (1987). Regulation of nitrite reductase. *Plant Physiology* **83**, 750–4.

Haynes, R. J. (1986). Uptake and assimilation of mineral nitrogen by plants. In *Mineral nitrogen in the plant–soil system* (ed. R. J. Haynes), pp. 303–58. Academic Press, Orlando.

Heidecker, G. and Messing, J. (1986). Structural analysis of plant genes. *Annual Review of Plant Physiology* **37**, 439–66.

Hewitt, E. J., Hucklesby, D. P., and Notton, B. A. (1976). Nitrate metabolism. In *Plant biochemistry* (eds J. Bonner and J. E. Varner), pp. 633–81. Academic Press, New York, London.

Hirasawa, M., Fukushima, K., Tamura, G., and Knaff, D. B. (1984). Immunochemical characterization of nitrite reductases from spinach leaves, spinach roots and other higher plants. *Biochemica et Biophysica Acta* **791**, 145–54.

Ho, C. H. and Tamura, G. (1973). Purification and properties of nitrite reductase from spinach leaves. *Agricultural and Biological Chemistry* **37**, 37–44.

Hucklesby, D. P., Dalling, M. J., and Hageman, R. H. (1972). Some properties of two forms of nitrite reductase from corn scutellum. *Planta* **104**, 220–33.

Ida, S. (1987). Immunological comparisons of ferredoxin-nitrite reductases from higher plants. *Plant Science* **49**, 111–16.

Ida, S. and Mikami, B. (1986). Spinach ferredoxin-nitrite reductase: a purification procedure and characterization of chemical properties. *Biochimica et Biophysica Acta* **871**, 167–76.

Ishiyama, Y. and Tamura, G. (1985). Isolation and partial characterization of homogeneous nitrite reductase from etiolated bean shoots (*Phaseolus angularis* W. F. Wight). *Plant Science Letters* **37**, 251–6.

Ishiyama, Y., Shinoda, I., Fukushima, K., and Tamura, G. (1985). Some properties of ferredoxin-nitrite reductase from green shoots of bean and an immunological comparison with nitrite reductase from roots and etiolated shoots. *Plant Science* **39**, 89–95.

Karlin-Neumann, G. A. and Tobin, E. M. (1986). Transit peptides of nuclear-encoded chloroplast proteins share a common amino acid framework. *EMBO Journal* **5**, 9–13.

Kelker, H. C. and Filner, P. (1971). Regulation of nitrite reductase and its relationship to the regulation of nitrate reductase in cultured tobacco cells. *Biochimica et Biophysica Acta* **252**, 69–82.

Lancaster, J. R., Vega, J. M., Kamin, H., Orme-Johnson, N. R., Orme-Johnson, W. H., Krueger, R. J., and Siegel, L. M. (1979). Identification of iron-sulfer center of spinach ferredoxin-nitrite reductase as a tetranuclear center and preliminary EPR studies of mechanism. *Journal of Biological Chemistry* **254**, 1268–72.

Lütcke, H. A., Chow, K. C., Mickel, F. S., Moss, K. A., Kern, H. F., and Scheele, G. A. (1987). Selection of AUG initiation codons differ in plants and animals. *EMBO Journal* **6**, 43–8.

Maniatis, T., Fritsch, E. F., and Sambrook, J. (1982). *Molecular cloning, A laboratory manual.* Cold Spring Harbor Laboratory Press, NY.

McRee, D. E., Richardson, D. C., Richardson, J. S., and Siegel, L. M. (1986). The heme and Fe^4S^4 cluster in the crystallographic structure of *Escherichia coli* sulfite reductase. *Journal of Biological Chemistry* **261**, 10277–81.

Mishkind, M. L., Wessler, S. R., and Schmidt, G. W. (1985). An intermediate in the maturation of the ribulose-1,5-bisphosphate carboxylase small subunit precursor. *Journal of Cell Biology* **100**, 226–34.

Murphy, M. J., Siegel, L. M., Tove, S. R., and Kamin, H. (1974). Siroheme: A new prothetic group participating in six-electron reduction reaction catalyzed by both sulphite and nitrite reductases. *Proceedings of the National Academy of Science, USA,* **71**, 612–16.

Ninomiya, Y. and Sato, S. (1984). A ferredoxin-like electron carrier from non-green cultured tobacco cells. *Plant and Cell Physiology* **25**, 453–8.

Oaks, A. and Hirel, B. (1985). Nitrogen metabolism in roots. *Annual Review of Plant Physiology* **36**, 345–65.

Oji, Y., Watanabe, M., Wakiuchi, N., and Okamoto, S. (1985). Nitrite reduction in barley root plastids: Dependence on NADPH coupled with glucose-6-phosphate and 6-phosphogluconate dehydrogenases, and possible involvement of an electron carrier and a diaphorase. *Planta* **165**, 85–90.

Ostrowski, J., Back, E. W., Madden, J., Siegel, L. M., and Kredich, N. M. (1987). Comparison of deduced amino acid sequences of α and β subunits of *S. Typhimurium* NADPH sulfite reductase with those of spinach nitrite reductase and cytochrome P450 oxidoreductase. *Federation Proceedings* **46**, 2231, Abst. 1780.

Rajasekhar, V. K. and Mohr, H. (1986). Appearance of nitrite reductase in cotyledons of the mustard (*Sinapis alba* L.) seedling as affected by nitrate, phytochrome and photooxidative damage of plastids. *Planta* **168**, 369–76.

Robinson, C. and Ellis, J. J. (1984*a*). Transport of proteins into chloroplasts: partial purification of a chloroplast protease involved in the processing of imported precursor polypeptides. *European Journal of Biochemistry* **142**, 337–42.

Robinson, C. and Ellis, R. J. (1984*b*). Transport of proteins into chloroplasts: the precursor of small subunit of ribulose bisphosphate carboxylase is processed to the mature size in two steps. *European Journal of Biochemistry* **142**, 343–6.

Sanger, F., Nicklen, S., and Coulson, A. R. (1977). DNA sequencing with chain terminating inhibitors. *Proceedings of the National Academy of Science, USA* **74**, 5463.

Schmidt, G. W., Bartlett, S. G., Grossman, A. R., Cashmore, A. R., and Chua, N. H. (1981). Biosynthetic pathways of two polypeptide subunits of the light-harvesting chlorophyll *a/b* protein complex. *Journal of Cell Biology* **91**, 468–78.

Schreier, P. H., Seftor, E. A., Schell, J., and Bohnert, H. J. (1985). The use of nuclear-encoded sequences to direct the light-regulated synthesis and transport of a foreign protein into plant chloroplasts. *EMBO Journal* **4**, 25–32.

Serra, J. L., Ibarlucea, J. M., Arizmendi, J. M., and Llama, M. J. (1982). Purification and properties of the assimilatory nitrite reductase from barley (*Hordeum vulgare*) leaves. *Biochemical Journal* **201**, 167–70.

Shah, D. M., Hightower, R. C., and Meagher, R. B. (1982). Complete nucleotide sequence of a soybean actin gene. *Proceedings of the National Academy of Science* **79**, 1022–6.

Siegel, L. M., Wilkerson, J. O., and Janick, P. A. (1987). Structural studies of the siroheme–[4Fe–4S] cluster active centers of spinach ferredoxin-nitrite reductase and *Escherichia coli* sulfite reductase. In *Inorganic nitrogen metabolism* (eds W. R. Ulrich, P. J. Aparicio, P. J. Syrett, and F. Castillo), pp. 118–22. Springer-Verlag, Heidelberg.

Small, I. S. and Gray, J. C. (1984). Synthesis of wheat leaf nitrite reductase *de novo* following induction with nitrate and light. *European Journal of Biochemistry* **145**, 291–7.

Smeekens, S., De Groot, M., Van Binsbergen, J., and Weisbeek, P. (1985). Sequence of the precursor of the chloroplast thylakoid lumen protein plastocyanin. *Nature* **317**, 456–8.

Smeekens, S., Bauerle, C., Hageman, J., Keegstra, K., and Weisbeek, P. (1986). The role of the transit peptide in the routing of precursors toward different chloroplast compartments. *Cell* **46**, 365–75.

Stewart, G. R. (1972). The regulation of nitrite reductase level in *Lemna minor*. *Journal of Experimental Botany* **23**, 171–83.

Suzuki, A., Oaks, A., Jacquot, J. P., Vidal, J., and Gadal, P. (1985). An electron transport system in maize *Zea mays* roots for reactions of glutamate synthase and nitrite reductase physiological and immunochemical properties of the electron carrier and pyridine nucleotide reductase. *Plant Physiology* **78**, 374–8.

Van den Broeck, G., Timko, M. P., Kausch, A. P., Cachmore, A. R., Van Montagu, M., and Herrera-Esterella, L. (1985). Targeting of foreign proteins to chloroplasts by fusion of the transit peptide from the small subunit of ribulose-1,5-biphosphate carboxylase. *Nature* **313**, 358–63.

Vega, J. M. and Kamin, H. (1977). Spinach nitrate reductase: purification and properties of a siroheme-containing iron–sulfer enzyme. *Journal of Biological Chemistry* **252**, 896–909.

Vega, J. M., Cardenas, J., and Losada, M. (1980). Ferredoxin-nitrite reductase. *Methods in Enzymology* **69**, 255–70.

Wood, W. I., Gitschier, J., Lasky, L. A., and Lawn, R. M. (1985). Base composition-independent hybridization in tetramethylammonium chloride: a method for oligonucleotide screening of highly complex gene libraries. *Proceedings of the National Academy of Science, USA* **82**, 1585–8.

Young, R. A. and Davis, R. W. (1983). Efficient isolation of genes using antibody probes. *Proceedings of the National Academy of Science, USA* **80**, 1194–8.

Regulation of Nitrate Assimilation

19. Regulation of nitrate assimilation in *Aspergillus nidulans*

Claudio Scazzocchio and Herbert N. Arst, Jr

Introduction

In *Aspergillus nidulans* most genes coding for activities involved in the utilization of nitrogen sources are subject to two forms of regulation, pathway-specific induction and nitrogen metabolite repression. Induction is usually mediated by a positive-acting pathway-specific gene, while nitrogen metabolite repression is mediated by the positive-acting wide domain *areA* gene (for a review see Arst and Scazzocchio (1985) and references therein).

Nitrate assimilation involves nitrate uptake and necessitates the conversion of nitrate to ammonium, catalysed sequentially by two enzymes, nitrate reductase (NR) and nitrite reductase (NiR). The three genes which code for these activities are clustered on chromosome VIII and are *crnA*, coding for a nitrate permease of physiological importance mainly in conidiospores and young germlings (Brownlee and Arst 1983), *niaD* encoding NR, and *niiA* encoding NiR (Cove 1979 and references therein; see Chapter 6 for recent molecular data on the nitrate assimilation cluster). The data of Brownlee and Arst (1983) imply the existence of at least one additional system for nitrate uptake.

While synthesis of Nr and NiR is induced by nitrate and nitrite and is under the specific control of the *nirA* gene (see below), the expression of *crnA* is independent of *nirA*. A functional product of the wide domain *areA* gene is necessary for the expression of all three genes of the nitrate assimilation cluster (Arst and Cove 1973; Brownlee and Arst 1983).

Nitrate assimilation and the *cnx* genes

Nitrate reductase activity requires the presence of a molybdenum-containing cofactor (MoCo) (Pateman *et al.* 1964; Cove 1979), synthesis of which depends upon the products of the *cnx* genes. MoCo is necessary for the activity of at least two other enzymes — purine hydroxylase I which is physiologically a hypoxanthine–xanthine dehydrogenase and purine hydroxylase II which is physiologically a nicotinate hydroxylase (Scazzocchio *et al.* 1973; Lewis *et al.* 1978; Scazzocchio 1980). The pathways of purine and nicotinate degradation have been studied in detail and their structural and regulatory genes and the physiological regulatory effectors have been identified (Scazzocchio and Darlington 1967, 1968; Scazzocchio *et al.* 1973; Scazzocchio 1973; Sealy-Lewis *et al.* 1978, 1979; Scazzocchio 1980; Arst and Scazzocchio 1985).

The function of each of the *Aspergillus cnx* genes in the synthesis of the cofactor is not yet clear, and it must be remembered that until recently the structure of the cofactor was unknown (Johnson *et al.* 1984; Chapter 14). Six *cnx* genes have been identified: *cnxABC, cnxE, cnxF, cnxG, cnxH* (Cove and Pateman 1963; Cove 1979), and *cnxJ* (Arst *et al.* 1982*a*). The *cnxABC* complex presents a complementation map typical of a gene coding for a bifunctional protein where B represents the non-complementing class (Cove and Pateman 1963; Cove 1979). A cryosensitive allele of *cnxC* has been described by Arst *et al.* (1982*a*). Mutations mapping in the *cnxE* gene have provided the first direct evidence for involvement of Mo in the cofactor. Strains carrying these mutations can be supplemented by high concentrations of molybdate (Arst *et al.* 1970), and show a clear intracistronic complementation pattern (Hartley 1970). Strains carrying mutations with a *cnx* phenotype which are supplemented by molybdate have since been found in a number of organisms. The *cnxH* gene is particularly interesting; thermosensitive alleles mapping in this gene result in a temperature-sensitive NR, in contrast to thermosensitive alleles mapping in the *cnxE* and *cnxF* genes. The stability of NR in strains carrying thermosensitive *cnxE* or *cnxF* mutations, grown at permissive temperature, cannot be distinguished from that of the wild-type enzyme (MacDonald and Cove 1974). When strains carrying thermosensitive *cnxF* or *cnxE* alleles are tested on media diagnostic for the presence of purine hydrogenase I or purine hydroxylase II, there is loss of growth at the non-permissive temperature and no effect on growth at the permissive temperature. This again contrasts with the behaviour of the thermosensitive *cnxH* alleles, which have a conditional phenotype for growth on nitrate but are non-conditional for the activity of the purine hydroxylases (Scazzocchio *et al.* 1973). The simplest interpretation is that the requirement for the *cnxH* product is more stringent for the purine hydroxylases than for NR. Moreover, intracistronic complementation tests with mutant *cnxH* alleles show that certain pairs complement for nitrate utilization whilst failing to complement for hypoxanthine utilization (D. J. Cove and C. Scazzocchio, unpublished results). There is no direct evidence for a *cnxH* structural role in the purine hydroxylases. In no case has more than one type of polypeptide chain been identified as a component of any of the three *Aspergillus* enzymes after purification and/or immunoprecipitation (Mehra and Coughlan 1984; Cooley and Tomsett 1985; H. M. Sealy-Lewis, F. Hanselman, and C. Scazzocchio, unpublished results). Thus the question of structural participation of the *cnxH* product in the molybdoproteins of *A. nidulans* remains open. No non-leaky mutations in *cnxJ* have ever been selected: Arst *et al.*

(1982*a*) have rationalized this finding by suggesting that loss of *cnxJ* function merely leads to a decreased concentration of MoCo. They postulated an unspecified regulatory role for *cnxJ*. One attractive hypothesis would be that *cnxJ* codes for a binding protein able to complex and protect the extremely labile MoCo. Recently, a protein with such a function has been found in other systems (Chapter 14).

The regulation of the *cnx* genes

Expression of the *cnx* genes is required in three different metabolic contexts, corresponding to growth on nitrate, purines or nicotinate as sole nitrogen source. An interesting question is whether their expression is constitutive or whether it can be elicited through the three pathway-specific, independent induction systems.

Two types of study indicate that the expression of the *cnx* genes is constitutive. Garrett and Cove (1976) monitored the concentration of MoCo using an *in vitro* complementation assay between *cnx* and *niaD* mutations and showed that the concentration of cofactor in a strain carrying a *niaD* mutation is not affected by nitrate, uric acid or nicotinate induction, but is strongly repressible by ammonium. Moreover, the concentration of cofactor does not depend on whether a $nirA^+$, $nirA^-$, or $nirA^c$ allele (see below) is present in the *niaD* mutant (Garrett and Cove 1976). This *in vitro* complementation test monitors the biosynthetically complete cofactor produced in the *niaD* strain, thus if any one of the *cnx* gene products is limiting under non-induced conditions (but not under induced conditions), the cofactor would have appeared inducible. The conclusion is that all *cnx* genes are expressed constitutively. The second line of evidence relevant to the problem of cofactor regulation is indirect and genetic. Cove (1977) has reported that mutations in *nirA*, the specific regulatory gene for the nitrate assimilation pathway (see below), complement very poorly in heterokaryons with mutations in *niaD* and *niiA*. These genes are under the control of *nirA* (see below and Chapter 6). Complementation in diploids is entirely normal. Scazzocchio and co-workers have observed the same mutant complementation pattern involving *uaY*, the regulatory gene of purine catabolism and *hxA* and *uaZ* encoding, respectively xanthine dehydrogenase (purine hydroxylase I) and urate oxidase (Scazzocchio, unpublished Thesis) 1966; Scazzocchio *et al.* 1982). Studies involving *in vitro* transcription and immunoprecipitation have shown that *uaY* controls expression of *hxA* and *uaZ*, probably at the level of transcription (unpublished results from the laboratory of C. Scazzocchio). Mutations in the structural genes (*niaD* and *niiA*; *hxA* and *uaZ*) complement each other readily in both heterokaryons and diploids. While the basis of the failure to complement in heterokaryons is controversial (Scazzocchio *et al.* 1973, 1982; Cove 1977) it provides a diagnostic test to establish whether any *cnx* genes are under the control of *nirA* or *uaY*. The answer is unequivocal; all mutations tested, representing all the *cnx* genes except *cnxJ* (where mutations are too leaky to test), complement completely with *nirA* mutations on nitrate and with *uaY* mutations on hypoxanthine as nitrogen sources.

These results acquire further significance when compared with the complementation pattern for mutations in the *hxB* gene. Mutations in this gene result in loss of substrate-specific enzyme activity for both purine hydroxylases. The two enzymes extracted from mutant strains have molecular weights, electrophoretic mobilities, immunological properties and NADH dehydrogenase ancillary activities identical to those of the wild-type enzyme (Scazzocchio 1980). Thus *hxB* probably codes for a post-transitional modification necessary for purine hydroxylases but not for NR. This modification might well involve the sulphide ion, which is a terminally inserted ligand of the molybdenum in the purine hydroxylases but not in nitrate reductase (Chapter 14). This function has been established for the *ma-1* (*maroon-like*) mutations of *Drosophila* (Wahl *et al.* 1982) which are in every way phenotypically equivalent to the *hxB* mutations of *Aspergillus* (Scazzocchio 1980). The activity of *hxB* is needed in the two contexts of purine and nicotinate utilization, when xanthine dehydrogenase (purine hydroxylase I) or nicotinate hydroxylase (purine hydroxylase II) is induced. Mutations in the *hxB* gene do not complement in heterokaryons but do complement in diploids with *uaY* mutations. There are two types of tightly linked mutations affecting nicotinate hydroxylase: a constitutive type mapping in *ap1A* and a non-inducible type mapping in *hxnR*. It is not clear whether *ap1A* and *hxnR* are indeed two separate genes, but this seems the most likely hypothesis. Mutations in *hxnR* do not complement with mutations in *hxB* in heterokaryons whilst complementing in diploids (L. Gorfinkiel and C. Scazzocchio, unpublished results). Constitutive *ap1A*[c] mutations must be in the same nucleus as a functional *hxB* allele for the constitutive phenotype to be detectable *in vivo* (Scazzocchio 1973). The conclusion is that, while the *cnx* genes seem to be expressed constitutively, *hxB* is under two alternative induction systems. Thus two different solutions have been adopted in the course of evolution to deal with very similar problems. The cloning of the *hxB* and *cnx* genes will allow a definitive analysis of their regulation. Transformation with an *A. nidulans* gene library has resulted in complementation of *hxB*, *cnxE*, *cnxG* and *cnxH* mutations, and cloning of these genes might be achieved in the near future (unpublished results from the laboratory of C. Scazzocchio).

The *nirA* gene and its role in the regulation of nitrate assimilation

The *nirA* gene has been defined by many hundreds of mutations of the non-inducible type (Cove 1979). The *nirA*[−] mutations are strictly co-dominant with *nirA*[+], implying a stringent dose effect (Cove 1969). Whilst deletion or polypeptide chain termination mutations have not been identified, the high frequency of appearance of *nirA⁻* mutations is only compatible with a positive control system. These mutations prevent the expression of *niaD* and *niiA* but not that of *crnA* (see below). Two other types of mutations (*nirA^c* and *nirA^d*) map in *nirA*: *nirA^d* mutations will be discussed later. The *nirA^c* mutations result in constitutivity for both NR and NiR and are strictly co-dominant to both *nirA⁺* and *nirA⁻*, implying, again, a stringent dose effect (Cove 1969, 1979). Whilst *nirA^c* mutations are rare, the fact that they can be obtained using rather cumbersome selection methods (Pateman and Cove 1967; Rand and Arst 1978) implies that they are far more frequent than would be expected for a purely positive induction system. Cove and Pateman (1969) have postulated that the native NR, [now known to be a dimer (Cooley and Tomsett 1985)], when complexed with the cofactor (Cove 1979), acts as a repressor of its own synthesis and that of NiR. The simplest hypothesis compatible with the data is that the *nirA* product is necessary for *niaD* and *niiA* expression and that the NR holoenzyme prevents activation by the *nirA* gene product by complexing it. It should be noted that this hypothesis necessitates only one nitrate binding site, the liganding of nitrate (or nitrite) to NR promoting the dissociation of the complex and allowing transcription of *niaD* and *niiA*. Constitutive mutations are easily accounted for as affecting the complexing of the *nirA* gene product with NR. This hypothesis was postulated on purely genetic grounds (Pateman and Cove 1967; Cove and Pateman 1969; Cove 1979). It is interesting that recent molecular work on other systems has shown similar mechanisms: in *Saccharomyces cerevisiae* the *GAL4* gene product is a positive regulator whose action can be prevented by complexation with the product of the *GAL80* gene (Lue *et al.* 1987 and references therein). In *Neurospora crassa* the product of the *qa-1F* gene is needed for expression of the quinate utilization gene cluster. The products of the *qa-1S* gene prevent its action, again probably by direct complexation (Giles *et al.* 1985). The co-inducer, quinate, would bind to the product of the *qa-1S* gene (Huiet and Giles 1986): it has recently been shown that the *qa-1S* gene has a striking isology with the *arom* cluster gene (Anton *et al.* 1987) which codes for a pentafunctional polypeptide which catalyses steps 2–6 of chorismic acid biosynthesis. Quinate degradation and aromatic amino acid biosynthesis share one enzymatic step, the dehydrogenation of quinate. The *qa-1S⁻* mutations described by Huiet and Giles (1986) lie in a region isologous with the biosynthetic quinate and shikimate dehydrogenase domain of *arom*, suggesting that they might lie in a region involved in inducer binding (Anton *et al.* 1987). These findings are important because they provide a bridge in the possible evolution of negative-acting regulatory proteins from their cognate (or related) enzymes. Returning to NR, a clear prediction can be made. It should be possible to isolate dominant mutations which would result in non-inducibility of both NR and NiR and which would map in *niaD*. These would define the nitrate binding site.

The *nirA* gene has recently been cloned by direct complementation of an *nirA⁻* mutation. A 5 kb fragment with *nirA⁻*-complementing activity has been isolated from a genomic library (J. Tilburn, G. Berger, and C. Scazzocchio, unpublished results). Thus all the elements are now available for a molecular approach to the mechanism of control of the nitrate assimilation gene cluster. Kinghorn (Chapter 6) reviews the progress made in *A. nidulans* thus far. Progress made with the *N. crassa* system is discussed in Chapter 20.

Nitrogen metabolite repression

In *A. nidulans* nitrogen metabolite repression is mediated by a positive-acting regulatory gene designated *areA* (Arst and Cove 1973). The loss-of-function phenotype (of alleles designated *areA*[r] for repressed) is inability to utilize nitrogen sources other than ammonium (and, to some degree, L-glutamine) and low to undetectable levels of enzymes and permeases necessary for utilization of such nitrogen sources (Arst and Cove 1973; Hynes 1975; Polya *et al.* 1975; Rand and Arst 1977; Arst *et al.* 1980, 1982*b*; Tollervey and Arst 1982; Spathas *et al.* 1982, 1983). The loss-of-function phenotype was originally identified on the basis of frequency and recessivity (Arst and Cove 1973) but has since been confirmed as the phenotype of putative polypeptide chain termination alleles, a translocation-based allele having a translocation breakpoint within the coding region and a major deletion allele present in duplication-deficiency strains derived from crosses heterozygous for a near terminal pericentric inversion (Rand and Arst 1977; Arst 1981, 1982; Al Taho *et al.* 1984; Caddick *et al.* 1986; S. Sibley, R. W. Davies, N. Martinez-Rossi, M. X. Caddick, and H. N. Arst, Jr., unpublished results). A much rarer class of *areA* alleles, designated *areA*[d], leads to nitrogen metabolite derepression of the synthesis of one or more enzymes or permeases under *areA* control (Cohen 1972; Arst and Cove 1973; Pateman *et al.* 1973; Hynes 1975; Polkinghorne and Hynes 1975, 1982; Arst and Scazzocchio 1975; Arst 1977; Shaffer and Arst 1984). The designation of a particular allele as *areA*[r] or *areA*[d] is made in the context of an individual activity. For example, *areA*-300 is derepressed for synthesis of thymine 7-hydroxylase but repressed for synthesis of at last one activity necessary for utilization of L-leucine or L-methionine as nitrogen source (Shaffer and Arst 1984). Table 19.1 illustrates *areA*[r] and *areA*[d] phenotypes with respect to nitrate reductase levels.

The occurrence of allele-specific genetic suppression of certain *areA*[r] mutations (i.e. the putative polypeptide chain termination mutations) is evidence that *areA* encodes a protein (Al Taho *et al.* 1984). There is very considerable heterogeneity of mutant *areA* phenotypes, easily uncovered by examination of conditional *areA*[r] alleles. For example, *areA*[r]-1 has a thermosensitive phenotype for expression of the nitrate assimilation pathway but a non-conditional phenotype for expression of a number of other activities whereas *areA*[r]-2 exhibits thermosensitivity for certain of these other activities but not for the nitrate assimilation pathway (Arst and Cove 1973). Similar variations in thermosensitivity involving the nitrate assimilation pathway can be seen when phenotypes of putative polypeptide chain termination mutations such as *areA*[r]-601 are compared in strains carrying

Table 19.1. Nitrate reductase levels in *A. nidulans* wild type and *areA* mutants

Relevant genotype	Relative nitrate reductase activity		
	Uninduced	Induced (NO_3^-)	Induced but repressed ($NO_3^- + NH_4^+$)
Wild type (*areA*$^+$)	8	100	13
areA^r-18	0	0	–
xprD-1 (*areA*^{d})	1	179	141

Data are collated from Arst and Cove (1973), Tollervey and Arst (1981) and Arst *et al.* (1982). Activities are expressed as a percentage of the induced, nitrogen metabolite-de-repressed activity of the wild type.

various allele-specific, broad specificity suppressor mutations (Al Taho *et al.* 1984). This heterogeneity of mutant *areA* phenotypes also includes instances, such as that noted above for *areA*-300, in which an *areA* mutation has an *areA*^{d} phenotype for expression of one structural gene and an *areA*^{r} phenotype for expression of another. It is even known for reversion of such an allele to lead to new mutant alleles of nearly opposite phenotype (Arst and Scazzocchio 1975; Wiame *et al.* 1985). Such non-hierarchical heterogeneity of mutant phenotypes indicates direct involvement of the *areA* product in the regulation of structural genes under its control as well as differences in structure of the various *areA* product receptor sites (which would provide a basis for differences in degrees of nitrogen metabolite repressibility) (Arst and Cove 1973; Arst and Bailey 1977; Arst and Scazzocchio 1985). There are also examples of genes indirectly involved in nitrogen metabolite repression. Mutations in *meaA* reduce ammonium uptake and result in nitrogen metabolite derepression (Arst and Cove 1969; Arst and Page 1973). Loss-of-function mutations in the structural genes for NADP-linked glutamate dehydrogenase (*gdhA*) and glutamine synthetase (*glnA*) lead to nitrogen metabolite de-repression because L-glutamine rather than ammonium is the probable repressing metabolite and effector for the *areA* product (Arst and MacDonald 1973; Pateman *et al.* 1973; Kinghorn and Pateman 1973,1975; Arst *et al.* 1982*b*; MacDonald 1982; Cornwell and MacDonald 1984; Arst and Scazzocchio 1985; Wiame *et al.* 1985). There is no evidence that any gene other than *areA* is directly involved in nitrogen metabolite repression in *A. nidulans* (Arst *et al.* 1982*b*; Wiame *et al.* 1985).

In addition to its relatively well-defined role in nitrogen metabolite repression, *areA* has been implicated in repression by atmospheric oxygen. Thymine 7-hydroxylase, a pyrimidine salvage pathway enzyme, is partially repressed by oxygen under nitrogen metabolite de-repressing growth conditions and the *areA*-300 mutation relieves oxygen as well as nitrogen metabolite repression of the synthesis of this enzyme (Shaffer and Arst 1984). The synthesis of at least one other *areA*-regulated activity is controlled in a similar manner but most activities under *areA* control tested are not (P. M. Shaffer and H. N. Arst, Jr., in preparation).

areA occupies an interesting chromosomal position in that no indispensable gene can lie between it and the telomere of the right arm of chromosome III (Arst 1982). This is shown by the viability of duplication-deficiency progeny from crosses heterozygous for a near terminal pericentric inversion (Arst 1982) or for a reciprocal translocation (Rand and Arst 1977; Arst 1981; Caddick *et al.* 1986). In both cases a breakpoint occurs within the *areA* gene and duplication-deficiency progeny lack a part of *areA* plus the whole of the region centromere distal to it (but the region of the genome which is duplicated differs in the two kinds of duplication-deficiency strains) (Caddick *et al.* 1986; S. Sibley, R. W. Davies, M. X. Caddick and H. N. Arst, Jr., unpublished results). A minimum size of 189 kb (nearly 0.75 per cent of the nuclear genome) can be deduced for the dispensable region distal to *areA* from clones isolated thus far but the whole of the region has yet to be cloned (Caddick *et al.* 1986; J. C. Connelly, S. Gnanasoorian, M. X. Caddick, and H. N. Arst, Jr., unpublished results).

The *areA* gene was cloned making novel use of a duplication-deficiency strain (Caddick *et al.* 1986). As noted above, crosses heterozygous for a mutant *areA* allele based upon a near terminal pericentric inversion yield (30–35%) duplication-deficiency progeny (Arst 1982). Clones were sought from a λ Charon gene library able to hybridize to (total) DNA from a wild-type strain but not to DNA from a duplication-deficiency strain. A number of clones obtained in this way contained DNA from entirely within the deficiency region but a clone able to hybridize weakly to DNA from the duplication-deficiency strain (because 2.8 kb of its 11.6 kb insert is contained in the DNA of such a strain) contained, as predicted, the *areA* gene. The cloning of *areA* was confirmed by the ability of the clone to transform an *areA*[r] strain to *areA*[+] and by hybridization of the clone in Southern blots to the genomic region containing breakpoints for two translocations and one inversion associated with different mutant *areA* alleles (Caddick *et al.* 1986).

The DNA sequence of *areA* has been largely determined (S. Sibley, R. W. Davies, N. Martinez-Rossi, M. X. Caddick, and H. N. Arst, Jr., unpublished results). The translated protein sequence reveals a 'helix-turn-helix' (Chou and Fasman 1978) likely to be a DNA-binding region (C. F. Bennett, M. X. Caddick, and H. N. Arst, Jr., unpublished results). Localization of the inversion breakpoint of an inversion-associated *areA*[d] allele and analysis of transformants in which defined parts of the *areA* gene have integrated heterologously suggest that a region where glutamine is able to interact with and inactivate the *areA* product is located towards the C-terminal end of the protein (C. F. Bennett, M. X. Craddick, and H. N. Arst, Jr., unpublished results).

One of the most surprising findings to emerge from classical genetic analysis of *areA* is that it is possible to obtain intra-cistronic partial revertants of *areA*[r]-18, an allele in which the coding region is split by a reciprocal translocation (Tollervey, unpublished thesis, 1981). Because duplication-deficiency progeny derived from crosses heterozygous for *areA*^{r}-18 contain the central and 3′ regions of *areA* but lack the 5′ coding and upstream sequences, the *areA* phenotype of such duplication-deficiency progeny clearly indicates whether the second-site reversion event has occurred in the central to 3′ region or in the 5′ region of the gene. Four different revertants have been analysed: in each case the reversion event is associated with the central to 3′ region (i.e. the duplication-deficiency progeny are partially revertant rather than wholly *areA*^{r} in phenotype) (H. N. Arst, Jr., unpublished results). This suggests that the second-site reversion events involve fusion of a functional promoter, ribosome binding site and 'in frame' initiation codon to the central and 3′ regions of the gene. Sequencing of the revertant alleles will provide a critical test of this hypothesis. Classical genetics has, however, provided some crucial support: three of the four reversion mutations are associated with an additional translocation (H. N. Arst, Jr., unpublished results). Although no additional translocation has been detected in the fourth case, the possibility of another chromosomal rearrangement such as an inversion, major deletion or insertion cannot be excluded. The possibility of a second translocation which, like *areA*^{r}-18, involves chromosomes III and IV also cannot be excluded, as it would go undetected upon haploidization of a diploid heterozygous for both translocations.

Selection of locus-specific extra-cistronic suppressors of *areA*^{r} mutations has proved an invaluable source of diverse regulatory mutations (reviewed by Arst and Scazzocchio 1985). One such mutation, *nis*-5 is an insertional translocation which probably fuses *niiA* to a promoter normally on chromosome II apparently in tandem to the *niiA* promoter/initiator region (Arst *et al.* 1979). All but one group of regulatory mutations obtained as locus-specific suppressors are able to suppress *areA*^{r} mutations only for utilization of one or a few nitrogen sources (with a correspondingly narrow regulatory domain when effects on enzyme and permease synthesis are determined). The exceptional group consists of six extremely rare mutations mapping to a locus designated *areB* (Tollervey and Arst 1982; H. N. Arst, Jr., unpublished results). These mutations suppress *areA*^{r} mutations, at least to a limited extent, on all nitrogen sources and therefore apparently cover the whole of the *areA* regulatory domain. All six mutations are associated with chromosomal rearrangements, five with translocations (Tollervey and Arst 1982), and one with a paracentric inversion (M. X. Caddick and H. N. Arst, Jr., unpublished results). Both of the translocation breakpoints of two of the translocation-associated alleles are in identical positions (by classical genetics), despite being selected in different parental strains in separate experiments and having distinct phenotypes (Tollervey and Arst 1982; Wiame *et al.* 1985). It therefore appears that one of a rather small number of chromosomal rearrangements is necessary for *areB* to substitute for *areA*. *areB* can be considered a cryptic regulatory gene, normally inactive under conditions in which *areA* is expressed but able to be activated by a chromosomal rearrangement fusing it to an appropriate promoter and/or expression sequence. This hypothesis is supported by the selection of loss-of-function *areB*^{r} mutations which abolish suppression of $areA^r$ mutations without affecting the *areB* translocation present in the parental strain (Wiame *et al.* 1985; H. N. Arst, Jr., unpublished results).

A possible interaction between the *nirA* and *areA* products

One group of mutations has very interesting characteristics: they map in *nirA*, but they bypass the *areA* functions, resulting in de-repression of NR and NiR. They have been obtained in both $nirA^+$ and $nirA^c$ genetic backgrounds. Standard genetic analysis showed that mutations to the constitutive and to the de-repressed phenotypes are separate and genetically separable, possibly defining different domains of the *nirA* gene (Rand and Arst 1978; Tollervey and Arst 1981). Table 19.2 summarizes the biochemical phenotypes of strains carrying these alleles.

Table 19.2. Nitrate and nitirite reductase levels in *A. nidulans* strains carrying various *nirA* alleles

Relevant genotype	Relative enzyme activities							
	Uninduced		Induced (NO_3^-)		Repressed (NH_4^+)		Induced but repressed ($NO_3^- + NH_4^+$)	
$nirA^+$ (wild type)	1	2	100	100	1	1	5	5
$nirA^c$	75	29	98	117	3	6	20	21
	111	40	94	79	13	9	5	7
$nirA^d$	3	5	120	134	10	6	70	61
$nirA^{c/d}$	94	34	108	124	85	49	74	90
	103	37	135	105	73	35	113	80
	85	27	143	111	85	40	79	63
$nirA^-$	2	2	2	2	–	–	0	4

Under each set of growth conditions the first column shows the nitrate reductase and the second column the nitrite reductase activities. Data are collated from Rand and Arst (1978) and Tollervey and Arst (1981) who give experimental details except for the $nirA^-$ data taken from the review by Cove (1979). Activities are expressed as a percentage of the induced, nitrogen metabolite de-repressed activity of the wild type under identical growth conditions. Data are shown for two $nirA^c$ alleles and three $nirA^{c/d}$ alleles, $nirA^{c/d}$ denoting $nirA^c$ $nirA^d$ double mutations.

In principle, three kinds of mechanism could account for these results, with differing implications for the site of action of the *areA* product. In the first (interacting) model the *areA* gene product would act together with *nirA* gene product on the *cis*-acting regions of *niaD* and *niiA*. It could be supposed that in the wild type only an *areA*-liganded *nirA* product binds to *cis*-binding regions adjacent to *niaD* and *niiA*. The *nirA*[d] mutations would, in this model, be interpreted as resulting in an *nirA* product which adopts a liganded-like conformation even in the absence of the *areA* product.

In the second model, the *nirA* and the *areA* products would both bind to *cis*-acting regions adjacent to *niaD* and *niiA* but without interacting directly with each other. If the *areA* product serves to facilitate binding of the *nirA* product, then *nirA*[d] mutations could be interpreted as increasing the affinity of the *nirA* product for its receptor sites.

In the third (cascade) model, the *areA* gene product would be necessary for *nirA* expression. The role of the wide domain regulatory gene would be to elicit transcription of the pathway-specific regulatory gene. In this model it would be predicted that the *nirA* mutations would lie in a *cis*-acting region adjacent to *nirA* rather than in the translated region of *nirA*. They would allow expression of *nirA* to be independent of *areA*.

Although these three models are not in principle mutually exclusive, it is extremely unlikely that any one *nirA*[d] mutation can act in more than one way. The fact that *nirA*[d] mutations can be obtained at all argues strongly against the *areA* product having more than one site of action per structural gene.

A cascade mechanism of the type discussed above has recently been described in another system of *A. nidulans*: carbon catabolite repression prevents the expression of *alcA* and *aldA* coding for alcohol dehydrogenase and aldehyde dehydrogenase, respectively. The negative-acting wide domain *creA* gene mediates carbon catabolite repression and, at least in this instance, it acts by preventing the transcription of the pathway-specific positive-acting regulatory gene (*alcR*) (Lockington *et al.* 1987). The recent cloning of *nirA* will facilitate distinguishing among the above three models by enabling it to be seen whether transcription of *nirA* is subject to nitrogen metabolite repression and *areA* control. More sophisticated experiments at the molecular level should subsequently allow definitive elucidation of the mechanism.

Acknowledgements

We thank our collaborators who kindly allowed us to discuss their unpublished data. Work in our laboratories described here has been supported over the years by the Royal Society (H.N.A.), the Science and Engineering Research Council (H.N.A. and C.S.), the Medical Research Council (H.N.A.) and the Centre National de la Recherche Scientifique (C.S.).

References

Al Taho, N., Sealy-Lewis, H. M., and Scazzocchio, C. (1984). Suppressible alleles in a wide domain regulatory gene in *Aspergillus nidulans*. *Current Genetics* **8**, 245–51.

Anton, I. A., Duncan, K., and Coggins, J. R. (1987). A eukaryotic repressor protein, the *qa-1S* gene product of *Neurospora crassa*, is homologous to part of *arom* multifunctional enzyme. *Journal of Molecular Biology* **197**, 367–71.

Arst, H. N., Jr. (1977). Some genetical aspects of ornithine metabolism in *Aspergillus nidulans*. *Molecular and General Genetics* **151**, 105–10.

Arst, H. N., Jr. (1981). Aspects of the control of gene expression in fungi. *Symposia of the Society for General Microbiology* **31**, 131–60.

Arst, H. N., Jr. (1982). A near terminal pericentric inversion leads to nitrogen metabolite derepression in *Aspergillus nidulans*. *Molecular and General Genetics* **188**, 490–3.

Arst, H. N., Jr. and Bailey, C. R. (1977). The regulation of carbon metabolism in *Aspergillus nidulans*. In *Genetics and physiology of Aspergillus* (eds. J. E. Smith and J. A. Pateman), pp. 131–46. Academic Press, London.

Arst, H. N., Jr. and Cove, D. J. (1969). Methylammonium resistance in *Aspergillus nidulans*. *Journal of Bacteriology* **98**, 1284–93.

Arst, H. N., Jr. and Cove, D. J. (1973). Nitrogen metabolite repression in *Aspergillus nidulans*. *Molecular and General Genetics* **126**, 111–41.

Arst, H. N., Jr. and MacDonald, D. W. (1973). A mutant of *Aspergillus nidulans* lacking NADP-linked glutamate dehydrogenase. *Molecular and General Genetics* **122**, 261–5.

Arst, H. N., Jr. and Page, M. M. (1973). Mutants of *Aspergillus nidulans* altered in the transport of methylammonium and ammonium. *Molecular and General Genetics* **121**, 239–45.

Arst, H. N., Jr. and Scazzocchio, C. (1975). Initiator constitutive mutation with an 'up-promoter' effect in *Aspergillus nidulans*. *Nature* **254**, 31–4.

Arst, H. N., Jr. and Scazzocchio, C. (1985). Formal genetics and molecular biology of the control of gene expression in *Aspergillus nidulans*. In *Gene manipulations in fungi* (ed. J. W. Bennett and L. L. Lasure), pp. 309–43. Academic Press, New York.

Arst, H. N., Jr., MacDonald, D. W., and Cove, D. J. (1970). Molybdate metabolism in *Aspergillus nidulans*. I. Mutations affecting nitrate reductase and/or xanthine dehydrogenase. *Molecular and General Genetics* **108**, 129–45.

Arst, H. N., Jr., Rand, K. N., and Bailey, C. R. (1979). Do the tightly linked structural genes for nitrate and nitrite reductases form an operon? Evidence from an insertional translocation which separates them. *Molecular and General Genetics* **174**, 89–100.

Arst, H. N., Jr., MacDonald, D. W., and Jones, S. A. (1980). Regulation of proline transport in *Aspergillus nidulans*. *Journal of General Microbiology* **116**, 285–94.

Arst, H. N., Jr., Tollervey, D. W., and Sealy-Lewis, H. M. (1982a). A possible regulatory gene for the molybdenum-containing cofactor in *Aspergillus nidulans*. *Journal of General Microbiology* **128**, 1083–93.

Arst, H. N., Jr., Brownlee, A. G., and Cousen, S. A. (1982b). Nitrogen metabolite represssion in *Aspergillus nidulans:* a farewell to *tamA*? *Current Genetics* **6**, 245–57.

Brownlee, A. G. and Arst, H. N., Jr. (1983). Nitrate uptake in *Aspergillus nidulans* and involvement of the third gene of the nitrate assimilation gene cluster. *Journal of Bacteriology* **155**, 1138–46.

Caddick, M. X., Arst, H. N., Jr., Taylor, L. H., Johnson, R. I., and Brownlee, A. G. (1986). Cloning of the regulatory gene *areA* mediating nitrogen metabolite repression in *Aspergillus nidulans*. *EMBO Journal* **5**, 1087–90.

Chou, P. Y. and Fasman, G. D. (1978). Empirical predictions of protein conformation. *Annual Review of Biochemistry* **47**, 251–76.

Cohen, B. L. (1972). Ammonium repression of extracellular protease in *Aspergillus nidulans*. *Journal of General Microbiology* **71**, 293–9.

Cooley, R. N. and Tomsett, A. B. (1985). Determination of the subunit size of NADPH nitrate reductase from *Aspergillus nidulans*. *Biochimica et Biophysica Acta* **831**, 89–93.

Cornwell, E. V. and MacDonald, D. W. (1984). *glnA* mutations define the structural gene for glutamine synthetase in *Aspergillus*. *Current Genetics* **8**, 33–6.

Cove, D. J. (1969). Evidence for a near-limiting intracellular concentration of a regulator substance. *Nature* **224**, 272–3.

Cove, D. J. (1977). The genetics of *Aspergillus nidulans*. In *Genetics and physiology of Aspergillus* (eds J. E. Smith and J. A. Pateman), pp. 81–95. Academic Press, London.

Cove, D. J. (1979). Genetic studies of nitrate assimilation in *Aspergillus nidulans*. *Biological Reviews* **54**, 291–327.

Cove, D. J. and Pateman, J. A. (1963). Independently segregating loci concerned with nitrate reductase activity in *Aspergillus nidulans*. *Nature* **198**, 262–3.

Cove, D. J. and Pateman, J. A. (1969). Autoregulation of the synthesis of nitrate reductase in *Aspergillus nidulans*. *Journal of Bacteriology* **97**, 1374–8.

Garrett, R. H. and Cove, D. J. (1976). Formation of NADPH-nitrate reductase activity *in vitro* from *Aspergillus nidulans niaD* and *cnx* mutants. *Molecular and General Genetics* **149**, 189–96.

Giles, N. H., Case, M. E., Baum, J., Geever, R., Huiet, L., Patel, V., and Tyler, B. (1985). Gene organisation and regulation in the *qa* (quinic acid) gene cluster of *Neurospora crassa*. *Microbiological Reviews* **49**, 338–58.

Hartley, M. J. (1970). Contrasting complementation patterns in *Aspergillus nidulans*. *Genetical Research* **16**, 123–5.

Huiet, L. and Giles, N. H. (1986). The *qa* repressor gene of *Neurospora crassa*: wild type and mutant nucleotide sequences. *Proceedings of the National Academy of Sciences*, **83**, 3381–5.

Hynes, M. J. (1975). Studies on the role of the *areA* gene in the regulation of nitrogen catabolism in *Aspergillus nidulans*. *Australian Journal of Biological Sciences* **28**, 301–13.

Johnson, J. L., Hainline, B. E., Rajagopalan, K. V., and Arison, B. H. (1984). The pterin component of the molybdenum cofactor. Structural characterization of two fluorescent derivatives. *Journal of Biological Chemistry* **259**, 5414–22.

Kinghorn, J. R. and Pateman, J. A. (1973). NAD and NADP λ-glutamate dehydrogenase activity and ammonium regulation in *Aspergillus nidulans*. *Journal of General Microbiology* **78**, 39–46.

Kinghorn, J. R. and Pateman, J. A. (1975). The structural gene for NADP L-glutamate dehydrogenase in *Aspergillus nidulans*. *Journal of General Microbiology* **86**, 294–300.

Lewis, N. J., Hurt, P., Sealy-Lewis, H. M., and Scazzocchio, C. (1978). The genetic control of molybdoflavoproteins in *Aspergillus nidulans*. IV. A comparison between purine hydroxylases I and II. *European Journal of Biochemistry* **91**, 311–16.

Lockington, R., Scazzocchio, C., Sequeval, D., Mathieu, M., and Felenbok, B. (1987). Regulation of *alcR*, the positive regulatory gene of the ethanol utilisation regulon of *Aspergillus nidulans*. *Molecular Microbiology*, **1**, 275–81.

Lue, N. F., Chasman, D. I., Buchman, A. R., and Kornberg, R. D. (1987). Interaction of GAL4 and GAL80 gene regulatory proteins *in vitro*. *Molecular and Cellular Biology* **7**, 3346–51.

MacDonald, D. W. (1982). A single mutation leads to loss of glutamine synthetase and relief of ammonium repression in *Aspergillus nidulans*. *Current Genetics* **6**, 203–8.

MacDonald, D. W. and Cove, D. J. (1974). Studies on temperature-sensitive mutants affecting the assimilatory nitrate reductase of *Aspergillus nidulans*. *European Journal of Biochemistry* **47**, 107–10.

Mehra, K. R. and Coughlan, M. P. (1984). Purification and properties of purine hydroxylase II from *Aspergillus nidulans*. *Archives of Biochemistry and Biophysics* **229**, 585–95.

Pateman, J. A. and Cove, D. J. (1967). Regulation of nitrate reduction in *Aspergillus nidulans*. *Nature* **215**, 1234–7.

Pateman, J. A., Cove, D. J., Rever, B. M., and Roberts, D. B. (1964). A common cofactor for nitrate reductase and xanthine dehydrogenase which also regulates the synthesis of nitrate reductase. *Nature* **201**, 58–60.

Pateman, J. A., Kinghorn, J. R., Dunn, E., and Forbes, E. (1973). Ammonium regulation in *Aspergillus nidulans*. *Journal of Bacteriology* **114**, 943–50.

Polkinghorne, M. and Hynes, M. J. (1975). Mutants affecting histidine utilization in *Aspergillus nidulans*. *Genetical Research* **25**, 119–35.

Polkinghorne, M. A. and Hynes, M. J. (1982). L-Histidine utilization in *Aspergillus nidulans*. *Journal of Bacteriology* **149**, 931–40.

Polya, G. M., Brownlee, A. G., and Hynes, M. J. (1975). Enzymology and genetic regulation of a cyclic nucleotide-binding phosphodiesterase-phosphomonoesterase from *Aspergillus nidulans*. *Journal of Bacteriology* **124**, 693–703.

Rand, K. N. and Arst, H. N., Jr. (1977). A mutation in *Aspergillus nidulans* which affects the regulation of nitrite reductase and is tightly linked to its structural gene. *Molecular and General Genetics* **155**, 67–75.

Rand, K. N. and Arst, H. N., Jr. (1978). Mutations in the *nirA* gene of *Aspergillus nidulans* and nitrogen metabolism. *Nature* **272**, 732–4.

Scazzocchio, C. (1966). Control of purine oxidation in *Aspergillus*. Unpublished Ph.D. Thesis. University of Cambridge.

Scazzocchio, C. (1973). Appendix: The induction of xanthine dehydrogenase II. In Scazzocchio, C., Holl, F. B. and Foguelman, A. I. The genetic control of molybdoflavoproteins in *Aspergillus nidulans*. Allopurinol-resistant mutants constitutive for xanthine dehydrogenase. *European Journal of Biochemistry* **36**, 428–45.

Scazzocchio, C. (1980). The genetics of the molybdenum containing enzymes. In *Molybdenum and molybdenum-containing enzymes* (ed. M. P. Coughlan), pp. 487–517. Pergamon Press, Oxford.

Scazzocchio, C. and Darlington, A. J. (1967). The genetic control of xanthine dehydrogenase and urate oxidase synthesis in *Aspergillus nidulans*. *Bulletin de la Société de Chimie Biologique* **49**, 1503–8.

Scazzocchio, C. and Darlington, A. J. (1968). The induction and repression of the enzymes of purine breakdown in *Aspergillus nidulans*. *Biochimica et Biophysica Acta* **166**, 557–68.

Scazzocchio, C., Holl, F. B., and Foguelman, A. (1973). The genetic control of molybdoflavoproteins in *Aspergillus nidulans*. Allopurinol-resistant mutants constitutive for xanthine dehydrogenase. *European Journal of Biochemistry* **36**, 428–45.

Scazzocchio, C., Sdrin, N., and Ong, G. (1982). Positive regulation in a eukaryote, a study of the *uaY* gene of *Aspergillus nidulans*: I. Characterisation of alleles, dominance and complementation studies and a fine structure map of the *uaY-oxpA* cluster. *Genetics* **100**, 185–208.

Sealy-Lewis, H. M., Scazzocchio, C., and Lee, S. (1978). A mutation defective in the xanthine alternative pathway of *Aspergillus nidulans*. Its use to investigate the specificity of *uaY* mediated induction. *Molecular and General Genetics* **164**, 303–8.

Sealy-Lewis, H. M., Lycan, D., and Scazzocchio, C. (1979). Product induction of purine hydroxylase II in *Aspergillus nidulans*. *Molecular and General Genetics* **174**, 105–6.

Shaffer, P. M. and Arst, H. N., Jr. (1984). Regulation of pyrimidine salvage in *Aspergillus nidulans*: a role for the major regulatory gene *areA* mediating nitrogen repression. *Molecular and General Genetics* **198**, 139–45.

Spathas, D. H., Pateman, J. A., and Clutterbuck, A. J. (1982). Polyamine transport in *Aspergillus nidulans*. *Journal of General Microbiology* **128**, 557–63.

Spathas, D. H., Clutterbuck, A. J., and Pateman, J. A. (1983). Putrescine as a nitrogen source for wild type and mutants of *Aspergillus nidulans. FEMS Microbiology Letters* **17**, 345–8.

Tollervey, D. W. (1981). Aspects of nitrogen metabolic regulation in *Aspergillus nidulans.* Unpublished Ph.D. Thesis. University of Cambridge.

Tollervey, D. W. and Arst, H. N., Jr. (1981). Mutations to constitutivity and depression are separate and separable in a regulatory gene of *Aspergillus nidulans. Current Genetics* **4**, 63–8.

Tollervey, D. W. and Arst, H. N., Jr. (1982). Domain-wide, locus-specific suppression of nitrogen metabolite repressed mutations in *Aspergillus nidulans. Current Genetics* **6**, 79–85.

Wahl, R. C., Warner, K. C., Finnerty, V., and Rajagopalan, K. V. (1982). *Drosophila melanogaster ma-1* mutants are defective in the sulfuration of desulfo Mo hydroxylases. *Journal of Biological Chemistry* **257**, 3958–62.

Wiame, J.-M., Grenson, M., and Arst, H. N., Jr. (1985). Nitrogen catabolite represssion in yeasts and filamentous fungi. *Advances in Microbial Physiology* **26**, 1–88.

20. Genetics, regulation, and molecular studies of nitrate assimilation in *Neurospora crassa*

George A. Marzluf and Ying-Hui Fu

Introduction

The assimilation of nitrate is a highly regulated process in *Neurospora crassa* and is subject to intricate metabolic and genetic controls (Marzluf 1981). Nitrate is converted to nitrite in a two-electron transfer reaction catalysed by nitrate reductase (NR). Nitrite is then converted to ammonium in a single but complex 6-electron transfer reaction catalysed by nitrite reductase (NiR). NR of *Neurospora* is a dimeric enzyme, composed of two identical polypeptide subunits, each of molecular weight 114 kDa (Bahns and Garrett 1980); it possesses a molybdenum-containing cofactor, which is also a component of other molybdenum enzymes, such as xanthine dehydrogenase (Scazzocchio 1979). NiR is a homodimeric protein of molecular weight 290 kDa (Lafferty and Garrett 1974). The synthesis of NR and NiR requires both a release from nitrogen catabolite repression and a specific induction by nitrate, plus the concerted action of various genes which specify structural components of these enzymes.

Genetic studies

Mutations in a number of unlinked genes result in the loss of the ability of *Neurospora* to utilize nitrate (Table 20.1). The NR enzyme subunit is encoded by the *nit-3* gene, which is unlinked to *nit-6*, the structural gene for NiR. At least four additional loci, *nit-1*, *nit-7*, *nit-8*, and *nit-9* are required for the synthesis and assembly of the molybdenum cofactor (MoCo) (Tomsett and Garrett 1980); mutants at any of these cofactor loci lack both NR and xanthine dehydrogenase activity, but possess NiR. Whilst the *nit-1*, *nit-7*, and *nit-8* loci appear to be single genes, the *nit-9* locus is complex and contains three complementation groups, suggesting that it may comprise three closely linked genes, all required for formation of MoCo (Tomsett and Garrett 1980). The cofactor is believed to consist of a molybdenum ligand and a pterin component (Chapter 14), both of which are necessary for the catalytic function of nitrate reductase.

Table 20.1. Genes involved in nitrate assimilation in *Neurospora crassa*

Locus	Linkage group	Function
nit-3	4R	Nitrate reductase structural gene
nit-6	6L	Nitrite reductase structural gene
nit-1	1R	Molybdenum cofactor biosynthetic gene
nit-7	3R	Molybdenum cofactor biosynthetic gene
nit-8	1R	Molybdenum cofactor biosynthetic gene
nit-9	4R	Molybdenum cofactor biosynthetic gene
nit-2	1L	Major regulatory gene mediating nitrogen catabolite repression
nit-4	4R	Pathway-specific regulatory gene mediating nitrate induction
nmr-1	5R	Involved in nitrogen repression
gln-1	5R	Glutamine synthetase structural gene

The products of two distinct and unlinked regulatory genes, designated *nit-2* and *nit-4* are required for nitrate assimilation. Mutants of either of these two regulatory genes lack both NiR and NR and cannot utilize nitrate or nitrite for growth (Marzluf 1981). Their null phenotype implies that each of these control genes acts in a positive manner and that their respective *trans*-acting products, almost certainly regulatory proteins, are required in order to switch on the expression of NR and NiR. Figure 20.1 presents a model to explain the action of *nit-2* and *nit-4*. Another nitrogen regulatory gene, *nmr* (for *n*itrogen *m*etabolic *r*egulation), appears to act in a negative fashion (Debusk and Ogilvie 1984) and *nmr* mutants constitutively express NiR (Premakumar *et al.* 1980*b*; Dunn-Coleman *et al.* 1981; Tomsett *et al.* 1981).

Nitrogen catabolite repression and nitrate induction

Although nitrate is an excellent nitrogen source for *Neurospora*, it is not utilized if certain favoured nitrogen sources, such as ammonium ion, glutamine or glutamate are present in adequate amounts. In the presence of any one of these preferred nitrogen sources, N-catabolite repression prevents the synthesis of NR and NiR. *nit-2* is a major nitrogen regulatory gene of *Neurospora* and appears to play a central role in nitrogen repression control. *nit-2* activity is required for expression of all of the enzymes within the nitrogen control circuit, including NR and NiR, and also various enzymes of purine metabolism, and protein and amino acid catabolism (Hanson and Marzluf 1975; Reinert and Marzluf 1975; Sikora and Marzluf 1982; Lindberg and Drucker 1984). Thus, nitrogen catabolite repression in *Neurospora* appears to be mediated by two separate regulatory genes, the positive-acting *nit-2* gene and the negative-acting *nmr* gene. In contrast, in *Aspergillus nidulans*, a single regulatory gene known as *areA* mediates

     Molecular and genetic aspects of nitrate assimilation

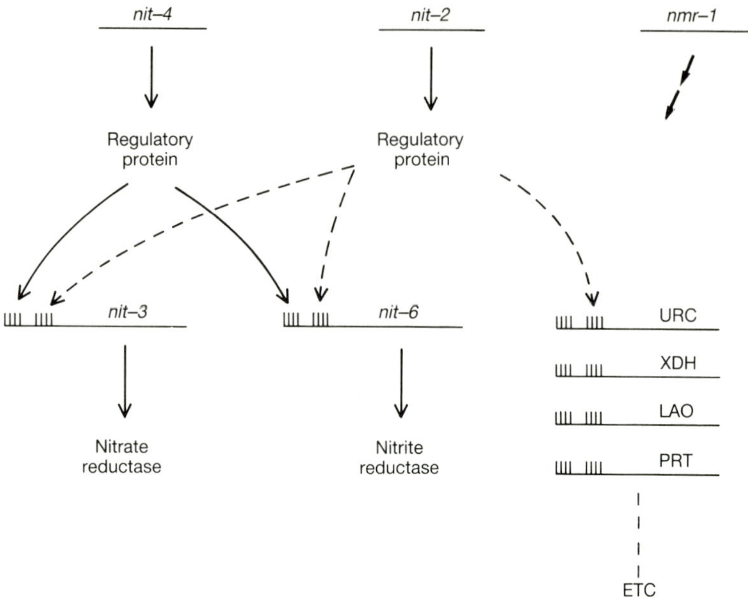

Fig. 20.1. Regulation of NR (*nit-3*) and NiR (*nit-6*) structural genes. The positive-acting major nitrogen regulatory gene, *nit-2*, and the pathway-specific regulatory gene, *nit-4*, are each presumed to encode a regulatory protein which is required for the expression of *nit-3* and *nit-6*. The *nit-2* protein is also required to activate various other nitrogen-related structural genes. The negative-acting *nmr* gene also is involved in control of these activities and could specify a protein which interacts directly with the *nit-2* protein. Other possible interactions are described in the text. URC = uricase, XDH = xanthine dehydrogenase, LAO = L-amino acid oxidase, PRT = protease.

N repression (Chapter 19). Several lines of evidence have demonstrated that glutamine is the effector molecule responsible for nitrogen catabolite repression of NR and NiR as well as the many other enzymes required to utilize various secondary nitrogen sources (Premakumar *et al.* 1979, 1980*a*). It is plausible that the effector function of glutamine may result from its binding to a regulatory protein specified by *nmr* or by *nit-2* or even a multimeric control protein comprised of *nmr* and *nit-2* subunits.

The second positive-acting regulatory gene, *nit-4*, is pathway-specific and mutants lack NR and NiR but are unaffected for other nitrogen-related enzymes (Fig. 20.1). The *nit-4* gene appears to mediate nitrate induction of NR and NiR, and is thought to specify a regulatory protein, which enters the nucleus and activates the *nit-3* and *nit-6* structural genes (Tomsett and Garrett 1980; Marzluf 1981). The *nit-4* protein might itself bind nitrate and thereby attain an active conformation; alternatively, it may be involved in a protein–protein interaction with NR in an autogenous control mechanism, as described below. Still another alternative that deserves mention is the possibility of sequential gene action in which *nit-2* controls the expression of *nit-4* which, in turn, is directly responsible for activation of the NR and NiR structural genes. The available genetic evidence cannot distinguish between these various possibilities, but molecular studies of *nit-2* and *nit-4* transcription should clarify the situation in the near future.

De novo synthesis of nitrate reductase

The step at which genetic and metabolic regulation of NR occurs is a question of long-standing interest. Bahns and Garrett (1980) presented convincing evidence that NR induction requires *de novo* enzyme synthesis. When uninduced cultures were transferred to medium containing 90 per cent deuterium oxide, induction via nitrate yielded NR of uniformly heavy density, demonstrating that enzyme induction was accompanied by *de novo* synthesis. In related studies, highly specific immunoelectrophoretic techniques were employed by Amy and Garrett (1979) to detect NR protein, including even partial chains which were devoid of all enzymatic activity. Nitrogen-repressed or uninduced wild-type cells lack any cross-reacting material (CRM) immunologically related to nitrate reductase. Induced *nit-1* and most *nit-3* mutants lack NR but, as would be expected, do contain CRM. The two regulatory mutants *nit-2* and *nit-4* were devoid of any detectable CRM under all conditions examined (Amy and Garrett 1979). These results clearly demonstrate that NR induction does not occur by activation of a pre-existing precursor protein or removal of an inhibitor to reveal a cryptic enzyme, but depends upon *de novo* enzyme synthesis.

A series of studies conducted by Sorger and his colleagues suggested that control of NR synthesis occurs at the level of transcription (Premakumar *et al.* 1979,1980*a*), based upon the apparent accumulation *in vivo* of specific mRNA under conditions that prevented active enzyme synthesis, followed by the subsequent translation of the accumulated mRNA to yield active NR. This work used a novel approach in which induction was carried out in the presence of tungsten, so that mRNA could accumulate but any NR synthesized would contain tungsten (in place of molybdenum) and thus lack activity. Translation of any accumulated mRNA was then allowed to occur in molybdenum-containing medium to form active NR, the amount synthesized reflecting the cellular content of NR mRNA. These studies were provocative and suggested that the regulation of NR reductase expression by both nitrogen repression and nitrate induction occurred at the transcriptional level. Because of the indirect nature of these experiments the results had to be viewed with caution. However, very recent work has provided direct evidence that NR expression is indeed subject to transcriptional control (see below).

Models of nitrogen catabolite repression and nitrate induction

It has been recognized for some time that the *nit-2* gene and perhaps also the *nmr* gene play major roles in nitrogen regulation, but it has not been possible to define exactly the operation of the nitrogen control circuit, although a number of different models have been suggested, two of which are displayed in Fig. 20.2. In model (a), it was proposed that the *nit-2* gene is expressed constitutively to yield a regulatory protein which activates the various N-related structural genes and whose activation function is inhibited upon binding glutamine (Grove and Marzluf 1981). The recent isolation of an amber nonsense mutant of *nit-2* that is

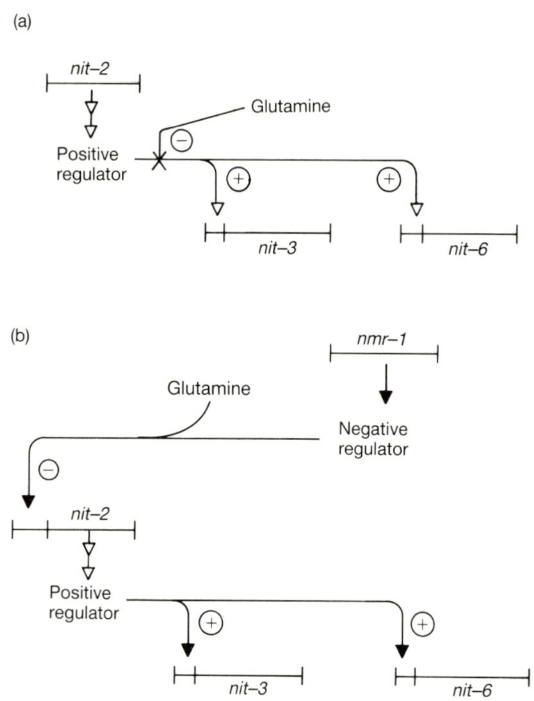

Fig. 20.2. Two models for operation of the nitrogen control circuit. (a) the *nit-2* gene is postulated to be expressed constitutively to yield a regulatory product active for turning on *nit-3* and *nit-6* unless it is inhibited by the metabolic repressor molecule, glutamine. (b) The *nmr* gene is visualized as producing a repressor protein which is activated by glutamine and precludes *nit-2* gene expression. In the absence of glutamine, *nit-2* is expressed and in turn activates *nit-3* and *nit-6*. Additional models are described in the text.

suppressible by *SSU-1*, a tRNA suppressor gene, implies that this control gene encodes a regulatory protein (Perrine and Marzluf 1986). Alternatively, expression of the *nit-2* control gene itself may be regulated, a feature present in two other models for operation of the nitrogen circuit. In model (b), the *nmr-1* gene was postulated to encode a repressor protein, which upon binding the effector glutamine, bound at a control site adjacent to the *nit-2* gene, thereby preventing its expression (DeBusk and Ogilvie 1984). Mutants of *nmr-1* are at least partially resistant to nitrogen catabolite repression of nitrate reductase, consistent with this proposed model since the positive acting *nit-2* protein would be synthesized and active in *nmr* mutant cells regardless of their nitrogen status. An entirely different mechanism for the operation of the nitrogen control circuit has also been proposed in which the enzyme glutamine synthetase carries out a primary regulatory role (Dunn-Coleman and Garrett 1980). Upon binding glutamine, glutamine synthetase is visualized as undergoing a change in oligomeric structure, enabling it to act as a repressor protein, binding at the *nit-2* gene and preventing its expression.

One further phenomenon of considerable interest is the possibility that NR expression is subject to autogenous control in which the enzyme itself plays a negative role in controlling its own synthesis and that of NiR (Tomsett and Garrett 1981; Cove 1976). Two classes of *nit-3* mutants have been described with respect to potential autogenous regulation. A few *nit-3* mutants require induction like the wild type, whereas the majority of *nit-3* mutants are found to be completely constitutive and synthesize NiR and NR partial activities or cross-reacting material in the absence of any added inducer (Tomsett and Garrett 1981). One possible autoregulatory role for NR protein suggests that it is the element that senses the presence of the inducer nitrate. In this proposed mechanism, NR protein binds to the *nit-4* control protein to stop its activation of the NR and NiR structural genes, perhaps by preventing the *nit-4* protein from entering the nucleus. In the presence of inducer, NR binds nitrate and releases the *nit-4* protein which can then switch on the NR and NiR structural genes. The fact that most *nit-3* mutants, including null and nonsense mutants, are constitutive at the level of enzyme activity or CRM is consistent with this model, but it would be of interest to determine whether autoregulation does in fact occur at the level of transcription (see below).

Molecular studies of *nit-2* gene expression

It recently became obvious that further progress in dissection of the *Neurospora* nitrogen regulatory circuit required the molecular cloning of representative structural and regulatory genes so that their interaction could be examined at the molecular level. Vollmer and Yanofsky (1986) have developed a cosmid library which permits the efficient cloning of *Neurospora* genes *via* complementation of mutant phenotypes. A cosmid carrying a *Neurospora* DNA insert of approximately 40 kb which complements a *nit-2* mutant was isolated from this library (Stewart and Vollmer 1986). A plasmid, designated pNit-2, that carries a 6 kb *Neurospora* *Eco*R1 DNA fragment subcloned from this cosmid, readily transformed the *nit-2* mutant (Fu and Marzluf 1987a). Additional subcloning experiments coupled with complementation studies revealed that the *nit-2* gene is contained in a unique *Neurospora* DNA fragment of approximately 3.5 kb. A restriction fragment length polymorphism analysis (Metzenberg *et al.* 1984) confirmed that it indeed corresponded to the *nit-2* gene.

The use of plasmid pNit-2 DNA as a probe provided a means to examine the expression of the *nit-2* gene in wild-type and mutant strains under various metabolic conditions (Fu and Marzluf 1987*a*). Northern blot analyses revealed that the *nit-2* gene is transcribed to give a single large messenger RNA of approximately 3.5 kb, which agrees well with the size of the functional gene as determined by deletion analysis. Moreover, the expression of the *nit-2* gene is itself regulated such that its transcript is present at a significantly higher level in cells which are limited for nitrogen in comparison with cells growing under nitrogen-repressed conditions (Fig. 20.3). The *nit-2* gene does not appear to be controlled by autogenous gene regulation since *nit-2* mutants gave the same RNA patterns during repressed and de-repressed conditions as was found with the wild-type strain. The *nit-2* RNA transcript was present at only the low level characteristic of

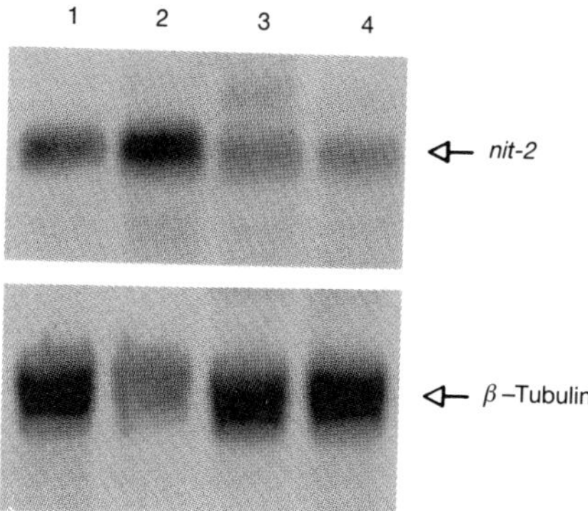

Fig. 20.3. Northern blot analysis of *nit-2* gene expression in the wild-type and an *nmr* (allele MS5) mutant. Lane 1, RNA from wild-type N-repressed cells; lane 2, RNA from wild-type de-repressed cells; lane 3, RNA from N-repressed *nmr* cells; lane 4, RNA from de-repressed *nmr* cells. The lower panel displays RNA hybridized to a *β*-tubulin probe to standardize the loading of each lane. An approximate 4-fold increase in the amount of *nit-2* transcript occurs upon de-repression of the wild-type strain, whereas no significant increase was detected with *nmr*. Flasks containing Vogel's minimal medium were inoculated with conidial suspensions and grown with shaking at 30°C for 24 h, when the mycelia were harvested and washed. The mycelia were then transferrd to fresh Vogel's minimal medium lacking nitrogen supplemented with 25 mM L-glutamine (N repression), or supplemented with 0.5 mM L-glutamine (N de-repression). After incubation in the second medium for 6 h, the mycelia were harvested and frozen at -80°C until used.

nitrogen-repressed wild-type cells in the *nmr-1* mutant strain under both nitrogen repressed and de-repressed conditions (Fig. 20.3). That is, the content of *nit-2* mRNA in the *nmr-1* mutant strain was similar whether the cells were N-repressed or N-de-repressed, but only a low basal amount was present in either case. This result indicates that *nmr* does not control *nit-2* gene transcription in a simple negative fashion as suggested in model (b) of Fig. 20.2. It remains possible that *nmr* might have both a negative and positive effect upon *nit-2* expression.

Similarly, it was found that *nit-2* expression was not constitutive in *gln-1b*, a structural gene mutant of the enzyme glutamine synthetase. The finding that the level of *nit-2* mRNA increased under N de-repression argues against model (a) of Fig. 20.2 which suggested that *nit-2* was expressed in a constitutive fashion; the *nit-2* gene is expressed even under nitrogen repressed conditions but the cellular level of its mRNA increases 3- to 4-fold upon de-repression (Fig. 20.3). Thus, it appears that the models proposed for nitrogen regulation in *Neurospora* will require careful revision in the face of new information that is now emerging.

nit-2 gene expression in *Aspergillus nidulans*

Another series of very recent experiments also implies that the *nit-2* gene of *Neurospora crassa* is itself regulated by the activity of other genes. It has been suspected for some time that the *areA* gene of *A. nidulans* was homologous to the *nit-2* gene of *Neurospora* and that each functioned similarly in nitrogen control. Davis and Hynes (1987) transformed *areA* mutants of *Apergillus* with plasmid pNit-2 and have found that the *Neurospora nit-2* gene can substitute for *areA* function in all *areA* mutants examined. Enzymes controlled by N catabolite repression in *Aspergillus* are not expressed in those *areA* mutants, leading to an inability to use many secondary nitrogen sources, such as nitrate, uric acid and acetamide. Upon transformation with the *Neurospora nit-2* gene, the *areA* mutant strains regained the ability to use acetamide, nitrate, and uric acid, and also possessed the relevant enzymes (Davis and Hynes 1987). This result implies that the *nit-2* gene is expressed in *Aspergillus* and that the regulatory protein it encodes is sufficiently similar to the *areA* gene product to act at 5′ regulatory recognition sites to activate the expression of *Aspergillus* genes controlled by N catabolite repression. One additional observation is of considerable interest: NR expression in the *Aspergillus* transformants containing the *Neurospora nit-2* gene was only weakly repressed by nitrogen, in contrast to the normal situation in *Aspergillus* where synthesis of this enzyme in cells with a functional *areA*$^+$ gene is strongly repressed by nitrogen. This result can be interpreted to indicate that the *nit-2* gene or its product is unregulated in *Aspergillus* because of differences between these fungi in additional genes which somehow modulate the activity of the major nitrogen regulatory gene. Thus, it appears possible that *nit-2* gene activity is controlled in *Neurospora* but that it becomes unregulated when it is introduced *via* transformation into *Aspergillus*. In this regard it is important to note that a negative-acting nitrogen control gene analogous to *nmr* has not been found in *Aspergillus*. One interesting possibility is that nitrogen regulation in *Aspergillus*, including both repression and activation aspects, is mediated entirely by the *areA* gene product. In contrast, in *Neurospora* the *nit-2* gene product might only be capable of *trans*-activation, and the *nmr* product, which could interact with the *nit-2* protein, could mediate nitrogen repression, perhaps by providing a glutamine-binding subunit. It would be interesting to determine whether the *areA* gene can substitute for the *nit-2* gene of *Neurospora*. It will likewise be of great interest to directly compare the *nit-2* and *areA* genes and the amino acid sequence of their corresponding protein products. This should be possible in the near future because the *areA* gene has also recently been cloned (Caddick *et al.* 1986) so that the DNA sequence of these two genes may soon be available for comparison. It will also be informative to compare possible *cis*-acting DNA sequences located upstream of the nitrate reductase genes and various other related structural genes in *Aspergillus* and in *Neurospora* since such sequences may comprise recognition elements for the *areA* protein and for the *nit-2* protein, respectively.

Molecular studies of nitrate reductase regulation

It has been known for some time that the induction of NR was dependent upon *de novo* enzyme synthesis (Bahns and Garrett 1980) as described above. Some indirect results also suggested that *nit-3*, the NR structural gene, was regulated at the transcriptional level (Premakumar *et al.* 1977,1979). We have recently isolated the *nit-3* gene by screening a *Neurospora* genomic library for a recombinant clone that complemented a *nit-3* mutant. The *nit-3* gene was found on an *EcoR1* fragment of approximately 3.8 kb (Fu and Marzluf 1987*b*). A restriction fragment length polymorphism analysis demonstrated that the cloned gene displayed linkage to the *cot-1* and 5 S gene 4, which are both located on linkage group 4 and which flank the *nit-3* gene. The cloned *nit-3* gene provides a molecular probe to directly examine the content of nitrate reductase mRNA in various wild-type and mutant cells grown under different metabolic conditions. The Northern blot analysis shown in Fig. 20.4 reveals that the *nit-3* gene is expressed to give a single poly(A)$^+$ RNA transcript of approximately 3.4 kb, the expected size required to encode the 114 kDa NR subunit. The results presented in Fig. 20.4 demonstrate that regulation of expression of the *nit-3* gene probably occurs at the level of transcription. Thus, wild-type cells did not accumulate detectable *nit-3* mRNA when subject to nitrogen repression or when they were not induced with nitrate. Upon simultaneous nitrate induction and N de-repression, the amount of *nit-3* mRNA present in wild-type cells increased dramatically, by a factor of at least 50-fold (Fig. 20.4). Moreover, neither *nit-2* mutant cells nor *nit-4* mutant cells expressed any detectable *nit-3* mRNA, even when de-repressed and induced. Thus, the activation of *nit-3* expression by the positive-acting *nit-2* and *nit-4* control genes occurs at the level of accumulation of mRNA, which is also the step affected by both of the metabolic signals, nitrogen de-repression and nitrate induction. These results strongly suggest that *nit-3* gene expression is controlled at the level of transcription, although it cannot yet be completely ruled out that a closely related step such as RNA processing or NR mRNA stability might also be involved.

The *nmr* mutant displays a partial constitutive synthesis of NR suggesting that *nmr* exerts negative regulation, either direct or indirect, over *nit-3* expression.

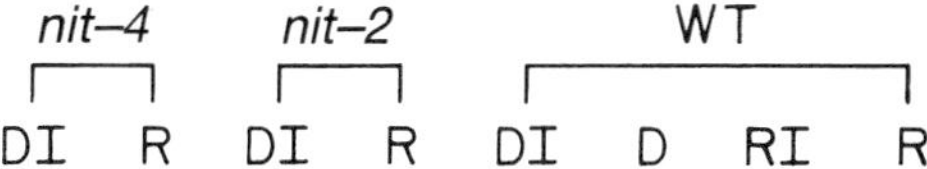

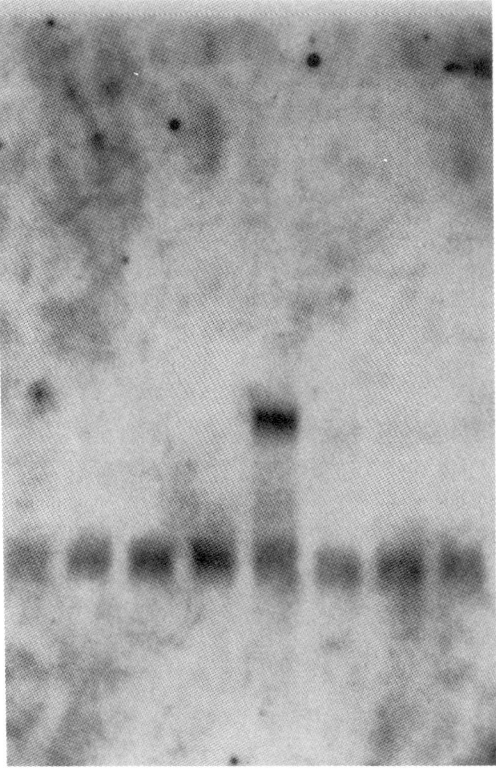

Fig. 20.4. Regulation of *nit-3* gene expression. A Northern analysis of nitrate reductase structural gene (*nit-3*) expression in wild-type, *nit-2*, and *nit-4* mutant cells under various metabolic conditions. The lower hybridization band represents a constitutive 0.9 kb RNA species which serves as an internal standard for RNA loading. The arrow identifies the position of the 3.4 kb *nit-3* transcript. DI, N-de-repressed, nitrate induced; D, de-repressed, uninduced; RI, N-repressed, nitrate induced; R, N-repressed. Mycelia were grown and harvested as described in the legend to Fig. 20.3, and then transferred to minimal medium lacking nitrogen and supplemented with 25 mM L-glutamine (N repression), 0.5 mM L-glutamine (N de-repression), 25 mM L-glutamine plus 50 mM sodium nitrate (N repression, induced) or 50 mM sodium nitrate (N de-repressed, induced). After incubation for 6 h in the second medium, the mycelia were harvested and frozen at $-80°C$ until used.

Thus, a similar Northern blot analysis was carried out to examine the expression of the *nit-3* gene in *nmr* mutant cells. The results shown in Fig. 20.5 demonstrate that *nmr* cells possessed a significant level of the 3.4 kb *nit-3* mRNA even in the presence of sufficient glutamine to fully repress expression in the wild type, i.e. *nit-3* expression was partially insensitive to nitrogen catabolite repression in the *nmr* mutant strain. The level of *nit-3* mRNA in nitrogen-repressed *nmr* cells was estimated to be approximately 20% of the fully de-repressed level. This amount of constitutive expression is significant because wild type cells maintained under identical conditions are completely devoid of any detectable *nit-3* mRNA. Nevertheless, the constitutive level of *nit-3* mRNA in *nmr* is less than the nearly total constitutive expression of NR enzyme activity. Thus, although *nmr* may have some direct role in controlling gene expression, it might instead act in some other fashion, e.g. in translational control; in this regard, it has been established that an increased rate of translation can lead to increased mRNA stability. Since the *nmr* gene has been very recently isolated *via* molecular cloning (Fu *et al.*, 1988), rapid progress can be expected in our understanding of how this interesting control gene exerts its regulatory function.

In the wild type, expression of *nit-3*, the NR structural gene, requires release from nitrogen catabolite repression and specific induction by nitrate: nitrogen de-repression alone does not lead to any detectable synthesis of *nit-3* mRNA in wild-type cells. In contrast, two different *nit-3* mutant strains, one a null mutant which lacks any NR CRM, the other an amber nonsense mutant, were fully constitutive in their expression of the *nit-3* transcript; i.e. each of these mutant strains maximally expressed the *nit-3* gene upon N de-repression, in the complete absence of any added nitrate as inducer (Fu and Marzluf 1988). The *nit-1* mutant produces a wild-type NR apoprotein, but completely lacks any enzyme activity because this strain lacks the MoCo which is required for catalysis. The *nit-1* mutant strain was also found to contain a fully constitutive level of *nit-3* mRNA under uninduced conditions. These results imply that the NR protein is indeed involved in autoregulation and that its control function occurs at the transcriptional level. As described above, the NR protein might carry out its postulated autogenous control by interacting with the *nit-4* regulatory gene product in direct protein–protein binding. Because the *nit-4* gene has now been cloned (Kneesi and Marzluf, unpublished results), direct studies of possible interaction of a *nit-4*-encoded protein with NR should soon be possible and will be invaluable in understanding any autoregulatory role of this remarkable enzyme.

Future directions

Our present understanding of nitrate assimilation in *Neurospora* has resulted from many parallel genetic and biochemical investigations which were carried out over the past two decades. This research has revealed that the structural genes which specify NR and NiR are not linked to one another nor to three distinct regulatory genes which govern their expression.The recent isolation and characterization of the *nit-2* and the *nit-3* genes has added greatly to our insight of the molecular mechanism of regulation in *Neurospora*, and the even more recent molecular cloning of the *nit-4* and *nmr* regulatory genes can only further amplify

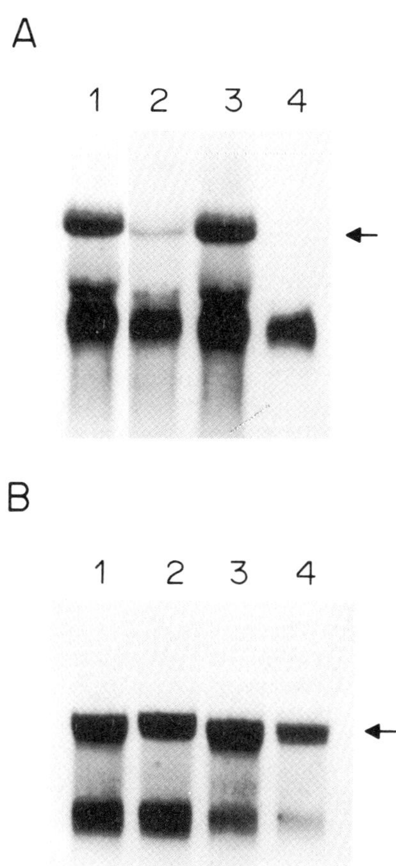

Fig. 20.5. Expression of the *nit-3* gene in wild-type, *nmr* and *nit-3* mutant cells. Upper panel: Northern analysis of wild-type and *nmr*. Lane 1, RNA from de-repressed, nitrate-induced *nmr* cells; lane 2, RNA from N-repressed, nitrate-induced cells; lane 3, RNA from de-repressed, nitrate-induced wild-type cells; lane 4, RNA from N-repressed, nitrate-induced cells. Lower panel: Northern analysis of two *nit-3* mutants. Lane 1, RNA from de-repressed and induced *nit-3* null mutant cells; lane 2, RNA from de-repressed, uninduced *nit-3* null mutant cells; lane 3, RNA from de-repressed and induced *nit-3* amber mutant cells; lane 4, RNA from de-repressed, uninduced *nit-3* amber mutant cells. In both panels, the arrow identifies the position of the 3.4 kb *nit-3* transcript and the lower band represents the 0.9 kb constitutively expressed RNA species that serves as an internal standard.

our understanding. Similar rapid advances are being accomplished with *Aspergillus* (Chapter 6, 19) which will permit very interesting comparisons of the homologous regulatory and structural genes of these two fungal species. Major advances in this area made possible by molecular approaches can be anticipated in the near future and promise to add an entirely new and exciting dimension to our understanding of the regulation of nitrate assimilation.

Acknowledgements

Research in the authors' laboratory was supported by Public Health Service Grant GM-23356 from the National Institutes of Health.

References

Amy, N. K. and Garrett, R. H. (1979). Immunoelectrophoretic determination of nitrate reductase in *Neurospora crassa*. *Analytical Biochemistry* **95**, 97–107.

Bahns, M. and Garrett, R. H. (1980). Demonstration of de novo synthesis of *Neurospora crassa* nitrate reductase during induction. *Journal of Biological Chemistry* **255**, 690–3.

Caddick, M. X., Arst, H. N., Jr., Taylor, L. H., Johnson, R. I., and Brownlee, A. G. (1986). Cloning of the regulatory gene *areA* mediating nitrogen metabolite repression in *Aspergillus nidulans*. *EMBO Journal* **5**, 1087–90.

Cove, D. J. (1976). Chlorate toxicity in *Aspergillus nidulans*. Study of mutants altered in nitrate assimilation. *Molecular and General Genetics* **145**, 147–59.

Davis, M. A. and Hynes, M. J. (1987). Complementation of *areA* regulatory gene mutations of *Aspergillus nidulans* by the heterologous regulatory gene *nit-2* of *Neurospora crassa*. *Proceedings of the National Academy of Sciences, USA* **84**, 3753–7.

DeBusk, R. M. and Ogilvie, S. (1984). Nitrogen regulation of amino acid utilization by *Neurospora crassa*. *Journal of Bacteriology* **160**, 656–61.

Dunn-Coleman, N. S. and Garrett, R. H. (1980). The role of glutamine synthetase and glutamine metabolism in nitrogen metabolite repression, a regulatory phenomenon in the lower eukaryote *Neurospora crassa*. *Molecular and General Genetics* **179**, 25–32.

Dunn-Coleman, N. S., Tomsett, A. B., and Garrett, R. H. (1981). The regulation of nitrate assimilation in *Neurospora crassa*: biochemical analysis of the *nmr-1* mutants. *Molecular and General Genetics* **182**, 234–39.

Fu, Y. H. and Marzluf, G. A. (1987a). Characterization of *nit-2*, the major nitrogen regulatory gene of *Neurospora crassa*. *Molecular and Cellular Biology* **7**, 1691–6.

Fu, Y. H. and Marzluf, G. A., (1987b). Molecular cloning and analysis of the regulation of NIT-3, the structural gene for nitrate reductase in *Neurospora crassa*. *Proceedings of the National Academy of Sciences, USA*, **84**, 8243–7.

Fu, Y. H. and Marzluf, G. A. (1988). Metabolic control and autogenous regulation of *nit-3*, the nitrate reductase structural gene, of *Neurospora crassa*. *Journal of Bacteriology* **170**, 657–61.

Fu, Y. H., Young, J. L., and Marzluf, G. A. (1988). Molecular cloning and characterization of a negative-acting nitrogen regulatory gene of *Neurospora crassa*. *Molecular and General Genetics* **214**, 74–9.

Grove, G. and Marzluf, G. A. (1981). Identification of the product of the major regulatory gene of the nitrogen control circuit of *Neurospora crassa* as a nuclear DNA-binding protein. *Journal of Biological Chemistry* **256**, 463–70.

Hanson, M. A. and Marzluf, G. A. (1975). Control of the synthesis of a single enzyme by multiple regulatory circuits in *Neurospora crassa*. *Proceedings of the National Academy of Sciences, USA* **72**, 1240–4.

Lafferty, M. A. and Garrett, R. H. (1974). Purification and properties of the *Neurospora crassa* assimilatory nitrite reductase. *Journal of Biological Chemistry* **249**, 7555–67.

Lindberg, R. A. and Drucker, H. (1984). Characterization and comparison of a *Neurospora crassa* RNase purified from cultures undergoing each of three different states of derepression. *Journal of Bacteriology* **157**, 375–9.

Marzluf, G. A. (1981). Regulation of nitrogen metabolism and gene expression in fungi. *Microbiological Reviews* **45**, 437–61.

Metzenberg, R. L., Stevens, J. N., Selker, E. U., and Morzycka-Wroblewska, E. (1984). A method for finding the genetic map position of cloned DNA fragments. *Neurospora Newsletter* **31**, 35–9.

Perrine, K. G. and Marzluf, G. A. (1986). Amber nonsense mutations in regulatory and structural genes of the nitrogen control circuit of *Neurospora crassa. Current Genetics* **10**, 677–84.

Premakumar, R., Sorger, G. J., and Gooden, D. (1979). Nitrogen metabolite repression of nitrate reductase in *Neurospora crassa. Journal of Bacteriology* **137**, 1119–26.

Premakumar, R., Sorger, G. J., and Gooden, D. (1980*a*). Repression of nitrate reductase in *Neurospora* studied by using L-methionine-DL-sulfoximine and glutamine auxotroph gln-1b. *Journal of Bacteriology* **143**, 411–15.

Premakumar, R., Sorger, G. J., and Gooden, D. (1980*b*). Physiological characterization of a *Neurospora crassa* mutant with impaired regulation of nitrate reductase. *Journal of Bacteriology* **144**, 542–51.

Reinert, W. J. and Marzluf, G. A. (1975). Genetic and metabolic control of the purine catabolic enzymes of *Neurospora crassa. Molecular and General Genetics* **139**, 39–55.

Scazzocchio, C. (1979). The genetics of molybdenum-containing enzymes. In *Molybdenum-containing enzymes* (ed. W. Coughlan), pp. 487–515. Pergamon Press, Oxford.

Sikora, L. and Marzluf, G. A. (1982). Regulation of L-amino acid oxidase and of D-amino acid oxidase in *Neurospora crassa. Molecular and General Genetics* **186**, 33–9.

Stewart, V. and Vollmer, S. J. (1986). Molecular cloning of *nit-2*, a regulatory gene required for nitrogen metabolite repression in *Neurospora crassa. Gene* **46**, 291–5.

Tomsett, A. B. and Garrett, R. H. (1980). The isolation and characterization of mutants defective in nitrate assimilation in *Neurospora crassa. Genetics* **95**, 649–60.

Tomsett, A. B. and Garrett, R. H. (1981). Biochemical analysis of mutants defective in nitrate assimilation in *Neurospora crassa*: Evidence for autogenous control by nitrate reductase. *Molecular and General Genetics* **184**, 183–90.

Tomsett, A. B., Dunn-Coleman, N. S., and Garrett, R. H. (1981). The regulation of nitrate assimilation in *Neurospora crassa*: the isolation and genetic analysis of *nmr-1* mutants. *Molecular and General Genetics* **182**, 229–35.

Vollmer, S. J. and Yanofsky, C. (1986). Efficient cloning of genes of *Neurospora crassa, Proceedings of the National Academy of Sciences, USA* **83**, 4869–73.

21. Molecular analysis of nitrate regulation of nitrate reductase in squash and *Arabidopsis*

Nigel M. Crawford and Ronald W. Davis

Introduction

Nitrate is an important source of nitrogen for plants. Its assimilation involves the uptake of nitrate by the roots followed by the reduction of nitrate to ammonium by the consecutive action of the enzymes nitrate reductase (NR) and nitrite reductase (NiR). Ammonium is then incorporated into the amino acid pool, primarily by the enzymes glutamine synthetase and glutamate synthase. The first step, catalysed by NR, plays a key role in the nitrate assimilation pathway because it is the rate-limiting step in the conversion of nitrate to ammonium and because it is tightly regulated. When a plant is exposed to nitrate, a dramatic increase in NR activity and protein occurs in both the roots and leaves (Somers *et al.* 1983; Remmler and Campbell 1986; Crawford *et al.* 1986). The response appears to be dependent on an increase of nitrate in a metabolic pool within the cell that is distinct from the storage compartment for nitrate, the vacuole (Ferrari *et al.* 1973). The molecular mechanisms responsible for regulation of NR in plants are not known.

The importance of NR in the metabolism of plants and the complexity of its regulation have attracted great interest, and much has been learned about the biochemistry and physiology of this protein (see Chapter 13). The enzyme is present in both the roots and leaves of plants, and several different forms have been identified: the best characterized form is the NADH-dependent enzyme, which has been purified from several different plants, including squash. The enzyme is a homodimer of 110 kDa subunits, and contains one each of two prosthetic groups (FAD, haem) and a Mo-pterin cofactor for each subunit of the enzyme (Redinbaugh and Campbell 1985). The protein is not abundant, comprising about 0.01 per cent of the total protein when maximally induced in leaves of plants. Antibodies have been prepared against NR from several different plants and have been used to isolate cDNA clones from barley (Cheng *et al.* 1986), squash (Crawford *et al.* 1986) and tobacco (Calza *et al.* 1987). The antibodies and clones have allowed a more thorough examination of the regulation of the enzyme (see also Chapters 9, 12, and 13).

The isolation of a cDNA clone is an excellent first step in the molecular analysis of the regulation of NR. To extend this work, we wish to isolate and characterize mutants which are impaired in the regulation of the enzyme. This genetic approach has been fruitful in microbial systems (see Chapters 6, 19, and 20) and is being applied to plants (see Chapters 11, 12, and 13). We wish to see if this approach will also be successful in the plant *Arabidopsis thaliana*, which is very useful for isolating mutants because its small genome size facilitates the isolation of genes and the molecular analysis of mutations, and it has a short generation time. To take advantage of these properties, we wished to isolate a NR cDNA clone from *Arabidopsis*. Unfortunately, this enzyme has not been purified from *Arabidopsis*; therefore, there are no antibodies or sequence information that would be useful for cloning. However, antibodies have been raised against NR purified from another dicotyledonous plant, squash (prepared by J. Remmler and W. Campbell), and we attempted to use these antibodies to isolate a cDNA clone from squash and then use this cDNA as a heterologous probe to isolate NR cDNA from *Arabidopsis*. This paper describes the cloning of NR mRNA from squash and the analysis of its regulation. For a more detailed description of this work, the reader may refer to the paper by Crawford *et al.* 1986. We also describe the cloning and analysis of NR mRNA from *Arabidopsis* in this paper.

The effect of nitrate on nitrate reductase levels in squash cotyledons

Before attempting to isolate a cDNA clone from squash, we wished to verify that the antibodies were mono-specific for NR and that the protein recognized by the antibodies was inducible by nitrate. Squash seedlings were grown in vermiculite in the absence of nitrate then treated with a defined nutrient medium containing nitrate. At certain time intervals thereafter, cotyledons were harvested to prepare protein extracts. The level of NR in the protein extracts was determined by measuring enzyme activity and by measuring NR protein levels with anti-NR antibody and immunoblots.

After irrigating squash seedlings with 35 mM nitrate, NR activity increased in the cotyledons (Fig. 21.1). No activity was detected before nitrate was applied (Fig. 21.1) or in control plants that were treated with chloride instead of nitrate (Table 21.1). The increase in activity was first detected after 1 h; the activity levelled off after approximately 12 h, but if a second application of nitrate was given at 12 h, a further 50 per cent increase was observed at 25 h. Therefore, extracts were usually prepared from cotyledons of plants that had been treated twice with nitrate over a 24–48 h period (Table 21.1).

The absence of NR activity in extracts from uninduced plants could be due to the presence of an inhibitor or due to the absence of NR protein. No *trans*-acting inhibitor was detected when extracts from nitrate-treated and control plants were

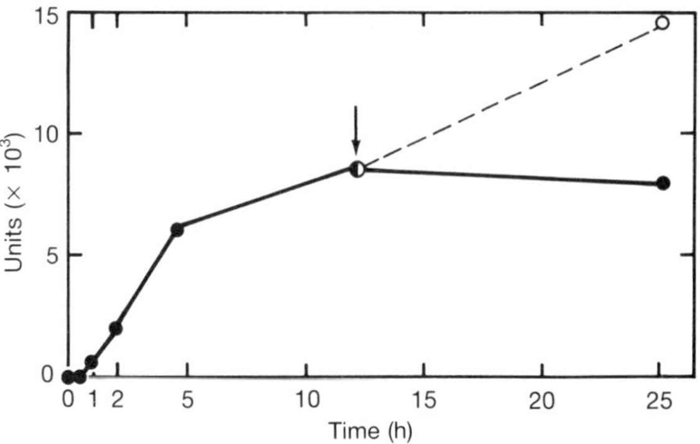

Fig. 21.1. Nitrate induction of NR activity. Squash plants (*Cucurbita maxima* L. cv Buttercup) were grown at 28°C with continuous light in vermiculite irrigated with water for 5 days and then (time zero) with 35 mM NH_4NO_3 in a nutrient medium containing 20 mM KPO_4, pH 5.5/1 mM $(NH_4)_2SO_4$/1 mM $CaCl_2$/1 mM $MgSO_4$/100 μM FeEDTA plus micronutrients. At the indicated times, crude protein extracts were prepared from cotyledons and assayed for NR activity (closed circles). At 12 h after the application of nitrate (arrow) some of the plants were irrigated again with 35 mM ammonium nitrate (open circles). Activity is expressed in units per mg of protein, one unit being defined as the production of 1 μmol of nitrite/min at 30°C.

Table 21.1. Nitrate induction of nitrate reductase activity

Tissue		Activity (units $\times 10^{-3}$)
Uninduced	(NH_4Cl)	<0.04
Induced	(NH_4NO_3)	2.1
Mix	$(I+U)$	2.1

Protein extracts were prepared from cotyledons of 6-day-old squash seedlings that had been treated twice with 35 mM NH_4Cl (uninduced) or 35 mM NH_4NO_3 (induced) 2 days prior to harvesting the tissue. Aliquots of these extracts (210 μg) were assayed for NR activity separately (rows 1 and 2) or together after a 15 min preincubation (row 3).

mixed and preincubated before measuring activity (Table 21.1). NR protein in the extracts was then analysed on Western blots using anti-NR antibody (mouse ascites fluid) prepared against gel-purified enzyme. A single band corresponding to a protein of the expected mass (110 kDa) was detected only in extracts from nitrate-treated plants (Fig. 21.2, lane 2) and not in extracts from control plants (lane 1). Thus, the antibodies appeared to recognize a single protein that was inducible by nitrate and precisely corresponded with the increase observed in nitrate reductase activity.

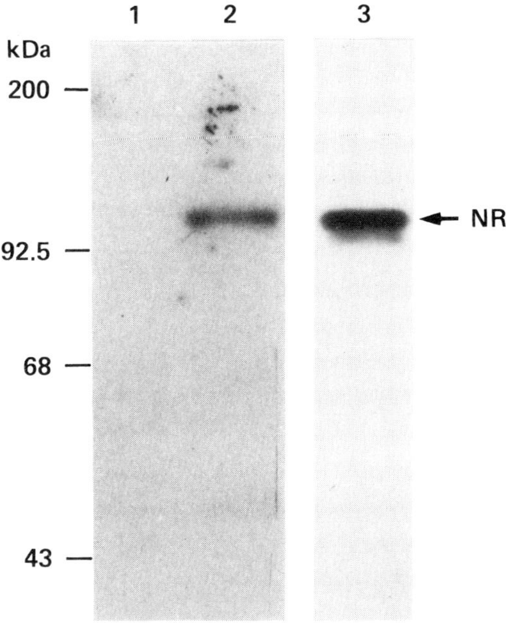

Fig. 21.2. Nitrate induction of NR protein. Crude protein extracts were prepared from cotyledons of squash seedlings irrigated with 35 mM ammonium chloride (lane 1) or with 35 mM ammonium nitrate (lane 2). Protein (100 μg) was separated by NaDodSO$_4$/ polyacrylamide gel electrophoresis then transferred to a nitrocellulase filter, which was incubated with anti-NR antibody (mouse ascites fluid provided by Jill Remmler and Wilbur Campbell) and ^{125}I-labelled Protein A then autoradiographed. In addition, 2.0 μg of protein enriched for NR 500-fold by Affi-Gel Blue chromatography was separated by electrophoresis, transferred to a filter, and incubated with affinity-purified antibody that bound to λ-Cmc1-encoded antigens (lane 3). No bands were seen when non-immune ascites fluid was used.

Isolation of a squash nitrate reductase cDNA clone

With the availability of mono-specific antibodies to NR, the most direct method for isolating a cDNA clone was an antibody screen of a λgt11 cDNA expression library. The anti-NR antibody appeared to bind to a single protein on the Western blot (Fig. 21.2) and was thus suitable for the screen. Poly(A)$^+$ RNA was prepared from cotyledons of nitrate-treated plants and was used to prepare a cDNA library (Huynh *et al.* 1985). From 250 000 recombinant phage screened with the antibody, one positive phage (λCmc1) was identified and purified. This phage had a 1.2 kb cDNA inserted into the *Eco*R1 site of λgt11.

An immunological test was performed to demonstrate that the λCmc1 cDNA encoded part of the NR protein. Antigens produced by the recombinant phage were used to affinity purify antibodies from the immune ascites fluid, and the purified antibodies were tested to see if they bound to NR. The purified antibodies bound to a 110 kDa protein on a Western blot of a highly enriched preparation of NR (Fig. 21.2, lane 3). A second test with the purified antibodies was also performed: immune ascites fluid inhibited NR activity in crude protein extracts while non-immune ascites fluid had little effect over the concentration tested (Fig. 21.3). The purified antibodies also inhibited NR activity by 90 per cent at the concentration tested (Table 21.2, row 4). Control antibodies [purified from the immune ascites fluid with antigens encoded by λgt11 (row 3), or antibodies purified from non-immune ascites fluid with antigens encoded by λCmc1, (row 2)] had little effect. Thus, λCmc1 encodes at least part of the NR protein.

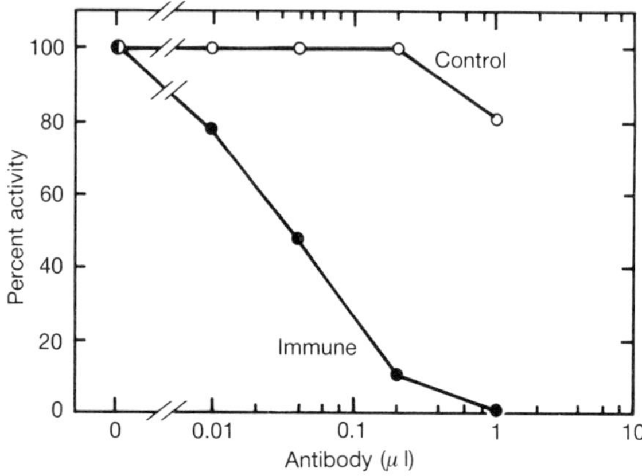

Fig. 21.3. Antibody inhibition of NR activity. Crude protein extracts (210 μg) from cotyledons of nitrate-treated squash seedlings were incubated with antibody (ascites fluid) for 30 min. on ice. Antibody–antigen complexes were precipitated with Protein A–Sepharose. The supernatant was then assayed for NR activity. 100% activity corresponds to 0.002 units.

Because the cDNA of λCmc1 is only ~ 30 per cent the expected size of the NR mRNA (a protein of 110 kDa would have approximately 3 kb of coding sequence), a longer cDNA was sought that would contain more of the coding sequence. The cDNA insert of λCmc1 was used to screen the λgt11 cDNA, and 13 positive phage were identified from a screening of 125,000 recombinant phage. Three phage had inserts of 2.8 kb with identical restriction maps (Fig. 21.4). The insert of one of the phage, λCmc2, was used for analysing nitrate reductase mRNA levels.

Effect of nitrate on nitrate reductase mRNA levels in squash cotyledons

It has been demonstrated that nitrate increases NR activity and protein levels in plants: this result was also confirmed for squash, as described above. To

Table 21.2. Inhibition of nitrate reductase activity with
affinity purified antibody

Purified antibodies	Activity	Remaining activity
None	2.1	100%
Control (λCmc1)	2.1	100%
Anti-NR (λgt11)	1.3	62%
Anti-NR (λCmc1)	0.32	12%

Protein extracts were prepared from cotyledons of 6-day-old squash
seedlings that had been treated twice with 35 mM NH_4NO_3 2 days prior
to harvesting the tissue. Aliquots of these extracts (210 μg) were assayed
for NR activity after incubation with no antibody (row 1) or with
antibody from non-immune ascites fluid (row 2) or from immune ascites
fluid (rows 3 and 4) that had been purified by binding to antigens
produced by bacteria infected with phage indicated in the paren-
theses. Before assaying NR activity, antibody was precipitated with
Protein A–Sepharose.

Squash cDNA clone

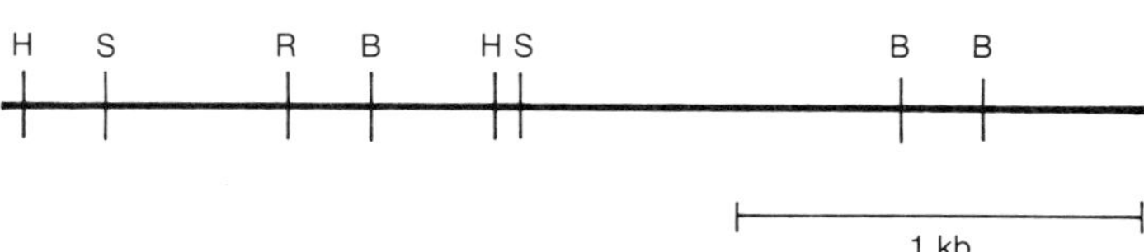

Fig. 21.4. Restriction map of λCmc2. The 2.8 kb cDNA insert of λCmc2 was isolated and
mapped with following restriction enzymes: H = *Hind*III, S = *Sal*1, B = *Bam*H1, R = *Eco*R1,
Sac = *Sac*1.

understand how this increase occurs it is important to know what is happening to
the NR mRNA levels. Poly(A)$^+$ RNA was prepared from squash cotyledons,
separated by electrophoresis and immobilized onto a filter which was hybridized
to radiolabelled λCmc2 cDNA. A single band, corresponding to RNA 3.2 kb in
length, was detected only in cotyledons of nitrate-treated plants and not in the
control plants (Fig. 21.5). Thus, the NR mRNA levels match the observed activity
and protein levels.

Cloning and analysis of a nitrate reductase mRNA from *Arabidopsis*

Immunological data indicate that NR antigenic sites are conserved in flowering
plants (Campbell and Smarrelli 1986; Kleinhofs *et al.* 1985); we therefore

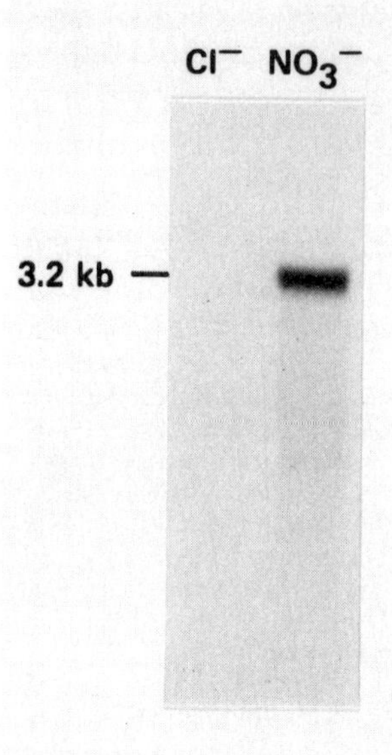

Fig. 21.5. Nitrate induction of squash NR mRNA. Poly(A)$^+$ RNA was prepared from cotyledons of squash seedlings 45 h after irrigation with 35 mM ammonium chloride (lane 1) or with 35 mM ammonium nitrate (lane 2). RNA (3 μg for each lane) was separated by agarose gel electrophoresis after treatment with glyoxal then transferred to a nylon filter which was hybridized with radiolabelled λCmc2 cDNA and washed as described (Crawford *et al.* 1986). After autoradiography, the filter was stripped and rehybridized with a control probe to demonstrate that the same amount of RNA had been transferred in each lane (see Crawford *et al.* 1986).

attempted to isolate a cDNA clone for NR from *Arabidopsis* using the λCmc1 cDNA from squash. Under low stringency conditions of hybridization, the squash cDNA was found to weakly hybridize to a 3.2 kb RNA in the leaves of nitrate-treated *Arabidopsis* plants (N. M. Crawford, unpublished data). A λgt10 library of size enriched cDNA was prepared from *Arabidopsis* mRNA (Huynh *et al.* 1985). From 4000 recombinant phage screened with λCmc1 cDNA at low stringency, 16 positive phage were identified and purified. Two phage had cDNA inserts 3.0 kb in length that had identical restriction maps, and the insert from one of these phage, λAtc22, was used for RNA analysis.

Poly(A)$^+$ RNA was prepared from the leaves of 3-week-old *Arabidopsis* plants that had been treated with different nitrogen regimens. NR levels were determined by enzymatic assays; NR mRNA levels were measured by Northern blot analysis using the λAtc22 cDNA as probe. The cDNA hybridized to a 3.2 kb RNA, the same size as that observed with squash (Fig. 21.6), the behaviour of which strongly indicated that it encodes NR. Plants grown on ammonium had little enzyme activity or NR mRNA (lane 1) whereas plants grown on nitrate had 15- to 30-fold higher levels of nitrate reductase activity and mRNA (lane 2). This increase was observed regardless of whether ammonium was present (lane 3). Nitrate, when applied to ammonium-grown plants, also resulted in an increase of NR activity and mRNA after 16 h (lanes 4 and 5). The cDNA also hybridized weakly to another mRNA that was constitutively expressed, highly abundant, and 0.8 kb in size (N. M. Crawford, unpublished data). What this RNA encodes and what relationship it has to NR is unknown.

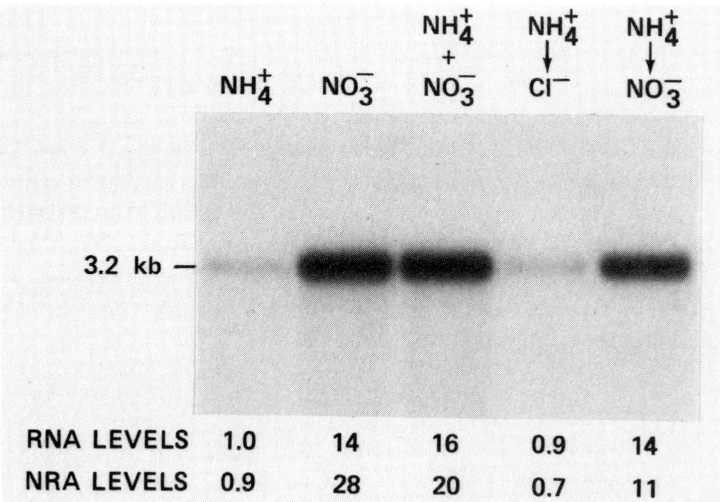

Fig. 21.6. Analysis of NR mRNA from *Arabidopsis thaliana.* Poly(A)$^+$ RNA was prepared from the leaves of 3-week-old plants that had been grown in a defined nutrient medium (Somerville and Ogren, 1982) with 2 mM (NH$_4$)$_2$SO$_4$ (lane 1), 5 mM KNO$_3$ (lane 2) or 3 mM NH$_4$NO$_3$ (lane 3). Ammonium-grown plants were also treated with 5 mM KCl (lane 4) or 5 mM KNO$_3$ (lane 5) 16 h before harvesting the leaves. Equal amounts of RNA (1 μg) were loaded onto each lane, separated by agarose gel electrophoresis after glyoxal treatment, and transferred to a nylon filter which was hybridized to radiolabelled λATc22 cDNA. After autoradiography, the filter was rehybridized with a control probe; identical signals were found in all lanes (data not shown). The hybridization signals were quantified with a densitometer (row marked 'RNA levels'). Before extracting the RNA, some of the leaf tissue was used to prepare protein extracts which were assayed for NR activity (row marked 'NRA levels' expressed in units/mg of protein).

Summary

We have described the isolation of cDNA clones from two plants, squash and *Arabidopsis.* Several lines of evidence indicate that these cDNAs encode NR: the antigens produced by the squash cDNA can be used to purify antibody that will bind and inhibit NR; the cDNAs hybridize to a mRNA that is of sufficient length to encode NR and that is regulated by nitrate. It is presumed that these clones encode the NADH-specific form of NR because no other form of the enzyme has been detected in these two plants (N. M. Crawford, unpublished data).

It has been known for over 30 years that nitrate increases the level of NR activity in plants (Tang and Wu 1957). It is now clear that in leaves this response is due to an increase in NR mRNA. In the absence of nitrate, there is little or no NR activity, protein or mRNA. When nitrate is added, the steady state level of NR activity, protein and mRNA increase. We have observed this response in two different species of plants and in two different types of leaves, cotyledons and mature leaves. It will be interesting to see in future experiments what role synthesis and degradation of NR mRNA play in the observed increase in steady-state level, although it is expected that activation of transcription of the NR gene will play a major role in nitrate regulation.

The molecular mechanisms responsible for nitrate regulation of NR are actively being investigated. The nitrate receptor needs to be identified, and regulatory genes that control the expression of nitrate assimilatory genes in plants need to be isolated and characterized. Both fungi and bacteria provide excellent models for regulatory networks that might exist in plants (Wray 1986), but more genetic and molecular work needs to be done before we can understand how plants control the vital process of nitrate assimilation.

Acknowledgements

We wish to thank Wilbur Campbell for anti-nitrate reductase antibody and for helpful discussion. This research was supported by National Science Foundation Research Grant DBC 8402390 and U.S. Department of Energy Research Grant DE-FG03-84ER13265. N.M.C. was supported by a National Science Foundation Postdoctoral Research Fellowship in Plant Biology.

References

Calza, R., Huttner, E., Vincentz, M., Rouze, P., Galangau, F., Vaucheret, H., Cherel, I., Meyer, C. Kronenberger, J., and Caboche, M. (1987). Cloning of DNA fragments complementary to tobacco nitrate reductase mRNA and encoding epitopes common to the nitrate reductases from higher plants. *Molecular and General Genetics* **209**, 552–62.

Campbell, W. H. and Smarrelli, J. (1986). Nitrate reductase: biochemistry and regulation. In *Biochemical basis of plant breeding* (ed. C. A. Neyra), vol. II, 1–39. CRC Press, Boca Raton, Florida.

Cheng, C.-L., Dewdney, J., Kleinhofs, A., and Goodman, H. M. (1986). Cloning and nitrate induction of nitrate reductase mRNA. *Proceedings of the National Academy of Sciences, USA* **83**, 6825–8.

Crawford, N., Campbell, W., and Davis, R. W. (1986). Nitrate reductase from squash: cDNA cloning and nitrate regulation. *Proceedings of the National Academy of Sciences, USA,* **83**, 8073–6.

Ferrari, T. E., Yoder, O., and Filner, P. (1973). Anaerobic nitrate production by plant cells and tissues: evidence for two nitrate pools. *Plant Physiology* **51**, 423–31.

Huynh, T., Young, R., and Davis, R. (1985). Constructing and screening cDNA libraries in λt10 and λt11. In *DNA cloning* (ed. D. Glover), vol. I, pp. 49–78. IRL Press, Oxford.

Kleinhofs, A., Warner, R., and Narayanan, K. (1985). Current progress towards an understanding of the genetics and molecular biology of nitrate reductase in higher plants. In *Oxford surveys of plant molecular and cell biology* (ed. B. J. Miflin), vol. 2, pp. 91–121. Oxford University Press.

Redinbaugh, M. G. and Campbell, W. H. (1985). Quaternary structure and composition of squash NADH: nitrate reductase. *Journal of Biological Chemistry* **260**, 3380–5.

Remmler, J. and Campbell, W. H. (1986). Regulation of corn leaf nitrate reductase. *Plant Physiology* **80**, 441–2.

Somers, D. A., Kuo, T.-M., Kleinhofs, A., Warner, R. L., and Oaks, A. (1983). Synthesis and degradation of barley nitrate reductase. *Plant Physiology* **72**, 949–52.

Sommerville, C. R. and Ogren, W. L. (1982). Isolation of photorespiration mutants in *Arabidopsis thaliana*. In *Methods in chloroplast molecular biology* (eds M. Edelman *et al.*), pp. 129–38. Elsevier, Amsterdam.

Tang, P. and Wu, H. (1957). Adaptive formation of nitrate reductase in rice seedlings. *Nature* **179**, 1355–6.

Wray, J. L. (1986). The molecular genetics of higher plant nitrate assimilation. In *A genetic approach to plant biochemistry* (ed. A. D. Blonstein and P. J. King), pp. 101–57. Springer-Verlag, New York.

Applied Aspects

22. Fungal biotechnology and the nitrate assimilation pathway

Shiela E. Unkles

Introduction

Biotechnology is often considered a concept of the late twentieth century but, in fact, as early as 6000 BC Middle Eastern peoples were using unicellular fungi, the yeasts, to produce alcohol and by around 4000 BC had developed technologies for wine making and for baking leavened bread. At least 2000 years ago, in the Far East, filamentous fungi were first used to produce fermented food, such as miso, as well as being cultivated for food themselves—the Japanese shiitake mushroom, for example. In the West, mould-ripened cheeses are said to have been eaten by the classical Romans, but it was not until the second half of the seventeenth century that the first fungal biotechnology industry developed in France around the cultivation of *Agaricus* mushrooms.

Following the work of van Leeuwenhoek, Pasteur, and others in the eighteenth and nineteenth centuries who demonstrated the involvement of micro-organisms in fermentation processes, came an understanding of the biochemistry involved in microbial fermentations, so that by the late nineteenth century, several industrially important compounds, for example organic acids, ethanol, and acetone, were being produced by micro-organisms on an increasingly large scale. The first process detailing the production of enzymes for industry, that of a starch hydrolysis enzyme preparation from *Aspergillus oryzae*, was patented in 1894. The 1940s saw the advent of the modern pharmaceutical industry with the industrial production of penicillin initially by *Penicillium notatum*, but latterly by *P. chrysogenum*.

In recent years, many more biotechnological processes have been introduced and today there is a multi-million pound industry built around the products of filamentous fungi. Not surprisingly, therefore, there is a great deal of interest in the possible introduction of useful, genetically engineered products as well as improvement of existing fermentations using recombinant DNA technology.

Current uses of filamentous fungi

The great variety of industrially important substances produced by filamentous fungi can be divided into several categories (Table 22.1). First, there are the primary metabolites, those substances produced during the course of normal cell growth. One of the most ubiquitous of these mould products must be citric acid, which is used extensively in the food and beverage industry as an acidulant and flavour enhancer as well as having various uses in the pharmaceutical and cosmetic industries (Kapoor *et al.* 1982).

Secondary metabolities are those compounds which are generally not essential to viability, although they may confer a selective advantage to the producer, and which are usually produced late in the growth phase of the organism. By far the most commercially important secondary metabolites are the β-lactam antibiotics— penicillins and cephalosporins—which together account for around 50 per cent of the world antibiotic market representing a value of US\$480 million (Pirt 1987). An interesting, although less lucrative, by-product of the antibiotic industry is the fungal mycelium itself which is used in some alpine regions as a fertilizer to rejuvenate the ski-slopes in summer. Many other secondary metabolites are made by filamentous fungi but few are produced commercially.

A wide variety of fungal enzymes is of primary importance in the food industry, particularly baking (α-amylase, proteases), dairy products (fungal rennin, catalase, lipases, proteases), fruit processing (pectinases, glucose oxidase), confectionary (lactase, invertase, pectinase, α-amylase) and brewing (α-amylase, glucoamylase, acid proteases, β-glucanase). Recent detailed reviews of the production and use of fungal enzymes have been published by Böing (1982), Lambert (1983), Montenecourt and Eveleigh (1985) and Alder-Nissen (1987). The cellulases, cellobiohydrolases and endoglucanases, although used only to a limited extent in the food industry at present, are, together with the hemicellulases and ligninases, the subjects of intensive research. It is hoped, eventually, to produce low cost, fermentable sugars from waste lignocellulose (Kirk 1983; Tsao and Chiang 1983; Knowles *et al.* 1987; Dale 1987).

A further economically important use of filamentous fungi is as a source of food, either as mushrooms or, potentially, as single cell protein (for example, Mycoprotein). Several types of cheeses are now available which use fungal inocula in the maturation (ripening) of curd by the action of proteases and lipases (Marth 1982).

Other processes in which fungi are used include the bioconversion of organic molecules to produce steroid hormones and the conversion of penicillin V to 6-aminopenicillanic acid, the precursor of many semi-synthetic β-lactam antibiotics. Additionally, there is growing interest in the use of moulds as biological insecticides as these are regarded as safer and more persistent than chemical insecticides (Lisansky and Hall 1983; Brown *et al.* 1987).

Although not of immediately obvious, direct biotechnological relevance, there exists, in addition, an array of animal and plant pathogenic fungi which world-wide can cause significant reduction of industrial man-hours and devastation of cash crops. *Phytophthora palmivora*, for example, has in the past destroyed as much as 60 per cent of the cocoa crop of West Africa. In the UK, *P. infestans* is the cause of potato late blight and is infamous as the agent responsible for the

Table 22.1. Industrial products of filamentous fungi

Category	Product	
Primary metabolites	Organic acids—citric, itaconic, gluconic	*Aspergillus niger, Aspergillus terreus*
	Vitamins—β-carotene, riboflavin	*Blakeslea trispora, Eremothecium ashbyii, Ashbya gossypii*
	Polysaccharides—pullulan, scleroglucan	*Aureobasidium pullulans, Sclerotium rolfsii*
Secondary metabolites	Antibiotics—penicillins, cephalosporins, griseofulvin, fusidanes	*Penicillium chrysogenum, Cephalosporium acremonium, Penicillium griseofulvin, Fusidium coccineum*
	Plant and animal growth hormones	*Gibberella fujikuroi, Gibberella zeae*
	Other drugs—alkaloids, cyclosporin	*Claviceps purpurea, Trichoderma polysporium*
Enzymes	α-Amylases	*Aspergillus niger, Aspergillus oryzae, Aspergillus awamori*
	Glucoamylase	*Aspergillus niger, Aspergillus awamori, Rhizopus sp.*
	β-glucanase	*Aspergillus niger*
	Cellulases	*Trichoderma viride, Trichoderma reesei*
	Acid proteases	*Aspergillus niger, Aspergillus oryzae, Mucor pusillus, Mucor miehei*
	Pectinase	*Aspergillus niger*
	Lipases	*Rhizopus sp., Mucor miehei, Aspergillus niger*
	Catalase	*Aspergillus niger*
	Glucose oxidase	*Aspergillus niger, Penicillium amagasakiense*
	Lactase	*Aspergillus niger, Aspergillus oryzae*
	Invertase	*Aspergillus niger, Aspergillus oryzae*
	Fungal rennin	*Mucor miehei, Mucor pusillus*
Food	Mushrooms	*Agaricus bisporus, Pleurotus ostreatus, Lentinus edodes, Volvariella volvacea*
	Mould-ripened cheeses	*Penicillium roquefortii, Penicillium camembertii*
	Oriental food fermentations	*Aspergillus oryzae, Rhizopus sp.*
	Single cell protein	*Fusarium, Paecilomyces*
Bioconversions	Steroid hormones	*Rhizopus sp. Fusarium solani, Aspergillus ochraceus, Curvularia lunata*
	6-Aminopenicillanic acid	*Penicillium chrysogenum, Cephalosporium sp., Fusarium sp.*
Insecticides	Colorado beetle control	*Beauveria bassiana*
	Aphid control	*Verticillium lecanii*
	Citrus mite control	*Hirsutella thompsonii*
	Spittle-bug control	*Metarrhizium anisopliae*

Compiled from Bigelis (1985), Bennett (1985), Montenecourt and Eveleigh (1985), Brown *et al.* (1987).

devastation of the potato crop and subsequent famine in eighteenth century Ireland. Equally important, and of growing interest to forestry and agriculture, are the mycorrhizal fungi which are found in symbiotic association with almost all plant roots, and without which many plants cannot survive. It has been shown that introduction of non-indigenous pine trees can fail without pre-inoculation of the roots or soil with the appropriate mycorrhizal fungus (Jackson and Mason 1984). While it could be argued that these fungi are of exclusively academic interest to medicine or botany, there is nevertheless the potential for applied, industrial or biotechnological exploitation.

Until now, strain improvement of industrial fungi has involved the selection of spontaneous and induced mutants as well as optimization of fermentation conditions, not without a considerable degree of success. For example, over the 40 years since the introduction of penicillin production, the concentration of the antibiotic which can be obtained in fermentations has increased 1000-fold (Pirt 1987). There is, however, a dearth of basic genetic understanding of many of the processes involved in fungal biotechnology and this ignorance is probably worse in respect to host–parasite or host–symbiont relationships. A major problem has been that classical genetic studies are impossible with many of the industrial strains and natural isolates because of the lack of sexual and parasexual cycles. Additionally, many of these strains can be diploid, polyploid, or even aneuploid, which further complicates mutant isolation and genetic crosses.

It is clear, therefore, that there are a great number of processes which could benefit from molecular genetic manipulation and the close scrutiny that modern recombinant DNA techniques afford. An obvious area of interest is the improvement of antibiotic biosynthesis pathways by, for example, increasing promoter activity, de-regulating repression control circuits and increasing gene copy number. Investigations have already shown that in high cephalosporin-producing *Cephalosporium acremonium*, two biosynthetic enzymes show greatly reduced sensitivity to glucose repression (Shen *et al.* 1986) and in a high penicillin-production strain of *P. chrysogenum*, the isopenicillin N synthetase gene is located on an at least 40 kb stretch of DNA which is amplified five times (Smith *et al.* 1987; G. Turner, personal communication).

Transformation in filamentous fungi

A primary requirement of genetic analysis and manipulation, using molecular genetic methods, is a gene-mediated transformation system. Several such systems have been developed for a number of biotechnologically relevant filamentous fungi from the oomycetes (*Achlya*) to the higher basidiomycetes (*Coprinus* and *Schizophyllum*) (Table 22.2). Not included are *A. nidulans* and *Neurospora crassa*, in which the original work on transformation in filamentous fungi was pioneered (for reviews see Johnstone 1985; Mishra 1985; Turner and Ballance 1985;

Hynes 1986). These organisms are considered to be of limited biotechnological value.

Transformation in filamentous fungi usually occurs by integrative recombination. Indeed, the only examples of recombinant plasmids existing by autonomous replication without demonstrable, simultaneous chromosomal integration are those of the zygomycetes, *Mucor circinelloides* (van Heeswijck 1986) and *Absidia glauca* (Wostemeyer *et al.* 1987), although autonomously maintained plasmids have also been described in *N. crassa* (Paietta and Marzluf 1985), *Phycomyces blakesleeanus* (Revuelta and Jayaram 1986), and *Podospora anserina* (Perrot *et al.* 1987). Integrative recombination can occur by insertion at a homologous site, insertion at a non-homologous site, or by conversion of the mutant allele by the transforming gene (Johnstone 1985; Hynes 1986; Ballance 1986). Integrative transformations are commonly less efficient than those resulting in autonomously replicating recombinant molecules, but this disadvantage is, to some degree, offset by the general stability of chromosomal integrations.

Fungal transformation systems can be placed in one of two broad categories. First, there are those systems based on the complementation of auxotrophic mutations by a corresponding cloned gene. Most, however, have certain limitations for use with poorly characterized organisms, the most obvious of which is the necessity for the isolation of the correct auxotrophic mutants. This usually involves random mutagenesis of strains, a procedure which could produce undesirable pleiotropic mutations in industrial strains, resulting in lower yields of product. Furthermore, the screening and subsequent characterization of mutants can be extremely difficult and time-consuming, particularly in non-haploid organisms. Random mutation can, of course, produce mutants in any of a number of genes in a biosynthetic pathway. An exception to this is selection based on reversal of uracil (or uridine) auxotrophy, for which mutations in the structural gene encoding orotidine-5′-monophosphate decarboxylase can be isolated by positive selection for fluoroorotic acid resistance in the presence of uracil or uridine. Such mutants have been isolated from *A. niger* (Goosen *et al.* 1987; van Hartingsveldt *et al.* 1987), *A. oryzae* (Mattern *et al.* 1987b) and *P. chrysogenum* (Cantoral *et al.* 1987). These studies have shown that mutants can be complemented by the homologous genes (*pyrG* or *pyrA* of *A. niger* and *pyrG* of *P. chrysogenum*), while *A. niger* and *P. chrysogenum*, but not *A. oryzae*, can be complemented by the *pyr-4* gene of *N. crassa*. *A. oryzae pyrG* mutants can however be complemented by the heterologous *A. niger pyrG* gene. The system has the disadvantage that, in addition to fluoroorotic acid itself being expensive, purine biosynthesis is of central metabolic importance. Thus mutation within this pathway, with subsequent introduction of a transforming gene, could conceivably cause disturbances to general metabolism affecting large scale cultivation of industrial organisms.

Positive transformation selection systems have been based on utilization of acetamide or on antibiotic resistance—the aminoglycosides hygromycin, G418, phleomycin, neomycin, or bleomycin, and oligomycin. The great advantage of

Table 22.2. Tranformation selection systems in filamentous fungi of potential commercial value

(A) COMPLEMENTATION OF AUXOTROPHY

Arginine	Tryptophan	Leucine	Uracil	Acetate	Methionine
Aspergillus niger	*Coprinus cinereus*	*Mucor circinelloides*	*Aspergillus niger*	*Coprinus cinereus*	*Ascobolus*
Aspergillus oryzae	*Penicillium*		*Aspergillus oryzae*		*immersus*
Magnaporthe	*chrysogenum*		*Penicillium*		
grisea	*Schizophyllum*		*chrysogenum*		
Trichoderma reesel	*commune*		*Podospora anserina*		
			Ustilago maydis		

(B) POSITIVE SELECTION

Acetamide	Aminoglycosides (hygromycin B/G418/phleomycin/bleomycin/neomycin)		Oligomycin	Benomyl
Aspergillus niger	*Absidia glauca*	*Nectria haematococca*	*Aspergillus niger*	*Aspergillus niger*
Aspergillus oryzae	*Aspergillus niger*	*Penicillium chrysogenum*		
Aspergillus terreus	*Aspergillus oryzae*	*Phycomyces blakesleeanus*		
Cochliobolus	*Achlya ambisexualis*	*Septoria nodorum*		
heterostrophus	*Cephalosporium acremonium*	*Ustilago maydis*		
Glomerella cingulata	*Cochliobolus heterostrophus*			
Penicillium chrysogenum	*Colletotrichum graminicola*			
Trichoderma reesei	*Fusarium sporotrichioides*			
	Fulvia fulva			
	Glomerella cingulata			
	Leptosphaeria maculans			

References cited in Gurr *et al.* (1987) Turgeon *et al.* (1987) and Abstracts from the 19th Lunteren Lectures on Molecular Genetics (1987).

these systems is that they do not require the isolation of recipient mutants. Acetamide utilization is permitted by the introduction of the *A. nidulans amdS* gene which encodes the enzyme acetamidase (Hynes *et al.* 1983). Some fungi, however, grow sufficiently well on acetamide to make background growth a significant problem in selection of transformants. In addition, while the gene is mitotically stable in *A. nidulans* (Kelly and Hynes 1987), this was found not to be the case in the plant pathogen *Glomerella cingulata* (Rodriguez and Yoder 1987). Selection for antibiotic resistance has been successful in a growing number of fungal species. Its usefulness, of course, depends on susceptibility of the recipient and this is species specific. *A. oryzae*, for example, is resistant to 5 mg/ml hygromycin B, but susceptible to 50 μg/ml phleomycin (Mattern *et al.* 1987*b*), while *M. circinelloides* is resistant to most useful antibiotics (van Heeswijck, personal communication). Furthermore, the use of a bacterial resistance gene often requires the isolation of a suitable fungal promoter in order to achieve optimum expression (Skatrud *et al.* 1987), although this may not be a difficult problem to overcome (Turgeon *et al.* 1987).

With positive selection systems, multiple integrations frequently occur (Kelly and Hynes 1985; Beri and Turner 1987; Skatrud *et al.* 1987), a feature which may be of considerable use when attempting over-production of an enzyme of interest, such as those involved in antibiotic biosynthesis. Conversely, the systems suffer the severe disadvantage that positive selection involves the introduction of a foreign gene. Generation of such recombinant strains would require expensive clinical trials before acceptance of a product to be used in the food or pharmaceutical industries.

Nitrate assimilation in filamentous fungi

Most filamentous fungi can grow on nitrate as a sole source of nitrogen (Lilly and Barnett 1951) although there are, nevertheless, exceptions in many genera, particularly in the zygomycetes *Rhizopus* and *Mucor* (Foster 1949; Lilly and Barnett 1951), and oomycetes e.g. *Phytophthora* (French 1953; Cameron and Milbrath 1965) but also in the ascomycetes *Aspergillus* and *Penicillium* (P. D. Bridge, personal communication).

The volume of information which exists on nitrate utilization by *A. nidulans* and *N. crassa* is greater by far than that available for any other fungi. Such information, biochemical and genetic, provides extremely valuable groundwork for extrapolation by analogy to less well characterized organisms. Detailed accounts of the systems and their regulation in these two 'model' species are given in Chapters 6, 19, and 20.

Briefly, extracellular nitrate is taken up by a permease (the product of *crnA* in *A. nidulans*) converted to nitrite by nitrate reductase (NR) (*A. nidulans niaD* and *N. crassa nit-3* gene product) and thence to ammonium by nitrite reductase (NiR) (*A. nidulans niiA* and *N. crassa nit-6* gene products). Both NR and NiR are under the control of two positively acting regulatory genes *nirA* or *nit-4* and *areA* or *nit-2* of *A. nidulans* (Chapter 19) or *N. crassa* (Chapter 20), respectively, and the products of both these genes are necessary for enzyme expression. In *A. nidulans* the nitrate assimilatory genes are clustered, with *niaD* and *niiA* divergently transcribed from an intergenic regulatory region (Chapter 6), while in *N. crassa*, the genes are unlinked (Chapter 20).

Studies of nitrate assimilation in other fungi are fragmentary. As with *A. nidulans* and *N. crassa*, Rippel (1931) demonstrated that ammonium ions were utilized in preference to nitrate by *A. niger* and *A. oryzae* and this phenomenon seems to be generally true among nitrate-assimilating fungi (Foster 1949). More detailed investigation of regulation in the basidiomycete *U. maydis* revealed that NR was induced when the organism was grown on nitrate as sole nitrogen source, but was completely repressed in the presence of ammonium (Lewis and Fincham 1970*a*).

Much information on the organization and regulation of nitrate assimilatory genes has been gained from the isolation and characterization of mutants unable to grow on nitrate. Indeed, a powerful advantage of the nitrate system for transformation of poorly characterized fungal species is the ease with which recipient mutants can be generated on the basis of resistance to chlorate. Nitrate assimilation gene mutants which have been identified in filamentous fungi, including *A. nidulans* and *N. crassa* for comparison, are shown in Table 22.3. Complementation analysis of nitrate non-utilizing, chlorate-resistant mutants has identified at least seven distinct loci in *A. amstelodami* (Bloomfield, unpublished Thesis, 1982), seven in *Fusarium oxysporum* and six in *Gibberella fujikuroi*, both pathogens of many agricultural crops (Klittich *et al.* 1986; Correll *et al.* 1987), nine in *P. chrysogenum* (Birkett and Rowlands 1981), seven in *Septoria nodorum*, an important pathogen of wheat (Newton and Caten 1988), and six in the maize pathogen *U. maydis* (Lewis and Fincham 1970*b*). Mutations in these loci can be assigned phenotypically, by simple growth tests, to (1) the structural gene for NR, *niaD* equivalent, which are unable to utilize nitrate; (2) control genes such as *nirA*, which show minimal growth on nitrate or nitrite (this is a 'leaky' phenotype) and (3) genes involved biosynthesis and assembly of the molybdenum cofactor (MoCo), *cnx* equivalents. The latter group of mutants also lack the enzyme purine hydroxylase 1 (xanthine dehydrogenase) which, like NR, requires the MoCo, and they can therefore easily be distinguished from *niaD* mutants by their inability to utilize hypoxanthine or adenine as sole sources of nitrogen.

A significant observation in terms of gene organization is that complementation groups displaying overlapping complementation patterns resembling those of *cnxB* mutations of *A. nidulans* have been described in both *P. chrysogenum* and *S. nodorum* (Birkett and Rowlands 1982; Newton and Caten 1988). This complex pattern probably reflects deletions spanning two closely linked *cnx* genes. Interestingly, a few chlorate-resistant mutants of *S. nodorum* have been described which are unable to use nitrogen sources other than ammonium (Newton and

Table 22.3. Nitrate assimilation genes identified in filamentous fungi

Organism	Nitrate reductase	Nitrite reductase	Nitrate uptake	Nitrate induction control	Nitrogen metabolite control	Molybdenum cofactor (loci)	Reference
Aspergillus amstelodami	*nia*	*nii*	*crn*	*nir*	—	*cnx* (5 loci)	Bloomfield (unpublished Thesis, 1982)
Aspergillus nidulans	*niaD*	*niiA*	*crnA*	*nirA*	*areA*	*cnx* (6 loci)	Chapter 6, 19
Aspergillus niger	*niaD*	—	—	—	—	*cnx*	Unkles *et al.* (1989)
Aspergillus oryzae	*niaD*	—	—	*nirA*	—	*cnx*	J. A. Macro and S. E. Unkles, unpublished.
Cephalosporium acremonium	*niaD*	—	(*crn*)[a]	*nir*	—	*cnx*	Whitehead *et al.* (1989*a*)
Fulvia fulva	*nia*	—	—	(*nir*)[a]	—	*cnx*	Talbot *et al.* (1988)
Fusarium oxysporum	*nit-1*	—	—	*nit-3*	—	*nit* (5 loci)	Correll *et al.* (1987)
Gibberella fujikuroi	*nit-1*	—	—	*nit-3*	—	*nit* (4 loci)	Klittich *et al.* (1986); Correll *et al.* (1987)
Neurospora crassa	*nit-3*	*nit-6*	—	*nit-4*	*nit-2*	*nit* (4 loci)	Chapter 20
Penicillium chrysogenum	*niaA*	*niiA*	—	*nir*	—	*cnx* (7 loci)	Birkett and Rowlands (1981)
Septoria nodorum	*Nia*	—	*ChlT*	*Nir*	*Nit*	*Cnx* (4 loci)	Newton and Caten (1988)
Ustilago maydis	*nar-A*	*nir-1*	—	*nar-C*	—	*cnx* (4 loci)	Lewis and Fincham (1970*b*); Holliday (1974)

[a] Tentative designation.

Caten 1988). This mutant locus, designated *Nit*, may correspond to the nitrogen metabolite control genes *areA* of *A. nidulans* and *nit-2* of *N. crassa* (Chapters 19, 20). Putative NR structural gene mutants have also been identified in *A. niger*, *A. oryzae* and *C. acremonium* as well as the plant pathogens *Fusarium oxysporum*, *G. fujikuroi*, and *Fulvia fulva*, and phenotypically distinguished from *nir* and *cnx* mutants (Table 22.3). Double mutants unable to utilize either nitrate or nitrite, probably the result of a deletion spanning *niiA* and *niaD*, have been isolated from *P. chrysogenum* and can be distinguished from *nirA* mutants by complementation studies (Birkett and Rowlands 1981). Preliminary evidence suggests that in *C. acremonium* the NR and NiR structural genes may be unlinked (Whitehead *et al.* 1989*a*).

Mutants deficient in nitrate uptake, analogous to *crnA* mutants of *A. nidulans*, which are chlorate-resistant but capable of growth on nitrate and nitrite, have been identified in *A. amstelodami* (Bloomfield, unpublished Thesis, 1982) and *S. nodorum* (Newton and Caten 1988) and possibly also in *C. acremonium* (Whitehead *et al.* 1989*a*). In addition, NiR structural gene mutants have been isolated, by screening for their inability to utilize nitrate or nitrite, from *A. amstelodami* (Bloomfield, unpublished Thesis, 1982) and *U. maydis* (Holliday 1974). These can be distinguished from *nir* mutants by complementation.

While in the majority of cases classification of chlorate-resistant mutants by growth properties is simple, a cautionary note must be struck by the example of *C. acremonium*. This organism is unable to grow on hypoxanthine alone as a nitrogen source and hence differentiation of NR structural gene mutants from MoCo mutants using the usual growth tests is impossible. In order to circumvent this problem, it was possible to demonstrate hypoxanthine dehydrogenase activity in polyacrylamide gels of protein extracts isolated from *C. acremonium* mutants grown in the presence of uric acid, an inducer of the enzyme in *A. nidulans* (Lewis *et al.* 1978). Mutants which did not display enzyme activity were classified as *cnx* mutants, while those which showed hypoxanthine dehydrogenase were designated NR structural gene mutants (Whitehead *et al.* 1989*a*). It was subsequently shown that growth could be obtained on medium containing inosine, a product of the hypoxanthine utilization pathway, when quinic acid was used as the carbon source rather than glucose but only with putative *niaD* and not *cnx* mutants. Presumably, inability of *C. acremonium* to grow on hypoxanthine with glucose as carbon source probably reflects carbon catabolite repression and possibly an inefficient hypoxanthine uptake system. Confirmation of the mutant designations was obtained by the demonstration of inducible cytochrome *c* reductase activity, an additional activity of the nitrate reductase apoprotein, in *cnx* mutants but not *niaD* mutants.

Finally, a number of chlorate-resistant, nitrate non-utilizing mutants have been isolated from several different strains of *A. flavus* (Papa 1986), *F. oxysporum* (Puhalla 1985) and *G. fujikuroi* (Puhalla and Spieth 1985). Although detailed phenotypic characterization of the mutants was not carried out, the studies illustrate the ease with which auxotrophic mutants can be generated in poorly characterized organisms using this system.

Cloning of fungal nitrate reductase structural genes

The NR structural genes of *A. nidulans* and *N. crassa* have been cloned, and the former has been sequenced (Chapters 6 and 20). Using radioactively labelled probes prepared from the *A. nidulans* gene, cross-hybridization at extremely low stringency to restricted DNA of a number of filamentous fungi can be demonstrated (Fig. 22.1). Hybridization to digested *U. maydis* DNA has not been detected using DNA probes but, using differential hybridization methods, *U. maydis* genomic sequences which encode a 4.2 kb mRNA have been isolated. The abundance of this transcript is markedly increased concomitant with NR activity, and it also shows cross-hybridization to the *A. nidulans niaD* gene (T. M. Smith, A. Spanos, and G. R. Banks, personal communication).

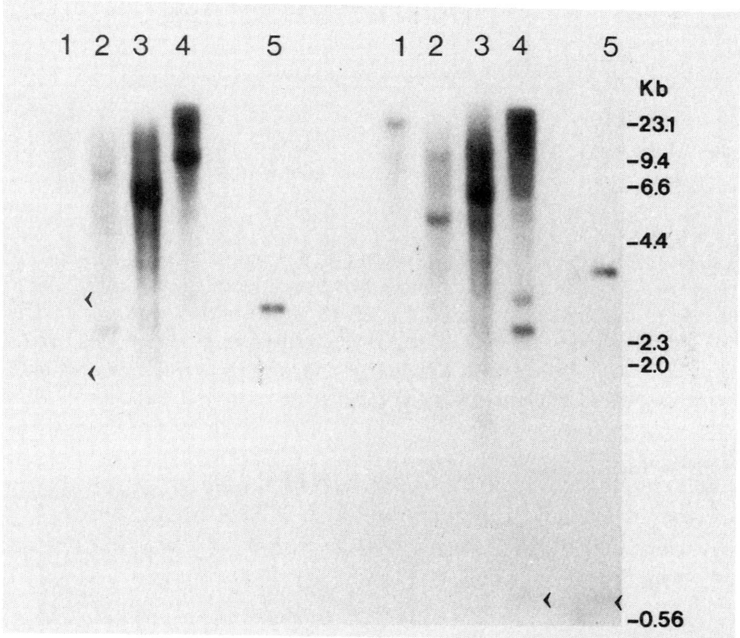

Fig. 22.1. Hybridization of the *Aspergillus nidulans niaD* gene to restricted fungal genomic DNA (S. E. Unkles, unpublished). Genomic DNA was digested and run on 0.8 per cent agarose and Southern blotted on to a nylon membrane. Blots were hybridized at very low stringency (54°C, washed once with $5 \times$ SSC) to the *A. nidulans niaD* gene. A. EcoRI-digested DNA. B. *Hind*III-digested DNA. 1. *Aspergillus niger* 3 μg; 2. *Aspergillus oryzae* 10 μg; 3. *Penicillium chrysogenum* 10 μg; 4. *Cephalosporium acremonium* 10 μg (DNA provided by M. Whitehead); 5. *Aspergillus nidulans* 0.5 μg (DNA provided by A. P. MacCabe). Arrows indicate the positions of faint bands.

Clones hybridizing strongly to *A. nidulans niaD* were isolated from an *A. niger* gene bank in pUC8. A detailed restriction map of one of these *niaD* clones is shown in Fig. 22.2, with that of *A. nidulans* for comparison. The approximate position of the coding sequence start has been inferred by hybridization to probes from the extreme 5′ end of the *A. nidulans* gene. Certainly, there is little similarity between the genes at the level of restriction sites. It has also been shown, however, using a probe prepared from the *A. nidulans* NiR gene, that the *niiA* and *niaD* genes of *A. niger* are contiguous in this clone and it is likely therefore that the arrangement of the genes will reflect that of *A. nidulans*.

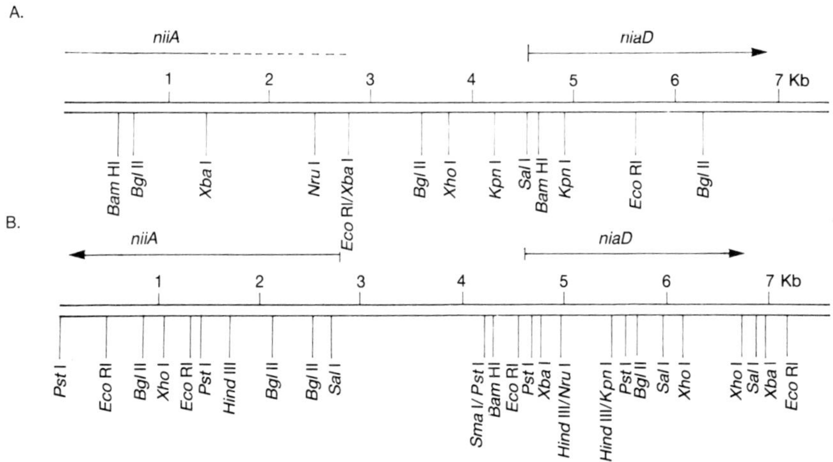

Fig. 22.2. Restriction maps of *Aspergillus niger* and *Aspergillus nidulans niaD* and *niiA* genes. A. *Aspergillus niger*. The position of the genes was inferred by hybridization to probes generated from the *A. nidulans* genes. The arrow indicates the direction of transcription following hybridization to extreme 5′- and 3′-*niaD* probes, respectively (Unkles *et al.* 1989). B. *Aspergillus nidulans*. The position of the genes and the direction of transcription indicated by the arrows was obtained from DNA sequence data (Johnstone *et al.* 1989).

In addition to the *U. maydis* and *A. niger* genes, the NR structural gene has been isolated from *A. oryzae* (J. A. Macro and S. E. Unkles, unpublished). As with *A. niger*, hybridization to the *A. nidulans niiA* gene has shown that the genes are probably also closely linked in *A. oryzae*.

Homologous transformation using *niaD*

Homologous transformation of *A. nidulans* has been demonstrated (Chapter 6). Similarly, introduction of the *A. niger niaD* gene has resulted in repair of the corresponding *A. niger* mutant (Fig. 22.3). Moreover, upon linearization of the transforming DNA close to the probable position of the NiR structural gene, frequencies of $\geqslant 100$ transformants/μg DNA have been observed (Unkles *et al.* 1989), approaching the range necessary for self-cloning experiments. Such frequency increases using linear DNA are not without precedent and have also been reported in *N. crassa* (Dhawale and Marzluf 1985) and *C. acremonium* (Skatrud *et al.* 1987). This is not, however, a universal phenomenon since linearization of a *trpC* plasmid had no effect on transformation frequency in *A. nidulans* (Yelton *et al.* 1984) or the rice blast disease fungus, *Magnaporthe grisea* (Parsons *et al.* 1987). Optimization of the transformation procedure itself may result in even higher frequencies. For example, manipulation of various parameters such as cell wall digestion, protoplast size fractionation, DNA concentration and polyethylene glycol treatment resulted in more efficient transformation and protoplast regeneration in *C. acremonium* (Skatrud *et al.* 1987). It has also been noted that spreading transformed protoplasts of *P. chrysogenum* rather than pour-plating significantly increased the transformation frequency in this organism (Whitehead *et al.* 1989*b*).

Several transformation systems have utilized vectors containing DNA sequences (ARS) which confer the ability for autonomous replication of integrative vectors in the yeast *Saccharomyces cerevisiae*. These experiments have met with varying success from showing no improvement as in *A. niger* (*pyr-4*/*ans-1* vector; van Hartingsveldt *et al.* 1987), *C. acremonium* (hygromycin resistance/ mitochondrial ARS; Skatrud *et al.* 1987) and *P. chrysogenum* (*trp*/mitochondrial ARS; Picknett *et al.* 1987) to increasing transformation frequency 3-fold and markedly increasing mitotic stability as in *P. chrysogenum* (*pyr-4*/*ans-1*; Cantoral *et al.* 1987). No example of integrative transformation has yet produced the dramatic increase in efficiency observed in *A. nidulans* using a vector containing the *N. crassa pyr-4* gene with the *A. nidulans ans-1* sequence. The effect of *ans-1*, however, has been shown to be location- and gene-specific (Wernars 1986), and may even be organism-specific. Thus it is possible that this or similar sequences could have a stimulating effect on transformation of *niaD* in at least some organisms. Preliminary evidence from *A. niger* transformations with a *niaD*/*ans-1* vector suggests that, at least in one location on the vector, *ans-1* has no effect (Unkles *et al.* 1989). The frequency of non-integrative transformation of *M. circinelloides*, however, has been demonstrated to be greatly influenced by the presence of a putative *Mucor ars* sequence (van Heeswijck 1986). Yet other studies have investigated the effect of inclusion of sequences homologous to host DNA, such as ribosomal DNA, without success in *A. nidulans* (Tilburn *et al.* 1983) and *C. acremonium* (Skatrud *et al.* 1987) but with improvement in *C. heterostrophus* (cited in Rodriguez and Yoder 1987). It would appear likely that this approach would have limited value, however, in a system which already includes homologous DNA.

In addition to *A. nidulans* and *A. niger*, homologous transformation of *niaD* has been successful in *A. oryzae* although in preliminary experiments, the transformation frequencies have not reached those achieved for *A. niger* (J. A. Macro, personal communication).

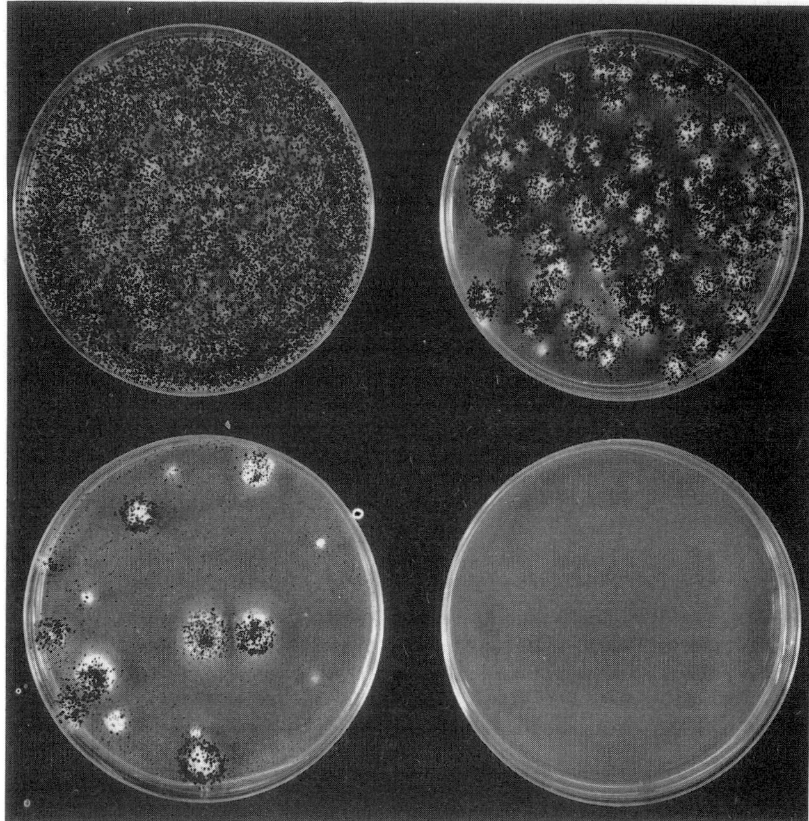

Fig. 22.3. Appearance of control and transformed *A. niger niaD*-5 strains on nitrate and ammonium. Upper left, untransformed on ammonium; upper right, transformed with linear *niaD* on nitrate; lower left, transformed with circular *niaD* on nitrate; lower right, untransformed on nitrate (Unkles *et al.* 1989).

Heterologous expression of fungal *niaD* genes

Although transformation based on the homologous system normally results in higher frequencies, heterologous transformation using *niaD* has also been achieved and such studies demonstrate the flexibility of recognition of control sequences within filamentous fungi. Complementation of NR structural gene mutants using heterologous *niaD* genes has been demonstrated by transformation of *F. oxysporum* with the *A. nidulans* gene (Malardier *et al.* 1987) and of *A. oryzae* with the *A. niger* gene (Mattern *et al.* 1987a). In both these cases, transformationf frequencies have been low, with < 10 transformants/μg DNA. In addition, it has been shown that the *A. nidulans* gene, which contains six introns

(Chapter 6), can also function in *A. niger*, *A. oryzae* and *P. chrysogenum*, indicating fidelity of intron splicing in these organisms. Likewise, the *A. niger niaD* gene is expressed not only in *A. oryzae* (J. A. Macro and S. E. Unkles, unpublished), but also in *A. nidulans* and *P. chrysogenum* (Whitehead *et al.* 1989*b*), and the *A. oryzae niaD* gene is functional in *P. chrysogenum* (M. Whitehead, personal communication). The exception to this relaxed accommodation of heterologous genes appears to be *C. acremonium* in which no successful transformation with either the *A. nidulans* or *A. niger* genes has been observed. Interestingly, assays of NR activity has shown that the regulation of both the *A. nidulans* and *A. niger niaD* genes introduced into *P. chrysogenum* is normal, i.e. repressed by ammonium and induced by nitrate (Whitehead *et al.* 1989*b*).

Expression of heterologous genes in filamentous fungi

One of the most exciting, and certainly novel, prospects for fungal biotechnology is the use of fungi for the expression of viral and mammalian genes. Although expression systems exist in bacteria, particularly *Escherichia coli*, and in yeast, particularly *Saccharomyces cerevisiae*, these suffer several disadvantages. The most obvious disadvantage of systems involving bacteria is the attempted expression of eukaryotic genes in a prokaryotic organism with the resultant lack of post-transcriptional processing (capping, polyadenylation, intron splicing) as well as post-translational modification (glycosylation, phosphorylation, acetylation). Furthermore, *E. coli* lacks an efficient secretion system and so, while several higher eukaryotic cDNA or intron-free genes can be expressed given appropriate bacterial transcription signals, these are accumulated within the cell resulting in insoluble aggregates (Emtage *et al.* 1983; Furman *et al.* 1987). Biological activity of such proteins can be restored but only after expensive extraction and renaturation procedures.

Perhaps surprisingly for a eukaryotic organism, *S. cerevisiae* shares many of the faults associated with *E. coli*. For example, *S. cerevisiae* is incapable of splicing accurately the introns of filamentous fungal and higher eukaryotic genes (Beggs *et al.* 1980; Innis *et al.* 1985; Pentilla *et al.* 1987) due to the absolute requirement for the sequence TACTAACA close to the 3′-end of yeast introns. Similarly, heterologous transcription signals function either poorly or not at all in yeast (Pentilla *et al.* 1984; Yoshizumi and Ashikari 1987) and so yeast require homologous transcription sequences for expression. *S. cerevisiae* will, however, carry out post-translational modification of heterologous proteins although glycosylation patterns may differ and hyperglycosylation is common (Pentilla *et al.* 1987; Kingsman *et al.* 1987; Yoshizumi and Ashikari 1987). A further disadvantage of yeast is the absence of efficient secretion systems. Those which have been used, such as the invertase gene signal sequence and that of the α mating-type factor, have had some limited success, particularly with small polypeptides (van Brunt 1986; Kingsman *et al.* 1987) but incorrect cleavage of the signal can occur (Vlasuk *et al.* 1986). Other proteins are expressed but not secreted and can be either soluble or insoluble, or may be associated with the cell wall.

Filamentous fungi are now becoming a focus of attention as microbial hosts for production of heterologous proteins. Of several possible attractions, not least are:

1. Large-scale fermentation technology based on filamentous fungi already exists.
2. The organisms in present use are recognized as being safe for production of pharmaceutical and industrial proteins.
3. Filamentous fungi are capable of secreting vast quantities of protein—for example *A. niger* production strains can secrete up to 20 g glucoamylase/l of medium (van Brunt 1986).

Several fungal and higher eukaryotic genes have been expressed and secreted in various filamentous fungi (Table 22.4). That genomic clones of glucoamylase and aspartic protease (fungal rennin) have been expressed in different fungal species, as well as fungi from widely different genera, illustrates the less rigid requirement of filamentous fungi for specific transcription sequences (Gurr *et al.* 1987). Interestingly, the glucoamylase gene of *A. niger* contains five introns which were efficiently spliced in *A. nidulans* (Berka *et al.*, cited in Cullen and Leong 1986). It is not yet known whether filamentous fungi would recognize mammalian intron splicing signals but this is an interesting possibility in view of the similarities of fungal introns to higher eukaryotes (Gurr *et al.* 1987). In addition, it has been shown that *Schizosaccharomyces pombe*, a yeast which is generally thought to be phylogenetically closer to *Aspergillus* than *Saccharomyces*, can accurately remove an intron of a mammalian transcript (Kaufer *et al.* 1985).

Current fungal expression and secretion systems use heterologous genes either with their own signal sequence or with a fungal signal sequence (such as that of glucoamylase) fused to a fungal promoter, with the exception of interferon-γ in *A. ambisexualis* for which the SV40 transcriptional regulatory sequence was used. These organisms have already been shown to be superior to yeast for the secretion of mammalian proteins such as the interferons α_2 and γ (Gwynne *et al.* 1987; Leung *et al.* 1987) and bovine chymosin (Cullen *et al.* 1987; Harkki *et al.* 1987) which is not secreted in bacteria or yeast. Over 90 per cent of the chymosin produced by *A. nidulans* and *Trichoderma reesei* was found to be extracellular. The most promising level of secretion yet, however, is that of aspartic protease by *A. oryzae* with some transformants approaching 3.3 g extracellular protease/l (Christensen *et al.* 1987). Recent work in *T. reesei* suggests that mechanisms of glycosylation in this group of organisms may resemble those of higher eukaryotes (Salovouri *et al.* 1987). Glycosylation of heterologous proteins by fungi has been reported (Leung *et al.* 1987), although hyperglycosylation of aspartic protease occurred in *A. oryzae* (Christensen *et al.* 1987). This did not, however, affect the specific activity of the enzyme.

One potentially serious problem for heterologous expression in organisms such

Table 22.4. Heterologous eukaryotic proteins secreted by filamentous fungi

Gene	Source	Host	Reference
Glucoamylase genomic	*A. niger*	*A. nidulans*	Berka *et al.*, cited in Cullen and Leong (1986)
Aspartic protease genomic	*M. miehei*	*A. nidulans*	Gray *et al.* (1986)
Aspartic protease genomic	*M. miehei*	*M. circinelloides*	Dickinson *et al.* (1987)
Chymosin cDNA	Calf	*A. nidulans*	Cullen *et al.* (1987)
Interferon-α_2 cDNA	Human	*A. nidulans*	Gwynne *et al.* (1987)
Tissue plasminogen activator cDNA	Human	*A. nidulans*	Upshall *et al.* (1987)
Aspartic protease cDNA	*M. miehei*	*A. oryzae*	Christensen *et al.* (1987)
Chymosin cDNA	Calf	*T. reesei*	Harrki *et al.* (1987)
Interferon gamma cDNA	Human	*A. ambisexualis*	Leung *et al.* (1987)
Pancreatic ribonuclease	Rat	*A. niger*	Finkelstein (1987)

as *A. niger* and *A. oryzae* is the destruction of secreted proteins by the action of extracellular proteases: both of these organisms produce copious amounts of several different proteases (Cohen, 1981). Manipulation of media and growth conditions may reduce the secretion of these proteases but an alternative is the isolation of mutants defective in protease production. To this end, a mutant of *A. oryzae* has been isolated which produces no detectable extracellular alkaline protease but which shows normal secretion of other enzymes such as α-amylase (J. A. Macro, personal communication).

The nitrate assimilation system offers a number of features which together make it extremely desirable for the development of expression/secretion vectors for filamentous fungi. First, there is the ease and economy of auxotrophic mutant isolation and characterization in industrial fungi. Second, the pathway for nitrate utilization is, under most growth conditions, entirely dispensable and so mutation within this pathway should not cause any secondary growth defects. Another advantage of the system is that transformant selection can be made with the corresponding homologous NR gene, thus avoiding potential problems, such as the need for clinical trials, in the production of commercially important proteins in recombinant organisms.

In order to obtain expression and secretion of heterologous proteins, it is, of course, necessary to place the gene of interest under the transcriptional control of a suitable promoter, possibly mammalian, but more likely fungal. In this regard, the NR promoter itself would be ideal as it is tightly regulated by the nitrogen source. In yeast, some proteins can be toxic or can reduce the growth rate of the organism (Kingsman *et al.* 1987; R. Contreras, personal communication). To circumvent such problems in filamentous fungi, it would be possible to grow the organism to a sufficient biomass with ammonium as sole nitrogen source, switch to nitrate and induce expression of the foreign gene. In order to obtain secretion, an appropriate signal sequence is required. This could be mammalian, although very efficient secretion has been demonstrated with the *A. niger* glucoamylase (Cullen *et al.* 1987) and *T. reesei* cellobiohydrolase signal sequences (Harrki *et al.* 1987). The glucoamylase gene of *A. oryzae* has also been isolated for this purpose (J. A. Macro and S. E. Unkles, unpublished). Constructs containing mammalian genes under the transcriptional control of the NR promoter are currently being tested and it is hoped that an evaluation of the usefulness of this system will be forthcoming in the near future.

Acknowledgement

I wish to thank colleagues in St Andrews and elsewhere who permitted me to include their unpublished work and to acknowledge support from the European Economic Community (BAP-00064-UK) for work on this research area. Thanks go to my colleague Dr J. R. Kinghorn for his encouragement, consideration, helpfulness, positive attitude, and kindness shown towards dumb animals.

References

Adler-Nissen, J. (1987). Newer uses of microbial enzymes in food processing. *Trends in Biotechnology* **5**, 170–4.

Ballance, D. J. (1986). Sequences important for gene expression in filamentous fungi. *Yeast* **2**, 229–36.

Beggs, J. D., van den Berg, J., van Ooyen, A., and Weissmann, C. (1980). Abnormal expression of chromosomal rabbit β-globin gene in *Saccharomyces cerevisiae*. *Nature* **283**, 835–40.

Bennett, J. W. (1985). Molds, manufacturing and molecular genetics. In *Molecular genetics of filamentous fungi* (ed. W. E. Timberlake), pp. 345–66. Alan R. Liss, New York.

Beri, R. K. and Turner, G. (1987). Transformation of *Penicillium chrysogenum* using the *Aspergillus nidulans amdS* gene as a dominant selective marker. *Current Genetics* **11**, 639–41.

Bigelis, R. (1985). Primary metabolism and industrial fermentations. In *Gene manipulations in fungi*, (eds. J. W. Bennett and L. L. Lasure), pp. 357–401. Academic Press, London.

Birkett, J. A. and Rowlands, R. T. (1981). Chlorate resistance and nitrate assimilation in industrual strains of *Penicillium chrysogenum*. *Journal of General Microbiology* **123**, 281–5.

Bloomfield, R. M. (1982). The genetics of nitrate assimilation in *Aspergillus amstelodami*. Ph.D. Thesis, University of Birmingham.

Böing, J. T. P. (1982). Enzyme production. In *Prescott and Dunn's industrial microbiology* (ed. G. Reed), pp. 634–708. Macmillan, Byfleet, Surrey.

Brown, C. M., Campbell, I., and Priest, F. G. (1987). *Introduction to biotechnology*, pp. 134–7. Blackwell, London.

Cameron, H. R. and Milbrath, G. M. (1965). Variability in the genus *Phytophthora*. Effects of nitrogen sources and pH on growth. *Phytopathology* **55**, 653–7.

Cantoral, J. M., Diez, B., Barredo, J. L., Alvarez, E., and Martin, J. F. (1987) High-frequency transformation of *Penicillium chrysogenum*. *Biotechnology* **5**, 494–7.

Christensen, T., Boel, E., Wöldike, H. F., Hjortshoj, K., Andreasen, F., Mortensen, S. B., Thim, L., and Hansen, M. T. (1987). Transformation of *Aspergillus oryzae* and expression of homologous and heterologous proteins in this organism. *Abstract from the 19th Lunteren Lectures on Molecular Genetics*, p. 33. TNO Corporate Communications Dept, The Hague, The Netherlands.

Cohen, B. L. (1981). Regulation of protease production in *Aspergillus*. *Transactions of the British Mycological Society* **76**, 447–50.

Correll, J. C., Klittich, C. J. R., and Leslie, J. F. (1987). Nitrate nonutilising mutants of *Fusarium oxysporum* and their use in vegetative compatibility tests. *Phytopathology* **77**, 1640–6.

Cullen, D. and Leong, S. (1986). Recent advances in the molecular genetics of industrial filamentous fungi. *Trends in Biotechnology* **4**, 285–8.

Cullen, D., Gray, G. L., Wilson, L. J., Hayenga, K. J., Lamsa, M. H., Rey, M. W., Norton, S., and Berka, R. M. (1987). Controlled expression and secretion of bovine chymosin in *Aspergillus nidulans*. *Biotechnology* **5**, 369–76.

Dale, B. E. (1987). Lignocellulose conversion and the future of fermentation biotechnology. *Trends in Biotechnology* **5**, 287–91.

Dhawale, S. S. and Marzluf, G. A. (1985). Transformation of *Neurospora crassa* with circular and linear DNA and analysis of the fate of the transforming DNA. *Current Genetics* **10**, 205–12.

Dickinson, L., Harboe, M., van Heeswijck, R., Stroman, P., and Jepsen, L. P. (1987). Expression of active *Mucor miehei* aspartic protease in *Mucor circinelloides*. *Carlsberg Research Communications* **52**, 243–52.

Emtage, J. S., Angal, S., Doel, M. T., Harris, T. J. R., Jenkins, B., Lilley, G., and Lowe, P. A. (1983). Synthesis of calf prochymosin (prorennin) in *Escherichia coli*. *Proceedings of the National Academy of Sciences, USA* **80**, 3671–5.

Finkelstein, D. B. (1987). Improvement of enzyme production in *Aspergillus*. *Antonie van Leeuwenhoek* **53**, 349–52.

Foster, J. W. (1949). *Chemical activities of fungi*, pp. 485–9. Academic Press, New York.

French, A. M. (1953). Physiologic differences between two physiologic races of *Phytophthora infestans*. *Phytopathology* **43**, 513–16.

Furman, T. C., Epp, J., Hsiung, H. M., Hoskins, J., Long, G. L., Mendelsohn, L. G., Schoner, B., Smith, D. P., and Smith, M. L. (1987). Recombinant human insulin-like growth factor II expressed in *Escherichia coli*. *Biotechnology* **5**, 1047–51.

Goosen, T., Bloemheuvel, G., Gysler, C., de Bie, D. A., van den Broek, H. W. J., and Swart, K. (1987) Transformation of *Aspergillus niger* using the homologous orotidine-5′-phosphate-decarboxylase gene. *Current Genetics* **11**, 499–503.

Gray, G. L., Hayenga, K., Cullen, D., Wilson, L. J., and Norton, S. (1986). Primary structure of *Mucor miehei* aspartic protease: evidence for a zymogen intermediate. *Gene* **48**, 41–53.

Gurr, S. J., Unkles, S. E., and Kinghorn, J. R. (1987). The structure and organization of nuclear genes of filamentous fungi. In *Gene structure in eukaryotic microbes* (ed. J. R. Kinghorn), pp. 93–139, IRL Press, Oxford.

Gwynne, D. I., Buxton, F. P., Williams, S. A., Garven, S., and Davies, R. W. (1987). Genetically engineered secretion of active human interferon and a bacterial endoglucanase from *Aspergillus nidulans*. *Biotechnology* **5**, 713–9.

Harkki, A., Bailey, M., Penttila, M., and Knowles, J. (1987). Active calf chymosin efficiently secreted from *Trichoderma reesei*. *Abstract from the 19th Lunteren Lectures on Molecular Genetics*, p. 42. TNO Corporate Communications Dept, The Hague, The Netherlands.

Holliday, R. (1974). *Ustilago maydis*. In *Handbook of genetics* (ed. R. C. King), pp. 575–95. Plenum, London.

Hynes, M. J. (1986). Transformation of filamentous fungi. *Experimental Mycology* **10**, 1–8.

Hynes, M. J., Corrick, C. M., and King, J. A. (1983). Isolation of genomic clones containing the *amdS* gene of *Aspergillus nidulans* and their use in the analysis of structural and regulatory mutations. *Molecular and Cellular Biology* **3**, 1430–9.

Innis, M. A., Holland, M. J., McCabe, P. C., Cole, G. E., Wittman, V. P., Tal, R., Watt, K. W. K., Gelfand, D. H., Holland, J. P., and Meade, J. H. (1985). Expression, glycosylation, and secretion of an *Aspergillus* glucoamylase by *Saccharomyces cerevisiae*. *Science* **228**, 21–6.

Jackson, R. M. and Mason, P. A. (1984). *Mycorrhiza*. Edward Arnold, London.

Johnstone, I. L. (1985). Transformation of *Aspergillus nidulans*. *Microbiological Sciences* **2**, 307–11.

Johnstone, I. L., Kinghorn, J. R., Gurr, S. J., Greaves, P., Innes, M. (1989). Cloning and characterisation of the *Aspergillus nidulans* gene cluster for nitrate assimilation. *EMBO Journal* (submitted).

Kapoor, K. K., Chaudhary, K., and Tauro, P. (1982). Citric acid. In *Prescott and Dunn's industrial microbiology* (ed. G. Reed), pp. 709–47. Macmillan, Byfleet, Surrey.

Kaufer, N. F., Simanis, V., and Nurse, P. (1985). Fission yeast *Schizosaccharomyces pombe* correctly excises a mammalian RNA transcript intervening sequence. *Nature* **318**, 78–80.

Kelly, J. M. and Hynes, M. J. (1985). Transformation of *Aspergillus niger* by the *amdS* gene of *Aspergillus nidulans*. *EMBO Journal* **4**, 475–9.

Kelly, J. M. and Hynes, M. J. (1987). Multiple copies of the *amdS* gene of *Aspergillus nidulans* cause titration of *trans*-acting regulatory proteins. *Current Genetics* **12**, 21–31.

Kingsman, S. M., Kingsman, A. J., and Mellor, J. (1987). The production of mammalian proteins in *Saccharomyces cerevisiae*. *Trends in Biotechnology* **5**, 53–6.

Kirk, T. K. (1983). Degradation and conversion of lignocelluloses. In *The filamentous fungi*. Vol. IV. Fungal technology (eds. J. E. Smith, D. R. Berry, and B. Kristiansen), pp. 266–95. Edward Arnold, London.

Klittich, C. J. R., Leslie, J. F., and Wilson, J. D. (1986). Spontaneous chlorate resistant mutants of *Gibberella fujikuroi* (*Fusarium moniliforme*). *Phytopathology* (Abstracts), **76**, 1142.

Knowles, J., Lehtovaara, P., and Teeri, J. (1987). Cellulase families and their genes. *Trends in Biotechnology* **5**, 255–61.

Lambert, P. W. (1983). Industrial enzyme production and recovery from filamentous fungi. In *The filamentous fungi*, Vol. IV. *Fungal technology* (eds. J. E. Smith, D. R. Berry and B. Kristiansen), pp. 210–37. Edward Arnold, London.

Leung, W.-C., Zwaagstra, J. C., and Leung, M. F. K. (1987). Use of filamentous fungus *Achlya ambisexualis* as host cell for protein engineering of human interferon gamma. *Abstract from the 19th Lunteren Lectures on Molecular Genetics*, p. 45. TNO Corporate Communications Dept, The Hague, The Netherlands.

Lewis, C. M. and Fincham, J. R. S. (1970a). Regulation of nitrate reductase in the basidiomycete *Ustilago maydis*. *Journal of Bacteriology* **103**, 55–61.

Lewis, C. M. and Fincham, J. R. S. (1970*b*). Genetics of nitrate reductase in *Ustilago maydis*. *Genetical Research, Cambridge* **16**, 151–63.

Lewis, N. J., Hurt, P., Sealy-Lewis, H. M., and Scazzocchio, C. (1978). The genetic control of the molybdoflavoproteins in *Aspergillus nidulans* IV. A comparison between purine hydroxylase I and II. *European Journal of Biochemistry* **91**, 311–16.

Lilly, V. G. and Barnett, H. L. (1951). *Physiology of the fungi*, pp. 97–102. McGraw-Hill, London.

Lisansky, S. G. and Hall, R. A. (1983). Fungal control of insects. In *The filamentous fungi. Vol. IV. Fungal technology* (eds. J. E. Smith, D. R. Berry and B. Kristiansen), pp. 327–45. Edward Arnold, London.

Malardier, L., Daboussi, M. J., Brygoo, Y., and Scazzocchio, C. (1987). Cloning the *niaD* gene of *Aspergillus nidulans* and its use for transformation in *Fusarium oxysporum*. *2nd International Symposium on Nitrate Assimilation*, M3. St Andrews, United Kingdom.

Marth, E. H. (1982). Cheese. In *Prescott and Dunn's industrial microbiology* (ed. G. Reed), pp. 65–112. Macmillan, Byfleet, Surrey.

Mattern, I. E., Punt, P. J., Unkles, S., Pouwels, P. H., and van den Hondel, C. A. M. J. J. (1987*a*). Transformation of *Aspergillus oryzae*. *Abstract from the 19th Lunteren Lectures on Molecular Genetics*, p. 34. TNO Corporate Communications Dept, The Hague, The Netherlands.

Mattern, I. E., Unkles, S., Kinghorn, J. R., Pouwels, P. H., and van den Hondel, C. A. M. J. J. (1987*b*). Transformation of *Aspergillus oryzae* using the *A. niger pyrG* gene. *Molecular and General Genetics* **210**, 460–1.

Mishra, N. C. (1985). Gene transfer in fungi. *Advances in Genetics* **23G**, 73–177.

Montenecourt, B. S. and Eveleigh, D. E. (1985). Fungal carbohydrases: Amylases and cellulases. In *Gene manipulations in fungi* (eds. J. W. Bennett and L. L. Lasure), pp. 491–512. Academic Press, London.

Newton, A. C. and Caten, C. E. (1988). Auxotrophic mutants of *Septoria nodorum* isolated by direct screening and by selection for resistance to chlorate. *Transactions of the British Mycological Society* **90**, 199–207.

Paietta, J. and Marzluf, G. A. (1985). Development of shuttle vectors and gene manipulation techniques for *Neurospora crassa*. In *Molecular genetics of filamentous fungi* (ed. W. E. Timberlake), pp. 1–13, Alan R. Liss, New York.

Papa, K. E. (1986). Heterokaryon incompatibility in *Aspergillus flavus*. *Mycologia* **78**, 98–101.

Parsons, K. A., Chumley, F. G., and Valent, B. (1987). Genetic transformation of the fungal pathogen responsible for rice blast disease. *Proceedings of the National Academy of Sciences (USA)* **84**, 4161–5.

Pentilla, M. E., Nevalainen, K. M. H., Raynal, A., and Knowles, J. K. C. (1984). Cloning of *Aspergillus niger* genes in yeast. Expression of the gene coding *Aspergillus* β-glucosidase. *Molecular and General Genetics* **194**, 494–9.

Pentilla, M. E., Suihko, M.-L., Lehtinen, U., Nikkola, M., and Knowles, J. K. C. (1987). Construction of brewer's yeast secreting fungal endo-β-glucanase. *Current Genetics* **12**, 413–20.

Perrot, M., Barreau, C., and Begueret, J. (1987). Non-integrative transformation in the filamentous fungus *Podospora anserina*: stabilisation of a linear vector by the chromosomal ends of *Tetrahymena thermophila*. *Molecular and Cellular Biology* **7**, 1725–30.

Picknett, T. M., Saunders, G., Ford, P., and Holt, G. (1987). Development of a gene transfer system for *Penicillium chrysogenum*. *Current Genetics* **12**, 449–55.

Pirt, S. J. (1987). Microbial physiology in the penicillin fermentation. *Trends in Biotechnology* **5**, 69–72.

Puhalla, J. E. (1985). Classification of strains of *Fusarium oxysporum* on the basis of vegetative compatibility. *Canadian Journal of Botany* **63**, 179–83.

Puhalla, J. E. and Spieth, P. T. (1985). A comparison of heterokaryosis and vegetative incompatibility among varieties of *Gibberella fujikuroi* (*Fusarium moniliforme*). *Experimental Mycology* **9**, 39–47.

Revuelta, J. L. and Jayaram, M. (1986). Transformation of *Phycomyces blackesleeanus* to G-418 resistance by an autonomously replicating plasmid. *Proceedings of the National Academy of Sciences (USA)* **83**, 7344–7.

Rippel, K. (1931). Qualititative Untersuchen uber die Abhangigkeit der Stickstoff. Assimilation von der Wasserstoffionenkonzentration bei einigen Pilzen. *Archives of Mikrobiology* **2**, 72–135.

Rodriguez, R. J., and Yoder, O. C. (1987). Selectable gene for transformation of the fungal plant pathogen *Glomerella cingulata* f.sp. *phaseoli* (*Colletotrichum lindemuthianum*). *Gene* **54**, 73–80.

Salovouri, I., Makarow, M., Rauvala, H., Knowles, J., and Kaariaihen, L. (1987). Low molecular weight high-mannose type glycans in a secreted protein of the filamentous fungus *Trichoderma reesei*. *Biotechnology* **5**, 152–6.

Shen, V.-Q., Wolfe, S., and Demain, A. L. (1986). Levels of isopenicillin N synthetase and deacetoxycephalosporin C synthetase in *Cephalosporium acremonium* producing high and low levels of cephalosporin C. *Biotechnology* **4**, 61–4.

Skatrud, P. L., Queener, S. W., Carr, L. G., and Fisher, D. L. (1987). Efficient integrative transformation of *Cephalosporium acremonium*. *Current Genetics* **12**, 337–48.

Smith, D. J., Bull, J. H., and Turner, G. (1987). Gene amplification in a strain of *Penicillium chrysogenum* producing high levels of penicillin. *Abstract from the 19th Lunteren Lectures on Molecular Genetics*, p. 37. TNO Corporate Communications Dept, The Hague, The Netherlands.

Stahl, U., Leitner, E., and Esser, K. (1987). Transformation of *Penicillium chrysogenum* by a vector containing a mitochondrial origin of replication. *Applied Microbiology and Biotechnology* **26**, 237–41.

Talbot, N. J., Coddington, A., Roberts, I. N., and Oliver, R. P. (1988). Diploid construction by protoplast fusion in *Fulvia fulva*: genetic analysis of an imperfect fungal plant pathogen. *Current Genetics* **14** (in press).

Tilburn, J., Scazzocchio, C., Taylor, G. G., Zabicky-Zissman, J. H., Lockington, R. A., and Davies, R. W. (1983). Transformation by integration in *Aspergillus nidulans*. *Gene* **26**, 205–21.

Tsao, G. T., and Chiang, L.-C. (1983). Cellulose and hemicellulose technology. In *The filamentous fungi. Vol. IV. Fungl technology* (ed. J. E. Smith, D. R. Berry and B. Kristiansen), pp. 296–326. Edward Arnold, London.

Turgeon, B. G., Garber, R. C., and Yoder, O. C. (1987). Development of a fungal transformation system based on selection of sequences with promoter activity. *Molecular and Cellular Biology* **7**, 3297–305.

Turner, G., and Ballance, D. J. (1985). Cloning and transformation in *Aspergillus*. In *Gene manipulations in fungi* (eds. J. W. Bennett and L. L. Lasure), pp. 259–78. Academic Press, London.

Unkles, S. E., Campbell, E. I., Carrez, D., Grieve, C., Contreras, R., Fiers, W., van den Hondel, C. A. M. J. J. and Kinghorn, J. R. (1989). Transformation of *Aspergillus niger* using the homologous *niaD* gene. *Gene* (in press).

Upshall, A., Kumar, A. A., Bailey, M. C., Parker, M. D., Favreau, M. A., Lewison, K. P.,

Joseph, M. L., Maraganore, J. M., and McKnight, G. L. (1987). Secretion of active human tissue plasminogen activator from the filamentous fungus *Aspergillus nidulans*. *Biotechnology* **5**, 1301–4.

van Brunt, J. (1986). Fungi: the perfect hosts? *Biotechnology* **12**, 1057–62.

van Hartingsveldt, W., Mattern, I. E., van Zeijl, C. M. J., Pouwels, P. H., and van den Hondel, C. A. M. J. J. (1987). Development of a homologous transformation system for *Aspergillus niger* based on the *pyrG* gene. *Molecular and General Genetics* **206**, 71–5.

van Heeswijck, R. (1986). Autonomous replication of plasmids in *Mucor* transformants. *Carlsberg Research Communications* **51**, 433–43.

Vlasuk, G. P., Bencen, G. H., Scarborough, R. M., Tsai, P.-K., Whang, J. L., Maack, T., Camargo, M. J. F., Kirsher, S. W., and Abraham, J. A. (1986). Expression and secretion of biologically active human atrial natriuretic peptide in *Saccharomyces cerevisiae*. *Journal of Biological Chemistry* **261**, 4789–96.

Wernars, K. (1986). DNA-mediated transformation of the filamentous fungus Aspergillus nidulans. Ph.D. Thesis, University of Wageningen.

Whitehead, M., Spence, D., Boyle, J., and Kinghorn, J. R. (1989*a*). Genetic studies of nitrate assimilation in *Cephalosporium acremonium*. (In preparation.)

Whitehead, M., Unkles, S. E., Ramsden, M., Campbell, E., Gurr, S. J., Spence, D., and Kinghorn, J. R. (1989*b*). Transformation of a *Penicillium chrysogenum niaD* mutant using the *Aspergillus niger* and *Aspergillus oryzae* nitrate reductase genes. *Molecular and General Genetics* (in press).

Wostemeyer, J., Burmester, A., and Weigtel, C. (1987). Neomycin resistance as a dominantly selectable marker for transformation of the zygomycete *Absidia glauca*. *Current Genetics* **12**, 625–7.

Yelton, M. M., Hamer, J. E., and Timberlake, W. E. (1984). Transformation of *Aspergillus nidulans* by using a *trpC* plasmid. *Proceedings of the National Academy of Sciences (USA)* **81**, 1470–4.

Yoshizumi, H. and Ashikari, T. (1987). Expression, glycosylation and secretion of fungal hydrolases in yeast. *Trends in Biotechnology* **5**, 277–81.

23. Plant biotechnology and nitrate assimilation

D. Cammaerts, R. Dirks, I. Negrutiu, I. Famelaer,
S. Hinnisdaels, D. Debeys, J. Veuskens, and M. Jacobs

Introduction

There are only a few reports of the successful selection of recessive, deficient biochemical mutants in plants. One of the most typical examples of this is the selection of nitrate assimilation variants based on the use of chlorate as a selective agent. In the case of nitrate assimilation we use a positive selection procedure which allows resistant cells, which are deficient in nitrate reductase (NR) as a consequence of mutation, to survive on a chlorate-containing medium while wild type cells die (Chapters 11, 12 and 13).

Selection, characterization and use of nitrate reductase-deficient mutants

NR-deficient mutants have been obtained in various species, at the cell/protoplast level as well as at the whole plant level (Chapters 11, 12 and 13).

We selected protoplast-derived colonies of haploid *N. plumbaginifolia* plants for the expression of chlorate resistance (Negrutiu *et al.* 1983). Eighty confirmed resistant lines were isolated from a total of 1.84×10^7 non-mutagenized haploid protoplasts; 50 of these were regenerated into plants. Only 30 were studied in more detail as losses or difficulties in maintaining the regenerants occurred; 26 were classified as apoprotein mutants (*nia* type) and four as Mo-cofactor (MoCo) mutants (*cnx* type).

Segregation tests have been performed with several of these lines (Dirks 1986). All progeny from selfed *nia* and *cnx* regenerants failed to grow on nitrate as sole nitrogen source. All F_1 progeny obtained by crossing the mutant with the wild type behaved as wild type (growth on nitrate and sensitivity to chlorate). The F_2 progeny showed a 3:1 growth/no growth segregation frequency. Such results indicated that loss of NR activity and resistance to chlorate as a consequence of *nia* and *cnx* mutations were inherited as single recessive nuclear genes.

Fusion of protoplasts and cross-analysis revealed four complementation groups, one including all *nia* mutants and three groups of *cnx* mutants: CNX20, CNX27, CNX82 and CNX103 (Dirks *et al.* 1985). CNX20 and CNX82 could be partially repaired when the culture medium was supplemented with unphysiologically high levels of molybdenum. CNX27 and CNX103 did not react to such addition.

The four CNX lines were fused with each other and in all necessary combinations with NX lines 1, 24, and 21 (Marton *et al.* 1982) previously isolated and characterized as representing respectively complementation group *cnxA*, *cnxB* and *cnxC* (Xuan *et al.* 1983). The results of somatic hybridization experiments showed that CNX20 and CNX82 are allelic and can be ascribed to the *cnxA* complementation group; that CNX27 belongs to complementation group *cnxB* and that CNX103 represents a new group, *cnxD* (Table 23.1).

Table 23.1. Somatic complementation of *cnx* mutants in *N. plumbaginifolia*

	CNX20	CNX27	CNX82	CNX103	NX 1 A	NX 21 C	NX 24 B
CNX20	−	+	+	+	−	+	+
CNX27	+	−	+	+	+	+	−
CNX82	−	+	−	NP[a]	−	NP[a]	NP[a]
CNX103	+	+	NP[a]	−	+	+	+

Complementation group		mutants
cnxA	Mo-repairable	CNX20, 82
cnxB	Mo-non repairable	CNX27
cnxC	Mo-non repairable	−
cnxD	Mo-non repairable	CNX103

The CNX lines were fused with each other and in all combinations with the NX lines, representing the established complementation groups *cnxA*, *cnxB*, *cnxC*. Four complementation groups were identified in the case of *cnx* mutants.

[a] These experiments were not performed.

In order to confirm such complementation groups, sexual crosses were carried out between distinct lines within and between the *nia* and *cnx* complementation groups (Table 23.2). All crosses between F_1 plants in the *nia* complementation group scored a 3:1 ratio. Identical results were obtained in a cross between F_1 plants of CNX20 and CNX82. When homozygous CNX20 was crossed with heterozygous NIA26, the progeny were completely green as a result of genetic complementation. Crosses between F_1 plants of CNX20 or CNX82 with CNX103 also resulted in completely green progeny.

Table 23.2. Genetic complementation between *nia* and *cnx* mutants, and among four independently isolated *cnx* mutants

Cross combination	Number of progeny		Ratio	P value
	Growing	Not growing		
CNX20/CNX20 × NIA26/+	150	0	1:0	
CNX103/+ × NIA26/+	150	0	1:0	
CNX20/+ × CNX82/+	418	150	3:1	0.3–0.5
CNX20/+ × CNX103/+	150	0	1:0	
CNX82/+ × CNX103/+	150	0	1:0	
CNX27/+ × CNX20/+	150	0	1:0	
CNX27/+ × CNX103/+	150	0	1:0	

Complementation groups were confirmed by carrying out sexual crosses between mutants belonging to *nia* and *cnx* complementation groups, *cnxA* (CNX20) and *cnxD* (CNX103) as well as within different *cnx* complementation groups, *cnxA* (CNX20, 82), *cnxB* (CNX27) and *cnxD* (CNX103).

Conditional lethal mutants such as NR deficient types can thus complement in sexual crosses. The results also confirmed that complementation occurred between apoenzyme and cofactor-deficient NR types and between molybdenum restorable and non-restorable *cnx* mutants (Dirks *et al.* 1985).

All of these NR-deficient mutants are of great importance for fundamental and applied research. They represent valuable tools for investigating genetic and biochemical processes as illustrated by the following topics.

Revertant studies

The essential suitability of some of the NR$^-$ mutants as recipients for genetic transformation experiments was established by setting up a screening method for revertant isolation (Dirks *et al.* 1986*a*). Auxotrophic mutants are excellent tools as recipients for fusion and transformation experiments provided that the selection conditions for recovery of the transformed phenotypes have been defined and that the spontaneous reversion frequency is sufficiently low. NR revertants themselves offer the possibility of learning more about the alterations induced in the biochemical functions that are damaged or absent in the forward mutations.

Protoplast cultures of four NR deficient mutants, NIA8, NIA26, CNX20, and CNX103 were used for selecting spontaneous revertants in selective nitrate medium (Table 23.3). Nitrate-utilizing colonies could be obtained 4–6 weeks after plating on selective medium. The reversion frequency for the *nia* mutants varies from 10^{-6} to 5×10^{-7}. For CNX20 no revertants were found, while for CNX103 further selection experiments need to be undertaken to determine a more meaningful reversion frequency.

Table 23.3. Selection of revertant cell lines of NR$^-$ mutants

NR$^-$ mutant	N° of colonies plated on		N° of colonies utilizing NO$_3^-$	Reversion frequency
	15 mM KNO$_3$	10 mM KNO$_3$		
NIA8	–	1.46×10^6	2	1.38×10^{-6}
NIA26	2.20×10^6		2	4.04×10^{-7}
		5.22×10^6	1	
CNX20	6.69×10^6		0	
		1.28×10^7	0	$<5 \times 10^{-8}$
CNX103	2.63×10^6	–	1	3.80×10^{-7}

Protoplast colonies were plated on medium containing 10 or 15 mM KNO$_3$, respectively, as sole source of nitrogen. Nitrate-utilizing colonies could be obtained 4–6 weeks after plating on selective medium.

Revertants of the *nia* type proved to be tetraploid and genetic analysis indicated that one out of the four NR$^-$ genes reverted to a functional allele (Table 23.4). To confirm this, several plants of the R$_1$ generation of RVNIA26.1, RVNIA26.2 and RVNIA26.3 were grown to maturity and individual segregations in the R$_2$ generation were analysed (Table 23.5). The progeny of some R$_1$ parents produced a 3:1 ratio (green growing:white not growing) while others segregated 35:1. A 3:1 ratio indicates that the R$_1$ parent contained only one functional NR gene; a 35:1 segregation indicates two functional NR genes.

Table 23.4. Segregation analysis of R$_0$[a] revertant NIA plants in the R$_1$[b] generation

Cross	Growing	Not growing	Ratio	P
RVNIA8.1	223	75	3:1	0.95–1.00
RVNIA8.2	237	75	3:1	0.70–0.90
RVNIA26.1	508	153	3:1	0.20–0.30
RVNIA26.2	521	204	3:1	0.05–0.10
RVNIA26.3	419	166	3:1	0.05–0.10

Seeds were germinated on plates containing nitrate as sole source of nitrogen: homozygous recessive NR$^-$ seedlings exhibited white cotyledons and died within 4 weeks of germination.
[a] Regenerated plants from protoplast culture.
[b] Generation produced after self-fertilization of R$_0$ plants.

Table 23.5. Segregation analysis in the R_2[a] generation of different R_1[b] NIA26 revertants

Cross	Growing	Not growing	Ratio	P
RVNIA26.1 R_2 2	429	157	3:1	0.30–0.50
RVNIA26.1 R_2 9	522	18	35:1	0.30–0.50
RVNIA26.2 R_2 1	805	22	35:1	0.90–0.95
RVNIA26.2 R_2 2	375	125	3:1	0.95–1.00
RVNIA26.3 R_2 1	308	99	3:1	0.70–0.90
RVNIA26.3 R_2 2	776	25	35:1	0.70–0.90

Experimental conditions were identical to those described in Table 23.4.
[a] Generation produced after selfing the R_1 generation.
[b] Generation produced after self-fertilization of the R_0 plants.

The low back-mutation frequency shows that the NR^- mutants so far studied are indeed suitable for transformation experiments.

Use of NR^- mutants to establish selection conditions for auxotrophy

NR-deficient mutants have also successfully been used in starvation–reconstruction experiments performed in order to define appropriate conditions for auxotroph isolation (Dirks *et al.* 1986*b*). Starvation and reconstruction experiments were carried out using protoplast cultures of NR^- mutants and a few previously isolated amino acid auxotrophs (Negrutiu *et al.* 1985).

The restoration of division activity of auxotrophic protoplasts was initially studied after starvation. Auxotrophic protoplasts were starved for increasing periods of time, after which the required metabolite was added. Fig. 23.1 shows the change of plating efficiency (PE%) as a function of time of starvation. After 6 days of starvation only the NR^- protoplasts showed a significant decrease, to 70 per cent and to 50 per cent when cultured on medium without organic acids and without nitrogen, respectively. For the other auxotrophs it is clear that up to 6 days, no significant differences in plating efficiency occur. After 9 days starvation however, all mutants under study dropped to a plating efficiency of 50 per cent.

In a second type of experiment the capacity to form colonies after dilution of starved protoplasts was studied. Protoplasts of auxotrophic mutants were cultured in unsupplemented medium and subsequently diluted to low densities (500 cells or colonies/ml) during 7 successive days in supplemented medium. The green colonies appearing on this medium were counted. The decrease in colony formation for the starved protoplasts is more drastic for NR^- than for other auxotrophic protoplasts.

The results showed that division activity of blocked cells (due to starvation) can be re-initiated without a significant decrease in plating efficiency up to 6 days

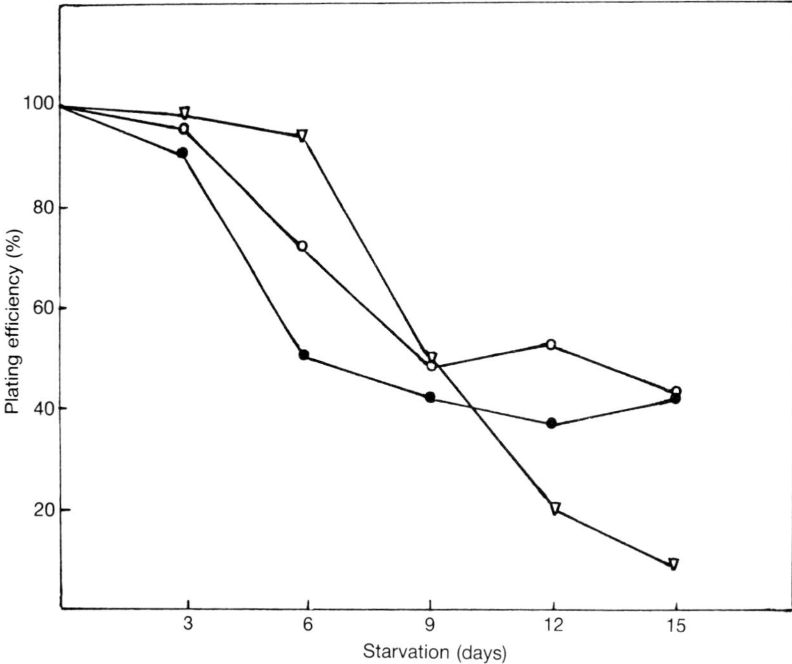

Fig. 23.1. Change in plating efficiency (PE) of auxotrophic mutants after different times of starvation. Isolated auxotrophic protoplasts were incubated in unsupplemented medium. After different periods of starvation the medium was supplemented with L-histidine or $(NH_4)_2$ succinate to ensure growth of respectively His$^-$ and NR$^-$ cells. Ten days after the addition of the required metabolite, the plating efficiency (PE%) was measured as the ratio between the amount of protoplast derived colonies that are counted and the amount of protoplasts that were initially inoculated. His$^-$, starved on medium without histidine ∇——∇; NR$^-$, starved on medium without organic acid $\bigcirc$——$\bigcirc$; NR$^-$, starved on medium without nitrogen ●——●. Results are presented in arbitrary units.

except for NR$^-$ protoplasts, which seem more sensitive to starvation. When unsupplemented protoplasts are cultured and diluted to low density, however, recovery of colonies in supplemented medium is low. Thus it is possible to re-initiate division activity of starved auxotrophic mutants, as long as this occurs at a sufficiently high density.

Somatic hybridization: symmetric and asymmetric fusion products

Somatic hybridization by protoplast fusion gives the opportunity of obtaining symmetric hybrids between different species. However, parasexual hybridization between remote species leads to morphologically abnormal and sterile hybrids. Spontaneous elimination of some or almost all chromosomes of one fusion partner results in so-called asymmetric hybrids (Gleba and Hoffmann 1980) which are morphologically less abnormal than symmetric hybrids and one can expect that further elimination of genetic material from one partner may lead to the production of fertile, partial hybrid plants. Methods for transferring part of the plant genome have been established: the use of irradiated protoplasts from one partner leads to the production of highly asymmetric hybrids which contain, in addition to the complete recipient genome, some chromosomes or chromosome fragments of the irradiated donor partner (Dudits *et al.* 1980; Gupta *et al.* 1984).

Symmetric hybrids. Due to the selective power of NR markers, studies were performed to optimize conditions for fusion and to identify parameters important in fusion protocols (Negrutiu *et al.* 1986). Different reports show that the NR⁻ phenotype is useful as a selection marker in the identification of somatic hybrids by complementation between mutants and also in plant transformation studies.

Somatic hybrids resulting from PEG fusion of protoplasts derived from *N. tabacum* NIA and CNX cell lines could be recovered on the basis of their growth on medium containing nitrate as sole nitrogen source. The possibility of reversion could be excluded (Glimelius *et al.* 1978). Wallin *et al.* (1979) used the NR⁻ cell lines in order to obtain proof for integration and expression of the genetic information of nuclei from microprotoplasts (nucleus surrounded by a plasma membrane and some cytoplasm). Others also reported the use of a NR⁻ cell line as one partner in somatic hybrid selection schemes (Glimelius and Bonnett 1981; Glimelius *et al.* 1981; Hein *et al.* 1983).

The NR⁻ mutants satisfy a number of criteria for effective use in somatic hybridization. One of the fusion partners, the NR⁻ mutant, can be excluded at the cellular level by transfer to a medium deficient in reduced nitrogen and, as already shown, reversion is very infrequent.

Asymmetric hybrids. In order to bypass the somatic incompatibility barriers between distantly related species of plants, several attempts have been made to transfer a limited amount of genetic material from one parent to another. γ-Fusion (fusion of recipient protoplasts with an irradiated donor partner) may provide new opportunities for somatic gene exchange.

The NR⁻ mutants can be used to obtain complementation at the interspecific and intergeneric level. Gupta *et al.* (1982) reported a stable correction of a nuclear genetic defect in NR activity in completely auxotrophic cells of *N. tabacum* by introducing the genes of two distantly related genera. The isozyme results confirmed that only a small amount of genetic material had been transferred from the highly irradiated protoplasts into the CNX line of *N. tabacum*. Somers *et al.* (1986) performed a protoplast fusion in which the selectable marker NR also served as a biochemical marker to provide direct evidence for intergeneric specific gene transfer. NR-deficient tobacco *nia* protoplasts were the recipients for the transfer of the NR structural gene from irradiated barley protoplasts.

In our laboratory, the transfer of genetic material from a γ-irradiated donor protoplast belonging to species with various phylogenetic relatedness has been attempted by means of fusion into NR-deficient mutant protoplasts from *N. plumbaginifolia*. Corrected phenotypes for NR were recovered at several doses of Krad in different fusion combinations (Famelaer *et al.* 1986). Different doses were tested in order to establish the effect of the irradiation on chromosome elimination.

In the fusion combination with *N. sylvestris* as donor partner, corrected phenotypes were recovered with a dose range of 10–200 Krad (Table 23.6). Karyological analysis of regenerated plants revealed the presence of whole chromosomes and chromosome fragments from the donor partner in somatic tissues (Fig. 23.2). Analysis of T_1 (first generation plants from T_0 plants) and T_2 (second generation) plants showed that chromosomes of the donor genome are still present in the progeny but have a tendency to be lost. In addition to the NR trait, several isozymes from the donor species were shown to be active in some fusion products. Shikimic dehydrogenase was active in the T_0 (regenerants obtained after fusion) and the T_1 generation (Fig. 23.3). Fusion combinations with other donor partners were only successful with *Atropa belladonna* and sugarbeet. In the fusion combination with *Atropa belladonna* corrected phenotypes for the NR deficiency were produced for each irradiation dose (Table 23.7). The introduction of genes from the donor into the recipient was shown by cytological analysis. Studies of xanthine dehydrogenase showed that the MoCo of *Atropa* complemented the apoprotein of *N. plumbaginifolia*, giving an active xanthine dehydrogenase. Also, a specific esterase isozyme characterizing *Atropa belladonna* was present in one fusion product (Fig. 23.4). Further experiments will concentrate on the screening of fertile regenerants, the study of meiotic behaviour of these plants and progeny analysis of the transmission of the NR trait.

Gene transfer

NR-deficient cell lines could be used as recipients in gene transfer experiments. However, systems of gene transfer such as those based on the use of the Ti-plasmid or direct gene transfer require the availability of cloned nitrate assimilation genes.

A possible approach for cloning nitrate assimilation genes consists of the disruption of a *nia* or *cnx* gene by insertion mutagenesis, accomplished by transformation with *Agrobacterium*, naked DNA or by use of a transposable element such as the Activator element (*Ac*) of maize. Indeed, the use of transposable elements for gene cloning is now being studied: Van Sluys *et al.* (1987) report that *Ac* transposes easily in different plant species. The disruption is identified by selecting for additional markers such as antibiotic resistance and defective NR activity. A genomic library can be constructed from the insertionally mutated NR⁻ cell lines and rescreened with the inserted sequence as a probe. In this way a NR gene fragment can be obtained. In a last step a probe can be

Table 23.6. Number of selected NR clones, regeneration response and chromosome number analysis in asymmetric fusion products

Fusion combination		Marker transferred	Dose (Krad)	Stable clones[a]	Correction frequency ($\times 10^{-5}$)	Number of regenerated clones	Number of plants analysed	Chromosome number range	
Recipient	Donor							T_0	T_1[b]
N. plumbaginifolia	*N. sylvestris*	*nia*[+]	10	3	4.3	2	1	46	39–42
NIA26	V-42, albino		20	3	4.6	2	3	44	42
			30	11	23	7	5	44	43
			50	5	8.3	5	9	46–49[c]	43
			100	3	9.5	3	3	46	48
			200	2	2.2	2	1	47	–

The transfer of genetic material from irradiated *N. sylvestris* donor protoplasts was done by fusion into *N. plumbaginifolia nia* mutants. Corrected phenotypes for NR were recovered at several doses of Krad.

[a] More than one regenerant was analysed per clone.

[b] One plant per dose.

[c] Higher degree of fragmentation of donor chromosome was observed.

T_0 = regenerated plants obtained after fusion.

T_1 = progeny of regenerated T_0 plants.

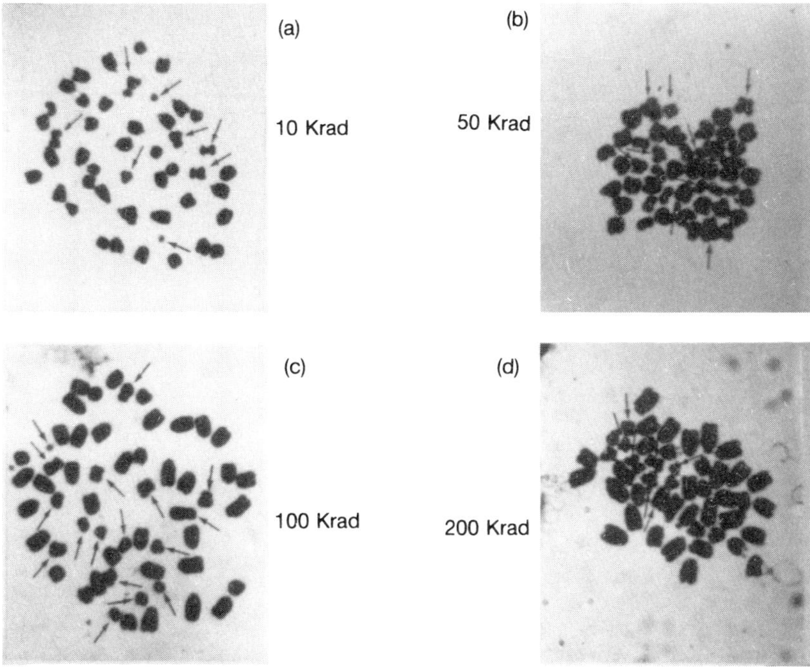

Fig. 23.2. Metaphase chromosomes of fusion products. *N. sylvestris* was used as irradiated donor partner in the fusion with nitrate reductase deficient mutant protoplasts from *N. plumbaginifolia.* The karyological analysis of the regenerated plants revealed the presence of whole chromosomes and chromosome fragments from the donor partner. *N. sylvestris* chromosomes are indicated by arrows. (a) 10 Krad; (b) 50 Krad; (c) 100 Krad; (d) 200 Krad.

constructed including the NR gene fragment to screen a wild-type genomic library.

Another method for cloning is based on a possible phenotypic complementation of NR⁻ cells using a cosmid genomic library comprising 10 000–20 000 clones based on *Arabidopsis thaliana,* which has the advantage of having a small genome. The DNA of the transformants can be isolated and digested. After self-ligation and transformation of *Escherichia coli* a plasmid with the NR plant gene or fragments can be rescued and can then be used as a probe to screen the cosmid library.

Uptake and regulatory mutants

Most of the mutants so far identified are the result of alterations in structural genes implicated in nitrate metabolism. Other categories, for instance mutants with an altered uptake of nitrate and mutants modified in regulatory loci might

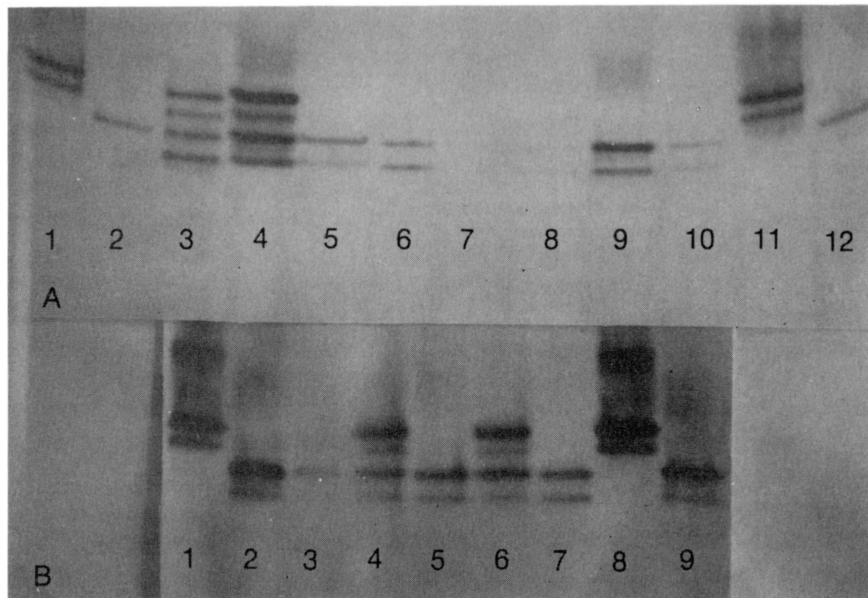

Fig. 23.3. Electrophoretic pattern of shikimic dehydrogenase showing the presence of isozymes from the donor species in asymmetric fusion products. *N. sylvestris* was used as irradiated donor and a *nia N. plumbaginifolia* mutant as recipient. Shikimic dehydrogenase isozymes from *N. sylvestris* were active in the fusion products. A. 1, 11 *N. sylvestris*; 2, 12 *N. plumbaginifolia*: 3–10 fusion products in T_0 generation (regenerated plants obtained after fusion). B. 1, 8 *N. sylvestris*; 2, 9 *N. plumbaginifolia*; 3–7 fusion products in T_1 generation (progeny of regenerated plants).

also be obtained. Putative nitrate uptake mutants have been described for *Arabidopsis thaliana* (Oostindier-Braaksma and Feenstra 1973) and *Aspergillus nidulans* (Chapters 6, 19, and 22).

Uptake mutants

A large number of mutants altered in nitrate reduction have been isolated, usually on the basis of chlorate resistance. Chlorate selection has led to the isolation of NR⁻ mutants. However, in addition to NR deficiency, another theoretical way to achieve chlorate resistance resides in a modified uptake system for nitrate (chlorate), but only a few types have been described (Oostindier-Braaksma and Feenstra 1973).

A more systematic study in our laboratory using *N. plumbaginifolia* protoplasts has led to the isolation of a few cell lines which are resistant to chlorate and are also NR positive (Table 23.8). The same isolation procedure was used as for the NR⁻ mutants. Three different irradiation doses have been used and the resistance to different chlorate concentrations was tested. From a total of 12×10^6 exposed

Table 23.7. Number of selected NR clones and regeneration response in asymmetric fusion products

Fusion combination		Marker transferred	Dose (Krad)	Stable clones	Correction frequency $(\times 10^{-3})$	Number of regenerated clones	Number of plants analysed
Recipient	Donor						
N. plumbaginifolia	Atropa	cnx$^+$	10	409	2.62	226	30
CNX20 Km[a]	belladonna		30	313	4.01	170	23
			50	80	5.29	37	7
			100	62	2.71	20	4

Atropa belladonna was used as donor partner in fusion experiments with a *N. plumbaginifolia cnx* mutant, which is also kanamycin resistant. Corrected phenotypes for NR were recovered for each irradiation dose.

[a] Kanamycin resistant.

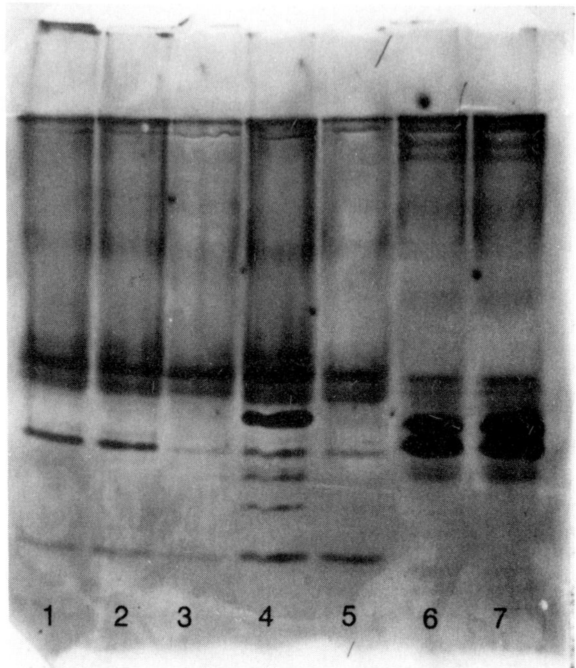

Fig. 23.4. Electrophoretic pattern of esterases showing the presence of an isozyme from the donor species in asymmetric hybrids. Irradiated protoplasts from *Atropa belladonna* were used as donors in the fusion with *N. plumbaginifolia cnx* mutant protoplasts. Specific esterase isozymes characterizing *Atropa belladonna* were present in one fusion product. 1, 2 *N. plumbaginifolia* CNX20; 3–5: fusion products; 6, 7 *Atropa belladonna*.

protoplasts, 441 resistant mutants were found from which four were confirmed as NR-positive and which probably have a deficient uptake mechanism for chlorate and possibly nitrate. Lowered nitrate uptake would have a positive effect on the nutritional value of the plants (see below). The availability of regenerated plants should allow the genetical analysis of this trait to be carried out.

Regulatory mutants

High soil nitrate concentrations can accumulate in plants and these are detrimental to health and also result in a lowered nutritional value. A very interesting type of nitrate assimilation variant would be those with changed kinetics: a higher nitrate turnover would accelerate the incorporation of nitrate into amino acids and other metabolites. As a result the free nitrate pool in the plants should be kept lower. Due to their faster nitrate assimilation such mutants could probably exploit nitrate-poor soils more efficiently.

To be able to distinguish a mutant with a high NR activity from a wild type a negative selection procedure has been proposed by Wang *et al.* (1986) using

Table 23.8. Selection for chlorate resistance in haploid protoplast culture of *N. plumbaginifolia*

	Selection on 75 mM KClO$_3$			Selection on 100 mM KClO$_3$	
UV treatment (erg/mm^2)	200	250	500	250	500
Number of exposed protoplasts	1.5×10^6	3×10^6	2×10^6	3×10^6	2.5×10^6
Number of resistant lines	62	150	96	65	68
Number of NR$^-$ lines	60	148	96	65	68
Number of NR lines	2	2	0	0	0

Three different irradiation doses were used and the resistance of the exposed protoplasts to two different chlorate concentrations was tested.

Arabidopsis thaliana (Table 23.9). Because such high activity mutants must convert chlorate into chlorite more rapidly than the wild type they become chlorotic sooner. At the first sign of chlorosis such hypersensitive plants were rescued and cultured for further analysis. One of the mutants, C-4, expressed an elevated NR activity throughout its life cycle. Inheritance studies of NR indicated that the F_1 hybrid enzyme activity level was very similar to that of the wild type and significantly lower than that of C-4. Consequently the wild type NR is genetically dominant and the elevated activity of C-4 is probably controlled by a single recessive allele.

Table 23.9. Nitrate reductase activity in shoots of mutants and wild-type plants and genetic analysis of mutants in reciprocal crosses between the mutant C-4 and the parental line (modified from Wang *et al.* 1986)

Nitrogen source	Genotype	NR activity (nmol NO_2^-/ mg protein/min)	Parents and cross	NR activity (nmol NO_2^-/ mg protein/min)
			1. Parents	
NO_3^-	WS	9.68	WS	9.62
	C-4	12.41	C-4	11.74
NH_4NO_3	WS	12.21		
	C-4	14.67		
			2. F_1 hybrids	
NH_4^+	WS	0.69	C-4 × WS	10.40
	C-4	1.28	WS × C-4	10.25

The race WS of *Arabidopsis thaliana* was used for mutagenic treatment and for comparison the resultant mutant. C-4 is a mutant line possessing elevated nitrate reductase activity. The inheritance studies of nitrate reductase activity in the parents and the mutant involving F_1 progeny were performed on plants growing on NH_4^+ medium.

We used a similar negative type of selection with chlorate to isolate hypersensitive mutants (Fig. 23.5). The media and cultures used in our selection procedure differed from those used by Wang *et al.* (1986). Wild-type seeds were treated with EMS and sown on soil in the greenhouse. The resulting plants were designated as the M_1 generation. M_2 seeds resulting from self-fertilization of M_1 plants were sown on agar with ammonium succinate as sole N source. Since ammonium does not induce NR activity possible mutants should be distinguished more easily from the wild type. 0.25 mM $KClO_3$ was added to the nutrient medium when the plants were 6 days old. Plants that exhibited early chlorosis (in less than 48 h) were assumed to be potentially chlorate-hypersensitive. These plants were immediately transferred to chlorate-free medium supplied with an excess of nitrate and after recovery to green plantlets they were transferred to soil.

From a total of 32 000 germinated M_2 seeds, 102 plantlets became chlorotic in less than 48 h. After transfer to a modified Hoagland medium (with higher KNO_3 concentration), 40 plantlets were saved. Unfortunately only five fertile plants could be obtained after transfer to soil, two of which showed a higher NR activity (an increase of 150 per cent) with ammonium as sole nitrogen source. The genetic basis of this mutation is now being determined in the M_4 generation by means of segregation analysis and back-crosses to the wild type.

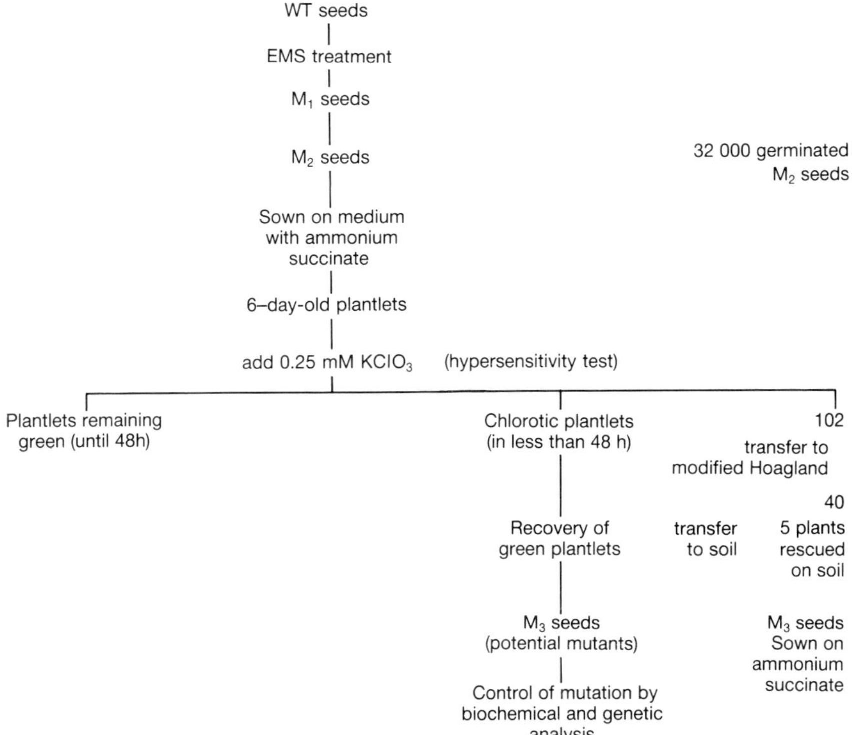

Fig. 23.5. Negative selection scheme for the isolation of a chlorate-hypersensitive mutant of *Arabidopsis thaliana*.

For the moment the high activity mutants of *Arabidopsis thaliana* are the only known higher plants with increased levels of NR activity. This selection may lead to the isolation of de-regulated types for NR activity, such as mutants which are permanently de-repressed for nitrate reduction. The recent isolation by different laboratories (Crawford *et al.* 1986; Crawford and Davies 1987; Calza *et al.*, 1987) of cDNA clones will allow the molecular study of such alterations and may improve the understanding of the processes controlling the assimilation of nitrate.

Acknowledgements

This work was supported by grants and a research contract from IWONL (4535 A and 4972 A).

References

Calza, R., Huttner, E., Vincentz, M., Ronzé, P., Galangau, F., Vaucheret, H., Chérel, I., Meyer, C., Kronenberger, J., and Caboche, M. (1987). Cloning of DNA fragments complementary to tobacco nitrate reductase in RNA and encoding epitopes common to the nitrate reductases from higher plants. *Molecular and General Genetics* **209**, 552–62.

Crawford, N. and Davis, R. (1987). Cloning of a nitrate-regulated gene from *Arabidopsis*. *3rd Intern. Meeting on Arabidopsis*, Abstract 44. Michigan State University, USA.

Crawford, N., Campbell, W., and Davis, R. (1986). Nitrate reductase from squash: cDNA cloning and nitrate regulation. *Proceedings of the National Academy of Science, USA,* **83**, 8073–6.

Dirks, R. (1986). Somatic cell genetics of *Nicotiana plumbaginifolia* Viviani: isolation and characterization of nitrate-reductase-deficient and amino acid auxotrophic mutants isolated in haploid protoplast cultures. Ph.D. Dissertation. Free University of Brussels.

Dirks, R., Negrutiu, I., Sidorov, V. and Jacobs, M. (1985). Complementation analysis by somatic hybridisation and genetic crosses of nitrate reductase-deficient *Nicotiana plumbaginifolia* protoplasts—evidence for a new category of *cnx* mutants. *Molecular and General Genetics* **201**, 339–43.

Dirks, R., Negrutiu, I., Heinderycks, M., and Jacobs, M. (1986*a*). Genetic analysis of revertants for the nitrate reductase function of *Nicotiana plumbaginifolia*. *Molecular and General Genetics* **202**, 309–11.

Dirks, R., Negrutiu, I., Jacobs, M., Sidorov, V. (1986*b*). Isolation of auxotrophic mutants based on reconstruction experiments with *Nicotiana plumbaginifolia* protoplasts. In *Genetic manipulation in plant breeding* (eds. Horn, Jensen, Odenbach, and Schieder), pp. 599–600. de Gruyter, Berlin–New York.

Dudits, D., Fejer, O., Hadlaczky, G., Koncz, C., Lazar, G., and Horvath, G. (1980). Intergeneric gene transfer mediated by plant protoplast fusion. *Molecular and General Genetics* **179**, 283–8.

Famelaer, I., Cammaerts, D., Karp, A., Sidorov, V., De Brouwer, D., Negrutiu, I., and Jacobs, M. (1986). Cellular engineering by 'gamma fusion' and 'egg transformation'. Recent experimental data and applications in plant breeding. In *Nuclear techniques and in vitro culture for plant improvement*, pp. 453–61. International Atomic Energy Agency, Vienna.

Gleba, Y. and Hoffmann, F. (1980). 'Arabidobrassica': a novel plant obtained by protoplast fusion. *Planta* **149**, 112–17.

Glimelius, K. and Bonnett, H. (1981). Somatic hybridization in *Nicotiana*: restoration of photo-autotrophy to an albino mutant with defective plastids. *Planta* **153**, 497–503.

Glimelius, K., Eriksson, T., Grafe, R., and Müller, A. (1978). Somatic hybridization of nitrate reductase-deficient mutants of *Nicotiana tabacum* by protoplast fusion. *Physiologia plantarum* **44**, 273–7.

Glimelius, K., Chen, K., and Bonnett, H. (1981). Somatic hybridisation in *Nicotiana*: segregation of organellar traits among hybrid and cybrid plants. *Planta* **153**, 504–10.

Gupta, P., Gupta, M., and Schieder, O. (1982). Correction of nitrate reductase defect in auxotrophic plant cells through protoplast mediated intergeneric gene transfers. *Molecular and General Genetics* **188**, 378–83.

Gupta, P., Schieder, O., and Gupta, M. (1984). Intergeneric nuclear gene transfer between somatically and sexually incompatible plants through asymmetric protoplast fusion. *Molecular and General Genetics* **197**, 30–5.

Hein, T., Przewozny, T., and Schieder, O. (1983). Culture and selection of somatic hybrids using an auxotrophic cell line. *Theoretical and Applied Genetics* **64**, 119–22.

Marton, L., Sidorov, V., Biasini, G., and Maliga, P. (1982). Complementation in somatic hybrids indicates four types of nitrate reduction deficient lines in *Nicotiana plumbaginifolia*. *Molecular and General Genetics* **187**, 1–3.

Negrutiu, I., Dirks, R., and Jacobs, M. (1983). Regeneration of fully nitrate reductase-deficient mutants from protoplast culture of *Nicotiana plumbaginifolia* (Viviani). *Theoretical and Applied Genetics* **66**, 341–7.

Negrutiu, I., De Brouwer, D., Dirks, R., and Jacobs, M. (1985). Amino acid auxotrophs from protoplast cultures of *Nicotiana plumbaginifolia*, Viviani. *Molecular and General Genetics* **199**, 330–7.

Negrutiu, I., De Brouwer, D., Watts, J., Sidorov, V., Dirks, R., and Jacobs, M. (1986). Fusion of plant protoplasts: a study using auxotrophic mutants of *Nicotiana plumbaginifolia* (Viviani). *Theoretical and Applied Genetics* **72**, 279–86.

Oostindier-Braaksma, F. and Feenstra, W. (1973). Isolation and characterization of chlorate-resistant mutants of *Arabidopsis thaliana*. *Mutation Research* **19**, 175–85.

Somers, D., Narayanan, K., Kleinhofs, A., Cooper-Bland, S., and Cocking, E. (1986). Immunological evidence for transfer of the barley nitrate reductase structural gene to *Nicotiana tabacum* by protoplast fusion. *Molecular and General Genetics* **204**, 276–301.

Van Sluys, M., Tempé, J., and Federoff, N. (1987). Studies on the introduction and mobility of the maize Activator element in *Arabidopsis thaliana* and *Daucus carota*. *EMBO Journal* **6**, 3881–9.

Wallin, A., Glimelius, K., and Eriksson, T. (1979). Formation of hybrid cells by transfer of nuclei via fusion of miniprotoplasts from cell lines of nitrate reductase deficient tobacco. *Zeitschrift für Pflanzenphysiologie* **91**, 89–94.

Wang, X., Scholl, R., and Feldmann, K. (1986). Characterization of a chlorate-hypersensitive high nitrate-reductase *Arabidopsis thaliana* mutant. *Theoretical and Applied Genetics* **72**, 328–36.

Xuan, L., Grafe, R., and Müller, A. (1983). Complementation of nitrate reductase deficient mutants in somatic hybrids between *Nicotiana* species. In *Protoplast 1983*, 6th International Protoplast Symposium Basel, pp. 75–6. Birkhäuser Verlag, Basel.

Comparative Aspects

24. Amino acid sequence relationships between bacterial, fungal, and plant nitrate reductase and nitrite reductase proteins

James R. Kinghorn and Edward I. Campbell

Introduction

The conversion of nitrate to ammonium has been extensively studied at both the biochemical and genetic level in a variety of organisms. Recently, genes encoding the nitrate reductase (NR) and nitrite reductase (NiR) apoproteins have been isolated by molecular cloning techniques (Chapters 3, 6, 12, 13, 15, 18, 20, 21, and 22). Comparison of the deduced primary amino acid sequence of these systems and other enzyme systems carrying out similar electron transfer reactions but otherwise unrelated physiological functions, reveals a pattern of conserved motifs. It is likely that at least some of these conserved motifs identify the functional domains of the enzymes.

Nitrate reductase

The properties of NR (EC 1.6.6.3) would appear to be generally similar in functional aspects (Chapters 4, 6, 7, 9, 11, and 14). The available biochemical evidence suggest that the enzyme possesses three domains, namely FAD (flavoreductase), haem (cytochrome b_{557}) and molybdenum cofactor (molybdo-reductase)—with a SH group functioning as the primary electron acceptor from NADH/NADPH.

In order to characterize similarities of protein sequence, the predicted amino acid sequence of the nitrate reductases of *Arabidopsis thaliana* (Crawford *et al.* 1988; N. M. Crawford, personal communication), *Nicotiana tabacum* (Calza *et al.* 1987; Chapter 12; M. Caboche, personal communication), *Escherichia coli* (*narG*) (J. C. Wootton, personal communication), and *Aspergillus nidulans* (*niaD*) (Johnstone *et al.* 1989) were compared by dot matrix (diagon) plots (Staden

A
917
A. thallana nitrate reductase
0
A. nidulans niaD
873
B
904
N. tabacum nitrate reductase
0
A. nidulans niaD
873

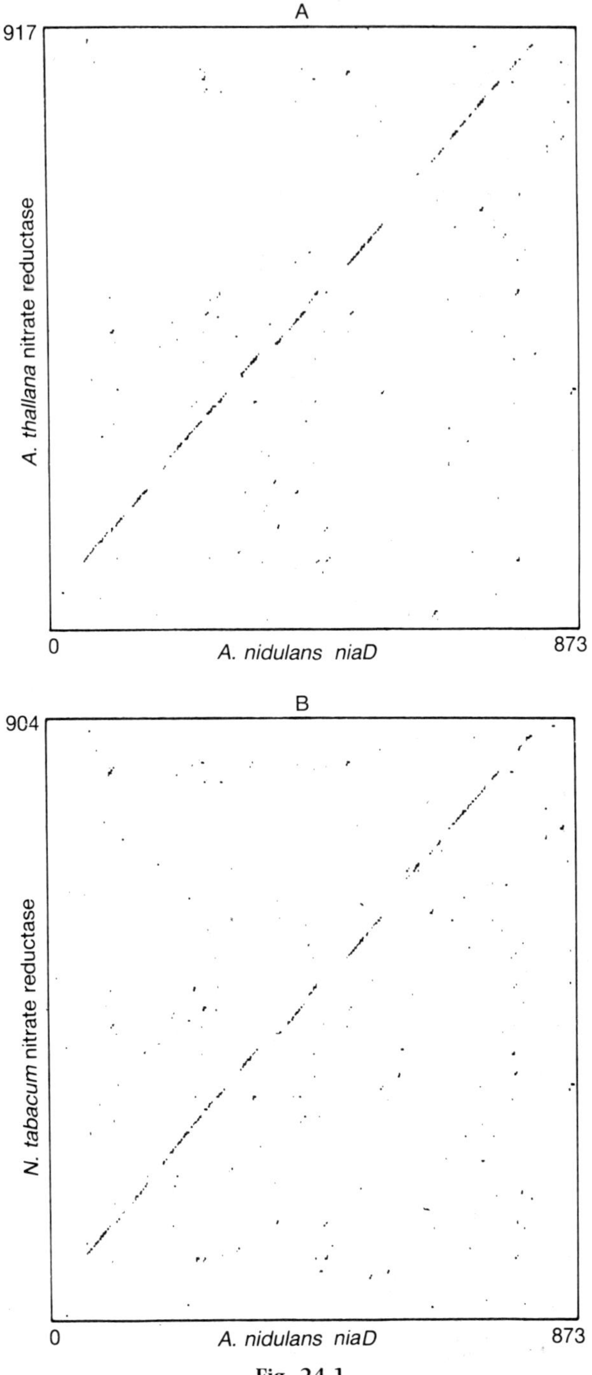

Fig. 24.1.

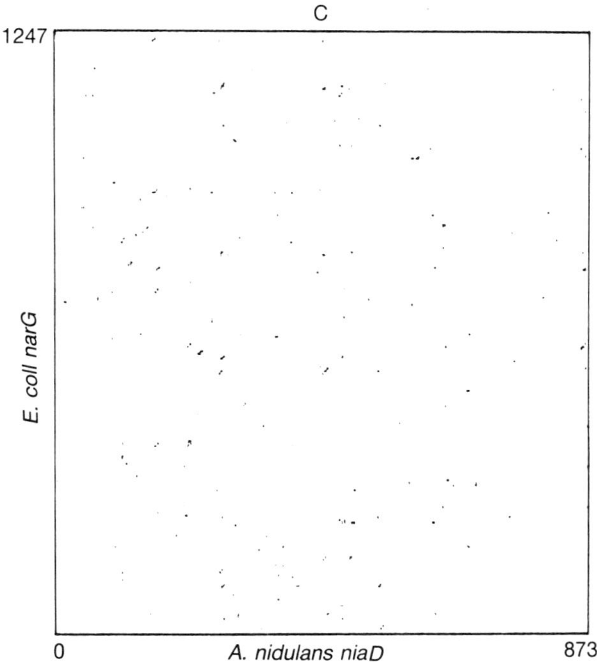

Fig. 24.1. Dot matrix comparison between the nitrate reductase amino acid sequence of *Aspergillus nidulans* and that of *Arabidopsis thaliana* (A), *Nicotiana tabacum* (B) and *Escherichia coli narG* (C).

1982, 1984). Even a casual perusal of Fig. 24.1 shows that there is extensive similarity between the fungal and plant proteins (Fig. 24.1A and 24.1B) suggesting evolutionary kinship and reflecting common ancestral sequences. In contrast, dot matrix analysis (Fig. 24.1C) shows that no obvious homology exists between the fungal NR and the *E. coli narG* protein encoding one of the subunits for respiratory NR (the only available *E. coli* subunit sequence so far). Fungal and plant NR amino acid residues were aligned by computer and refined by eye for maximum homology. Inspection of the three sequences shows a pattern of conserved regions of motifs (Table 24.1).

Comparison of these sequences with other enzymes which are likely to contain FAD, haem or molybdenum cofactor (MoCo) approximately locates these motifs to domains on the NR protein molecule (Fig. 24.2; Table 24.1). The *A. nidulans* amino acid sequence was compared with FAD (bovine NAD cytochrome b_5 reductase, Fig. 24.2A), haem (yeast flavocytochrome b_2 Fig. 24.2B) and MoCo (fruitfly xanthine dehydrogenase, Fig. 24.2C)-containing enzymes by dot matrix analysis. Significant similarity exists between NR and bovine and human NAD cytochrome b_5 reductase (Ozols *et al.* 1985) at the C-terminus of the NR protein (Fig. 24.2A). No obvious similarity with other flavoproteins, including glutath-

Table 24.1. Alignment of the deduced amino acid sequences of NR from *Arabidopsis thaliana* (Crawford *et al.* 1989), *Nicotiana tabacum* (Calza *et al.* 1987; M. Caboche, personal communication) and *Aspergillus nidulans* (Johnstone *et al.* 1989). Protein sequences are aligned for maximum homology by computer analysis using the program BEST FIT from the University of Wisconsin sequence analysis and software package (Devereux *et al.* 1984) and subsequently refined by eye

```
                                                *                   *                *        **       *  *
A. thaliana   MAASVDNRQYARLEPGLNGVVRSYKPPVPGRSDSPKAHQNQTTNQTVFLKPAKVHDDDEDVSSEDENETHNSNAVYYKEMIRKSNAELEPSVLDPRDEYT   100
N. tabacum    MAASVENRQFSHLEAGLS---RSFK---PRSDSPVRGCNFPSPNSTNFQKKPNSTIYLDYSSSEDDDDDDEKNE--YLQMIKKGNSELEPSVHDTRDEGT    92
A. nidulans                         MSTTVTQVRTGSIPKTLKTSQIRVEEQEITE-LDTADIPLPPPSKEPTEVLSLDKTT    56


                *       *    ****  *    *  *  *      **  **    *  ******  **              *   *        *           **********
A. thaliana   ADSWIERNPSMVRLTGKHPFNSEAPLNRLMHHGFITPVPLHYVRNHGHVPKAQWAE---WTVEVTGFVKRPMKFTMDQLVSEFAYREFAATLVCAGNRRK   197
N. tabacum    ADNWIERNFSMIRLTGKHPFNSEPPLNRLMHHGFITPVPLHYVRNHGPVPKGTWDD---WTVEVTGLVKRPMKFTMDQLVNEFPCRELPVTLVCAGNRRK   189
A. nidulans   PDSHVPRDPRLIRLTGVHPFNVEPPLTALFQQGFLTPPELFYVRNHGPVPHVRDEDIPNWELRIEGLVEKPITLSFKQILQNYDQITAPITLVCAGNRRK   156
                 LRINSQRPFNAEPPP
                 sul. oxid.        intron 1                                       intron 2


              *** *      **  *   *    *         *              *   **  **    **  *      *******            *  *  **        *  *  *  *
A. thaliana   EQNMVKKSKGFNWGSAGVSTSVWRGVPLCDVLRRCGIFSRKGGALNVCFEGSEDLPGGAGTAGSKYGTSIKKEYAMDPSRRIILAYMQNGEYLTPDHGFP   297
N. tabacum    EQNMVKQTIGFNWGAAAVSTTIWRGVPLRALLKRCGVFSKNKGALMVCFEGADVLPGG---GGSKYGTSIKKEFAMDPARDIIVAYMQNGEKLAPDHGFP   286
A. nidulans   EQNTVRKSKGFSWGSAALSTALFTGPMMADIIKSAKPLRR---AKYVCMEGADNLPNG------NYGTSIKLNWAMDPNRGIMLAHKMNGEDLRPDHGRP   247
                 G-AVVPT                                                      AMDPQAEVLLAYEMNGQPLPRDSGFP
                 xan. dehyd.                              intron 3                    sul. oxid.


                *      **  **  *  ***  *   *  *     **   **   ******   *          *    *  *  **  **    *  *  *  *     **   **  *  *
A. thaliana   VRIIIPGFIGGRMVKWLKRIIVTTKESDNFYHFKDNRVLPSLVDAELADEEGWWYKPE-YIINELNINSVITTPCHEEILPINAFTTQRPYTLKGYAYSG   396
N. tabacum    VRMIIPGFIGGRMVKWIKRIIVTTQESDSYYHFKDNRVLPPHVDAELANTEAWWYKPE-YIINELNINSVITTPCHEEILPINAWTTQRPYTLRGYSYSG   385
A. nidulans   LRAVVPGQIGGRSVKWLKKLIITDAPSDNWYHIYDNRVLPTMVTPDMSSQNPSWWRDERYAIYDLNVNSAAVYPQHKETLDLAA--ARPFYTAKGYAYAG   345
              VRV
                                                             intron 4


                *       ****    *  *   *       *     **               **  ****     *   *  *     ***  **  *  **      *  *****
A. thaliana   GGKKVTRVEVTVDGGETWNVCALDHQEKPNKY----------------GKFWCWCFWSLEVEVLDLLSAKEIAVRAWDETLNTQPEKMIWNLMGMMNNC   479
N. tabacum    GGKKVTRVEVTLDGGETWQVSTLDHPEKPTKY----------------GKYWCWCFWSLEVEVLDLLSAKEIAVRAWDETLNTQPEKLIWNVMGMMNNC   468
A. nidulans   GGRRITRVEISLDKGKSWRLARIEYAED--KYRDFEGTLYGGRVDMAWREACFCWSFWSLDIPVSELASSDALLVRAMDEALSLQPKDMYWSVLGMMNNP   443
              GGKESRGISVAL
              xan. dehyd.    intron 5
```

```
              **▩*           *    *           *                  *                           *   *   *     ▩  * *  *▩ ▩ ▩
A. thaliana   WFRVK-----TNVCKPHKGEIGIVFEHPTLPGNESGGWMAKERHLEKSADAPPSLKKSVSTPFMNTT-AKMYSMSEVKKHNSADSCWIIVHGHIYDCTRF   573
N. tabacum    WFRVK-----MNVCKPHKGEIGIVFEHPTQPGNQSGGWMAKERHLEISAEAPQTLKKSISTPFMNTA-SKMYSMSEVRKHSSADSAWIIVHGHIYDATRF   562
A. nidulans   WFRVKITNENGRLLFEHPTDITGSSGWMEQIKKAGGDLTNGNWGERQEGEEPVEAEPVVEVNMKKEGVTRIIDLEEFKKNSSDERPWFVVNGEVYDGTAF   543
              RVG-----GWVCK
              xan. dehyd.

              *      ▩       **   *▩ *   **▩  *▩*  ***    *   *▩ *                     *                        ▩            ▩   * *
A. thaliana   LMDHPGGSDSILINAGTDCTEEFEAIHSDKAKKMLEDYRIGELITTGYSSDSSSPNNSVHGSSAVFSLLAPIGEATPVRNLALVNPRAKVPVQLVEKTSI   673
N. tabacum    LKDHPGGTDSILINAGTDCTEEFDAIHSDKAKKLLEDFRIGELITTGYTSD-SPGNSVHGSSS-FSSFLAPIKELVPAQRSVALIPREKIPCKLIDKQSI   660
A. nidulans   LEGHPGGAQSIISAAGTDASEEFLEIHSETAKKMMPDYHIGTLDKASLEALRKGNADTTDSS--------------SDPRPTFLTPKAWTKATLTKKTSV   629
                                     ▲                      bovine NAD cyt b5 reductase   LENPDIKYPLRLIDKEVI   52
                                  intron 6                human NAD cyt b5 reductase   LESPDIKYPLRLIDREII   27

              ▩▩ ▩  ▩  ▩          ▩▩▩▩ ▩▩ ▩ ▩ ▩  ▩       ▩ ▩▩ ▩▩▩ ▩ ▩         ▩*▩          ▩▩ ▩▩ ▩*  * ▩*    *▩▩ ▩
A. thaliana   SHDVRKFRFALPVEDMVLGLPVGKHIFLCATINDKLC----LRAYTPSSTVDVVGYFELVVKIYFGGVHPRFPNGGLMSQYLDSLPIGSTLEIKGPLGHV   769
N. tabacum    SHDVRKFRFALPSEDQVLGLPVGKHIFLCAVIDDKLC----MRAYTPTSTIDEVGYFELVVKIYFKGIHPKFPNGGQMSQYLDSMPLGSFLDVKGPLGHI   756
A. nidulans   SSDTHIFTLSLEHPSQALGLPTGQHLMLKTPDPKSSSSGSIIRSYTPISPSDQLGMVDILIKIY---AETPSIPGGKMTTALDTLPLGSVIECKGPTGRF   726
Bov b5 red    SHDTRRFRFALPSPEHILGLPVGQHIYLSARIDGNLV----IRPYTPVSSDDDKGFVDLVIKVYFKDTHPKFPAGGKMSQYLESMGIGDTIEFRGPNGLL   148
Hum b5 red    SHDTRRFRFALPSPQHILGLPVGQHIYLSARIDGNLV----VRPYTPISSDDDKGFVDLVIKVYFKDTHPKFPAGGKMSQYLESMQIGDTIEFRGPSGLL   123

              *▩    ▩            **  *    ▩    ▩▩▩▩▩▩▩* ▩  ▩   ▩   ▩  ▩▩     ▩* ▩ ▩▩▩▩  ▩▩*        ▩              ▩*▩▩
A. thaliana   EYLGLGSF------TVHGKPKFADKLAMLAGGTGITPVYQIIQAILKDPEDETEMYVIYANRTEEDILLREELDGWAEQYPDRLKVWYV-V-ESAKEGWA   861
N. tabacum    EYQGKGNF------LVHGKQKFAKKLAMIAGGTGITPVYQVMQAILKDPEDDTEMYVVYANRTEDDILLKEELDSWAEKIPERVKVWYV-VQDSIKEGWK   849
A. nidulans   EYLDRGRV------LISGKERFVKSFVMICGGTGITPVFQVLRAVMQDEQDETKCVMLDGNRLEEDILLKNELDEFEALAGKKEKCKIVHTLTKGSESWT   820
Bov b5 red    VYQGKGKFAIRPDSKSDPVIKTVKSVGMIAGGTGITPMLQVIRAIMKDPDDHTVCHLLFANQTEKDILLRPELEELRDEHSARFKLWYT-V-DKAPEAWD   246
Hum b5 red    VYQGKGKFAIRPDKKSNPIIRTVKSVGMIAGGTGITPMLQVIRAIMKDPDDHTVCHLLFANQTEKDILLRPELEELRNKHSARFKLWYT-L-DRAPEAWD   221

              ▩ * ▩  ▩▩          ▩▩▩▩ ▩ ▩  ▩  ▩            ▩
A. thaliana   YSTGFISEAIMREHIPDGLDGSALAMACGPPPMIQFAVQPNLEKMQYNIKEDFLIF   917
N. tabacum    YSIGFITEAILREHIPEPSHTT-LALACGPPPMIQFAVNPNLEKMGYDIKDGLLVF   904
A. nidulans   GRRGRIDEELIRQH-AGTPDFGTMVLVCGPEAMEK-ASKKILLSLGWKEENLHY-F   873
Bov b5 red    YSQGFVNEEMIRDHLPPP-EEEPLVLMCGPPPMIQYACLPNLDRVG-HPKERCFAF   300
Hum b5 red    YDQGFVNEEMIRDHLPPP-EEEPLVLMCGPPPMIQYACLPNLDHVG-HPTERCFVF   275
```

*Denotes identical amino acids in NR in all three species. Where such identities are additionally shared with other physiologically unrelated enzyme systems the * is replaced be a vertical bar (▩). Short arrows (▲) denote exon boundaries on the *Aspergillus nidulans* NR protein (intron positions in the *A. thaliana* and *N. tabacum* genes are unknown).

FAD domain: Alignment of NR amino acid sequence residues with the FAD/NADH domain (*A. thaliana* residues ~659–917) human NAD cytochrome b_5 reductase (Yabisui *et al.* 1984) and bovine NAD cytochrome b_5 reductase (Crawford *et al.* 1989) with NR.

Haem domain: Alignment of NR amino acid sequence residues with the haem domain region (residues ~540–612) of seven members of the cytochrome b_5 family as well as yeast flavocytochrome b_2 and chicken sulphite oxidase (Calza *et al.* 1987; Chapter 12 and references therein).

Molybdenum cofactor domain: Alignment of homologous sequence of two peptides of rat liver sulphite oxidase (Crawford *et al.* 1989) (residues 113–126, ~272–299) and homologous stretches of xanthine dehydrogenase sequences (Keith *et al.* 1987) (residues 211–217, 397–408, 481–488) with NR proteins.

ione reductase (Krauth-Siegel *et al.* 1982), pig D-amino acid oxidase (Ronchi *et al.* 1982), *Pseudomonas fluorescens* p-hydroxy benzyloate hydroxylase (Weier *et al.* 1982), dogfish lactate dehydrogenase (Eventoff *et al.* 1977), *E. coli* respiratory lactate dehydrogenase (Campbell *et al.* 1984), or *Clostridium* MP flavodoxin (Tanaka *et al.* 1974), was observed by dot matrix analysis (E. I. Campbell and J. R. Kinghorn, unpublished). A second group of proteins, haem-binding proteins, as exemplified by yeast flavocytochrome b_2 shows a small region of homology near the mid-region of the NR gene (Fig. 24.2B). Finally, when the fruitfly *Drosphila melanogaster* MoCo-containing xanthine dehydrogenase protein (Keith *et al.*, 1987) was compared no clear similarity is seen by dot matrix analyses (Fig. 24.2C).

These sequences were aligned with NR by computer and refined by eye for maximum homology (Tables 24.1, 24.2) to more precisely locate functional domains.

FAD domain (A. thaliana residues ∼659–917)

Close inspection of Table 24.1 shows that as well as complete consensus amino acid blocks, there are additional but not complete similarities. For instance the FRFALP block (*A. thaliana* residues 681–687) is conserved faithfully between plant NR and mammalian NAD b_5 reductase, but reads FTLSLE in fungal NR (residues 636–742). There are 40 further shared residues between plant and mammalian genes, suggesting that these systems may be more closely related to each other than to the fungal enzyme. The analogous SH residue of bovine b_5 reductase (at residue 273 on the mammalian sequence) for acceptance of NADH reducing equivalents would appear to be conserved in all three NR proteins and occurs at position 889 in the *A. thaliana* sequence (Crawford *et al.* 1989).

Haem domain (A. thaliana residues ∼540–612)

The haem-containing group includes the cytochrome b_5 superfamily (seven members sequenced), yeast flavocytochrome b_2 and chicken sulphite oxidase (Chapter 12 and references therein). No less than 13 'full house' or complete identities are seen between these and the three NR in this region (Table 24.1). Alignment of *A. thaliana* (residues 538–617), *N. tabacum*, *A. nidulans*, and *N. crassa* with other haem-binding proteins (Table 24.2) shows that significant identities occur between NR and these proteins at positions other than 'full house' identities. This, together with a number of conservative changes, only serves to increase awareness of homology in this region. Calza *et al.* (1987) observed that the tobacco NR haem domain is closer to chicken cytochrome b_5 than to any others.

MoCo-binding domain (A. thaliana residues ∼110–490)

Aligning sequences between NR and fruitfly xanthine dehydrogenase (Keith *et al.* 1987) reveals three short stretches of amino acid residue similarities (Table 24.1)

Table 24.2. Alignment of inferred primary sequences in the haem domain in the region of highest homology (*A. thaliana* residues 538–617). Modified after Calza *et al.* (1987; and Chapter 12)

A. thaliana	NTTAKMYSMSEVKKHNSADS-CWIIVHGHIYDCTRFLMDHPGGSDSILINAGTDCTEEFEAI---HST-KAKKMLEDYR--IGELI
N. tabacum	NTASKMYSMSEVRKHSSADS-AWIIVHGHIYDATRFLKDHPGGTDSILINAGTDCTEEFDAI---HSD-KAKKLLEDFR--IGELI
A. nidulans	EGVTRIIDLEEFKKNSSDER-PWFVVNGEVYDGTAFLEGHPGGAQSIISAAGTDASEEFLEI---HSE-TAKKMMPDYH--IGTLD
N. crassa	--AEMDLEYEIKQYHNNSKS-TWLILHYKVYDLTKFLEEFPGGQEVLCR

Rat mitochondrial cyt b_5	DPAVTYYRLEEVALRMTAEE-TWMVIHGRVYDITRFLSEHPGGEEVLLEQAGADATESFEDV--GHS-PDAREMLKQY--YIGDVH
Rat microsomal cyt b_5	DKDVKYYTLEEIQKHKDSKS-TWVILHHKVYDLTKFLEEHPGGEEVLREQAGGDATENFEDV--GHS-TDARELSKTYLLIIGELH
Horse microsomal cyt b_5	SKAVKYYTLEEIKKHNHSKS-TWLILHHKVYDLTKFLEDHPGGEEVLREQAGGDATENFEDI--GHS-TDARELSLTF--IIGELH
Rabbit microsomal cyt b_5	DKDVKYYTLEEIKKHNHSKS-TWLILHHKVYDLTKFLEEHPGGEEVLREQAGGDATEEFEDV--GHS-TDARELSKTF--IIGELH
Bovine microsomal cyt b_5	SKAVKYYTLEQIEKHNNSKS-TWLILHYKVYDLTKFLEEHPGGEEVLREQAGGDATEDFEDV--GHS-TDARELSKTF--IIGELH
Human microsomal cyt b_5	DEAVKYYTLQEIEKHNHSKS-TWLILHHKVYDLTKFLEEHPGGEEVLREQAGGDATEEFEDV--GHS-TDAREMSKTF--IIGELH
Chicken microsomal cyt b_2	SKAGRYYRLEEVQKHNZABS-TWIIVHHRIYDITKFLDEHPGGEEVLREQAGGDATEDFEDV--GHS-TDARALSETF--IIGELH

Yeast flavo cytochrome b_2	DMNKQKISPAEVAKHNKPDD-CWVVINHYVYDLTRFLPNHPGGQDVIKFNAGKDVTAIFEPL---HA-PNVIDKYIAPEKKLGPLE

Chicken sulphite oxidase	APSYPRYTREEVGRHRSPEERVWVTHGTDVFDVTDFVELHPGGPDKILLAAGGAL-EPFWALYAVHGEPHVLELLQQY--KVGELS

'Full house' amino acids identities are for all 13 genes denoted by a vertical bar (▒).

A
300
Bovine NADH cytochrome b_5 reductase
0
A. nidulans niaD
873
B
511
S. cerevisiae flavocytochrome b_2
0
A.nidulans niaD
873

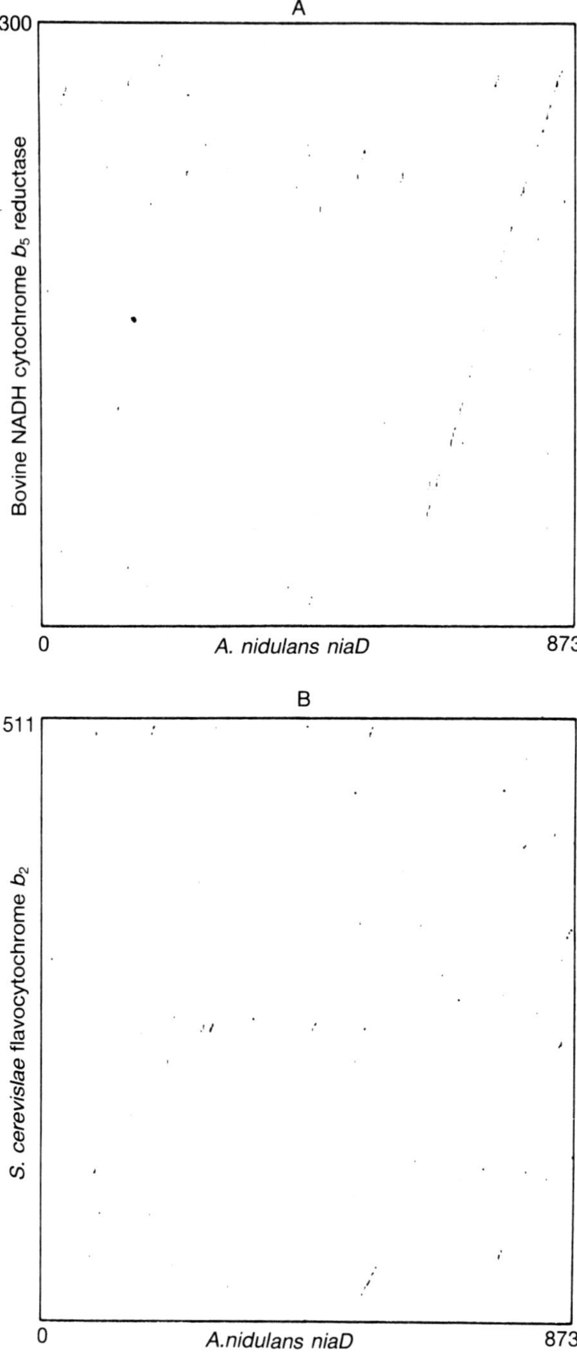

Fig. 24.2.

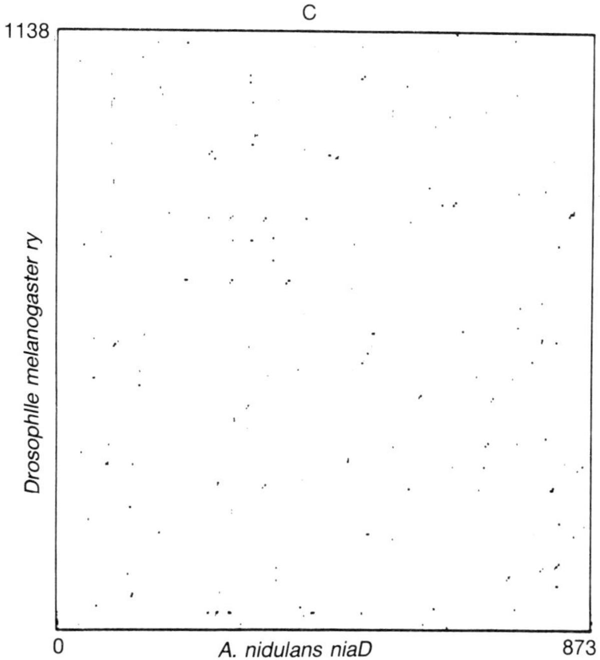

Fig. 24.2. Dot matrix comparisons between the nitrate reductase amino acid of *Aspergillus nidulans* and bovine cytochrome b_5 reductase (A), yeast flavocytochrome b_5 (B) and fruitfly xanthine dehydrogenase (C).

which correspond to *A. thaliana* residues 211–217, 397–408 and 481–488. Exact matches are represented in all three regions. Furthermore, examination of conserved (but not 'full house') identities indicate that the plant sequences are closer to fruitfly xanthine dehydrogenase in these regions than to the fungal NR. Two other regions (*A. thaliana* residues 113–126 and 272–299) show strong similarity with the molybdenum-binding domain of rat liver sulphite oxidase (Crawford *et al.* 1989). Finally, it may be appropriate to restate here that there appears to be little homology (Fig. 24.1C) between the assimilatory NR proteins and the *E. coli narG* protein which contains the molybdenum domain of respiratory NR (J. C. Wootton, personal communication). Other blocks of strong homology between the plant and fungal proteins exist, for example *A. thaliana* residues 140–146.

Intron positions

The position on the *A. nidulans* amino acid sequence at which introns are spliced out to give the mature mRNA (Johnstone *et al.* 1989) is shown in Fig. 24.1. It is generally considered that individual exons may correspond to the functional domains of a given protein (Gilbert 1985). With regard to assessing whether or not exons encode independent domains of the *A. nidulans* proteins it would appear that exon 7 carries the FAD domain (~ 659–917) in its entirety. Most of the haem domain, as well as part of the region showing homology with xanthine dehydrogenase and the connecting dissimilar region, is found together within exon 6. The putative MoCo region, i.e. the two motifs which show homology with rat liver sulphite oxidase, are found on exons 1 and 4. The three regions showing homology with xanthine dehydrogenase are found within exons 3, 5, and 6 (the latter exon also includes the haem domain).

In summary, it would appear from such comparison studies that the MoCo domain is located towards the N-terminus of the NR protein between *A. thaliana* residues 110 and 490. This is separated from haem (residues ~ 540–612) by a region which is dissimilar between plant and fungal systems. Finally the FAD domain is found towards the C-terminus of the protein (from *A. thaliana* residue ~ 659) separated again by a relatively dissimilar amino acid stretch (residues 620–659). Identification of the precise location of functional components of the electron transfer sequence on the NR protein await further analysis. Dissimilar amino acid sequences connecting the three domains are unlikely to be involved in electron transfer, catalytic activity or the maintenance of 'higher order' protein structure and probably function as interdomain 'hinge' regions.

Nitrite reductase

NiR (EC 1.6.6.4) catalyses a six-step electron reduction of nitrite to ammonium, the reaction being linked to intermediaries which serve in the transfer of electrons. Whilst plant assimilatory NiR utilizes ferredoxin, filamentous fungal assimilatory NiR, as well as *Escherichia coli* respiratory NiR, utilize NADH/NADPH as electron donors and contain a FAD domain (Chapter 18). These three sets of enzymes contain the prosthetic group sirohaem in addition to iron sulphur centres.

E. coli/fungi   NADH/NADPH→$\boxed{\text{SH}\rightarrow\text{FAD}\rightarrow\text{Fe}_4\text{S}_4 \rightarrow\text{sirohaem}}$──────→$\text{NO}_2^-$

plants   ferredoxin   ───────→$\boxed{\text{Fe}_4\text{S}_4\rightarrow\text{sirohaem}}$──────→$\text{NO}_2^-$

Fe_4S_4 denotes tetranuclear iron–sulphur centres while sirohaem is a tetrahydroporphyrin (Chapters 17, 18). Four cysteine residues of the NiR protein are thought to bind the tetranuclear iron cluster whilst sirohaem is also thought to be bound to one of these cysteine residues (Chapter 18).

The predicted NiR amino acid sequences of *A. nidulans* (Chapter 6, Johnstone *et al.* 1989), *S. oleracea* (Chapter 18, E. Back, personal communication), and *E. coli* (Chapter 15, J. A. Cole, personal communication) were compared for sequence similarity (Fig. 24.3) to identify functional domains of the enzyme.

Iron–sulphur centre and sirohaem domain

It has been suggested by Back *et al.* (Chapter 18) that cysteine residues observed at positions 473, 479, 514, and 518 (Table 24.3 marked ◆) on the spinach sequence, are responsible for binding the tetranuclear iron–sulphur centre (and sirohaem) to the protein, as these residues are conserved between NADPH-sulphite reductase of *Salmonella typhimurium* and spinach NiR (Chapters 17 and 18). Significantly the *A. nidulans* nitrite reductase protein has a similar cysteine block but at positions 720, 726, 760, and 764. Although the number of residues between these are similar, i.e. N-terminal cys-X-X-X-X-X-cys-X_n-cys-X-X-X-cys C-terminal; interestingly amino acids between these cysteine residues are largely dissimilar. Significantly perhaps, this domain may be contained within exon 7 of the *A. nidulans* protein. The cysteine residues apart, there is therefore little obvious homology between fungal and spinach primary sequences (Fig. 24.3A) except in the short region succeeding the cysteine residues where there are five identities (Table 24.3). As expected there is no similarity observed between haem-containing enzymes such as yeast flavocytochrome b_2 and the Fe_4S_4/sirohaem domain of *A. nidulans* nitrite reductase (E. I. Campbell and J. R. Kinghorn, unpublished).

Comparison of the *A. nidulans* NiR with that of *E. coli* by dot matrix analysis shows that there is striking homology between these systems (Fig. 24.3B) which implies similarity of structure or function. Computer alignment of the two sequences (Table 24.3) reveals that there are an extensive number of shared motifs, including the second (C-terminal) cysteine-containing sequence i.e. Gly-Cys-X-Arg-Glu-Cys-X-Glu-Ala. In contrast, however, the earlier (N-terminal) cysteine-containing sequence is not observed in the *E. coli* NiR sequence (Table 24.3). This might support the notion that the *E. coli* NiR is of the Fe_2S_2 type (Cammack *et al.* 1982) as opposed to the more common Fe_4S_4 form found in fungi and higher plants (Chapter 17).

FAD domain

A major difference between fungal and plant enzyme is that the plant polypeptide is much smaller, 566 amino acids, while the fungal enzyme is 1104 amino acids long. This additional stretch of sequence in the lower organism may be due to the presence of a FAD domain. With respect to the identity of the FAD domain, no extended similarity is observed when the NiR sequence of *A. nidulans* is compared by dot matrix analysis with the sequences of certain other flavoproteins (E. I. Campbell and J. R. Kinghorn, unpublished) such as human glutathione reductase (Krauth-Siegel *et al.* 1982), pig D-amino acid oxidase (Ronchi *et al.* 1982), *Pseudomonas fluorescens* p-hydroxybenzoate hydroxylase (Weijer *et al.* 1982), dogfish lactate dehydrogenase (Eventoff *et al.* 1977), *E. coli* respiratory lactate dehydrogenase (Campbell *et al.* 1984), and *Clostridium* MP flavodoxin (Tanaka *et al.* 1974). Some very short runs of amino acid matches were observed; for

Table 24.3. Alignment of the deduced amino acid sequences of nitrite reductase from *A. nidulans* (Johnstone *et al.* 1988) *E. coli* (J. A. Cole, personal communication), and *S. oleracea* (E. Back *et al.* 1987; Chapter 18; Back, personal communication) as well as sulphite reductase of *S. typhimurium* (Chapter 17)

```
                                        * ***    *** *    *      ** *  *** *** ** ***** *    *
E. coli                                 MSKVRLAIIGNGMVGHRFIEDLLD-KSDAANFDITVFCEEPRIAYDRVHLSSYFSHHTAE    59
A. nidulans    MPLLDGPRNGETVTASAHNGIPIIDGVDPSTLRGDIDQDPNRRQKIVVVGLGMVAVAFIEKLVKLDSERRKYDIVVIGEEPHIAYNRVGLSSYFEHRKIE   100
                                                           intron 1


               * *      *          *       * *      * *  * ** *  ****     *      * *   ****** **       *      * ***
E. coli        ELSLVREGFYE---KHGIKVLVGERAITINRQEKVIHSSAGRTVFYDKLIMATGSYPWIP-PIKGSDTQDCFVYRTIEDLNAIESCA--RRSKRGAVVGG   153
A. nidulans    DLYLNPKEWYGSFKDRSFDYYLNTRVTDVFPQHKTVKTSTGDIVSYDILVLATGSDAVLPTSTPGHDAKGIFVYRTISDLERLMEFAANHKGQTGVTVGG   200
                          intron 2


               * ******* *    *        *      * **      **  *   *   *         * ** **     *      ****** *
E. coli        GPLGLEAAGALKNLGI--ETHVIESAPMLMAERLDRMDDERLRRKIESMGVRVHTSKNTLGIVQEGVEARKTMRFADGSELEVDFIVFSTGIRPRDKLAT   251
A. nidulans    GLLGLEAAKAMTDLEDFGSVKLIDRNKWVLARQLDGDAGSLVTKKIRDLGLEVLHEKRVAKIHTDDDNNVTGILFEDGQELDCCCICFAIGIRPRDELGG   300
                                                                                               intron 3


                  *    * *** **    *  **   *********** *  **  *** ** *          *   *** ********   **
E. coli        QCGLDVAPRGGIVINDSCQTSDPDIYAIGECASWNNRVFGLVAPGYKMAQVAVDHILGSENA---FEGADLSAKLKLLGVDVGGIGD------------   335
A. nidulans    STGIQCAKRGGFVIDESLRTSVNDIYAIGECASWENQTFGIIAPGIEMADVLSFNLTNPDKEPKRFNRPDLSTKLKLLGVDVASFGDFFADRDGPKFLPG   400


                               *         * *    ***      * * ***    *** **  * * *      *    *    ** * **        * * **
E. coli        -------------AHGRTPGARSYVYLDESKEIYKRLIVSEDNKTLLGAVLVGDTSDYGNLLQLVLNAIELPENPDSLILPAHSGSGKPSIGVDKLPDSA   422
A. nidulans    QRPSAESIGAADPNREEEPQVKALTYRDPFGGVYKKYLFTMDGKYLLGGMMIGDTKDYVKLNQMVKSQKPLEVPPSEFILGAQSGGEE---NADDLDDST   497
                                                                                                        intron 4


               *****  *****        * * *  * *********** ***       *      * ** **** * *** *  *      **        **
E. coli        QICSCFDVTKGDLIAAINKA-CHTVAALKAETKAGTGCGGCIPLVTQVLNAELAKQGIEVNNNLCEHFAYSRQELFHLIRVEGIKTFEELLAKHGK---G   518
A. nidulans    QICSCHNVTKGDVVESVKSGTCKTIADVKSCTKAGTGCGGCMPLVQSIFNKTMLDMGQEVSNNLCVHIPYSRADLYNVIAIRQLRTFDDVMKSAGKCPDS   597
                                                              intron 5
```

```
               ***  ***    *  *  *   *    *   **  ** ******  ** ** ** ******   *  *  *  *   *** **** **   *** * **  **
E. coli      YGCEVCKPTVGSLLASCWNEYILKPEHTPLQDSNDNFLANIQKDGTYSVIPRSPGGEITPEGLMAVGRIAREFNLYTKITGSQRLAMFGAQKDDLPEIWR  618
A. nidulans  LGCEICKPAIASILSSLFNPHLMDKEYHELQETNDRFLANIQRNGTFSVVPRVPGGEITADKLIAIGQVAKKYNLYCKITGGQRIDMFGARKQDLLDIWT  697


                                   ◆       ◆     *       * ■ *** ** *** *  **■◆ **◆ **  *■ ■ ****        ** *
E. coli      QRLKPASKPVMPMRKHCVWRKPAWVALVPLRRWRQLG----LGVEL-ENRYKGIRTPHKMKFGVSGCTRECSEARGKDVGIIATE--------KGWNLY   704
S. oleracea  LLNEPLLKERYSPEPPILMKGLV-ACTGSQFCGQAIIETKARALKVTEEVQRLVSVTRPVRMHWTGCPNSCGQVQVADIGFMGCMTRDENGKPCEGADVF
S. typhimurium  sulphite reductase       ACVSFPTCPLAMAE                   MRVTGCPNGCG
A. nidulans  ELVDAGMESG---HRYAKSLRTVKSCVGTTWCRFGVGDSVGMAIRL-EQRYKSIRAPHKFKGAVSGCVRECAEAQNKDFGLIATE--------KGFNIF   784
                                 ▲                                              ▲
                              intron 6                                      intron 7


              ■ ■**■  ***   *** *    *  *  ***   ********* ***   ** * *** *** *    ****  *  *  *  *    ** ****   *
E. coli      VCGNGGMKPRHADLLAADIDRETLIKYLDRFMMFYIRTADKLTRTAPWLENLEGGIDYLKAVIIDDKLGLNAHLEEEMARLREAVLCEWTETVNTPSAQT   804
S. oleracea  VGGRIGSDSHLGDIYKKAVPCKDLVPVVAEILINQFGAVPREREEAE
A. nidulans  VGGNGGAKPRHSELLAKDVPPEEVIPILDRYVIFYIRTADKLQRTARWLESLPGGIEYLKDVVLNDKLGIAAEMERQMQELVDSYFCEWTETVRNPKRRK   884


                *     *  *    **   *  **** **
E. coli      RFKHFINSDKRDPNVQMVPEREQHRPATPVNMSQJ 839
A. nidulans  YFQQFANTDETVENVEIVKEREQVRPTYWPKDGANEDFKGHQWSSLSWQPVIKADYFSDGPPAISSANIKRGDTQLAIFKVKGKYYATQQMCPHKRTFVL   984


A. nidulans  SDGLIGDDDNGKYWVSCPYHKRNFELNGEQAGRCQNDEAMNIATFPVEEREDGWIYMKLPPVEELDSVLGTEKWKVKKGEAVDPFEAYDKKYSGMKGKRA  1084


A. nidulans  GAKGIEGSKPTRSPSNTIDW   1104
```

Symbols: *Denotes identical amino acids in NiR proteins from *A. nidulans* and *E. coli*. Where these matched amino acids are additionally shared with *S. oleraceae* a box (■) is used. The cysteine residues that are thought responsible for binding the iron sulphur centre and sirohaem are denoted by a diamond (◆).

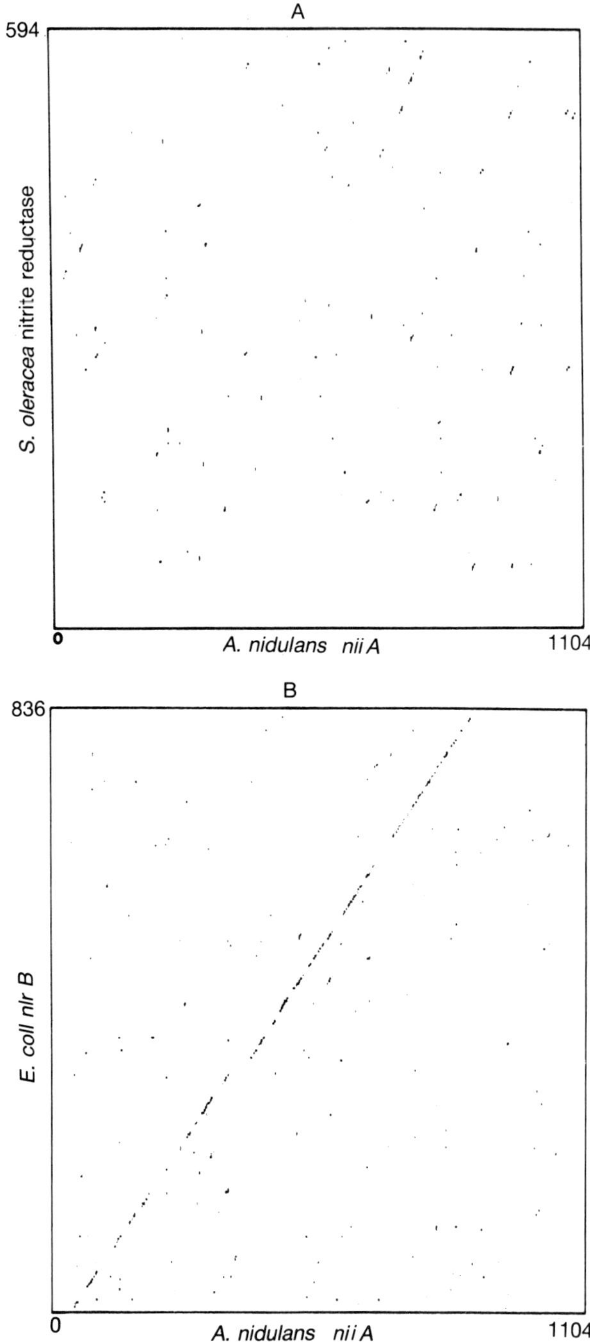

Fig. 24.3. Dot matrix comparisons between the nitrite reductase amino acid sequence of *Aspergillus nidulans* and the of *S. oleracea* (A) and *E. coli* (*narB*) (B) nitrite reductase.

example, human glutathione reductase gave five amino acid matches i.e. Tyr-Asp-X-X-Val-Ile-Gly with the *A. nidulans* NiR sequence at residues 70–77 (*A. nidulans* sequence). However, it is unlikely that any functional significance can be attached to this since pentapeptide homologies or less can occur in very different proteins (Kabsch and Sander 1984). None of the above proteins show extended matches with each other and no clear similarity is observed between NiR of *A. nidulans* and FAD-containing human or bovine NAD cytochrome b_5 reductase (E. I. Campbell and J. R. Kinghorn, unpublished) or indeed with *A. nidulans* NR itself (Fig. 24.4).

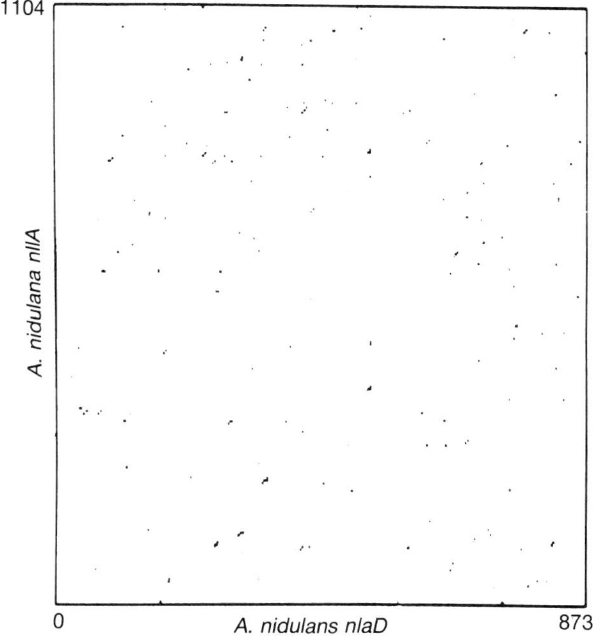

Fig. 24.4. Dot matrix comparison between the nitrate reductase and nitrite reductase amino acid sequence of *Aspergillus nidulans*.

In summary, as no meaningful homology has been demonstrated when comparing *A. nidulans* NiR with various FAD-containing proteins of unrelated physiological function, we were unable to identify a FAD-containing domain.

Codon preference

A summary of codon usage is shown in Table 24.4 (NR) and Table 24.5 (NiR). From these data, it would appear that assimilatory NR and NiR proteins are not especially biased in their codon usage.

Table 24.4. Codon usage in *A. nidulans* (A) and *A. thaliana* (B) assimilatory nitrate reductase proteins

		A	B			A	B			A	B			A	B
Phe	TTT	2 (8)	13(37)	Ser	TCT	6(10)	11(18)	Tyr	TAT	8(36)	12(35)	Cys	TGT	3(33)	6(38)
Phe	TTC	23(92)	22(63)	Ser	TCC	15(25)	20(33)	Tyr	TAC	14(64)	22(65)	Cys	TGC	6(67)	10(63)
Leu	TTA	1 (1)	7(10)	Ser	TCA	7(11)	8(13)	Ter	TAA	0	0	Ter	TGA	0	0
Leu	TTG	10(13)	19(28)	Ser	TCG	14(23)	11(18)	Ter	TAG	1	1	Trp	TGG	19	20
Leu	CTT	11(14)	12(18)	Pro	CCT	15(26)	16(29)	His	CAT	9(50)	10(36)	Arg	CGT	4 (8)	14(35)
Leu	CTC	26(33)	19(28)	Pro	CCC	18(32)	8(14)	His	CAC	9(50)	18(64)	Arg	CGC	12(24)	4(10)
Leu	CTA	13(17)	7(10)	Pro	CCA	13(23)	17(30)	Gln	CAA	11(48)	12(63)	Arg	CGA	4 (8)	4(10)
Leu	CTG	17(22)	3 (4)	Pro	CCG	11(19)	15(27)	Gln	CAG	12(52)	7(37)	Arg	CCG	11(22)	8(20)
Ile	ATT	17(34)	22(44)	Thr	ACT	13(21)	12(24)	Asn	AAT	8(27)	10(23)	Ser	AGT	7(11)	4 (7)
Ile	ATC	29(58)	21(42)	Thr	ACC	16(25)	15(31)	Asn	AAC	22(73)	33(77)	Ser	AGC	12(20)	7(11)
Ile	ATA	4 (8)	7(14)	Thr	ACA	20(32)	9(18)	Lys	AAA	23(41)	23(38)	Arg	AGA	8(16)	7(18)
Met	ATG	26	28	Thr	ACG	14(22)	13(27)	Lys	ACG	33(59)	37(62)	Arg	AGG	11(22)	3 (8)
Val	GTT	9(19)	27(39)	Ala	GCT	10(19)	20(34)	Asp	GAT	28(54)	24(51)	Gly	GGT	7(10)	27(39)
Val	GTC	28(58)	18(26)	Ala	GCC	25(46)	19(33)	Asp	GAC	24(46)	23(49)	Gly	GGC	29(43)	13(19)
Val	GTA	4 (8)	7(10)	Ala	GCA	12(22)	7(12)	Glu	GAA	28(43)	23(34)	Gly	GGA	16(24)	19(27)
Val	GTG	7(15)	17(25)	Ala	GCG	7(13)	12(21)	Glu	GAG	37(57)	44(66)	Glu	GGG	15(22)	11(16)

Numbers without bracket represent number of codons used whilst numbers inside bracket is frequency expressed as a percentage.

Table 24.5. Codon usage in *A. nidulans* (A) and *S. oleracea* assimilatory nitrite reductase proteins

		A	B			A	B			A	B			A	B
Phe	TTT	16(37)	11(58)	Ser	TCT	5 (8)	4(11)	Tyr	TAT	7(19)	3(33)	Cys	TGT	14(54)	9(69)
Phe	TTC	27(63)	8(42)	Ser	TCC	10(16)	5(13)	Tyr	TAC	29(81)	6(67)	Cys	TGC	12(46)	4(31)
Leu	TTA	5 (6)	5(10)	Ser	TCA	11(18)	13(34)	Ter	TAA	0	0	Ter	TGA	0	0
Leu	TTG	10(12)	9(18)	Ser	TCG	13(21)	5(13)	Ter	TAG	1	1	Trp	TGG	14	8
Leu	CTT	21(25)	15(29)	Pro	CCT	16(31)	17(50)	His	CAT	7(39)	8(67)	Arg	CGT	9(15)	7(17)
Leu	CTC	23(28)	4 (8)	Pro	CCC	9(18)	2 (6)	His	CAC	11(61)	4(33)	Arg	CGC	9(15)	1 (2)
Leu	CTA	7 (8)	8(16)	Pro	CCA	11(22)	12(35)	Gln	CAA	19(51)	13(65)	Arg	CGA	15(25)	3 (7)
Leu	CTG	17(20)	10(20)	Pro	CCG	15(29)	3 (9)	Gln	CAG	18(49)	7(35)	Arg	CCG	10(17)	2 (5)
Ile	ATT	36(54)	17(49)	Thr	ACT	17(30)	3(14)	Asn	AAT	15(35)	12(41)	Ser	AGT	8(13)	6(16)
Ile	ATC	26(39)	12(34)	Thr	ACC	21(37)	6(27)	Asn	AAC	28(65)	17(59)	Ser	AGC	15(24)	5(13)
Ile	ATA	5 (7)	6(17)	Thr	ACA	12(21)	9(41)	Lys	AAA	19(23)	12(34)	Arg	AGA	13(22)	15(36)
Met	ATG	23	17	Thr	ACG	7(12)	4(18)	Lys	ACG	64(77)	23(66)	Arg	AGG	3 (5)	14(33)
Val	GTT	23(30)	20(40)	Ala	GCT	20(29)	15(42)	Asp	GAT	33(39)	20(65)	Gly	GGT	34(34)	8(18)
Val	GTC	26(34)	4 (8)	Ala	GCC	32(46)	4(11)	Asp	GAC	52(61)	11(35)	Gly	GGC	35(35)	2 (5)
Val	GTA	7 (9)	11(22)	Ala	GCA	11(16)	12(33)	Glu	GAA	30(42)	19(39)	Gly	GGA	23(23)	18(41)
Val	GTG	21(27)	15(30)	Ala	GCG	6 (9)	5(14)	Glu	GAG	41(58)	30(61)	Glu	GGG	8 (8)	16(36)

Numbers without bracket represent number of codons used whilst numbers inside bracket is frequency expressed as a percentage.

Acknowledgements

We wish to thank Professor M. Caboche, Drs E. Back, J. Cole, and J. C. Wootton, for making their unpublished data available to us. Receipt of the *A. thaliana* manuscript from Dr N. Crawford during the later stages of preparation of this chapter is acknowledged. Thanks go to Dr A. Coulson (University of Edinburgh) and Ms C. Toogood (University of St Andrews) for advice on computing aspects. Support from the European Economic Community (Grant BAP-0039-UK) is acknowledged.

References

Calza, R., Hutter, E., Vincentre, M., Rouze, P., Galangue, F., Vaucheret, H., Chevel, I., Mezer, C., Kronenberger J., and Caboche, M. (1987). Cloning of DNA fragments complemetary to tobacco nitrate reductase mRNA and encoding epitopes common to the nitrate reductases from higher plants. *Molecular and General Genetics* **209**, 552–62.

Cammack, R., Jackson, R. H., Cornish-Bowden, A., and Cole, J. A. (1982). Electron spin resonance studies of the NADH-dependent nitrite reductase from *Escherichia coli* K12. *Biochemical Journal* **107**, 333–9.

Campbell, H. D., Rogers, B. L., and Young, I. G. (1984). Nucleotide structure of the respiratory D-lactate dehydrogenase gene of *Escherichia coli*. *European Journal of Biochemistry* **144**, 367–73.

Crawford, N. G., Smith, M., Bellissimo, D., and Davis, R. W. (1989). Sequence and nitrate regulation of the *Arabidopsis thaliana* mRNA encoding nitrate reductase: a metalloflavo-protein with three functional domains. *Proceeding of the National Academy of Science, USA* (in press).

Devereux, J. R., Haeherli, P. and Smithies, O. (1984). A comprehensive set of sequence analysis programs for VAX. *Nucleic Acids Research* **12**, 387–95.

Eventoff, W., Rossmann, M. G., Taylor, S. S., Torff, H.-J., Meyer, H., Keil, W., and Kiltz, H. H. (1977). *Proceedings of the National Academy of Science, USA* **74**, 2677–81.

Gilbert, W. (1985). Genes-in-pieces, revisited. *Science* **228**, 823–4.

Johnstone, I. L., Greaves, P., Gurr, S. J., Kinghorn, J. R., and Innes, M. (1989). Cloning and characterisation of the *Aspergillus nidulans* gene cluster for nitrate assimilation. *EMBO Journal* (submitted).

Kabsch, W. and Sander, C. (1984). On the use of sequence homologies to predict protein structure: identical pentapeptides can have completely different conformations. *Proceedings of the National Academy of Science, USA* **81**, 1075–8.

Keith, R. P., Riley, M. A., Kreitman, M., Lewontin, R. C., Curtis, D., and Chambers, G. (1987). Sequence of the structural gene for xanthine dehydrogenase (rosy locus) in *Drosophila melanogaster*. *Genetics* **116**, 67–73.

Krauth-Siegel, R. L., Blatterspiel, R., Saleh, M., Schiltz, E., Schirmer, R. H., and Untucht-Grau, R. (1982). Glutathione reductase from human erythrocytes. *European Journal of Biochemistry* **121**, 259–67.

Ozols, J., Korza, G., Heinemann, F. S., Hediger, M. A., and Strittmatter, P. (1985). Complete amino acid sequence of steer liver microsomal NADH-cytochrome b_5 reductase. *Journal of Biological Chemistry* **260**, 11953–61.

Ronchi, S., Minchiotti, L., Galliano, M., Curtis, B., Swenson, R. P., Williams, C. H., Jr., and Massey, V. (1982). The primary structure of D-amino acid oxidase from pig kidney. *Journal of Biological Chemistry* **257**, 8824–34.

Tanaka, M., Haniu, M., Yasunobu, K. T., and Mayhew, S. G. (1974). The amino acid sequence of the *Clostridium* MP flavodoxin. *Journal of Biological Chemistry* **249**, 4393–6.

Staden, R. (1982). An interactive graphics programme for comparing and aligning nucleic acid and amino acid sequences. *Nucleic Acids Research* **10**, 2951–61.

Staden, R. (1984). Graphic methods to determine the function of nucleic acid sequences. *Nucleic Acids Research* **12**, 521–38.

Weijer, W. J., Hofsteenge, J., Vereijken, J. M., Jekel, P. A., and Beintema, J. J. (1982). Primary structure of *p*-hydroxybenzyoate hydroxylase from *Pseudomonas fluorescens*. *Biochimica et Biophysica Acta* **704**, 385–8.

Yabisui, T., Miyata, T., Iwanaga, S., Tamura, M., Yosheda, S., Takeshita, M., and Nakajima, H. (1984). Amino acid sequence of NADH-cytochrome b_5 reductase of human erythrocytes. *Journal of Biochemistry* **96**, 579–82.

Index

algal nitrate assimilation
  genetics 101–8
  regulatory aspects 108–16
algal nitrate reductase
  effectors 94, 95
  functional domains 91–3
  functional size 90, 91
  genetics 101–8
  kinetic properties 93, 94
  prosthetic groups 89–90
  regulation 111–16
  structure–function relationships 88–97
  thermodynamic properties 95–7
amino acid sequence
  nitrate reductase 191, 385–94
  nitrite reductase 287–90, 394–9
ammonium repression
  nitrate reductase 45, 63, 113, 114
  nitrate uptake 42
  nitrite reductase 45, 116
ammonium uptake
  cyanobacterial 42–3
  light 110
  nitrogen starvation 110
Anabaena cycadeae
  nitrate uptake 109
Anabaena cylindrica
  nitrate reductase 43
  nitrite reductase 44
Anabaena sp. 7119
  nitrite reductase 44, 45
Anacystis nidulans
  nitrate reductase 43, 46, 47, 88
  nitrite reductase 44, 45
Ankistrodesmus braunii
  nitrate reductase 91, 93, 102, 111, 113
  nitrate uptake 110
Arabidopsis thaliana
  chlorate-hypersensitive mutants 378–81
  chlorate-resistant mutants 16
  molecular analysis of nitrate
      regulation 328–37
  nitrate reductase 132, 328–37, 388–92
  nitrate uptake 4, 9, 16, 373
  nitrate uptake mutants 16, 17, 374
  regulatory mutants 376–9
Aspergillus amstelodami
  chlorate-resistant mutants 348
  nitrate assimilation genes 349
Aspergillus nidulans
  areA gene 71, 72, 304–9

chlorate-resistant mutants 70
cnx genes 70, 71, 299–301
crn gene 70, 71, 73
crnA niiA niaD gene cluster 69–87
  homologous transformation 352
  molybdenum cofactor 213, 299–301
  nirA gene 71–3, 304–6
  nitrate assimilation mutants 348, 349
  nitrate reductase 70, 71, 93, 155, 384–94
  nit-2 gene expression 321, 322
  nitrite reductase 70, 71, 394–401
  nitrogen metabolite repression 304–8
  regulation of nitrate assimilation 299–313
Aspergillus niger
  cloning of nitrate reductase structural
      gene 352
  homologous transformation 352
  nitrate assimilation genes 349
Aspergillus oryzae
  cloning of nitrate reductase structural
      gene 352
  nitrate assimilation genes 349
autoregulation (autogenous control) 73, 115,
      182, 207, 316, 319, 320

Barley (*Hordeum vulgare*)
  molybdenum cofactor 144
  molybdenum cofactor mutants 198, 202–6,
      212, 213
  NADPH-nitrate reductase 200–2, 206, 207
  nitrate reductase 126, 127, 132, 146, 157,
      197–208
  nitrate reductase structural gene 198–202,
      204
  nitrate reductase mutants 157, 197–207
  nitrate uptake 7, 9
  nitrate uptake mutants 17–23
  nitrite reductase 244, 247, 248, 250, 251,
      253–56
  nitrite reduction mutants 253–6
biotechnology
  fungal 341–63
  higher plant 364–81

Candida nitratophila
  nitrate assimilation 53, 58–63
  nitrate reductase 54–7, 59–63
  nitrate uptake 53

406   *Index*

Candida nitratophila—cont.
  nitrite reductase 57
  nitrite uptake 54
Candida utilis
  nitrate assimilation 52, 58–60, 63
  nitrate reductase 54, 59, 60
  nitrite reductase 54
Chara corallina
  nitrate uptake 108, 110
Cephalosporium acremonium
  nitrate assimilation genes 349
Chenopodium album
  nitrite reductase 245
Chlamydomonas reinhardtii
  nitrate reductase 102, 111, 112, 114–16
  nitrite reductase 113, 116, 266, 267
  mutants 103–8, 111
chl mutants 28–39, 73
chlorate-hypersensitive mutants 378, 379
chlorate resistance 16, 17, 29–35, 45, 46, 64,
    70, 73, 75, 102, 166, 167, 169–70, 187,
    237, 239, 348, 364, 374
Chlorella fusca
  nitrite reductase 102
Chlorella sorokiniana
  mutants 102
  nitrate reductase 112
Chlorella vulgaris
  molybdenum cofactor 89
  molybdopterin 89
  NADH-dehydrogenase 89
  nitrate reductase 55, 88–100, 102, 112,
    113, 156
Citrobacter freundii
  molybdenum cofactor mutants 212
Cryptococcus
  nitrate assimilation 51
Cyanidium caldarium
  nitrate reductase 112
  nitrate uptake 110
cyanobacterial nitrate assimilation
  biochemistry 41–4
  genetics 45–6
  regulatory aspects 40–7
cytochrome *c* reductase 76, 77, 89, 93, 101,
    131, 170, 198–203

Debaromyces hansenii
  nitrite assimilation 51, 52
Debaromyces pseudopolymorpha
  nitrite assimilation 51, 52
Drosophila melanogaster
  molybdenum cofactor mutants 212
Dunaliella
  nitrate reductase mRNA 113
Dunaliella tertiolecta
  nitrate assimilation mutants 102

EPR spectroscopy
  nitrate reductase 93
  nitrite reductase 270
  sulphite reductase 270
Escherichia coli
  biosynthesis of nitrate reductase 33–5
  chlorate-resistant mutants 28–39
  cytochrome c_{552} and the electrogenic
    formate–nitrite oxidoreductase
    complex 235–7
  FNR protein 237, 238
  genetics 233, 237–9
  molybdenum cofactor 30–9, 45, 212
  NADH-dependent nitrite reductase 234, 235
  NADPH-dependent sulphite reductase 234
  nitrate reductase genetics 28–9
  nitrate respiration 28
  nitrite reductase 264, 394–9
  nitrite reductase genetics 237–9
  physiology of nitrite reduction 229–34
  protein FA 32–34
  protein PA 32–35
  respiratory nitrate reductase 27–39, 155,
    385, 387
  structure of nitrate reductase 27, 28
Eudorina elegans
  mutants 102

FNR protein 237, 238
Formate–nitrite oxidoreductase
  complex 235–7
Fulvia fulva
  nitrate assimilation genes 349
Fungal biotechnology
  heterologous expression of *nia*D gene 354–5
  nitrate assimilation 341–63
  nitrate assimilation genes 349, 351–2
  transformation 344–7
Fusarium oxysporum
  chlorate resistant mutants 348

Galium aparine
  nitrite reductase 245
Gibberella fujikuroi
  chlorate resistant mutants 348
  nitrate assimilation genes 349

Hansenula anomala
  nitrate assimilation 52, 59
  nitrate reductase 54–60
  nitrogen starvation 60
Hansenula wingei
  chlorate-resistant mutants 64
Haplopappus gracilis
  nitrite reduction mutants 252

higher plant nitrate assimilation
  biotechnology 364–81
higher plant nitrate reductase
  biochemical and somatic cell
     genetics 166–86
  immunology 155–65
  molecular genetics 186–211
  nitrate regulation 328–37
  structure and regulation 125–54
higher plant nitrate uptake
  molecular aspects 3–14
  genetics 15–24
higher plant nitrite reductase
  molecular cloning 284–96
  molecular and genetic aspects 244–62
  structure and function 263–83
Hydrodictyon africanum
  nitrate uptake 110
Hyoscyamus muticus
  molybdenum cofactor mutants 212

immunology
  nitrate reductase 76, 126–8, 142, 143,
    145, 146, 155–65, 171–3, 186, 187,
    190, 199–203, 328–32
  nitrite reductase 248, 251, 285, 286

Klebsiella aerogenes
  molybdenum cofactor mutants 212, 213

lacZ gene fusions 31, 82
Lemna minor
  nitrate uptake 5, 9
  nitrite reductase 250
light
  ammonium uptake 110
  molybdenum cofactor 144
  nitrate reductase 113, 141–3, 145, 147,
    190–3
  nitrate uptake 110, 113
  nitrite reductase 245, 250–2, 285, 291
  nitrite uptake 110

maize (corn)
  molybdenum cofactor 144
  nitrate reductase 126–32, 138, 140, 141,
    143, 145–7, 161, 162
  nitrate uptake 5, 6, 9
  nitrite reductase 245, 247, 250, 251
man
  molybdenum cofactor mutants 213
molecular cloning
  chl genes 30
  molybdate transport gene 31

nitrate reductase structural gene 77–82,
  147, 193, 197, 322–4, 331–5, 351,
  352
nitrate uptake structural gene 77–9
nitrite reductase structural gene 77–9,
  285–7
regulatory genes 303, 306, 319–22
molybdate transport 31, 34
molybdenum cofactor
  assay 144, 177, 213–15
  biosynthesis 33–5, 144, 176–80, 182, 183
  chemistry and biology 212–26
  genetics 28–35, 64, 102–8, 167–70,
    176–80, 182, 183, 187, 188, 198,
    202–5, 212, 213, 299–302, 314, 315,
    348, 364–7
  light 144
  regulation 144, 182, 183
  structure 218, 219
molybdopterin 215–19
monoclonal antibodies
  nitrate reductase 157–62, 186, 187
mRNA
  crnA 81
  in vitro translation 60
  niaD 81
  niiA 81
  nitrate reductase 60–3, 113, 114, 115,
    147, 188–94, 204, 205, 207, 317,
    322, 332–5
  nitrite reductase 251, 285, 291, 292
  transcriptional control 60
mustard
  nitrite reductase 250, 251
mutants
  chlorate-resistant 29–33, 46, 64, 70–3,
    102, 167, 169, 187, 237, 239, 348
  molybdenum cofactor 64, 103–6, 167–9,
    176–80, 182, 187, 188, 198, 202,
    212, 213, 299–301
  nitrate assimilation 64, 70, 102–8
  nitrate reductase 70–5, 103–6, 137, 167,
    170–6, 181, 187, 188, 198–202, 314,
    315, 364–72
  nitrate reduction 45, 46, 167, 168
  nitrate uptake 16–23, 70–5, 374–6
  nitrite reductase 70–5
  nitrite reduction 252–6
  regulatory 73, 106, 302–4, 308, 309,
    319–22, 373–9
  revertant studies 174, 366–8

Neurospora crassa
  genetic studies 314, 315
  nitrate assimilation 314–27
  nitrate assimilation genes 315–6
  nitrate assimilation mutants 348, 349

Neurospora crassa—cont.
  molybdopterin 217, 218
  nitrogen catabolite repression and nitrate
      induction 315–19
  nit-1 reconstitution assay 213, 214
  nit-2 gene 318–22
  nit-3 gene 322–4
  nit-4 gene 316, 317, 319
  nmr-1 gene 315, 316, 320–6
  nitrate reductase 55, 156
  nitrate reductase regulation 322–4
Nicotiana, see tobacco
nitrate reductase
  amino acid sequence homologies 132,
      385–93
  ammonium repression 45, 63, 113, 114
  autoregulation 73, 115, 182, 195, 207,
      317, 319, 320
  cDNA 147, 188–90, 197, 202, 331, 332,
      333
  domains 91–3, 130–2
  effectors 94
  EPR potentiometry 95
  EPR spectroscopy 93
  functional size 90
  gene dosage effects 181, 182
  genetics 28–35, 64, 70–76, 102–8, 137,
      170–6, 187, 188, 198–204, 314, 315,
      364–73
  immunology 76, 126–28, 142, 143, 145,
      146, 155–65, 171–3, 187, 188,
      199–202, 329–32
  internal electron transfer 136
  kinetics 54–6, 93, 94, 134, 135
  light 113, 141–3, 145, 147, 190–3
  limited proteolysis 92
  mRNA 60–3, 113, 114, 115, 147, 188–94,
      204, 205, 207, 317, 322–4, 332–5
  partial reactions 133–5
  prosthetic groups 54, 89–90, 129–30, 138,
      170
  purification 43–4, 54–7, 76, 88, 126–8,
      160
  radiation inactivation analysis 91
  regulation 45, 58–63, 73, 81, 111–16,
      141–7, 180–3, 190–3, 302–9, 322–5,
      328–37, 376–9
  structural gene 170–6, 187, 188, 190–4,
      198–202, 322–4, 351, 352, 354, 355
  structure 27, 28, 54, 76, 89–93, 102,
      128–32
  structure–function model 92
  structure–function relationships 88–100
  thermodynamic properties 95–7
nitrate respiration 28
nitrate uptake
  ammonium 41, 42, 108, 109

crnA gene 70, 71, 73
  distribution of plasmalemma transporters 5,
      6
  effects of nitrate shortage 7, 8
  energy requirement 110
  genetics 15–24, 70–5, 374–6
  intracellular transmembrane transport 6, 7
  kinetics 3, 4, 52
  light 41, 110, 113
  mechanism 4, 5
  molecular aspects 3–14
  nitrogen starvation 110
  regulation 8, 9, 41, 42, 58, 108–11
nitrite reductase
  amino acid composition 247
  amino acid sequence 287–91
  amino acid sequence homologies 394–9
  catalytic properties 265, 266
  cDNA 249, 251, 285–7
  domains 395–9
  EPR spectroscopy 270
  formate–nitrite oxidoreductase
      complex 235–7
  genetics 70–6, 237–9, 252–6, 314, 315
  hexahaem 264, 278
  immunology 248, 251, 285, 286
  light 245, 250–2, 285, 291
  mechanism of catalysis 276
  molecular cloning 77, 284–96
  mRNA 251, 285, 291–3
  NADH-dependent 234, 235
  prosthetic groups 235, 245, 264, 265,
      267–71
  purification 44, 102, 244, 245, 266, 285,
      286
  pyridine nucleotide specificity 56, 57
  regulation 45, 59–63, 73, 112, 116, 237,
      239, 248–52, 291–3
  sirohaem (4Fe–4S) 264–78
  structural gene 70, 287
  structure and function 263–83
  subcellular location 245
  transit peptide 249, 290–1
nitrite uptake
  ammonium 42, 109
  light 110
  mechanism 42, 53, 54
  nitrogen starvation 110
nitrogen metabolite (catabolite) repression
  nitrate assimilation 304–8, 315–19
  nitrate reductase 73
  nitrite reductase 73
nitrogen starvation 60, 61, 110
Nostoc muscorum
  chlorate-resistant mutants 45, 46
  molybdenum cofactor mutants 213
  nitrate reduction mutants 45, 46

oats
  nitrate uptake 5

Paul's Scarlet rose
  nitrite reductase 244
pea
  nitrite reductase 244, 245, 248, 251
Penicillium chrysogenum
  chlorate-resistant mutants 348
  nitrate assimilation genes 349
pentose phosphate pathway 63
Phaeodactylum tricornutum
  nitrate uptake 108
Phaseolus angularis
  nitrite reductase 245, 248
Phaseolus vulgaris
  nitrate uptake 4
physical map
  crnA, niiA, niaD gene cluster 75
  nitrate reductase of tobacco 194
phytochrome 142, 251, 252
Pisum
  nitrate uptake 7
plant biotechnology
  gene transfer 371–3
  nitrate assimilation 364–81
  somatic hybridization 369–71
Platymonas striata
  nitrite reductase 116
Plectonema boryanum
  nitrate reductase 43, 46, 47
Porphyra yezoensis
  nitrite reductase 102
protein FA 32–34
protein PA 32–35
Proteus vulgaris
  molybdenum cofactor mutants 213
Pseudomonas aeruginosa
  molybdenum cofactor mutants 213
  nitrate assimilation 232
  respiration 232

regulatory aspects
  ammonium repression 42, 45, 63, 113,
    114, 116
  ammonium uptake 42, 43, 110
  areA gene 71, 73, 304–309
  autoregulation 73, 115, 182, 207, 317,
    319, 320
  cnx genes 176–180, 301, 302
  molybdenum cofactor synthesis 144, 182
  nitrate assimilation 40–50, 57–63, 108–16,
    299–313, 315–26
  nitrate reductase 45, 57–63, 73, 111–16,
    141–7, 180–2, 190–3, 302–9, 322–4,
    328–37, 374–8
  nitrate uptake 8, 9, 41, 42, 58, 108–11

nitrite reductase 45, 59–63, 73, 112, 116,
  237, 239, 248–52, 291–3
  nirA gene 71–3, 302, 303, 308
  nit-2 gene 318–22
  nit-4 gene 316, 317, 319, 322
  nitrite uptake 42, 108–10,
  nitrogen metabolite (catabolite)
    repression 73, 304–8, 315–19
  nmr-1 gene 315, 316, 320–6
regulatory genes
  areA gene 71–3, 304–9
  nirA gene 71–3, 302–4, 308
  nit-2 gene 318–22
  nit-4 gene 316, 317, 319, 322
  nmr-1 gene 315, 316, 320–6
respiratory nitrate reductase 27–39, 155,
  385, 387
Rhodotorula glutinis
  nitrate assimilation mutants 64
  nitrate reductase 54, 55, 58
rice
  nitrate reductase 140

Saccharomyces cerevisiae
  nitrate assimilation 51
Salmonella typhimurium
  NADPH-sulphite reductase 395, 397
  molybdenum cofactor mutants 212
Septoria nodorum
  chlorate-resistant mutants 348
  nitrate assimilation genes 349
sequence homologies
  nitrate reductase 191, 385–94
  nitrite reductase 394–9
sirohaem 234, 235, 244, 267–74, 276
  redox properties 273, 274
  structure 264, 265
sirohaem (4Fe–4S) active centre
  nitrite reductase 269–72
  possible ligands 271, 272
  redox properties 273, 274
  sulphite reductase 267–9
Skeletonema costatum
  nitrate uptake 108
Sorghum bicolor
  nitrite reductase 250
soybean
  nitrate reductase 126, 127, 129, 136–40,
    147
  nitrate reductase deficient mutant 137
  nitrate reductase mRNA 147
  nitrate uptake 5
spinach
  nitrate reductase 55, 129, 135, 155, 156,
    159
  nitrite reductase 244, 245, 248, 249, 251,
    263–83, 285–96, 394, 397, 398

Sporobolomyces roseus
  nitrate assimilation 58
  nitrate reductase 56
  nitrate uptake 53
Sporopachydermia cereana
  nitrate assimilation 51, 52
Sporopachydermia lactativora
  nitrate assimilation 51, 52
squash
  nitrate reductase 126, 127, 129–32, 138,
    147, 156, 162, 328–36
  sirohaem 4Fe–4S centre 274
sulphite reductase
  EPR spectroscopy 270
  possible ligands to the 4Fe–4S centres 271
  sirohaem (4Fe–4S) active centre 267–9,
    271, 272

tobacco (*Nicotiana*)
  autogenous regulation 182
  biochemical genetics of nitrate
    reduction 166–83
  chlorate resistance 166, 167, 169, 187
  immunology 187, 188
  molecular genetics of nitrate
    reduction 186–96
  molybdenum cofactor mutants 167–9,
    176–180, 182, 213, 364–7
  nitrate reductase apoprotein mutants 167,
    170–6, 181, 188, 364–7
  nitrate reductase cDNA 188–90
  nitrate reductase structural gene 190–4
  nitrate reduction 166–83
  nitrate uptake mutants 373–6
  nitrite reductase 250
  somatic cell genetics of nitrate
    reduction 166–83

tomato
  nitrite reductase 245
Torulopsis nitratophila
  nitrate assimilation 52, 57
transformation
  filamentous fungi 344–7
  higher plants 193
  homologous transformation 352–4
transit peptide 249, 290
tungstate 32–4, 107, 113, 250, 317

Ustilago maydis
  chlorate-resistant mutants 348
  cloning of nitrate reductase structural
    gene 352
  nitrate assimilation genes 349

Volvox
  mutants 102

wheat
  nitrite reductase 244, 245, 248, 250, 251
wild oat
  nitrite reductase 245

yeast nitrate assimilation
  generation of reduced pyridine
    nucleotide 63, 64
  genetics 64
  nitrate uptake 53, 54
  nitrate reductase 54–7
  nitrite uptake 53, 54
  nitrite reductase 57
  occurrence 51, 52
  regulation 57–63